Essential Plant Cell Biology

Essential Plant Cell Biology

Edited by Agatha Wilson

SYRAWOOD
PUBLISHING HOUSE

New York

Published by Syrawood Publishing House,
750 Third Avenue, 9th Floor,
New York, NY 10017, USA
www.syrawoodpublishinghouse.com

Essential Plant Cell Biology
Edited by Agatha Wilson

International Standard Book Number: 978-1-68286-399-2 (Hardback)

Cataloging-in-publication Data

Essential plant cell biology / edited by Agatha Wilson.
p. cm.
Includes bibliographical references and index.
ISBN 978-1-68286-399-2
1. Plants--Cytology. 2. Plant cells and tissues. I. Wilson, Agatha.
QK725 .E87 2017
571.6--dc23

TABLE OF CONTENTS

PREFACE

Plant cell biology is defined as the study of the eukaryotic cells. Plant cells may seem similar to animal cell but in size the former is larger than the latter. This book presents the complex subject of plant cell biology and its varied branches in the most comprehensible and easy to understand language. It aims to provide a comprehensive understanding of plant cells through explaining the cell structures and other aspects. The text strives to provide its reader the essentials of plant cell biology through presenting the key components of cell biology. Scientists and students actively engaged in this field will find this book full of crucial and unexplored concepts. Coherent flow of topics, student-friendly language and extensive use of examples make this book an invaluable source of knowledge.

The main aim of this book is to educate learners and enhance their research focus by presenting diverse topics covering this vast field. This is an advanced book which compiles significant studies by distinguished experts. This book addresses successive solutions to the challenges arising in the area of application, along with it; the book provides scope for future developments.

It was a great honour to edit this book, though there were challenges, as it involved a lot of communication and networking between me and the editorial team. However, the end result was this all-inclusive book covering diverse themes in the field.

Finally, it is important to acknowledge the efforts of the contributors for their excellent chapters, through which a wide variety of issues have been addressed. I would also like to thank my colleagues for their valuable feedback during the making of this book.

Editor

Ocean-Scale Patterns in Community Respiration Rates along Continuous Transects across the Pacific Ocean

Jesse M. Wilson*, Rodney Severson, J. Michael Beman

Life and Environmental Sciences, Environmental Systems, and Sierra Nevada Research Institute, University of California Merced, Merced, California, United States of America

Abstract

Community respiration (CR) of organic material to carbon dioxide plays a fundamental role in ecosystems and ocean biogeochemical cycles, as it dictates the amount of production available to higher trophic levels and for export to the deep ocean. Yet how CR varies across large oceanographic gradients is not well-known: CR is measured infrequently and cannot be easily sensed from space. We used continuous oxygen measurements collected by autonomous gliders to quantify surface CR rates across the Pacific Ocean. CR rates were calculated from changes in apparent oxygen utilization and six different estimates of oxygen flux based on wind speed. CR showed substantial spatial variation: rates were lowest in ocean gyres (mean of 6.93 mmol m^{-3} d^{-1}±8.0 mmol m^{-3} d^{-1} standard deviation in the North Pacific Subtropical Gyre) and were more rapid and more variable near the equator (8.69 mmol m^{-3} d^{-1}±7.32 mmol m^{-3} d^{-1} between 10°N and 10°S) and near shore (e.g., 5.62 mmol m^{-3} d^{-1}±45.6 mmol m^{-3} d^{-1} between the coast of California and 124°W, and 17.0 mmol m^{-3} d^{-1}±13.9 mmol m^{-3} d^{-1} between 156°E and the Australian coast). We examined how CR varied with coincident measurements of temperature, turbidity, and chlorophyll concentrations (a proxy for phytoplankton biomass), and found that CR was weakly related to different explanatory variables across the Pacific, but more strongly related to particular variables in different biogeographical areas. Our results indicate that CR is not a simple linear function of chlorophyll or temperature, and that at the scale of the Pacific, the coupling between primary production, ocean warming, and CR is complex and variable. We suggest that this stems from substantial spatial variation in CR captured by high-resolution autonomous measurements.

Editor: Jack Anthony Gilbert, Argonne National Laboratory, United States of America

Funding: Data were supplied by gliders built and deployed by Liquid Robotics (http://liquidr.com/). The funders had no role in study design, data analysis, decision to publish, or preparation of the manuscript. The authors have no additional sources of funding to report.

Competing Interests: The authors have declared that no competing interests exist.

* Email: jwilson34@ucmerced.edu

Introduction

Microorganisms have profound effects on their surrounding environment, chemically modifying habitat and affecting other organisms through their biogeochemical activity [1], [2], [3]. Some of the most fundamental metabolic functions—in terms of both cell function and environmental importance—involve the production and consumption of oxygen. Oxygenic photosynthesis and aerobic respiration are basic measures of ecosystem function because the production, respiration, cycling, and overall availability of carbon affect everything from the number of trophic levels, to the types of organisms present [4]. Oxygenic photosynthesis dictates the amount of oxygen available for aerobic organisms (including both microbes and larger organisms), and photosynthesis in the ocean is almost exclusively microbial, accounting for approximately half of global oxygenic photosynthesis [5]. The subsequent oxidation of organic carbon with oxygen yields tremendous amounts of free energy, making aerobic respiration a preferred method of obtaining energy, and an indication of the total amount of production and activity in an ecosystem.

In many open ocean systems, photosynthesis and aerobic respiration rates tend to be tightly coupled and close to balanced [6], [7], [8], [9], [10], [11]. However, the biological carbon pump depends in part on decoupling between photosynthesis and the

respiration of organic carbon, leading to net export of carbon in areas where gross primary production (GPP) ultimately exceeds community respiration (CR). Importantly, GPP and CR are expected to differ in their sensitivity to ocean warming: metabolic theory predicts that both CR and GPP will increase as global temperatures rise [12], but that CR should increase more rapidly than GPP [13], [14], [15], [16], [17], [18]. This scenario leads to a positive feedback for climate change, as a larger fraction of primary production would be channeled through the microbial food web and respired to carbon dioxide [13], [16], [17]. Because this has global implications for marine ecosystems and biogeochemical cycles, understanding how CR is affected by temperature and other environmental variables is critical for our understanding of oceanic carbon cycling.

Recent results indicate that CR does scale with temperature, but the majority of these measurements have been collected in the Mediterranean Sea, and Arctic, Atlantic, Indian, and Southern Oceans [18], [19]. Most of the limited Pacific Ocean measurements are restricted to the vicinity of station ALOHA north of the Hawaiian Islands [20], [21], [22], [23], [24]. While these data are a powerful tool for understanding open ocean processes in the North Pacific Subtropical Gyre (NPSG), microbial community composition and nutrient and carbon sources differ significantly between different open-ocean and nearshore sites, such that relationships between GPP and CR are inconstant [20]. The Pacific also covers a substantially greater area than the Atlantic, and CR in the Pacific Ocean consequently may have a greater influence on global carbon cycling. However, the general lack of information regarding variations in CR across the Pacific means that this is essentially unknown. Of particular interest is how CR varies with natural oceanographic features (like the California Current and Equatorial Upwelling) and proxies for production (such as chlorophyll a) compared with temperature-driven changes in CR; understanding and quantifying this variability will lead to a better understanding of the factors driving CR in different regions of the ocean [15], and how they may change. Finally, most of our understanding of large-scale patterns in CR is drawn from syntheses of existing datasets (e.g., [19])—systematic investigations of CR are extremely rare.

We used data continuously collected by autonomous Wave Gliders[TM] [25] to assess how CR changes according to latitude, temperature, phytoplankton biomass (chlorophyll a), and turbidity across the Pacific Ocean. Quantifying how CR varies across Earth's largest ocean will further our understanding of how climate change may affect CR and ocean carbon cycling in the near future, and we focus on CR alone because of the general disagreement concerning the ideal method for also calculating GPP and net community production (NCP; see below). Our approach dramatically expands the number of CR measurements from the Pacific, and our findings indicate that CR rates are driven by different processes in different regions of the Pacific Ocean.

Materials and Methods

Autonomous sampling and data

The *Piccard Maru* and *Benjamin* Wave Gliders were deployed by Liquid Robotics [25] on November 17th, 2011 off the coast of San Francisco, California as part of the Pacific Crossing or 'PacX.' *Benjamin* traveled to Hawaii and turned south, crossing the equator on August 3rd, 2012 and arriving in Brisbane, Australia on February 14th, 2013 (Figure 1). *Piccard Maru* also traveled to Hawaii and then continued west towards Japan. The glider did not complete the journey to Japan and contact was lost in the Western Pacific. These two datasets are the most spatially- and temporally-extensive of the datasets collected during PacX.

Conductivity, dissolved oxygen concentration (mL L^{-1}), oxygen solubility (mL L^{-1}), salinity (psu), water temperature (°C), air temperature (°C), and average wind speed (knots), chlorophyll a (mg m^{-3}), and turbidity (NTU) were recorded by the wave gliders using a Glider Payload CTD, SBE 43F DO sensor, PB200 WeatherStation and C3 submersible fluorometer; data were obtained from the PacX data retrieval site (http://data.liquidr.com/fetch/). Oxygen, salinity, and water temperature were measured by the CTD at 10 seconds intervals, while chlorophyll and turbidity were measured every 2 minutes, and wind speed and air temperature were measured every 10 minutes. Underwater data were collected just below the sea surface (~0.2 m) while wind speed was measured 1 m above the sea surface. Only data from days when both CTD and weather data are available were used when calculating CR, such that the flux of oxygen from the atmosphere to the ocean or from the ocean to the atmosphere could be accounted for based on wind speed.

Respiration rate calculations

To quantify respiration rates, we adapted the approach of Needoba, Peterson, and Johnson [26] and calculated (1) the nighttime drawdown of oxygen, as well as (2) the expected flux of oxygen into or out of the mixed layer. We directly followed the

Needoba et al. [26] approach for one set of calculations; in these calculations, 'biological demand of oxygen' (BDO) [26] is calculated as the difference in oxygen concentrations between any two time points, and then all of these BDO values are summed over each dark period. We performed additional calculations where we modified this approach in two ways. First, in addition to summing changes in dissolved oxygen (DO) over time, we calculated nighttime drawdown of DO based on regressions between apparent oxygen utilization (AOU) and time (AOU slope approach). Second, we used more than one relationship with wind speed to calculate oxygen flux; this allowed us to account for a variety of physical processes that affect dissolved oxygen concentrations, including breaking waves and bubble entrainment [27; see below]. We applied all of these windspeed relationships to both the BDO approach of Needoba et al. [26] and to our AOU slope approach.

For the first part of these calculations, DO and oxygen solubility (OS) were used to compute AOU as the difference between OS and DO for ten minute averages of high frequency measurements. AOU measures the cumulative effects of biological activity that have occurred in a water sample and is negative when oxygen is supersaturated, and positive when oxygen is undersaturated (due to consumption of DO). DO drawdown was calculated during nighttime (when incoming solar radiation was zero) based on regressions between AOU and time, which were tested for significance. Nighttime DO drawdown may differ significantly from daytime DO drawdown [28], meaning that nighttime CR rates may differ slightly from actual rates. This is an unavoidable limitation of using autonomously collected DO data to calculate CR.

We calculated gas exchange based on the air-sea concentration gradient and established parameterizations based on wind speed, where the diffusive flux (F) of oxygen equals the difference between the observed oxygen concentration of surface water and expected saturation oxygen concentration (which depends on temperature and salinity), multiplied by the gas transfer velocity (or coefficient of gas exchange; k$_{O2}$) for oxygen at a given temperature (eq. 1) [29], [30], [31].

$$F = k_{O2}([DO]\text{-}[OS])\Delta time \qquad \text{(eq.1)}$$

Oxygen is transferred from the water to the atmosphere when F (umol cm^{-2}) is positive, while oxygen is transferred from the atmosphere into the water when F is negative. Flux was calculated every ten minutes based on the measured wind speed, and was integrated over the dark period to obtain the total nighttime flux for each night. Nighttime integrated flux was subtracted from the entire nighttime AOU respiration estimate while each ten minute flux estimate was subtracted from each interval calculated following Needoba et al. [26].

k$_{O2}$ was obtained by first computing k$_{660}$ (cm hr^{-1}) based on wind speed: this is the k value for CO_2 at 20°C and has been measured empirically for various wind speeds in the ocean. There are multiple mathematical equations relating wind speed to k$_{660}$ and we used six relationships to ultimately calculate oxygen flux and CR; these equations vary in the formulation of k as function of wind speed, especially whether relationships are linear [32], quadratic [30], [33], cubic [34], or power functions [32]. Needoba et al. [26] recommend using CR1, while CR2 and CR3 explicitly consider the higher k values resulting from breaking waves and bubble entrainment [27], [33], [34]. CR4-CR6 were specifically designed to capture lower wind speeds. We include all six approaches given the wide variation in environmental conditions

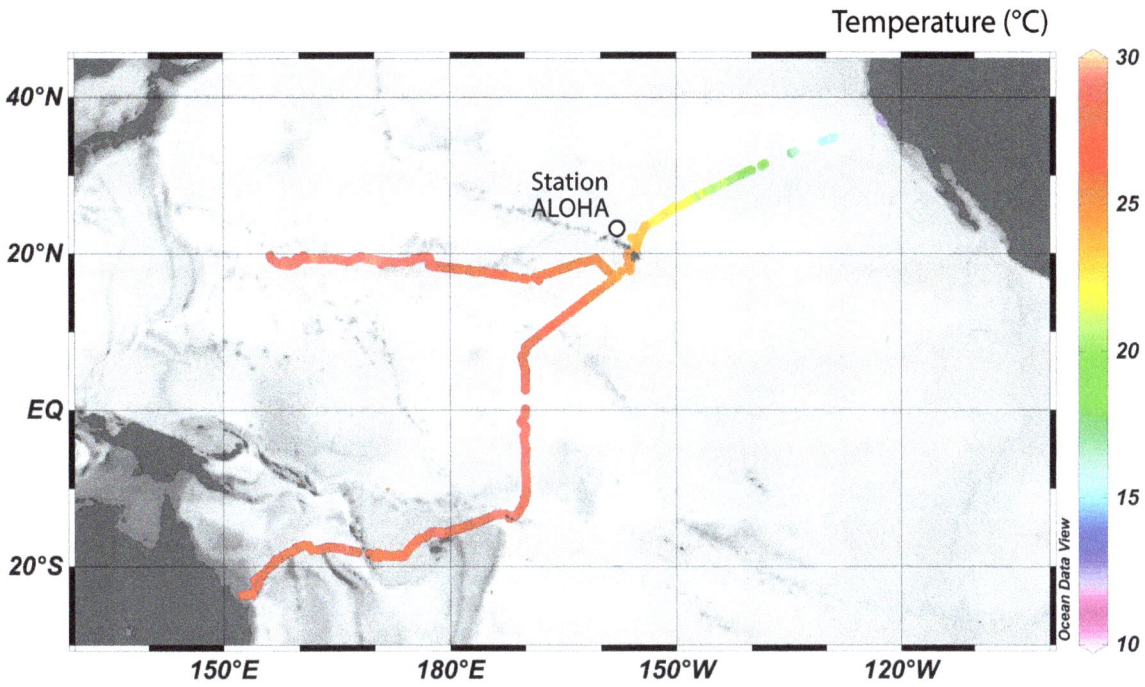

Figure 1. Map of the transpacific paths of the California to Australia and the California to Japan gliders and sea surface temperature as measured by the gliders. Eleven datapoints are not displayed due to obvious GPS errors. Ocean Data View [59] was used to plot and visualize data across the Pacific.

across the Pacific and to provide broader context. All of the equations were originally derived using chemical tracers such as sulfur hexafluoride.

CR1) $\quad k_{660} = 0.31(U_{10})^2$ $\quad\quad$ (eq.2) [30]

CR2) $\quad k_{660} = 0.0283(U_{10})^3$ $\quad\quad$ (eq.3) [34]

CR3) $\quad k_{660} = 0.222(U_{10})^2 + 0.333(U_{10})$ $\quad$ (eq.4) [33]

CR4) $\quad k_{660} = 0.72(U_{10})\ for\ U_{10} < 3.7 m\ s^{-1}$ $\quad$ (eq.5) [32]

$\quad\quad\quad k_{660} = 4.33(U_{10})\ for\ U_{10} \geq 3.7$

CR5) $\quad k_{660} = 0.228(U_{10})^{2.2} + 0.168\ cm\ hr^{-1}$ $\quad$ (eq.6) [32]

CR6) $\quad k_{660} = 1.0 cm\ hr^{-1}\ for\ U_{10} < 3.7\ m\ s^{-1}$ (eq.7) [32]

$$k_{660} = 5.41(U_{10}) - 17.9\ cm\ hr^{-1}\ for\ U_{10} \geq 3.7\ m\ s^{-1}$$

In all of these equations, wind speed (U_{10}) is measured in m s^{-1} at 10 m above the water's surface whereas our data were collected at 1 m; wind speeds were therefore scaled to 10 m (eq. 8) [35]:

$$U_{10} = U_z[1 + [(C_{d10}^{0.5})/K]ln(10/z)] \quad\quad (eq.8)$$

Where C_{d10}—the surface drag coefficient for wind above water at 10 m—is 1.3×10^{-3} [36], K refers to the von Karman constant of 0.41 [37], and z is the height above the sea surface where wind speed was measured. The Crusius & Wanninkhof [32] relationships were specifically developed for low wind speeds that may periodically occur over the ocean. In our data, wind speed was greater than 3.7 m s^{-1} 96.5% of the time.

Schmidt numbers (Sc) were used to covert from the modeled k_{660} value in equations 2–6 to a k_{O2} at each recorded surface temperature. A Schmidt number is a dimensionless number that characterizes fluid flow, is unique for each dissolved gas, and is defined as the kinematic viscosity of water divided by the diffusion coefficient of the gas at a given temperature. The Schmidt number for O_2 in saltwater can be calculated with the following equation, in which T is temperature in Celsius (eq. 9) [30]:

$$Sc_{O2} = 1953.4 - 128T + 3.9918T^2 - 0.050091T^3 \quad (eq.9)$$

The ratio of k values (for each CO_2 and O_2) equals the ratio of Schmidt numbers raised to negative n (eq. 10) [38]:

$$k_{gas1}/k_{gas2} = \left(Sc_{gas1}/Sc_{gas2}\right)^{-n} \qquad \text{(eq.10)}$$

n depends on the processes that dominate gas transfer including the friction velocity and the mean square slope of the waves [39]. Based on the laboratory experiments of Jahne et al. [39], [40] we set n to 2/3 when $U_{10} < 2$ m s^{-1} and n at 1/2 when $U_{10} \geq 2$ m s^{-1}.

Computed oxygen fluxes were then divided by the mixed layer depth, and subtracted from the nighttime slopes of AOU versus time to obtain respiration rates. (Note that fluxes are defined relative to the atmosphere, and a negative flux represents transfer of oxygen into the ocean from the atmosphere.) Positive CR values indicate a higher rate of respiration while a 'negative' CR value indicates that DO increased during nighttime hours. Negative CR can be explained by daytime primary production exceeding and masking the amount of nighttime CR that occurred, with isopycnal mixing also potentially playing a role, as these measurements were only made at the surface, whereas photosynthesis and respiration likely varied with depth throughout the mixed layer.

Data analysis

Two outliers (3-9-2012, and 12-31-2012) were removed from the California to Australia dataset due to extremely high calculated respiration values (>2 standard deviations from the mean). One outlier was removed from the California to Japan dataset (3-6-2012) due to extremely high chl a and turbidity values (>2 standard deviations from the means).

Relationships between respiration, latitude, temperature, chl a, and turbidity were explored using multiple regression analysis in R. Regression analysis was completed for all data, for each individual glider, and for latitudinal and longitudinal zones for the California to Australia dataset. These zones follow Longhurst's biogeographical zones [41], [42] and ranged from 20°N to 38°N (California Current province), 10°N to 20°N (North Pacific [Sub]Tropical Gyre province), 0°N to 10°N (North Pacific Equatorial Countercurrent province), 15°S to 0°S (Pacific Equatorial Divergence province), and 23°S to 15°S (South Pacific Subtropical Gyre province). The natural log was taken of chlorophyll and turbidity in order to correct slight non-linearity. Due to several slightly 'negative' values (see Discussion for an explanation), turbidity was offset by 1.38 so that the minimum value was 1 before transformation.

Results

Pacific-scale Patterns in CR

Data from autonomous gliders provide ocean-scale patterns in temperature, chlorophyll, turbidity, and respiration; all showed substantial variation across the Pacific Ocean. For the California to Australia transect, water temperature increased as latitude decreased, and was warmest just south of the equator (29.3°C) (Figs. 1, 2A and 2B). Water temperatures increased from the eastern to western Pacific along the transect from California to Japan. Chlorophyll concentrations were greatest off coastal California (above 30°N latitude) and were generally low in the open Pacific, with the exception of the elevated levels observed near the equator (Fig. 2E and 2F). Chlorophyll also increased near 18–19°N in the West Pacific, and turbidity was extremely variable throughout the Pacific (Fig. 2). Several of these variables were correlated with one another across the Pacific, along the different transects, or within different biogeographical zones (Table 1).

In total, we generated 341 independent measurements of CR from California to Australia and 264 independent measurements

from California to Japan. All of our computational approaches demonstrated that respiration rates were variable across the Pacific. The different windspeed-based approaches to calculate CR ashowed strong agreement, with r^2 values ranging from 0.810–0.999 (all $P < 0.001$) between different CR datasets within a given method for calculating respiration (AOU change over time, or summed changes in DO based on Needoba et al. [26]) (Tables 2 and 3). Independent of the windspeed-based flux estimates, the AOU and BDO approaches for estimating oxygen consumption were significantly correlated ($P < 0.001$) for each transect, but the California to Japan dataset had an r^2 value of 0.701 while the California to Australia dataset had an r^2 value of 0.3. The BDO approach typically generated more instances of negative CR rates (Figure S1)—most likely due to summing many small changes in DO—and so our analysis is focused on the AOU slope approach. These six CR datasets varied slightly in terms of maximum, minimum, mean, and median values (Figs. 3 and 4) but were generally comparable: for example, CR1 rates ranged from −88.9 to 75.4 mmol m^{-3} d^{-1}, CR2 ranged from −88.6 to 77.2 mmol m^{-3} d^{-1}, CR3 ranged from −89.0 to 75.0 mmol m^{-3} d^{-1}, CR4 ranged from −89.2 to 74.7 mmol m^{-3} d^{-1}, CR5 ranged from −88.8 to 76.0 mmol m^{-3} d^{-1}, and CR6 rates ranged from −89.1 to 74.9 mmol m^{-3} d^{-1} for the California to Australia dataset. We focus on CR2 because it considers breaking waves and bubble entrainment, the method used to obtain this relationship utilizes long-standing protocols, and it performs better statistically (see below). CR2 ranged from −120 to 84.4 mmol m^{-3} d^{-1} from California to Japan using our AOU slope approach (both for CR2; Fig. 3), while using Needoba et al.'s BDO approach [26], CR2 ranged from −58.9 to 61.9 mmol m^{-3} d^{-1} for the California to Australia dataset, and from −84.8 to 93.9 mmol m^{-3} d^{-1} for the California to Japan dataset.

The gliders traveled nearly identical paths from California to the Hawaiian Islands over slightly different time periods, yet they showed highly similar patterns in CR: both were variable and reached maximum and minimum values near the California Coast, and each was more consistent beyond 128°W. Several local maxima were recorded at various points within the northeastern portion of the North Pacific Gyre, particularly near 150°W. This included high CR rates measured at 148.7°W on 2012-02-09 and 156.2°W on 2012-03-13 for the California to Australia transect, and at 151°W on 2012-02-19 and 153.1°W 2012-03-09 for the California to Japan transect (for CR2). These local maxima were slightly greater for the California to Japan transect (39.7 mmol m^{-3} d^{-1} and 37.2 mmol m^{-3} d^{-1}) than the California to Australia transect (35.3 mmol m^{-3} d^{-1} and 17.9 mmol m^{-3} d^{-1}).

The east-to-west, California to Japan transect then crossed a large swath of the oligotrophic North Pacific Subtropical Gyre (NPSG), whereas the northeast-to-southwest, California to Australia transect crossed multiple ocean provinces (Figs. 3 and 4). We expected that the east-west transect would therefore show relatively low rates and little variation, but this was not the case: CR2 averaged 8.38 mmol m^{-3} d^{-1} with a standard deviation of 8.44 mmol m^{-3} d^{-1}. Many of the high CR rates were observed within the NPSG—particularly in the central Pacific to the west of Hawaii—as well as closer to mainland Asia. The overall range of CR values was similar for the California to Australia transect, and also showed substantial spatial variability. CR above 20°N latitude was variable between days, latitude, and longitude with no obvious increasing or decreasing trends, but became less variable after 128°W longitude and remained below 10 mmol m^{-3} d^{-1} on all but 6 occasions after leaving the California Coast. CR rates were generally low from Hawaii to the equator, but a small peak in CR occurred near 10°N latitude. CR rates exhibited several obvious

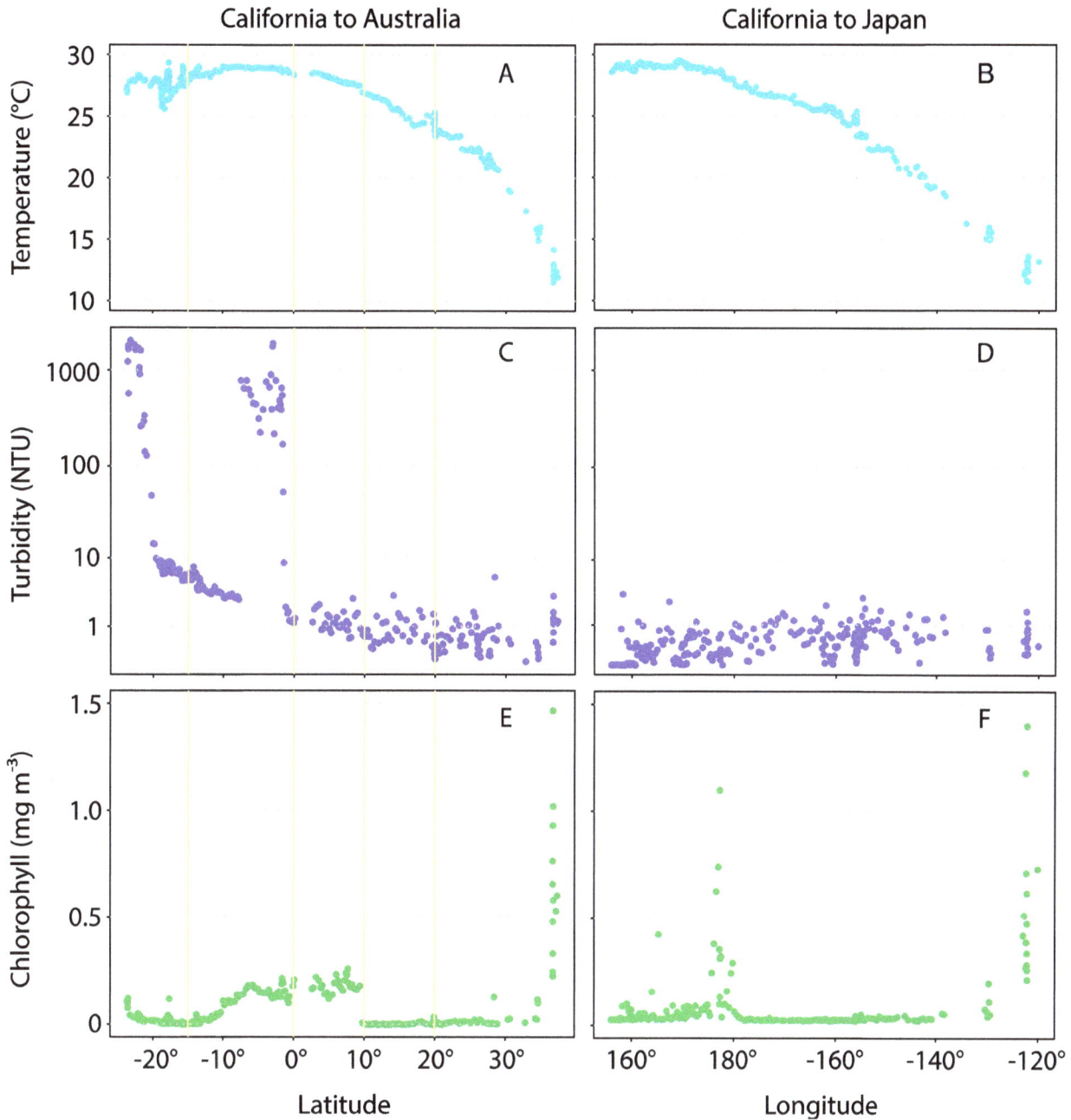

Figure 2. Environmental data collected by the gliders along the two transects. Panels depict temperature (°C) according to latitude for California to Australia (A) and according to longitude for California to Japan (B); turbidity (NTU) from California to Australia (C) and California to Japan (D); and Chlorophyll (RFU) from California to Australia (E) and California to Japan (F).

peaks below the equator and were consistently elevated from the equator to 10°S. CR2 averaged 6.49 mmol m^{-3} d^{-1} with a standard deviation of 11.0 mmol m^{-3} d^{-1} for the entire transect, but reached 26.2 mmol m^{-3} d^{-1} at 6.36°S latitude; CR was generally elevated within several degrees of latitude of this peak. CR again increased around 18–19°S, and the greatest values south of the equator were observed near the coast of Australia (46.5 mmol m^{-3} d^{-1} at 22.5°S latitude).

Relationships to environmental data

We used several statistical approaches to analyze relationships between possible explanatory variables and CR. No approach to

calculating CR was uniformly most significantly related to environmental data, however CR2 generally outperformed the other approaches for both datasets, and for both univariate and multiple linear regression by producing more significant relationships and stronger significant relationships. Most of our analysis therefore focuses on CR2, however all of the CR datasets were correlated and yielded similar results despite differences in their underlying assumptions. For these analyses, we examined both transects together as well as separately, and also analyzed the California to Australia dataset based on multiple latitudinal zones. This stems from the large expected and observed differences in oceanographic conditions across these regions, as well as the fact

Table 1. r values for relationships among environmental datasets in different regions of the Pacific.

	Pacific Ocean	California to Japan	California to Australia	20 to 38°N	10 to 20°N	0 to 10°N	−15 to 0°S	−23 to −15°S
Chlorophyll and temperature	−0.418	−0.423	−0.410	−0.795	0.341	Not significant	0.592	Not significant
Temperature and turbidity	0.276	0.454	−0.288	Not significant	−0.270	Not significant	0.458	Not significant
Chlorophyll and turbidity	Not significant	Not significant	Not significant	Not significant	Not significant	Not significant	0.639	0.647

that datasets from different latitudes exhibit different relationships between temperature and CR (e.g., [15]).

Across the long oceanographic transects, CR2 varied significantly with turbidity from California to Australia ($r^2 = 0.085$ and $P<0.001$), and increasing the number of possible explanatory variables in stepwise multiple linear regression did not increase the predictive strength of the model. For the California to Japan data, CR again varied most strongly with turbidity ($r^2 = 0.112$, $P<0.001$); the most descriptive multiple regression model included chlorophyll and turbidity as explanatory variables ($r^2 = 0.167$ and $P<0.001$) using CR2. When California to Australia data were assessed according to latitudinal zones, different explanatory variables were important in different areas. From north to south, no significant relationships were observed from 20°N to 38°N, 10°N to 20°N, and 0°N to 10°N, whereas chlorophyll was most significantly related to CR from 15°S to 0°S ($r^2 = 0.166$, $P<0.001$ using CR2), and turbidity was the best predictor from 23°S to 15°S ($r^2 = 0.325$, $P<0.001$ using CR2). Multiple linear regression did not yield any significant relationships north of the equator, and only increased the explanatory power of the model from 23°S to 15°S slightly, such that water temperature and turbidity explained 37.3% of the variation in CR ($P<0.001$).

Chlorophyll and turbidity were also log-transformed to fix slight non-linearity but produced highly similar results (Figure 5): CR2 had the most significant models; no significant relationships occurred north of the equator; chlorophyll was the best predictor from 15°S to 0°S ($r^2 = 0.165$, $P<0.001$); and water temperature and turbidity were collectively the best predictors from 15°S to 23°S ($r^2 = 0.347$, $P<0.001$). Allowing second or third order interactions to the log-transformed variable models for CR2 increased the explanatory power of all the models except from 20°N to 38°N (Fig. 5). For example, there were no significant relationships between CR and any of the variables between 10°N to 20°N, but adding an interaction between chlorophyll and turbidity resulted in a significant relationship ($r^2 = 0.391$, $P<0.001$). Adding a third order interaction between water temperature, chlorophyll, and turbidity strengthened this relationship ($r^2 = 0.518$, $P<0.001$).

All data were ultimately pooled together to analyze cross-Pacific patterns in respiration as a function of chlorophyll, turbidity, and temperature. Across the dataset, the strongest relationship was between the natural log of turbidity and CR2 ($r^2 = 0.036$, $P<0.001$). Using multiple explanatory variables only incrementally increased the amount of variation explained, with the natural logs of chlorophyll and turbidity producing the best model ($r^2 = 0.043$, $P<0.01$ for CR2). Allowing interactions between explanatory variables likewise incrementally increased the explanatory power of the models (e.g., $r^2 = 0.077$, $P<0.001$ for CR2 allowing interactions between temperature, chlorophyll, and turbidity). A third order interaction term between water temperature, the natural log of chlorophyll, and the natural log of turbidity only explained 8.54% of the data ($P<0.001$ for CR2). We also used multivariate adaptive regression splines (MARS) [43] to model CR; this approach includes interactions among variables, and identifies regions of the dataset where different basis functions may be applied. The MARS model explained 7.5% of the variation in CR across the Pacific ($P<0.001$) as a function of temperature, chlorophyll, and turbidity.

For these pooled data, we found stronger relationships within the NPSG than across the entire Pacific. The NPSG is Earth's largest ecosystem [44], [45] and different regions of the NPSG were covered extensively by the two gliders. The natural log of turbidity was the most significant explanatory variable for 4 of the 6 methods of calculating CR ($r^2 = 0.110$, $P<0.001$ for CR2), with

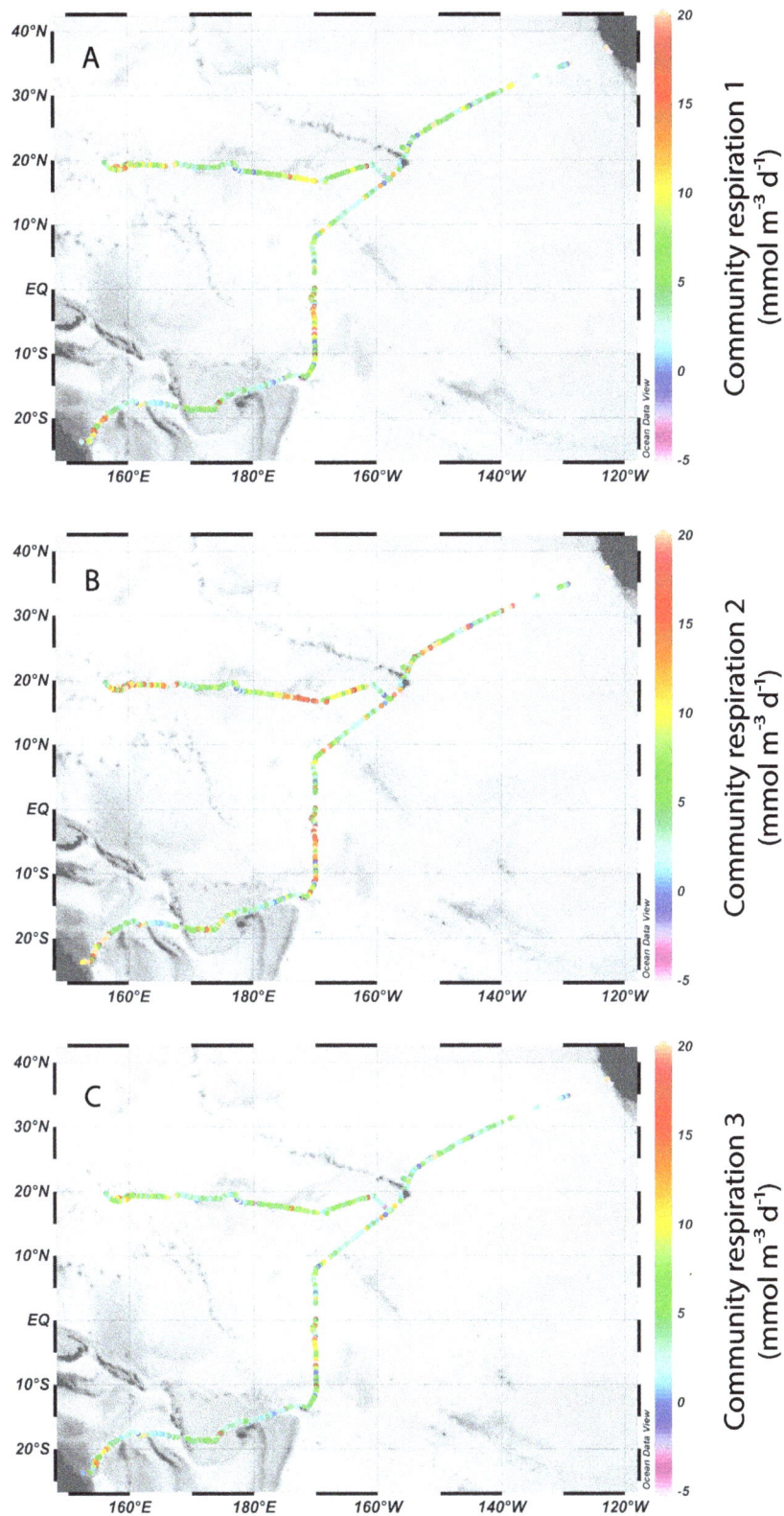

Figure 3. Computed community respiration rates along the two transects. Color shading shows (A) community respiration 1 (mmol m^{-3} d^{-1}) (B) community respiration 2 (mmol m^{-3} d^{-1}) and (C) community respiration 3 (mmol m^{-3} d^{-1}) across the Pacific Ocean. 14 datapoints, 24 datapoints, and 15 datapoints exceed 20 mmol m^{-3} d^{-1} and appear pink for CR1, CR2, and CR3, respectively; 19 datapoints, 15 datapoints, and 19 datapoints fall below −5 mmol m^{-3} d^{-1} and appear light purple for CR1, CR2, and CR3, respectively. Data displayed using Ocean Data View [59].

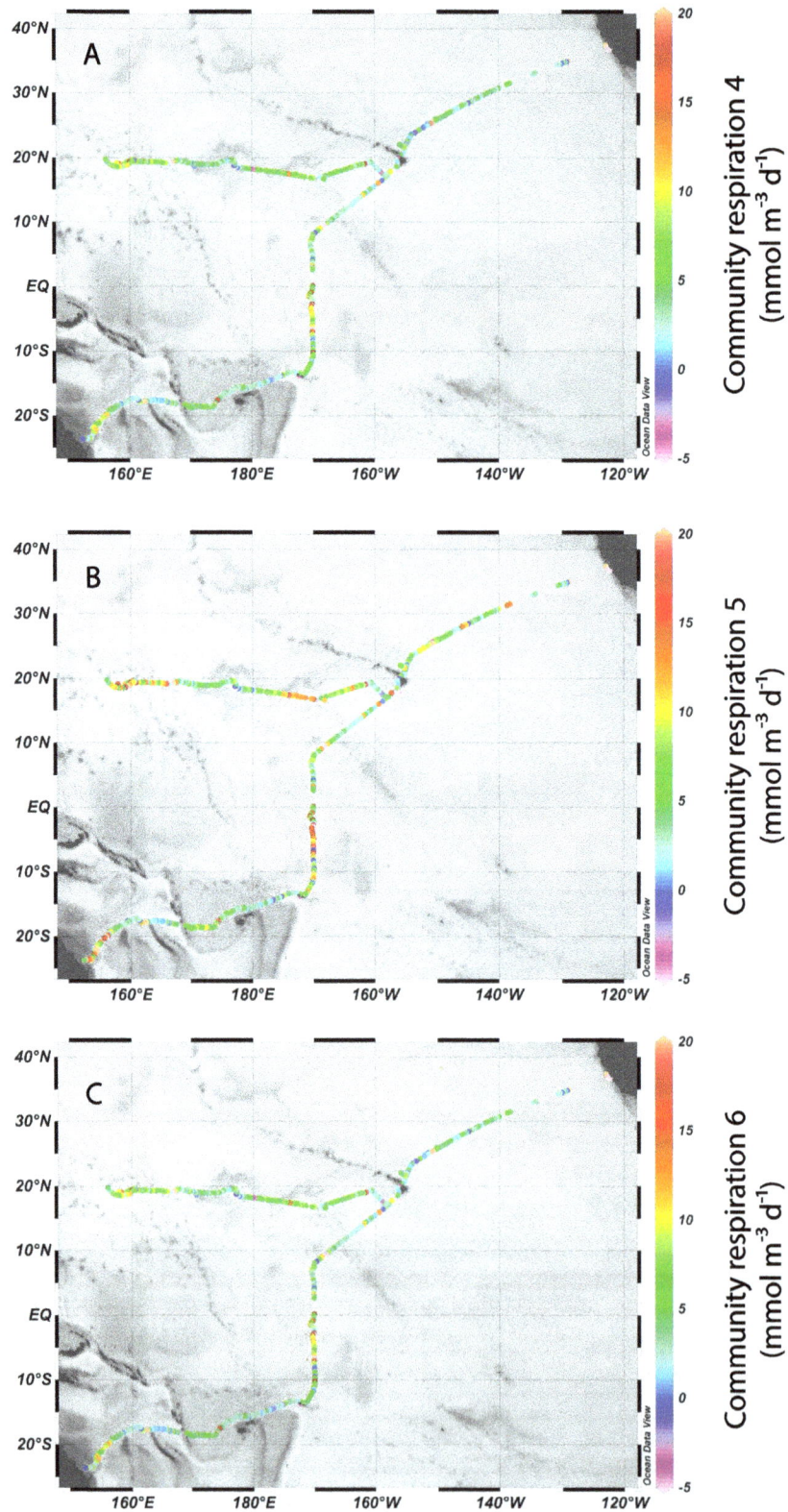

Figure 4. Computed community respiration rates along the two transects [58]. Color shading shows (A) community respiration 4 (mmol m^{-3} d^{-1}) (B) community respiration 5 (mmol m^{-3} d^{-1}) and (C) community respiration 6 (mmol m^{-3} d^{-1}) across the Pacific Ocean. 14 datapoints, 16 datapoints, and 39 datapoints exceed 20 mmol m^{-3} d^{-1} and appear pink for CR4, CR5, and CR6, respectively; 19 datapoints, 18 datapoints, and 14 datapoints fall below -5 mmol m^{-3} d^{-1} and appear light purple for CR4, CR5, and CR6, respectively. Data displayed using Ocean Data View [59].

Table 2. r^2 for comparisons among different AOU approaches to calculating CR for the California to Australia (above the diagonal) and California to Japan (below the diagonal datasets).

	CR1	CR2	CR3	CR4	CR5	CR6	Raw
CR1		0.905	0.993	0.979	0.986	0.986	0.842
CR2	0.917		0.856	0.810	0.957	0.826	0.601
CR3	0.996	0.881		0.988	0.959	0.994	0.899
CR4	0.985	0.839	0.995		0.940	0.999	0.883
CR5	0.990	0.961	0.973	0.953		0.948	0.750
CR6	0.990	0.853	0.998	0.999	0.961		0.891
Raw	0.921	0.718	0.952	0.965	0.860	0.961	

$P<0.001$ in all instances.

Table 3. r^2 for comparisons among different BDO approaches to calculating CR for the California to Australia (above the diagonal) and California to Japan (below the diagonal datasets).

	CR1	CR2	CR3	CR4	CR5	CR6	Raw
CR1		0.923	0.993	0.984	0.987	0.989	0.870
CR2	0.914		0.881	0.848	0.968	0.860	0.665
CR3	0.996	0.876		0.990	0.963	0.996	0.919
CR4	0.985	0.834	0.995		0.951	0.999	0.903
CR5	0.990	0.960	0.972	0.952		0.956	0.788
CR6	0.989	0.848	0.998	0.999	0.959		0.911
Raw	0.921	0.712	0.952	0.965	0.859	0.961	

$P<0.001$ in all instances.

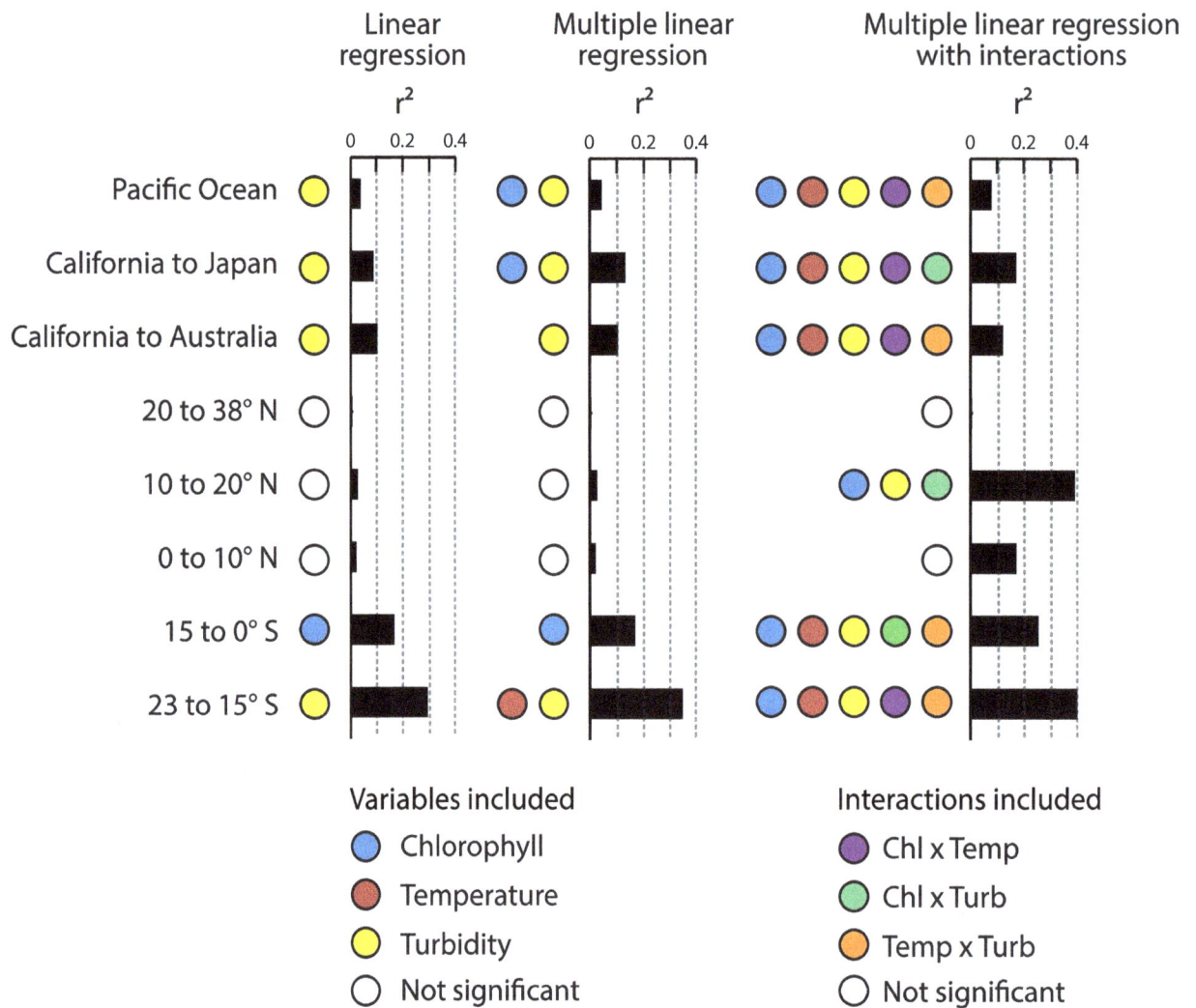

Figure 5. Regression statistics for best AIC derived relationships between AOU-slope-based CR2 data and environmental data. Bar graphs show regression r^2 values for univariate linear regression, multiple linear regression, and multiple linear regression with interactions. Color-coded symbols next to the bar graphs indicate which variables or interactions yielded significant relationships; a white circle indicates that no significant relationships were found. These model statistics represent log-transformed turbidity and chlorophyll data.

water temperature being more strongly correlated with CR4 ($r^2 = 0.041$, $P<0.002$) and CR6 ($r^2 = 0.033$, $P<0.005$), from $10°N$ to $20°N$ in the NPSG. Including both of these variables in multiple regression ($r^2 = 0.202$, $P<0.001$ for CR2) and their interactions ($r^2 = 0.245$, $P<0.001$ for CR2) produced stronger relationships.

Discussion

The CR rate data presented here are unique in both spatial coverage and the frequency with which measurements were taken, as no previous study has calculated CR continuously across an entire ocean basin. While CR has been measured across a few latitudinal and longitudinal transects, these consist largely of discrete measurements made at a limited number of locations [20], [23], [24]. Our data also provide a large number of measurements from the undersampled Pacific Ocean: in the most recent compilation of CR measurements, Regaudie-de-Gioux and Duarte [46] report 3854 measurements, with only 296 made in the Pacific Ocean (some of which are unpublished). We report 341 measurements from north to south, and 264 measurements from

east to west, across the Pacific, more than tripling the total number of CR measurements made in Earth's largest ocean. Our data and findings are in broad agreement with the limited information available from the Pacific and other large-scale studies, display several interesting spatial patterns, and provide new insight into the environmental drivers of CR.

Most measurements of CR in the Pacific Ocean have been conducted at Station ALOHA (A Long Term Oligotrophic Habitat Assessment) of the Hawaii Ocean Time-series (HOT) program, and both gliders were in the vicinity of Station ALOHA for several days—approaching from the east and passing to the south (Figure 1). This allows for comparison of CR between the glider dataset and the extensive HOT dataset. In a year-long study at ALOHA from 2001–2002, Williams et al. [22] found that surface CR at station ALOHA ranged from 0.63 to 1.15 mmol O_2 m^{-3} d^{-1}; unpublished CMORE data from more recent HOT cruises show that surface respiration varies from 0.18 to 2.8 mmol O_2 m^{-3} d^{-1} [46], [http://cmore.soest.hawaii.edu/cmoredata/ Church/GPP_NCP_R_results_summary.xls]. These data illustrate the fact that even stratified, oligotrophic, open ocean sites exhibit order-of-magnitude variations in CR over time [22]. For the gliders, mean CR values ranged from 2.38-4.64 mmol m^{-3} d^{-1} near station ALOHA. These values overlap the upper end of CR rates from dark bottle incubations conducted at station ALOHA.

Glider-based CR data are also consistent with other datasets collected in the equatorial Pacific. Viviani et al. [23] measured CR, GPP, and NCP along a transect from 14.3°S, 169.2°W to Station ALOHA that lies to the west of our CA-Australia transect; they observed increased CR near the equator, with a maximum around 10°S. In our data, CR varied widely near the equator and in association with high chlorophyll and turbidity that are likely indicative of a phytoplankton bloom; CR appears to have responded by increasing between the equator and 10°S latitude. NCP and GPP also varied substantially over relatively short distances (~50 km) along the equator in the Western Equatorial Pacific [47]. While Stanley et al. 's data were primarily collected along an east-west transect, and they did not report volumetric CR rates, they followed 170°W for ±2 degrees north and south of the equator. NCP decreased from north to south along 170°W, which could reflect decreased GPP, increased CR, or both. In our data, the glider *Benjamin* detected an increase in CR from north to south also while tracking 170°W, which is consistent with but not directly comparable to Stanley et al. [47].

As a whole, our datasets substantially increase the number of CR measurements available for the Pacific, and our approach may be useful in constraining the metabolic balance of the oligotrophic ocean [24], [48], [49], [50], [51], [52]. *In situ* studies of oxygen production and consumption tend to indicate that the open ocean is net autotrophic while *in vitro* studies indicate net heterotrophy (see [24]). Setting aside issues regarding the depth of integration, the discrepancies most likely have to do with an underestimation/ overestimation of photosynthesis rather than mistakes calculating respiration [24], [53]. This view is supported by comparisons of in vitro and *in situ* data from HOT, which found general agreement between respiratory rates for the two approaches [22], [53], [54]. While most of the global CR database is comprised of light-dark bottle measurements [19], [46], our data represent a large collection of *in situ* measurements that provide a useful point of comparison. Clearly some uncertainty lies in gas exchange, which leads to variations in absolute magnitudes among our CR datasets—however the variation in gas exchange as a function of windspeed (typical r^2 values of 0.8±0.1) is a long-running issue without clear resolution (reviewed by [27]). Regardless of the approach we used to calculate CR, the majority of the data fell within observations reported elsewhere in the ocean [19], with > 90% falling within the range of surface CR rates reported in Regaudie-de-Gioux and Duarte's 2013 dataset [46]. The remaining values are almost entirely 'negative' CR rates. Our approach could therefore be adapted to daytime increases in oxygen and used to calculate patterns in surface NCP across the Pacific.

Based on these data we observed several large-scale patterns in CR across the Pacific Ocean, including high CR rates near the equator and the coast of Australia, and surprising variation in CR within the NPSG. While rates were low throughout much of the NPSG, high CR rates may occur at least transiently in portions of the NPSG west of Hawaii. CR was also highly variable in the South Pacific, exhibiting a wide range of values and sharp changes within zones. These data represent the first extensive series of measurements made in the highly undersampled South Pacific Subtropical Gyre [19], [42], [46] and indicate that large gradients in CR may occur in this ocean. For the most part, these variations were not explained by coincident turbidity, temperature, and chlorophyll measurements made by the gliders. Previous work in

the Atlantic Ocean has shown significant relationships with beam attenuation [55], which is consistent with the presence of particles and aggregates in the water column that may be subsequently remineralized and respired. Turbidity was frequently related with CR, including significant relationships across the whole dataset and for each transect. However, there are two main issues with the turbidity data collected by the PacX gliders. First, there were many low, but negative, turbidity values that likely reflect sensor drift over the extended deployment at sea (Fig. 2C). Second, microbubbles can potentially interfere with the turbidity sensor: as outlined by Villareal and Wilson [56], this likely explains the highly elevated turbidity values observed south of 18°S, as the glider *Benjamin* traversed Tropical Storm Freda and may have had bubbles entrained in the sensor. While the significant relationships observed in other biogeographic regions and for the other glider correspond with much lower turbidity values, we cannot exclude the possibility that these stem from artifacts present in the data.

Temperature was generally weakly correlated with CR—if at all—and was only a significant predictor in multiple linear regression. The slopes of these relationships were variable, ranging from negative to positive. This contrasts with some previous work [18], [19], but supports the idea that temperature has mixed effects on CR in different regions of the ocean, and specifically the idea that temperature effects are weak outside of high latitude regions [15]. Kirchman et al. [15] convincingly argue that temperature is not a strong regulating factor in other parts of the ocean because the availability of organic C substrates and nutrients is more important. This likely applies to much of our dataset, as outside of the region that extends from California to 20°N, temperatures fell between a relatively narrow range of 23– 30°C (representing >84% of the temperature data). No significant relationship between CR and temperature was observed from 20°–38°N—despite a wider range of temperatures—but no significant relationships were observed between CR and any of the *in situ* measurements in this region. This is due to the localized hotpots in CR observed near 150°W and 25°N, which are not associated with distinct changes in turbidity, temperature, or chlorophyll.

Compilations of CR datasets have also shown that CR can be positively correlated with chlorophyll concentrations [18], [19], [46]. Production ultimately sets the pace for CR rates, and it seems likely that coupling between CR and either the biomass (chlorophyll) or photosynthetic activity (GPP) of phytoplankton would co-vary across the ocean and through time. For example, several studies have found that chlorophyll consistently explains ca. 30% of the variance in CR—whether in estuaries, throughout the Atlantic Ocean, or across oceans [19], [55], [57], [58]. However, Regaudie-de-Gioux and Duarte [46] found that both net community metabolism and CR were each only weakly correlated with chlorophyll while 30% of the variability in GPP was explained by chlorophyll. Robinson and Williams [19] showed that, overall, CR is not strongly correlated with chlorophyll concentrations, while stronger relationships are observed in regions of the ocean with sharp chlorophyll and CR gradients. Our data from the region south of the equator in the Pacific are consistent with this: high CR rates observed south of the equator correspond with high chlorophyll concentrations (Fig. 2E). These latter data capture a phytoplankton bloom associated with equatorial upwelling, and CR was significantly related to chlorophyll in the latitudinal band just south of the equator for all CR datasets (Fig. 5). An increase in CR near the equator appears to be a general feature in the Atlantic Ocean [20], [55], [57] that was also observed by Viviani et al. [23] in Pacific Ocean, and our data extend these observations. These variations in CR are likely driven by upwelling in the equatorial oceans—either by directly affecting CR through nutrient supply, or by fueling GPP and increasing the supply of organic C to CR.

We also observed high CR rates where chlorophyll showed little change—especially high CR rates in the oligotrophic NPSG west of Hawaii, and low CR rates where chlorophyll was elevated near the International Date Line. Chlorophyll and CR were in fact inversely related along the California to Japan transect. This reflects decoupling between phytoplankton biomass and CR and is in line with the intensive measurements made by Williams et al. [22] at Station ALOHA: they argue that primary production occurs in intermittent bursts—which they may not have detected even with comprehensive sampling—whereas CR is less variable, and more integrative, over time. Our cross-Pacific data suggest that another distinguishing feature of different oceanic provinces may be differences in the coupling between primary production and CR, with strong coupling in productive provinces, and decoupling in less productive regions. Sampling discrete stations may obscure this pattern, because our data demonstrate that isolated hot spots or moments of CR occur in oligotrophic regions. This finding has implications for understanding carbon and metabolic balance in the sea, as it reinforces the importance of integrating CR over time and depth, but also lateral space.

The net effects of ocean warming and other forms of global change on CR will ultimately depend on a complex series of responses among communities of phytoplankton and heterotrophic microbes in different ocean provinces. For the first time, we provide basin-scale measurements of CR conducted at high spatial and temporal resolution. Like previous CR datasets, our data capture patterns over limited periods of time, and temporal variation may be intertwined with spatial differences. Unlike primary production, CR has not been measured regularly or systematically and cannot be easily sensed from space. Instead, use of spatially-distributed, high-quality, *in situ* biogeochemical measurements made at high frequency by autonomous gliders, floats, and moorings seems a promising approach for regular measurements of CR across the ocean. Our work provides CR data that are consistent in magnitude and pattern with previous data, demonstrating the feasibility of using autonomous platforms to measure CR over large scales. Such data allow us to identify broad biogeographic and biogeochemical patterns that are otherwise undetectable by isolated measurements. This reveals a more nuanced view of the environmental controls on CR, highlighting weak temperature effects in warm waters of the Pacific, and the coupling and decoupling between phytoplankton biomass and CR in different regions. Of particular relevance are hot spots and moments of CR, which may affect carbon flux estimates and represent interesting oceanographic and ecological phenomena. The novel dataset and approach provides hundreds of new measurements from the under-sampled Pacific, yielding new insight into variation in CR across the world's largest ocean.

Supporting Information

Figure S1   BDO computed community respiration rates along the two transects. Color shading shows (A) community respiration 1 (mmol m^{-3} d^{-1}) (B) community respiration 2 (mmol m^{-3} d^{-1}) (C) community respiration 3 (mmol m^{-3} d^{-1}) (D) community respiration 4 (mmol m^{-3} d^{-1}) (E) community respiration 5 (mmol m^{-3} d^{-1}) and (F) community respiration 6 (mmol m^{-3} d^{-1}) across the Pacific Ocean. Data displayed using Ocean Data View [59].

Acknowledgments

We thank Dr. Michael Colvin for his help with data management and R code.

Author Contributions

Conceived and designed the experiments: JMW JMB. Performed the experiments: JMW RS JMB. Analyzed the data: JMW RS JMB. Contributed reagents/materials/analysis tools: JMW RS JMB. Wrote the paper: JMW RS JMB.

References

1. Breitburg DL, Hondorp DW, Davias LA, Diaz RJ (2009) Hypoxia, nitrogen, and fisheries: Integrating effects across local and global landscapes. Marine Science 1: 349.

2. Breitburg DL, Crump BC, Dabiri JO, Gallegos CL (2010) Ecosystem engineers in the pelagic realm: Alteration of habitat by species ranging from microbes to jellyfish. Integrative and Comparative Biology 50: 188–200.

3. Wright JJ, Konwar KM, Hallam SJ (2012) Microbial ecology of expanding oxygen minimum zones. Nature Reviews Microbiology 10: 381–394.

4. Odum HT (1956) Primary production in flowing waters. Limnology and Oceanography 1: 103–117.

5. Falkowski PG (1994) The role of phytoplankton photosynthesis in global biogeochemical cycles. Photosynthesis Research 39: 235–258.

6. Spitzer WS, Jenkins WJ (1989) Rates of vertical mixing, gas exchange and new production: Estimates from seasonal gas cycles in the upper ocean near bermuda. J Mar Res 47: 169–196.

7. Williams P, Purdie D (1991) *In vitro* and *in situ* derived rates of gross production, net community production and respiration of oxygen in the oligotrophic subtropical gyre of the north pacific ocean. Deep Sea Research Part A.Oceanographic Research Papers 38: 891–910.

8. González N, Anadón R, Marañón E (2002) Large-scale variability of planktonic net community metabolism in the atlantic ocean: Importance of temporal changes in oligotrophic subtropical waters. Mar Ecol Prog Ser 233: 21–30.

9. Hamme RC, Emerson SR (2006) Constraining bubble dynamics and mixing with dissolved gases: Implications for productivity measurements by oxygen mass balance. J Mar Res 64: 73–95.

10. Marañón E, Pérez V, Fernández E, Anadón R, Bode A, et al. (2007) Planktonic carbon budget in the eastern subtropical north atlantic. Aquat Microb Ecol 48: 261–275.

11. Luz B, Barkan E (2009) Net and gross oxygen production from O 2/Ar, 17 O/ 16 O and 18 O/16 O ratios. Aquat Microb Ecol 56: 133–145.

12. Brown JH, Gillooly JF, Allen AP, Savage VM, West GB (2004) Toward a metabolic theory of ecology. Ecology, 85: 1771–1789.

13. Pomeroy LR, Deibel D (1986) Temperature regulation of bacterial activity during the spring bloom in newfoundland coastal waters. Science 233: 359–361.

14. López-Urrutia Á, San Martin E, Harris RP, Irigoien X (2006) Scaling the metabolic balance of the oceans. Proceedings of the National Academy of Sciences 103: 8739–8744.

15. Kirchman DL, Moran XA, Ducklow HW (2009) Microbial growth in the polar oceans-role of temperature and potential impact of climate change. Nature Reviews Microbiology 7: 451–459.

16. Riebesell U, Körtzinger A, Oschlies A (2009) Sensitivities of marine carbon fluxes to ocean change. Proceedings of the National Academy of Sciences 106: 20602–20609.

17. Wohlers J, Engel A, Zöllner E, Breithaupt P, Jürgens K, et al. (2009) Changes in biogenic carbon flow in response to sea surface warming. Proceedings of the National Academy of Sciences 106: 7067–7072.

18. Regaudie-de-Gioux A, Duarte CM (2012) Temperature dependence of planktonic metabolism in the ocean. Global Biogeochem Cycles 26: n/a-n/a. 10.1029/2010GB003907.

19. Robinson C, Williams PJ le B (2005) Respiration and its measurement in surface marine waters. Respiration in aquatic ecosystems: 148–181.

20. Serret P, Fernández E, Robinson C (2002) Biogeographic differences in the net ecosystem metabolism of the open ocean. Ecology 83: 3225–3234.

21. Quay P, Stutsman J (2003) Surface layer carbon budget for the subtropical N. pacific: Δ13C constraints at station ALOHA. Deep Sea Research Part I: Oceanographic Research Papers 50: 1045–1061. 10.1016/S0967-0637(03) 00116-X.

22. Williams PJ le B, Morris PJ, Karl DM (2004) Net community production and metabolic balance at the oligotrophic ocean site, station ALOHA. Deep Sea Research Part I: Oceanographic Research Papers 51: 1563–1578. 10.1016/ j.dsr.2004.07.001.

23. Viviani DA, Björkman KM, Karl DM, Church MJ (2011) Plankton metabolism in surface waters of the tropical and subtropical pacific ocean. Aquat Microb Ecol 62: 1–12.
24. Williams PJ le B, Quay PD, Westberry TK, Behrenfeld MJ (2013) The oligotrophic ocean is autotrophic*. Annu Rev Marine Sci 5: 535–549. 10.1146/annurev-marine-121211-172335. Available: http://dx.doi.org/10.1146/annurev-marine-121211-172335. Accessed 30 June 2014.
25. Liquid Robotics (2012) Pacx. Available: http://liquidr.com/. Accessed 1 February 2013.
26. Needoba JA, Peterson TD, Johnson KS (2012) Method for the quantification of aquatic primary production and net ecosystem metabolism using in situ dissolved oxygen sensors. In: Anonymous Molecular Biological Technologies for Ocean Sensing. Springer. pp. 73–101.
27. Wanninkhof R, Asher WE, Ho DT, Sweeney C, McGillis WR (2009) Advances in quantifying air-sea gas exchange and environmental forcing. Marine Science 1.
28. Teira E, Martínez-García S, Fernández E, Calvo-Díaz A, Morán XAG (2010) Lagrangian study of microbial plankton respiration in the subtropical North Atlantic Ocean: bacterial contribution and short-term temporal variability. Aquat Microb Ecol 61:31–43.
29. Liss PS, Slater PG (1974) Flux of gases across the air-sea interface. Nature 247: 181–184.
30. Wanninkhof R (1992) Relationship between wind speed and gas exchange over the ocean. Journal of Geophysical Research: Oceans (1978–2012) 97: 7373–7382.
31. MacIntyre S, Wanninkhof R, Chanton J (1995) Trace gas exchange across the air-water interface in freshwater and coastal marine environments. Biogenic trace gases: Measuring emissions from soil and water. Methods in ecology. Blackwell. 52–97.
32. Crusius J, Wanninkhof R (2003) Gas transfer velocities measured at low wind speed over a lake. Limnol Oceanogr 48(3): 1010–1017.
33. Nightingale PD, Malin G, Law CS, Watson AJ, Liss PS, et al. (2000) In situ evaluation of air-sea gas exchange parameterizations using novel conservative and volatile tracers. Global Biogeochem Cycles 14: 373–387.
34. Wanninkhof R, McGillis WR (1999) A cubic relationship between air-sea CO2 exchange and wind speed. Geophys Res Lett 26: 1889–1892.
35. Donelan M (1990) Air-sea interaction. The sea 9: 239–292.
36. Stauffer RE (1980) Windpower time series above a temperate lake. Limnol Oceanogr 25(3): 513–528.
37. Engle D, Melack JM (2000) Methane emissions from an amazon floodplain lake: Enhanced release during episodic mixing and during falling water. Biogeochemistry 51: 71–90.
38. Holmén K, Liss P (1984) Models for air–water gas transfer: An experimental investigation. Tellus B 36: 92–100.
39. Jähne B, Münnich KO, Dutzi RBA, Huber W, Libner P (1987) On the parameters influencing air-water gas exchange. Journal of Geophysical research 92: 1937–1949.
40. Jähne B, Libner P, Fischer R, Billen T, Plate E (1989) Investigating the transfer processes across the free aqueous viscous boundary layer by the controlled flux method. Tellus B 41: 177–195.
41. Longhurst A (1995) Seasonal cycles of pelagic production and consumption. Prog. Oceanogr. 36: 77–167.
42. Longhurst A (2010) Ecological Geography of the Sea. Academic Press.
43. Friedman JH (1991) Multivariate adaptive regression splines. The annals of statistics, 1–67.
44. Sverdrup HU, Johnson MW, Fleming RH (1946) The oceans, their physics, chemistry and general biology. New York: Prentice-Hall.
45. Karl DM (1999) A sea of change: Biogeochemical variability in the north pacific subtropical gyre. Ecosystems 2: 181–214.
46. Regaudie-de-Giouxa A, Duarte CM (2013) Global patterns in oceanic planktonic metabolism. Limnol Oceanogr 58: 977–986.
47. Stanley RH, Kirkpatrick JB, Cassar N, Barnett BA, Bender ML (2010) Net community production and gross primary production rates in the western equatorial pacific. Global Biogeochem Cycles 24: GB4001.
48. Duarte CM, Agustí S, del Giorgio PA, Cole JJ (1999) Regional carbon imbalances in the oceans. Science 284: 1735.
49. Williams PJ le B, Bowers DG (1999) Regional carbon imbalances in the oceans. Science 284: 1735–1735. 10.1126/science.284.5421.1735b.
50. del Giorgio PA, Duarte CM (2002) Respiration in the open ocean. Nature 420: 379–384.
51. Karl DM, Laws EA, Morris P, Williams PJlB, Emerson S (2003) Metabolic balance of the open sea. Nature 426: 32.
52. Duarte CM, Regaudie-de-Gioux A, Arrieta JM, Delgado-Huertas A, Agustí S (2013) The oligotrophic ocean is heterotrophic. Annu Rev Marine Sci 5: 551–569. 10.1146/annurev-marine-121211-172337. Available: http://dx.doi.org/10.1146/annurev-marine-121211-172337. Accessed 30 June 2014.
53. Westberry TK, Williams PJ le B, Behrenfeld MJ (2012) Global net community production and the putative net heterotrophy of the oligotrophic oceans. Global Biogeochem Cycles 26: n/a-n/a. 10.1029/2011GB004094.
54. Quay PD, Peacock C, Björkman K, Karl DM (2010) Measuring primary production rates in the ocean: Enigmatic results between incubation and non-incubation methods at station ALOHA. Global Biogeochem Cycles 24: n/a-n/a. 10.1029/2009GB003665.
55. Robinson C, Serret P, Tilstone G, Teira E, Zubkov MV, et al. (2002) Plankton respiration in the eastern atlantic ocean. Deep Sea Research Part I: Oceanographic Research Papers 49: 787–813.
56. Villareal TA, Wilson C (2014) A comparison of the pac-X trans-pacific wave glider data and satellite data (MODIS, aquarius, TRMM and VIIRS). PloS one 9: e92280.
57. Pérez V, Fernández E, Marañón E, Serret P, Varela R, et al. (2005) Latitudinal distribution of microbial plankton abundance, production, and respiration in the equatorial atlantic in autumn 2000. Deep Sea Research Part I: Oceanographic Research Papers 52: 861–880.
58. Hopkinson CS Jr, Smith EM (2005) Estuarine respiration: An overview of benthic pelagic, and whole system respiration. Respiration in aquatic ecosystems. Oxford University Press. pp. 122–146.
59. Schlitzer R (2013) Ocean Data View. Available: http://odv.awi.de. Accessed 30 June 2014.

2

A Global View of Transcriptome Dynamics during *Sporisorium scitamineum* Challenge in Sugarcane by RNA-seq

Youxiong Que[¶], **Yachun Su**[¶], **Jinlong Guo, Qibin Wu, Liping Xu***

Key Laboratory of Sugarcane Biology and Genetic Breeding, Ministry of Agriculture, Fujian Agriculture and Forestry University, Fuzhou, Fujian, China

Abstract

Sugarcane smut caused by *Sporisorium scitamineum* is a critical fungal disease in the sugarcane industry. However, molecular mechanistic studies of pathological response of sugarcane to *S. scitamineum* are scarce and preliminary. Here, transcriptome analysis of sugarcane disease induced by *S. scitamineum* at 24, 48 and 120 h was conducted, using an *S. scitamineum*-resistant and -susceptible genotype (Yacheng05-179 and "ROC"22). The reliability of Illumina data was confirmed by real-time quantitative PCR. In total, transcriptome sequencing of eight samples revealed gene annotations of 65,852 unigenes. Correlation analysis of differentially expressed genes indicated that after *S. scitamineum* infection, most differentially expressed genes and related metabolic pathways in both sugarcane genotypes were common, covering most biological activities. However, expression of resistance-associated genes in Yacheng05-179 (24–48 h) occurred earlier than those in "ROC"22 (48–120 h), and more transcript expressions were observed in the former, suggesting resistance specificity and early timing of these genes in non-affinity sugarcane and *S. scitamineum* interactions. Obtained unigenes were related to cellular components, molecular functions and biological processes. From these data, functional annotations associated with resistance were obtained, including signal transduction mechanisms, energy production and conversion, inorganic ion transport and metabolism, and defense mechanisms. Pathway enrichment analysis revealed that differentially expressed genes are involved in plant hormone signal transduction, flavonoid biosynthesis, plant-pathogen interaction, cell wall fortification pathway and other resistance-associated metabolic pathways. Disease inoculation experiments and the validation of *in vitro* antibacterial activity of the chitinase gene *ScChi* show that this sugarcane chitinase gene identified through RNA-Seq analysis is relevant to plant-pathogen interactions. In conclusion, expression data here represent the most comprehensive dataset available for sugarcane smut induced by *S. scitamineum* and will serve as a resource for finally unraveling the molecular mechanisms of sugarcane responses to *S. scitamineum*.

Editor: Binying Fu, Institute of Crop Sciences, China

Funding: This work was funded by National Natural Science Foundation of China (31101196 and 31340060), the earmarked fund for the Modern Agriculture Technology of China (CARS-20), Program for New Century Excellent Talents in Fujian Province University (2014) and Research Funds for Distinguished Young Scientists in Fujian Agriculture and Forestry University (xjq201202). The funders had no role in study design, data collection and analysis, decision to publish, or preparation of the manuscript.

Competing Interests: The authors have declared that no competing interests exist.

* Email: xlpmail@126.com

¶ These authors are co-first authors on this work.

Introduction

Sugarcane (*Saccharum officinarum*) is an important sugar crop, and disease within this commodity affects cane yield and sugar content. Sugarcane smut, or sugarcane whip smut, is an airborne fungal disease first discovered in South Africa's Natal in 1877 [1]. The disease commonly manifests after infection with *Sporisorium scitamineum*, presenting as a black growth from the tip ("smut whip") of the diseased sugarcane stalk. Infected sugarcanes sprout early, and tiller more than normal with slender stems and leaves. Also, smut whips grow on tillers, reducing sugarcane yield and sugar quality. Currently, sugarcane smut has emerged as a globally important disease, and prevalence is increasing annually. When infection is severe, it can cause a 20–50% loss in sugarcane production [2,3]. Thus, replacing susceptible with resistant cultivars is a cost-effective measure for controlling sugarcane smut [4].

Currently, studies of the molecular mechanisms of resistance to sugarcane smut are still few and only preliminary. Raboin's group used amplified fragment length polymorphism (AFLP) markers to analyze genetic maps of hybrids of resistant cultivar R570 and susceptible cultivar MQ76/53 [5]. Xu and colleagues developed random amplified polymorphic DNA (RAPD) markers for genes associated with resistance to sugarcane smut [6]. Thokoane and Rutherford applied cDNA-AFLP to investigate differentially expressed genes in sugarcane exposed to *S. scitamineum*. Sequence homology analysis revealed that with *S. scitamineum* stress, resistant cultivars differentially expressed putative serine/threonine protein kinase, chitin receptor kinase and long terminal repeat retrotransposon (LTR). In addition, 7 days after *S.*

scitamineum infection, expression of phenylpropanoid, flavonoid genes and chitinase protein family members were induced [7]. Heinze and colleagues analyzed different gene sequences expressed in sugarcane after *S. scitamineum* infection, reporting that these genes involved transcription factors and signal receptors associated with disease resistance and proteases associated with the phenylpropane-flavonoid metabolic pathway [8]. Borrás-Hidalgo's laboratory used a cDNA-AFLP technique for screening and obtained 62 genes that were differentially expressed after sugarcane infection with *S. scitamineum*. Among these, expression of 10 genes was down-regulated and 52 was up-regulated and of these, 19 were associated with defense and signal transduction. For example, sugarcane genes encoding nucleotide binding site-leucine-rich repeats, a nucleotide-binding site and a leucine-rich region (NBS-LRR), protein kinases and proteins associated with auxin and ethylene signaling pathways were found to be important to sugarcane smut resistance stability [9]. Que et al. applied both cDNA-AFLP and silver staining methods, and obtained 136 transcript-derived fragments (TDFs) differentially expressed in *S. officinarum* in response to *S. scitamineum* infection. Of these 40 TDFs (including 34 newly induced TDFs and 6 significantly up-regulated TDFs) were sequenced and data were confirmed using reverse transcription PCR (RT-PCR) [10].

Wu and colleagues applied a Solexa high-throughput sequencing technique to analyze differential gene expression after *S. scitamineum* infection, and obtained 2,015 differentially expressed sequence tags. Among these, 1,125 up-regulated and 890 down-regulated ESTs were identified, including 3 up-regulated ESTs associated with the MAPK signaling cascade pathway [11]. Que et al. examined sugarcane smut-resistant cultivar NCo376 and susceptible cultivar F134 using differential display PCR (DDRT-PCR) and identified 7 differentially expressed genes after *S. scitamineum* inoculation, and RT-PCR was applied to measure gene expression patterns in roots, stems and leaves after *S. scitamineum*, salicylic acid (SA) or hydrogen peroxide stress [12]. Su and colleagues used infected sugarcane buds to clone pathogenicity-associated β-1,3-glucanase genes *ScGluA* and *ScGluD* [13] with real-time quantitative PCR (RT-qPCR) examined gene expression, reporting that TDFs of target genes *ScGluA* and *ScGluD* were up-regulated under the stress of *S. scitamineum*. Moreover, compared to the susceptible cultivar, gene up-regulation in the resistant cultivar were faster, longer-lasting, and occurred in response to SA, methyl jasmonate (MeJA) or abscisic acid (ABA) induction, as well as NaCl or $CdCl_2$ stress. Different expression patterns of *ScGluA* and *ScGluD* genes under biotic and abiotic stresses were also documented. Que and co-workers applied two dimensional electrophoresis (2-DE) to measure protein expression of sugarcane after *S. scitamineum* inoculation [14]. Using matrix-assisted laser desorption/ionization time of flight mass spectrometry (MALDI-TOF-TOF/MS), 23 differentially expressed proteins were identified and bioinformatic analysis revealed that 20 of these proteins were associated with photosynthesis, signal transduction or disease resistance and 3 proteins had an uncertain function. It can be deduced that, after *S. scitamineum* infection, various disease-resistant pathways are activated in sugarcane, and studies suggest that sugarcane and *S. scitamineum* interactions involve complex biological processes. Further in-depth research is needed to study the mechanism behind these observations.

RNA-Seq is an emerging transcriptomic technology utilizing high-throughput sequencing to analyze tissue or cell cDNA libraries obtained via reverse transcription of total RNA. After counting read numbers, RNA expression alterations were calculated to identify new TDFs. Until now, many transcriptomic studies have been conducted on stressed plants and many pathogen stress-response genes have been identified from *Arabidopsis thaliana*, *Oryza sativa*, *Zea mays*, and *Triticum aestivum*, and pathogen resistance mechanisms have been explored. Wu and colleagues [15] used Solexa sequencing to analyze mixed *Vitis vinifera* leaf samples collected 4–8 days after *Plasmopara viticola* inoculation, and obtained 15, 249 differentially expressed candidate genes. Ward and co-workers [16] applied RNA-Seq to obtain transcriptome expression profiles of red raspberry cultivars resistant and susceptible to *Phytophthora rubi*. Data indicated that expression of genes associated with lignin synthesis and the citric acid cycle, as well as genes encoding pathogenesis-related proteins and WRKY family transcription factors were all increased. Strau and colleagues [17] applied RNA-Seq to isolate one *Xanthomonas vesicatoria*-resistant gene-Bs4C-from *Capsicum annuum* which can regulate AvrBs4, a transcription activator-like effector of *Xanthomonas*. Li's laboratory [18] used Solexa sequencing to analyze a transcriptome from an early interaction between *O. sativa* and *Magnaporthe grisea*, to provide a basis for investigating genes encoding *M. grisea* effector proteins and their functions. Thus, high-throughput techniques to examine the response of sugarcane inoculated with *S. scitamineum* at the transcriptome level may reveal metabolic pathways and molecular regulation networks involved, as well as to define the characteristics of transcriptional regulation and identify key genes involved in sugarcane smut resistance.

In the present study, a *S. scitamineum*-resistant sugarcane genotype (Yacheng05-179) and a susceptible genotype ("ROC"22) were analyzed 24, 48 and 120 h after *S. scitamineum* inoculation, and Illumina RNA-Seq sequencing, bioinformatics and RT-qPCR, transcriptome expression was performed to identify differentially expressed genes and offered detail of how sugarcane responds to *S. scitamineum* stress.

Materials and Methods

Ethics Statement

We confirm that no specific permits were required for the described locations/activities. We also confirm that the field studies did not involve endangered or protected species.

Plant Materials and Pathogen Inoculation

The source of *S. scitamineum* inoculum was collected from the most popular cultivar "ROC"22 in the Key Laboratory of Sugarcane Biology and Genetic Breeding, Ministry of Agriculture (Fuzhou, China), and stored at 4°C. Two cultivars of sugarcane, *S. scitamineum*-resistant Yacheng05-179 and -susceptible "ROC"22, were also maintained in our laboratory. Robust stems were collected from both genotypes after soaking in water for 24 h. Stems were placed in a light incubator (12-h light-dark cycle, 32°C) for germination. When buds grew to 2 cm, 5×10^6/mL *S. scitamineum* spore suspension (containing 0.01% volume ratio of Tween-20) was used to inoculate the sugarcane buds via puncture. Control buds received water inoculations. Next, sugarcane stems were cultured at 28°C and (12-h light-dark cycle) [19]. At 24, 48 and 120 h after inoculation, five single buds were randomly selected from each group, and immediately fixed with liquid nitrogen before being stored at −80°C. Each experiment was repeated three times.

Total RNA Extraction, Construction of cDNA Library and Illumina Sequencing

The above five buds from Yacheng05-179, 24 h after water inoculation (T1) and 24, 48 and 120 h after *S. scitamineum* inoculation (T2–T4), and "ROC"22, 24 h after water inoculation (T5) and 24, 48 and 120 h after *S. scitamineum* inoculation (T6–T8), were collected for total RNA extraction using Trizol reagent (Invitrogen, Shanghai, China), respectively. At least 20 μL extracted total RNA was then sent to Beijing Biomarker Technologies Inc. for cDNA library construction and Illumina sequencing (HiSeqTM 2000, Illumina Inc., San Diego, CA, USA).

Basic Data Processing and Analysis

Raw reads (double-ended sequences) obtained from sequencing were evaluated and a unigene library for sugarcane was obtained. Based on this library, gene structure annotation, expression analysis and function annotations were performed. The subroutine Getorf in the EMBOSS software package (http://emboss. sourceforge.net/apps/cvs/emboss/apps/getorf.html/) was used to predict open reading frames (ORFs). Comparing T2 *vs.* T1, T3 *vs.* T1, T4 *vs.* T1, T6 *vs.* T5, T7 *vs.* T5 and T8 *vs.* T5, unigene expression in both cultivars 24, 48 and 120 h after *S. scitamineum* inoculation were conducted. IDEG6 software (http://telethon.bio.unipd.it/bioinfo/IDEG6/) was used for a generalized Chi-square test, and obtained P values were corrected for multiple hypotheses testing using a false discovery rate (FDR). After correction, unigenes with false discovery rate (FDR) no greater than 0.01 and reads per kb per million reads (RPKM) between samples of no less than 2 (fold-change (FD) ≥ 2) were considered to be differentially expressed genes.

For gene function annotation, obtained unigene sequences were annotated by searching in various protein databases, including the National Center for Biotechnology Information (NCBI) non-redundant protein (Nr) database, the NCBI non-redundant nucleotide sequence (Nt) database, Swiss-Prot, TrEMBL, Cluster of Orthologous Groups (of proteins) (COG), Gene Ontology (GO) and the Kyoto Encyclopedia of Genes and Genomes (KEGG). Annotation information of homologous genes in these databases was used to represent annotations of obtained unigenes. In addition, information for differentially expressed genes was collected from unigene annotations, and these genes were subjected to GO and KEGG significant enrichment analyses to identify biological functions and metabolic pathways in which these genes participate.

Customized Data Analysis

To identify dynamic changes in differentially expressed genes in sugarcane after *S. scitamineum* stress, expression of infected cultivars and controls at different time points (two groups), and between different time points of the same cultivar (multiple groups) were analyzed. Data for gene roles were then analyzed.

Two-group analysis. Specifically, two-group analysis was used to study differentially expressed genes of both cultivars at 24, 48 and 120 h after *S. scitamineum* inoculation and corresponding controls at 24 h. Then, up/down-regulated genes were counted. Differentially expressed genes were subjected to COG functional annotation, GO classification analysis, and KEGG enrichment analysis, to obtain information about gene function and relevant regulation networks at different time points.

Multi-group analysis. To analyze differential gene expression of one genotype at different time points (multi-group analysis), genes with sustained differential expression in both genotypes collected 24 h after water inoculation (control) and 24, 48 and 120 h after *S. scitamineum* inoculation were investigated to find gene intersections among the four time points. Moreover, differentially co-expressed genes in both cultivars at the same time point, and differentially co-expressed genes in both cultivars at different time points were also counted. Comparisons of T1, T2 *vs.* T1, T3 *vs.*

T1 and T4 *vs.* T1, or T5, T6 *vs.* T5, T7 *vs.* T5 and T8 *vs.* T5, which underwent classification analysis in both cultivars and dynamic gene expression patterns, were obtained. For certain dynamic expression patterns, GO significance analysis and pathway enrichment analysis were performed.

The Role of Chitinase Genes in Response to Pathogen Infection

Based on the RNA-Seq data, the unigenes encoding sugarcane chitinases were differently expressed in sugarcane after inoculation with *S. scitamineum*. The chitinase gene *ScChi* (unigene ID: gi36003099) was cloned and identified. Expression profiles of *ScChi* during Yacheng05-179-*S. scitamineum* interaction and "ROC"22-*S. scitamineum* interaction at 0 h, 24 h, 48 h and 120 h, as well as mock plants were investigated by RT-qPCR. The *ScChi* transcript was calculated by subtracting mock plant expression from inoculated sample at each corresponding time points.

For transient expression of *ScChi* in *Nicotiana benthamiana*, we constructed an overexpression vector pCAMBIA 1301-*ScChi* to analyze its defense response. The *Agrobacterium strain* EHA105 carrying the recombinant vector was grown overnight in LB liquid medium containing 35 μg/mL rifampicin and 50 μg/mL kanamycin at 28°C. Culture cells were collected and resuspended in MS liquid medium containing 200 μM acetosyringone at $OD_{600} = 0.8$. Then, cells were infiltrated into eight-leaf stage-old *N. benthamiana* leaves. For comparison, the *Agrobacterium strain* containing the pCAMBIA 1301 vector alone was also transiently expressed in *N. benthamiana* leaves. The materials were incubated at 28°C for 24 h (16 h light/8 h darkness). Then a dilution ($OD_{600} = 0.5$) of *Fusarium solani* var. *coeruleum* or *Botrytis cinerea* suspended in 10 mmol/L $MgCl_2$ was infiltrated into the main vein of the infected leaves. Tested plants were cultured at 28°C (16 h light/8 h darkness) for 20 d and photographed.

To validate antifungal activity, *N. benthamiana* plants were infected with *Agrobacterium strain* EHA105 carrying pCAMBIA 1301-*ScChi* or pCAMBIA 1301 vector by the leaf disc method. The initial transgenic *N. benthamiana* lines (T_0) was selected with 35 mg/mL hygromycin and were further identified by PCR and RT-PCR. The mycelium of the *Fusarium solani* var. *coeruleum* was inoculated in the middle of the petri plates containing potato dextrose agar (PDA). Four days after inoculation, filter papers ~1 cm distance from hyphae were filled with chitinase from three different T_0 generation of *ScChi* transgenic *N. benthamiana* plants, and controls were filled with chitinase from the T_0 generation of pCAMBIA 1301 transgenic *N. benthamiana* or untransgenic *N. benthamiana* plants, or 0.1 mol/L sodium acetate buffer (pH 5.0). The antibacterial effect was observed after cultivation at 28°C for 2 d.

RT-qPCR Validation

To validate the reliability of differentially expressed genes obtained from Illumina RNA-Seq sequencing, six co-expressed, up-regulated genes from both cultivars: sugar cane_unigene_BMK.40387 (metacaspase-1-like, Q1), sugar cane_unigene_BMK.49302 (ribonuclease 3-like, Q2), sugar cane_unigene_BMK.51436 (pathogenesis-related protein PR-10, Q3), sugar cane_unigene_BMK.57924 (sucrose transporter SUT1, Q4), sugar cane_unigene_BMK.63074 (vacuolar amino acid transporter 1-like, Q5) and sugar cane_unigene_BMK.63784 (heat shock protein-like, Q6) were subjected to RT-qPCR. First-strand cDNAs (10-fold dilution) of sugarcane buds collected from both cultivars 24 h after water inoculation (control) and 24, 48 and 120 h after *S. scitamineum* inoculation were used as templates, and specific primers were designed according to differential gene sequences [20] (Table S1 in File S1). Glyceraldehyde 3-phosphate dehydrogenase (*GAPDH*) [21] served as the internal reference gene. SYBR Green staining was applied for RT-qPCR using the ABI 7500 fast real-time PCR system (Applied Biosystems, Foster, CA, USA). The total reaction volume was 25 µL, including 12.5 µL FastStart Universal SYBR Green PCR Master (ROX Medical, Shanghai, China), 0.4 µmol/L primer and 2.0 µL template. Reaction conditions were: 50°C, 2 min; 95°C, 10 min; 95°C, 15 s, 60°C, 1 min, and 40 cycles and three replicates were performed for each. PCR using distilled water as the template was used as a blank control. A $2^{-\Delta\Delta Ct}$ algorithm was applied for quantitative gene expression analysis [22].

Results

RNA-Seq Results

Illumina RNA-Seq of eight samples yielded 36.68 Gb of data and 181,603,016 read pairs. Trinity software was used to assemble data for T1–T8. Data indicate that the assembled eight-sample unigene library included 148,605 unigenes. Among these, 46,525 exceeded 500 bp, accounting for 31.31% of all unigenes, and 20,798 exceeded 1.0 kb, accounting for 14% of all unigenes.

Merged data were assembled and clustered according to similarity to 15,394 sugarcane unigene sequences downloaded from the National Center for Biotechnology Information (NCBI) website to construct the merged unigene database (Merge_Unigene) and for subsequent analyses. As shown in Table 1, compared to unigenes obtained via simple assembling, Merge_Unigene had greater gene comparison efficiency. Merge_Unigene contains 99,824 unigenes. Among these, 47,345 exceeded 500 bp, accounting for 47.43% of all unigenes in Merge_Unigene, and 22,091 exceeded 1.0 kb, accounting for 22.13% of all unigenes.

The 99,824 sugarcane unigenes were annotated by searching various protein databases, including Nr, Nt, Swiss-Prot, TrEMBL, COG, GO and KEGG. In total, transcriptome sequencing of all the eight samples revealed gene annotations of 65,852 unigenes.

Differential Gene Analysis of Both Genotypes at Different Time Points (Two-group Analysis)

Screening Two-group Genes after S scitamineum Inoculation. Two-group analysis yielded data about differentially expressed sugarcane genes at different time points after *S. scitamineum* inoculation (see Table 2). After *S. scitamineum* induction, up-regulated genes exceeded down-regulated genes. Also, as *S. scitamineum* inoculation was extended, differentially expressed genes in "ROC"22 gradually increased. After 120 h interaction, induced differentially expressed genes in "ROC"22 were 1.85 times greater than those of Yacheng05-179 (incompatible interaction). Also, differentially expressed genes in Yacheng05-179 48 h after *S. scitamineum* inoculation were greater than that at 24 and 120 h after inoculation, and exceeded the number of differentially expressed genes in "ROC"22 at the same time point (48 h). Data suggest that after *S. scitamineum* stress, differential gene expression was induced in the smut-resistant cultivar (24 and 48 h) earlier than that in the smut-susceptible cultivar (120 h).

Functional Annotation of Differentially Expressed Genes. COG functional annotation revealed that after *S. scitamineum* stress, in "ROC"22, 22 differentially expressed general function genes were predicted at 24 h, 47 at 48 h, and 174 at 120 h. Functional annotation information associated with smut resistance, such as signal transduction mechanisms (16 genes distributed at 24 h, 19 genes distributed at 48 h, and 88 genes distributed at 120 h), energy production and conversion (5, 11, and 28), inorganic ion transport and metabolism (2, 8, and 33), and defense mechanisms (0, 2, and 14) were observed. Table S2 in File S1 showed the GO classification of up/down regulated genes (p≤0.05) in "ROC"22 after *S. scitamineum* inoculation. Some differentially expressed genes appeared to be related to smut resistance, including metabolic process (299, 460, and 1176), response to stimulus (236, 353, and 958), biological regulation (218, 297, and 838), immune system process (58, 80, and 213), and antioxidant activity (9, 15, and 21).

COG analysis of genes in Yacheng05-179 after *S. scitamineum* inoculation revealed that differentially expressed general function genes predicted at 24, 48 and 120 h were 61, 102 and 80, respectively. We also found some functional annotation information associated with smut resistance which involved in signal transduction mechanisms (21 genes distributed at 24 h, 40 genes distributed at 48 h, and 26 genes distributed at 120 h), energy production and conversion (17, 27, and 32), inorganic ion transport and metabolism (10, 24, and 19), and defense mechanisms (3, 7, and 3). The differentially expressed genes distributed to the metabolic process (505, 758, and 634), response to stimulus (400, 570, and 473), biological regulation (351, 511, and 424), immune system process (61, 102, and 72), and antioxidant activity (9, 18, and 19) were found according to GO analysis (Table S3 in File S1).

Table 1. Assembly results of sugarcane transcriptome using trinity software.

Length range	Unigene	Merge_Unigene
200 bp–300 bp	28,473 (30.85%)	27,684 (27.73%)
300 bp–500 bp	24,852 (26.93%)	24,795 (24.84%)
500 bp–1,000 bp	18,488 (20.03%)	25,254 (25.30%)
1,000 bp–2,000 bp	12,792 (13.86%)	14,383 (14.41%)
2,000+bp	7,692 (8.33%)	7,708 (7.72%)
Total number	92,297	99,824
Total length	70,251,430 bp	77,293,229 bp
N50 length	1,300 bp	1,143 bp
Mean length	761.1453 bp	774.2950 bp

Notes: N50 length is an indicator of measuring assembly effect, which is calculated by the accumulated length of the assembled fragments from long to short. When the sum is greater than or equal to 50% of the total length, the final accumulated fragment length is the N50 value. Mean length = for the average assembly length.

Table 2. Statistics of differentially expressed genes.

Combination	DEG set name	Up regulated	Down regulated	All DEG
T2 vs. T1	DR24	323	185	508
T3 vs. T1	DR48	520	230	750
T4 vs. T1	DR120	1,270	535	1,805
T6 vs. T5	DY24	536	219	755
T7 vs. T5	DY48	832	265	1,097
T8 vs. T5	DY120	727	250	977

Notes: T2 vs. T1, T3 vs. T1 and T4 vs. T1 represent the combination of "ROC"22 under *S. scitamineum* stress for 24, 48, or 120 h and "ROC"22 under sterile water stress after 24 h, respectively. T6 vs. T5, T7 vs. T5 and T8 vs. T5 refer to the combination of Yacheng05-179 under *S. scitamineum* stress for 24, 48, or 120 h and Yacheng05-179 under sterile water stress after 24 h, respectively. Unigenes with false discovery rate (FDR) no greater than 0.01 and reads per kb per million reads (RPKM) between samples of no less than 2 (fold-change (FD) ≥2) were considered to be differentially expressed genes (DEG).

Data for functional annotation of differentially expressed genes obtained from two-group analysis reveal the overall transcriptome of sugarcane in response to *S. scitamineum* infection. TDF expression involves all aspects of biological activity. As the time of *S. scitamineum* inoculation extended, differentially expressed genes in "ROC"22 gradually increased. In addition, differentially expressed resistance-associated genes in Yacheng05-179 48 h after *S. scitamineum* inoculation were higher than those at 24 and 120 h after inoculation, and were also greater than differentially expressed genes with the corresponding COG function and GO classification in the susceptible cultivar "ROC"22 at 24 and 48 h. However, at 120 h, differentially expressed resistance-associated genes in the susceptible cultivar exceeded those in the resistant cultivar, suggesting that both cultivars have specific responses to *S. scitamineum*, and that the timing of induced gene expression in the smut-resistant cultivar was earlier.

Differential Gene Analysis of a Sugarcane Genotype at Different Time Points (Multi-group Analysis)

Here, both cultivar controls and samples at 24, 48 and 120 h after *S. scitamineum* inoculation were subjected to multi-group differential analysis to identify differentially expressed genes in sugarcane associated with response to *S. scitamineum* infection.

Analysis of Differentially Co-expressed Genes at the Same Time Point in Both Genotypes. Differentially expressed gene sets at the same time points of both cultivars (DR24–DY24, DR48–DY48 and DR120–DY120) were analyzed (see Fig. 1). At 24 h after *S. scitamineum* inoculation, 48 differentially co-expressed genes were identified in both cultivars (38 up-; 3 down-regulated). At 48 h, 115 differentially co-expressed genes were identified (103 up-; 5 down-regulated). Finally, at 120 h after inoculation, 246 differentially co-expressed genes were identified (218 up-; 18 down-regulated). Overall, after *S. scitamineum* infection of both cultivars, with increasing time, the number of induced differentially co-expressed genes increased.

Analysis of Sustained Differentially Expressed Genes in Both Genotypes Across Different Time Points. Differentially expressed gene sets at different time points of the same cultivar (DR24–DR48–DR120 h or DY24–DY48–DY120) were analyzed (see Fig. 2) and indicate that in "ROC"22 177 genes had sustained differential expression at 24, 48 and 120 h after inoculation (88 up-; 88 down-regulated). At 24 h 171 specific differentially expressed genes were identified (121 up-; 52 down-regulated). At 48 h, 289 were identified (200 up-; 89 down-regulated), and at 120 h, 1372 were identified (986 up-; 389 down-regulated). In Yacheng05-179, 328 genes had sustained differential

expression at 24, 48 and 120 h after inoculation (218 up-; 110 down-regulated). At 24 h 247 specific differentially expressed genes were identified (182 up-; 67 down-regulated). At 48 h, 425 were identified (338 up-; 87 down-regulated) and at 120 h, 377 were identified (289 up-; 90 down-regulated). Thus from 24 to 120 h after *S. scitamineum* infection, continuously differentially expressed genes in Yacheng05-179 is greater than those in "ROC"22 (1.85 times more; 2.48 times more for up-regulated genes and 1.25 times more for down-regulated genes). Genes with sustained differential expression in both cultivars are depicted in Fig. 2.

Analysis of Genes with Sustained Differential Co-expression in Both Cultivars at Different Time Points. Differential genes with sustained expression in both cultivars at all time points were analyzed (see Fig. 3) and data suggest that the co-expressed, up-regulated genes are associated with plant resistance, and can be used as candidate resistance genes in future studies.

Analysis of Dynamic Gene Expression Pattern of Both Genotypes at Different Time Points. Pathogen invasion into the host cell is a dynamic process and its growth and development in the host is a prerequisite for causing plant disease. During the invasion process, differentially expressed genes associated with disease resistance undergo changes in expression and genes with the same expression pattern for the same biological activity usually have similar functions. Here, we studied dynamic gene expression in both cultivars at different time points, and multi-group expression pattern clustering analysis was performed. After statistical analysis, cluster heatmaps of the dynamic expression patterns of differentially co-expressed genes in "ROC"22 or Yacheng05-179 at different time points (control→24 h→48 h→120 h, i.e. T1→T2→T3→T4 or T5→T6→T7→T8) (Fig. 4) and different dynamic expression models (Figs. 5 and 6) were plotted. As illustrated in Tables 3 or 4, in the four treated samples, 9 dynamic expression patterns of differentially expressed genes were identified. Combined with our knowledge about biological processes that *S. scitamineum* spores undergo after inoculation, we obtained three models of distinct and notable dynamic expression patterns of (0, 1, 2, 3), (0, −1, −2, −3) and (0, −1, −1, −1) in both genotypes.

In the pattern of (0, 1, 2, 3), Cluster No. 1 in "ROC"22 and Cluster No. 4 in Yacheng05-179, expression of differentially expressed genes continuously increased as inoculation time increased, and peaked at 120 h. The sustained accumulation of this gene type in response to *S. scitamineum* stress suggests important biological significance. Based on the different top 10

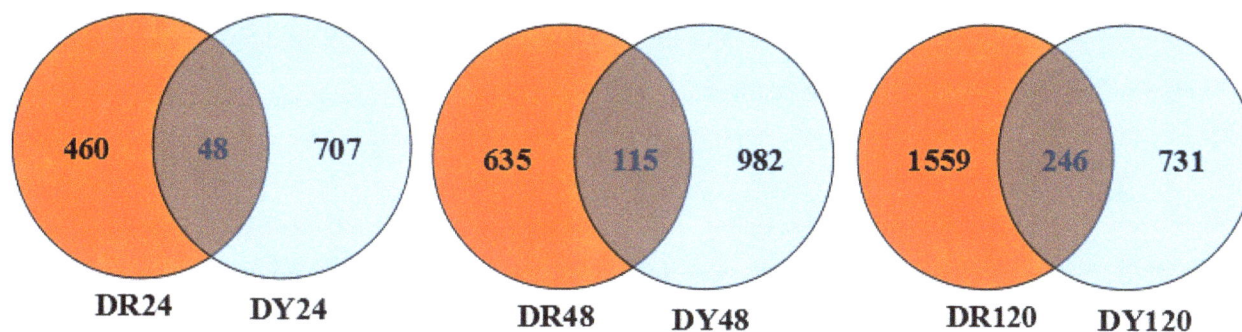

Figure 1. Venn diagram of differentially co-expressed genes in both sugarcane cultivars after *S. scitamineum* inoculation at the same time points. DR24, DR48 and DR120 denote differentially expressed gene sets obtained from "ROC"22 samples at 24, 48, and 120 h after *S. scitamineum* inoculation compared to controls 24 h after water inoculation, respectively; DY24, DY48 and DY120 denote differentially expressed gene sets obtained from Yacheng05-179 samples at 24, 48 and 120 h after *S. scitamineum* inoculation compared to control sample 24 h after water inoculation, respectively. All DEGs: differentially expressed genes.

GO classifications, the number of differentially co-expressed genes in the resistant Yacheng05-179 cultivar was greater than that in "ROC"22 (Table 5). As for the pattern of $(0, -1, -2, -3)$, Cluster No. 3 in "ROC"22 and Cluster No. 8 in Yacheng05-179, expression of differentially expressed genes continuously decreased as inoculation time increased, and were minimal at 120 h of plant-pathogen interaction. In the pattern of $(0, -1, -1, -1)$, Cluster No. 6 in "ROC"22 and Cluster No. 3 in Yacheng05-179, differentially expressed genes were expressed at a low levels in controls, and maintained even lower expression at different time points after *S. scitamineum* stress, suggesting sustained down-regulation of these gene types after pathogen induction. Also, based on the different top 10 GO classifications in these two dynamic expression patterns, the number of differentially co-expressed genes in resistant cultivar were less than that in susceptible one (Table 5). It is interesting that the genes of Cluster No. 3 in the susceptible cultivar "ROC"22 were continuously down-regulated after *S. scitamineum* inoculation, and these were involved in the DNA replication pathways (14), pyrimidine metabolism (8), purine metabolism (7), ribosome (6), nucleotide excision repair (6), mismatch repair (6), homologous recombination (6), ubiquitin mediated proteolysis (5), glutathione metabolism (3) and base excision repair (3).

Analysis of Metabolic Pathways

Analysis of metabolic pathways in which differentially expressed genes may be involved or may participate are shown in Tables 6 and 7. After *S. scitamineum* inoculation in both cultivars, differential gene expression involved in resistance-associated metabolic pathways were induced (Tables 6, 7 and 8). In addition, as infection time increased (24 to 120 h), differentially expressed genes involved in metabolic pathways gradually increased. At 24, 48 and 120 h after infection, differentially expressed genes involved in metabolic pathways in "ROC"22 were 112, 142 and 287, respectively. Differentially expressed genes involved in metabolic pathways in Yacheng05-179 were 138, 217 and 202, respectively. Overall, after *S. scitamineum* inoculation, during early and middle stages of infection (24 and 48 h) differentially expressed genes involved in resistance-associated metabolic pathways in Yacheng05-179 were greater than those of the susceptible cultivar. Resistance-associated metabolic pathways of significant enrichment were also greater than those in "ROC"22. Thus, further exploration of differentially expressed genes in metabolic pathways is needed to understand the mechanism underlying sugarcane smut resistance.

The Role of Chitinase Genes in Response to Pathogen Infection

Chitinases (EC 3.2.2.14), which can catalyze poly chitin are present in the cell walls of most fungi, and homologues in plant typical pathogenesis-related (PR) proteins. Our results indicated that 26 unigenes (gi36003099, Sugarcane_Unigene_BMK.51590, Sugarcane_Unigene_BMK.47839, Sugarcane_Unigene_BMK.34637, gi34957207, Sugarcane_Unigene_BMK.38981, gi35081719, gi35238203, Sugarcane_Unigene_BMK.38726, gi35992663, Sugarcane_Unigene_BMK.69934, Sugarcane_Unigene_BMK.49423, gi35980761, gi36002588, Sugarcane_Unigene_BMK.44826, Sugarcane_Unigene_BMK.56580, gi32815041, Sugarcane_Unigene_BMK.60969, Sugarcane_Unigene_BMK.40091, gi35045219, gi36066432, Sugarcane_Unigene_BMK.60821, gi36021860, Sugarcane_Unigene_BMK.48857, Sugarcane_Unigene_BMK.68059 and Sugarcane_Unigene_BMK.64954) encoding chitinases were differently expressed in sugarcane after inoculation with *S. scitamineum* (Table 8). The transcript of an acidic class III chitinase *ScChi* (gi36003099) was triggered during challenge with *S. scitamineum* in both resistant (Yacheng05-179) and susceptible ("ROC"22) cultivars, but gene expression was greater and maintained longer in the resistant cultivar (Fig. 7A). To determine whether *ScChi* (GenBank Accession No. KF664180) affects resistance to fungi, over-expressing pCAMBIA 1301-*ScChi* helped improve *N. benthamiana* in defending *Fusarium solani* var. *coeruleum* and *Botrytis cinerea* after inoculation for 20 d (Fig. 7B) which had significantly greater disease resistance than controls. Meanwhile, chitinase from plant 3 of the T_0 generation of *ScChi* transgenic *N. benthamiana* could inhibit hyphal growth of *Fusarium solani* var. *coeruleum* (Fig. 7C).

RT-qPCR Validation

To validate sequencing reliability, 6 differentially co-expressed genes in both genotypes were subjected to RT-qPCR analysis (metacaspase-1-like gene (Q1), ribonuclease 3-like gene (Q2), pathogenesis-related protein (PR-10) gene (Q3), sucrose transporter (SUT1) gene (Q4), vacuolar amino acid transporter 1-like gene (Q5) and heat shock protein-like gene (Q6)). RT-qPCR results (Fig. 8) for these differentially expressed genes were similar to Illumina sequencing results, but bias in the degree of differential expression between the two data sets, likely because the sensitivity of Illumina sequencing is greater than that of RT-qPCR [23]. In general, RT-qPCR data depicted up/down regulation patterns of differential TDFs that were consistent with Illumina sequencing results, suggesting that Illumina data are relatively reliable.

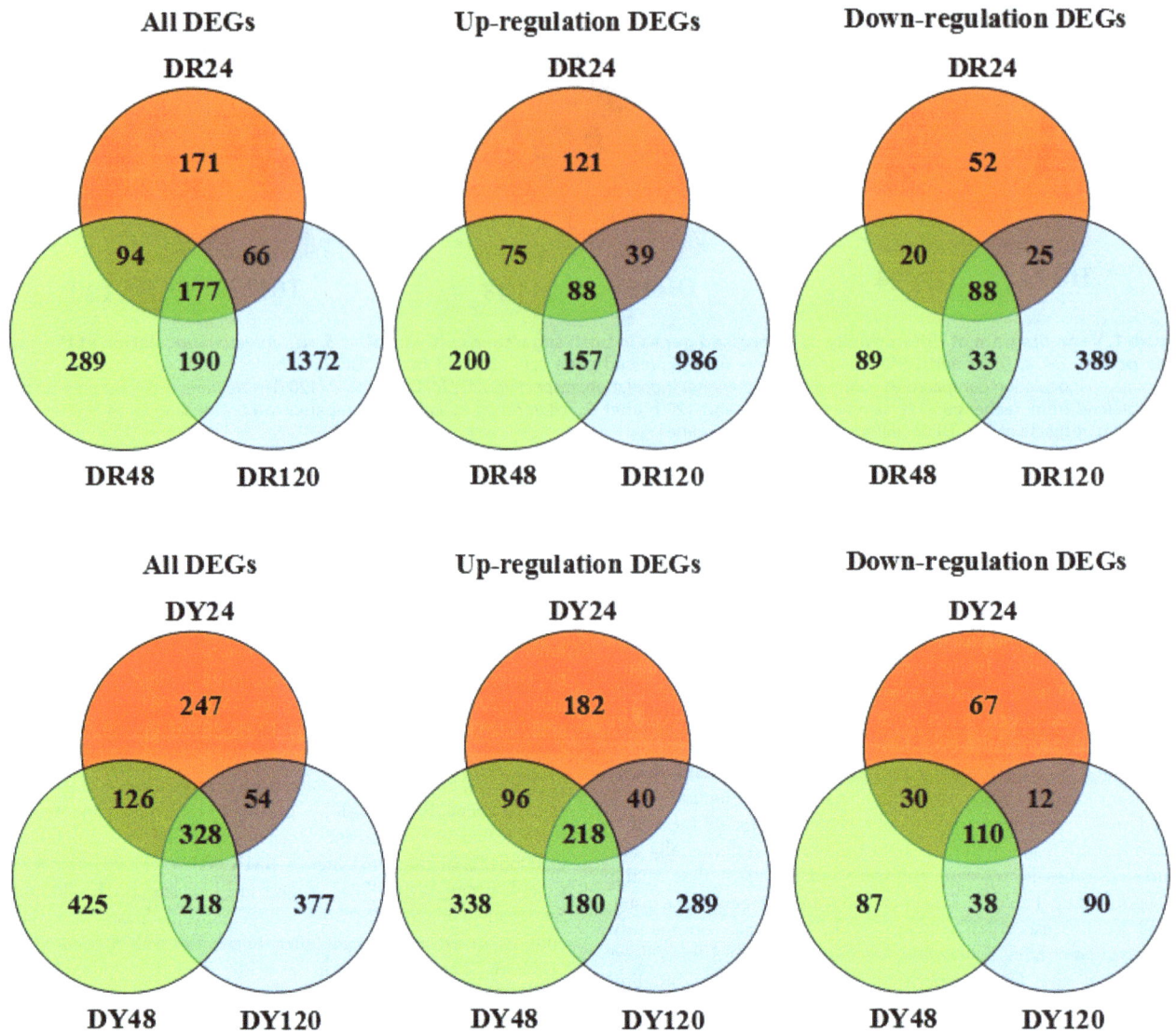

Figure 2. Venn diagram of differentially co-expressed genes in both sugarcane cultivar after *S. scitamineum* inoculation at different time points. DR24, DR48 and DR120 denote differentially expressed gene sets obtained from "ROC"22 samples at 24, 48 and 120 h after *S. scitamineum* inoculation compared to control sample at 24 h after water inoculation, respectively; DY24, DY48 and DY120 denote differentially expressed gene sets obtained from Yacheng05-179 samples at 24, 48 and 120 h after *S. scitamineum* inoculation compared to control sample at 24 h after water inoculation, respectively; All DEGs, all differentially expressed genes; Up-regulation DEGs, up-regulated genes; Down-regulation DEGs, down-regulated genes.

Discussion

Application of Illumina RNA-Seq in Transcriptome Studies

With RNA-Seq technology, no prior assumption is needed to obtain transcriptional activity of tissues or cells under particular conditions (including coding and non-coding RNAs). Illumina technology can combine reads of different lengths, single and double-terminal sequencing, strand specificity, and obtain billions of reads. This method permits not only annotation of encoding simple sequence repeats (SSR) and single nucleotide polymorphisms (SNP), as well as identification of alternative splicing, but also allows discovery of new and rare TDFs, while identifying regulatory RNAs and determining the expression abundance of TDFs. Thus, accurate and complete gene function maps can be obtained. Transcriptome sequencing technology has widely used in many studies-such as bilberry fruit transcriptome library

sequencing and investigating expression of genes associated with anthocyanin biosynthesis [24], transcriptome analysis of eucalyptus under scorch viral stress [23], and transcriptome analysis of interactions between tomato and powdery mildew [25].

Here, RNA-Seq of eight sugarcane samples yielded 36.68 GB data and 181,603,016 pairs of reads. By assembling data from each cultivar (Table 1), a Unigene library was constructed. Unigenes of sugarcane published online were merged with sequencing data to construct the Merge_Unigene library. In the Unigene library and the Merge_Unigene library, there are 148,605 and 99,824 unigenes, respectively but unigenes exceeding 500 bp and 1.0 kb in both libraries do not differ much. TDFs of rather low expression were also assembled relatively completely in Merge_Unigene. Thus, for subsequent analyses, the Merge_Unigene database was used. Searches indicate that gene numbers of *Z. mays*, *Sorghum bicolor*, and other species related to *S. officinarum*

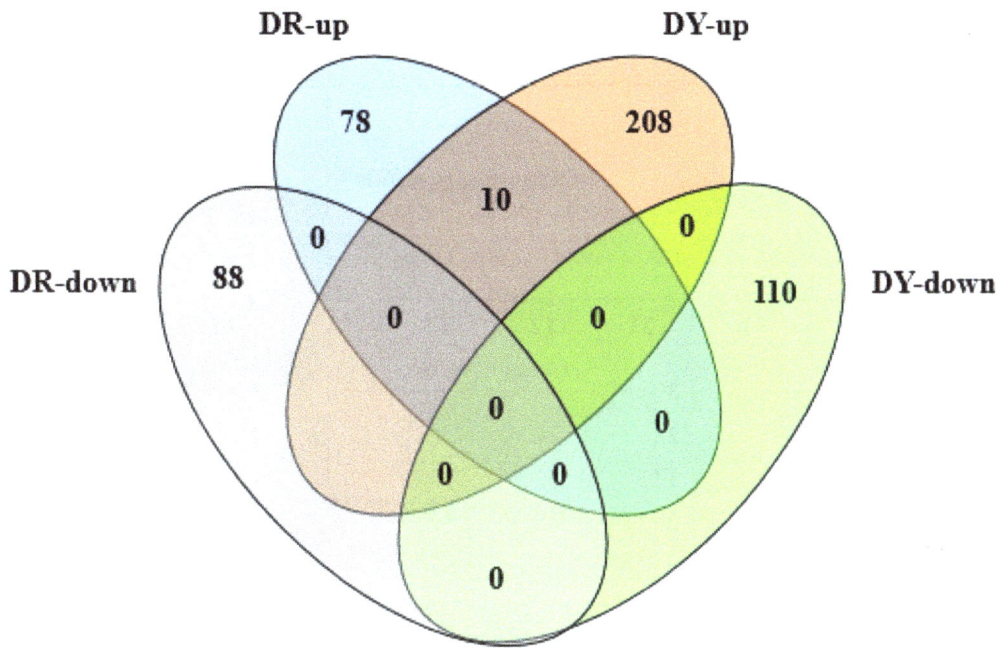

Figure 3. Venn diagram showing the number of genes with sustained differential co-expression between both sugarcane cultivars. DR-up and DR-down denote continuously up-regulated/down-regulated gene sets in "ROC"22 samples at 24, 48 and 120 h after *S. scitamineum* inoculation compared to control sample 24 h after water inoculation, respectively; DY-up and DY-down denote continuously up-regulated/down-regulated gene sets in Yacheng05-179 samples at 24, 48 and 120 h after *S. scitamineum* inoculation compared to control sample 24 h after water inoculation, respectively.

fall into the range of 30,000–40,000, a number similar to the number of relatively long unigenes (99,824) after merged assembling, suggesting a satisfactory outcome of merged assembling. Subsequent comparisons of unigenes with various databases showed that for most unigenes, especially those longer than 1.0 kb, annotation information on homologous sequences could be obtained. This also indicates the accuracy of the unigenes we obtained after assembling. The Merge_Unigene library constructed provided sufficient information for subsequent analysis and can be a reference for future sugarcane studies.

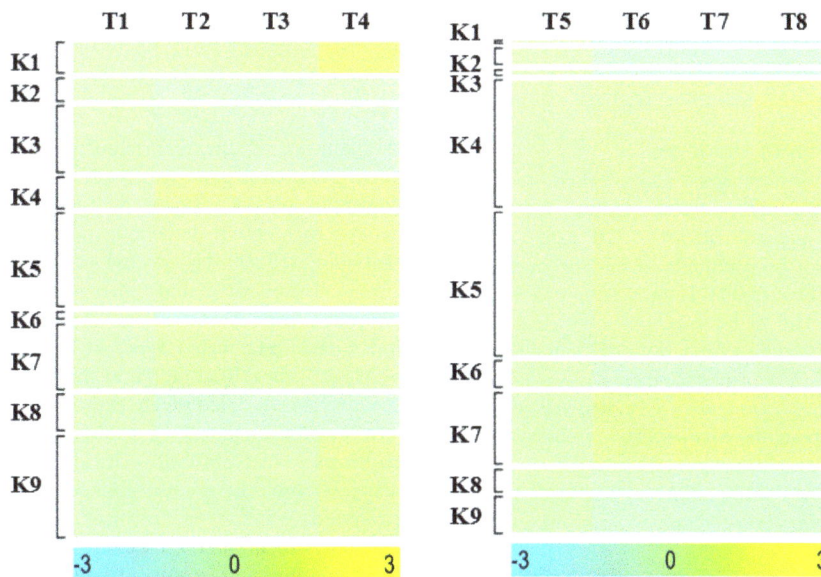

Figure 4. Cluster heatmap of expression patterns of differentially co-expressed genes in both sugarcane cultivars at different time points after *S. scitamineum* inoculation. T1, T2, T3 and T4 denote "ROC"22 at 24 h after water inoculation, and at 24, 48 and 120 h after *S. scitamineum* inoculation, respectively; T5, T6, T7 and T8 denote Yacheng05-179 at 24 h after water inoculation, and at 24, 48 and 120 h after *S. scitamineum* inoculation, respectively; k1~k9 indicate nine distinct expression patterns of differentially co-expressed genes.

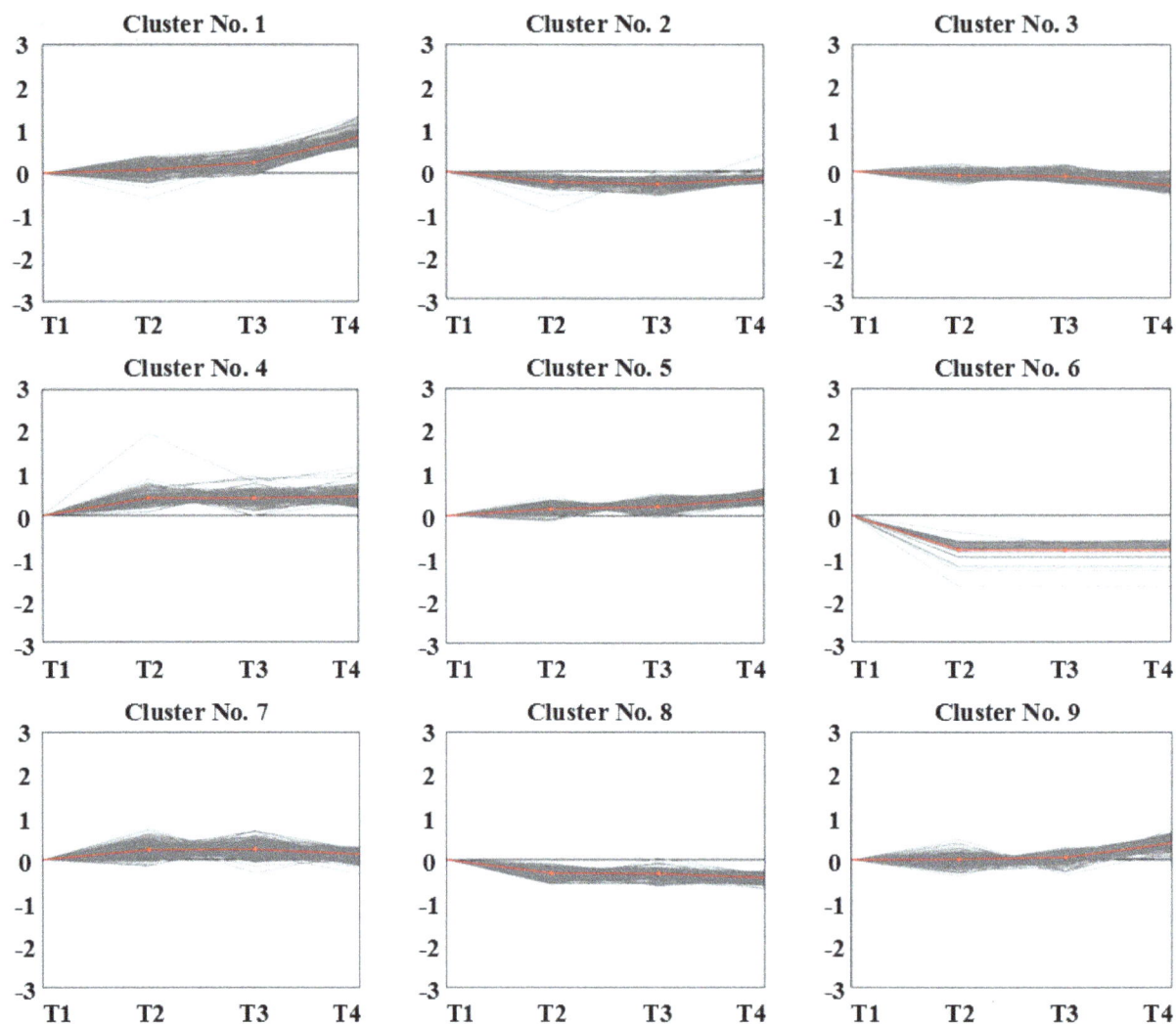

Figure 5. Dynamic expression models of differentially expressed genes in "ROC"22 after *S. scitamineum* inoculation (T1→T2→T3→T4). T1, T2, T3 and T4 denote "ROC"22 at 24 h after water inoculation, and at 24, 48 and 120 h after *S. scitamineum* inoculation, respectively.

Bioinformatic Annotation of Unigenes

Sugarcane is a highly heterozygous allopolyploid or highly heterozygous aneuploid crop and sugarcane hybrids have more than 120 chromosomes (genome sizes = 10 GB [26]). Whole genome sequencing has not been completed but the number of unigenes published by NCBI is 15,394. Here, we observed that the Merge_Unigene database had better comparison efficiency than the Unigene library; however, the comparison efficiency between the Merge_Unigene and *S. bicolor* unigene database was not high. Although *S. bicolor* and *S. officinarum* are related species, their gene sequences differ. Merging the *S. officinarum* unigene library with that of *S. bicolor* will introduce many unigenes that are useless to *S. officinarum*, and may likely affect subsequent analyses. Hence, the obtained *S. officinarum* library was not merged with the *S. bicolor* unigene library.

Bioinformatic annotations of the unigenes obtained 65,852 unigene annotations involved in cell parts, molecular functions and biological processes, among other functions. From this analysis, functional annotations associated with resistance were obtained (signal transduction mechanisms, energy production and conversion, inorganic ion transport and metabolism and defense

mechanisms). Pathway enrichment analysis revealed that differentially expressed genes are involved in plant-pathogen interaction (ko04626), plant hormone signal transduction (ko04075), phenylalanine metabolism (ko00360), peroxisome (ko04146), flavonoid biosynthesis (ko00941), phenylpropanoid biosynthesis (ko00940), ribosome (ko03010) and other resistance-associated metabolic pathways. As illustrated in Table 2, after *S. scitamineum* induction, up-regulated genes in both cultivars exceeded down-regulated genes. In addition, as the *S. scitamineum* infection time was prolonged, differentially expressed genes gradually increased. Transcriptome sequencing analysis showed that most genes and pathways induced in both cultivars were similar. *S. scitamineum* activated multiple smut-resistance pathways, and differentially expressed genes were involved in defense response, signal transduction and other processes. The response to *S. scitamineum* involved almost all aspects of biological activities, suggesting the pathogen response is regulated by multi-gene networks, a finding consistent with previous data which suggest that after pathogens infect plants, many metabolic pathways are affected, and gene expression in the transcription network is disturbed [27–29].

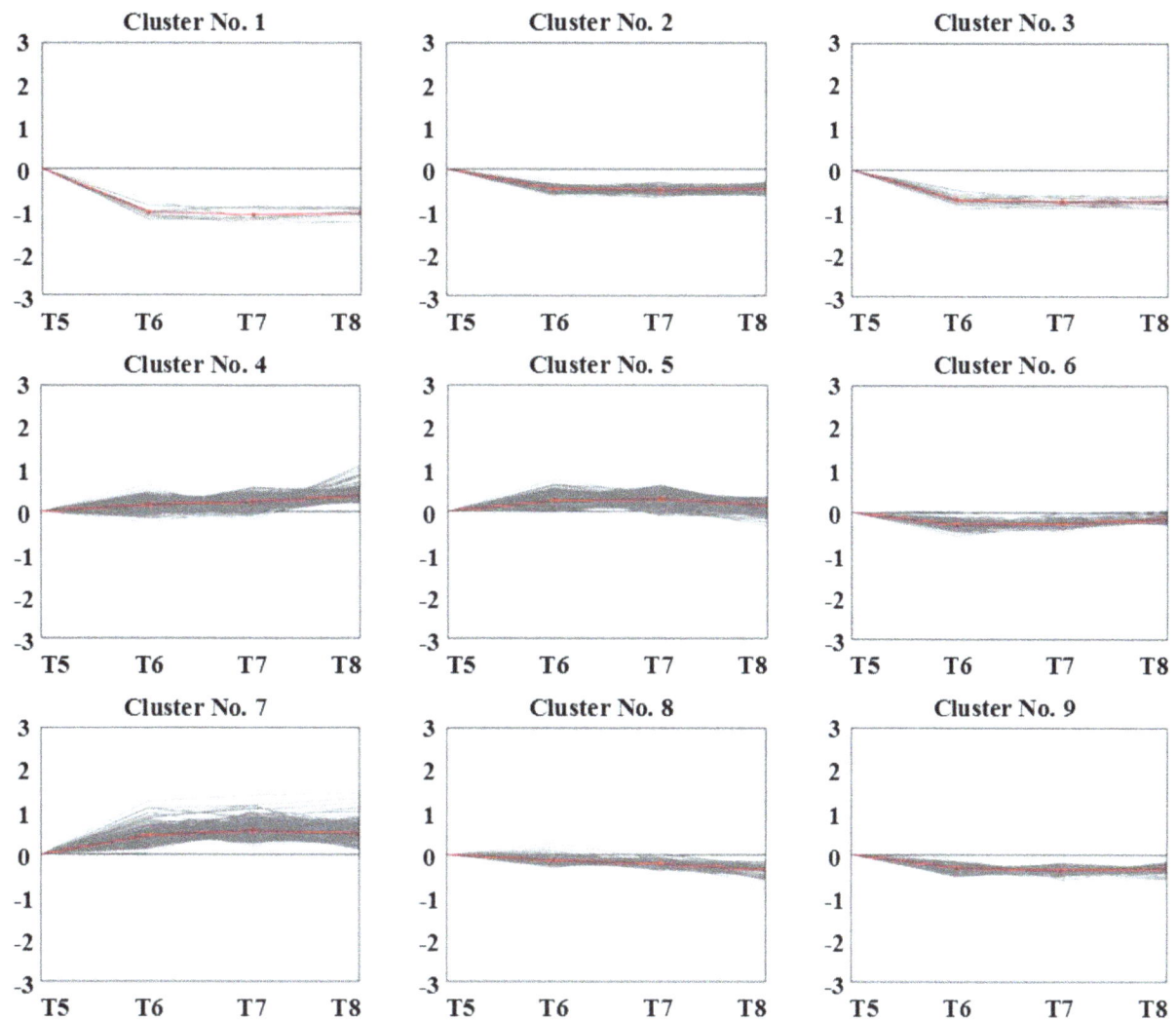

Figure 6. Dynamic expression models of differentially expressed genes in Yacheng05-179 after *S. scitamineum* inoculation (T5→T6→T7→T8). T5, T6, T7 and T8 denote Yacheng05-179 at 24 h after water inoculation, and at 24, 48 and 120 h after *S. scitamineum* inoculation, respectively.

Table 3. Dynamic expression patterns of differentially expressed genes in "ROC"22 after *S. scitamineum* inoculation.

Cluster No.	T1	T2	T3	T4	Gene Number
1	0	1	2	3	162
2	0	−2	−3	−1	118
3	0	−1	−2	−3	344
4	0	1	1	1	160
5	0	1	1	2	482
6	0	−1	−1	−1	39
7	0	2	2	1	339
8	0	−1	−1	−2	188
9	0	0	1	2	527

Notes: 0, 1, 2, 3, −1, −2 and −3 do not refer to the actual expression of the differentially expressed genes, but for the classification mark of gene dynamic changes. Gene numbers represent the actual number of dynamic expression patterns of differentially expressed genes. T1, "ROC"22 sample under sterile water stress after 24 h; T2–T4, "ROC"22 sample under *S. scitamineum* stress for 24, 48, and 120 h, respectively.

Table 4. Dynamic expression patterns of differentially expressed genes in Yacheng05-179 after *S. scitamineum* inoculation.

Cluster No.	T5	T6	T7	T8	Gene Number
1	0	−1	−3	−2	12
2	0	−1	−2	−1	64
3	0	−1	−1	−1	22
4	0	1	2	3	497
5	0	2	3	1	564
6	0	−3	−2	−1	104
7	0	1	3	2	280
8	0	−1	−2	−3	90
9	0	−1	−2	−1	142

Notes: 0, 1, 2, 3, −1, −2 and −3 do not refer to the actual expression of the differentially expressed genes, but for the classification mark of gene dynamic changes. Gene numbers represent the actual number of dynamic expression patterns of differentially expressed genes. T5, Yacheng05-179 sample under sterile water stress after 24 h; T6–T8, Yacheng05-179 samples under *S. scitamineum* stress for 24, 48, and 120 h, respectively.

What should also be stressed here is that, when comparing the samples at 48 h and 120 h post *S. scitamineum* inoculation with the samples at 24 h post water inoculation, they do have some differentially expressed genes which are involved in the senescing process. In the present study, as for this issue, firstly, only unigenes with no less than 2 (fold-change (FD) ≥2) were considered to be differentially expressed genes, which should largely decrease or even avoid the number of genes related to senescing process. Secondly, before we decide to further investigate the function of a certain differentially expressed gene, we need to confirm again that the differential expression is due to the challenged by *S. scitamineum*, but not only because of the senescing process, such as by RT-qPCR or Northern blot.

In the present studies, 33,972 unigenes (34.03%) were not annotated, likely due to incomplete whole genome sequencing of sugarcane, suggesting that a whole-genome database with functional annotations has not been established. Also, the number of sugarcane ESTs with known functions is limited and some short sequences obtained affected data comparisons. These unannotated unigenes may be new TDFs, and future experiments are needed to confirm this.

Analysis of Resistance-Associated Metabolic Pathways and Genes

Mechanisms underlying plant resistance to unique pathogens are diverse involving many molecular processes that could be slow or weak in susceptible plants. Such molecular changes affect later

Table 5. Analysis of GO classifications involving differentially co-expressed genes in three models of distinct and notable dynamic expression patterns in both sugarcane genotypes.

GO classifications	Pattern (0, 1, 2, 3)		Pattern (0, −1, −2, −3)		Pattern (0, −1, −1, −1)	
	"ROC"22	Yacheng05-179	"ROC"22	Yacheng05-179	"ROC"22	Yacheng05-179
cell	110	372	284	65	21	10
cell part	115	379	285	65	22	10
organelle	103	340	281	64	20	10
organelle part	0	0	0	34	0	0
membrane	54	203	0	42	17	0
cellular process	91	324	267	59	20	11
cellular component organization or biogenesis	0	0	180	0	0	0
metabolic process	96	0	241	62	0	10
developmental process	0	317	165	0	0	7
response to stimulus	87	0	179	31	20	10
immune system process	0	246	0	0	15	0
biological regulation	54	210	199	0	19	10
multicellular organismal process	0	0	0	0	0	7
catalytic activity	72	202	0	48	0	0
binding	77	274	247	49	17	10
localization	0	0	0	0	16	0

Table 6. Analysis of pathways involving differentially expressed genes after S. *scitamineum* inoculation in "ROC"22.

Pathway	ko_id	T2 vs. T1		T3 vs. T1		T4 vs. T1	
		Cluster_frequency	P-value	Cluster_frequency	P-value	Cluster_frequency	P-value
Plant-pathogen interaction	ko04626	7 out of 112 6.25%	0.0494	6 out of 142 4.24%	0.1354	4 out of 287 1.39%	0.8759
Phenylalanine metabolism	ko00360	3 out of 112 2.68%	0.3411	2 out of 142 1.41%	0.6382	5 out of 287 1.74%	0.3392
Phenylpropanoid biosynthesis	ko00940	3 out of 112 2.68%	0.4184	2 out of 142 1.41%	0.7069	6 out of 287 2.09%	0.2714
Plant hormone signal transduction	ko04075	3 out of 112 2.68%	0.9346	6 out of 142 4.22%	0.5623	8 out of 287 2.79%	0.8507
Flavonoid biosynthesis	ko00941	2 out of 112 1.79%	0.2076	3 out of 142 2.11%	0.0573	3 out of 287 1.05%	0.2081
Terpenoid backbone biosynthesis	ko00900	1 out of 112 0.89%	0.7516	3 out of 142 2.11%	0.1702	2 out of 287 0.70%	0.7300
Peroxisome	ko04146	1 out of 112 0.89%	0.9322	2 out of 142 1.41%	0.7644	1 out of 287 0.35%	0.9935
Ribosome	ko03010	10 out of 112 8.93%	0.3227	8 out of 142 5.63%	0.6471	10 out of 287 3.48%	0.9534

Notes: T2 *vs.* T1, T3 *vs.* T1 and T4 *vs.* T1 = combination of "ROC"22 sample under S. *scitamineum* stress for 24, 48, or 120 h and Yacheng05-179 under sterile water stress after 24 h, respectively.

Table 7. Analysis of pathways involving differentially expressed genes after S. *scitamineum* inoculation in Yacheng05-179.

Pathway	ko_id	T6 vs. T5		T7 vs. T5		T8 vs. T5	
		Cluster_frequency	P-value	Cluster_frequency	P-value	Cluster_frequency	P-value
Plant-pathogen interaction	ko04626	3 out of 138 2.17%	0.8646	10 out of 217 4.61%	0.0437	3 out of 202 1.49%	0.8842
Phenylalanine metabolism	ko00360	12 out of 138 8.70%	-	7 out of 217 3.23%	0.0486	8 out of 202 3.96%	0.0119
Phenylpropanoid biosynthesis	ko00940	12 out of 138 8.70%	0.0001	7 out of 217 3.23%	0.0844	8 out of 202 3.96%	0.0241
Plant hormone signal transduction	ko04075	8 out of 138 5.80%	0.5909	15 out of 217 6.91%	0.0463	9 out of 202 4.46%	0.5035
Flavonoid biosynthesis	ko00941	2 out of 138 1.45%	0.3429	2 out of 217 0.92%	0.3999	2 out of 202 0.99%	0.3647
Terpenoid backbone biosynthesis	ko00900	2 out of 138 1.45%	0.5882	3 out of 217 1.38%	0.3814	3 out of 202 1.49%	0.3380
Peroxisome	ko04146	0 out of 138 -	-	3 out of 217 1.38%	0.8038	2 out of 202 0.99%	0.9100
Ribosome	ko03010	0 out of 138 -	-	0 out of 217 -	-	1 out of 202 0.50%	1.0000

Notes: T6 *vs.* T5, T7 *vs.* T5 and T8 *vs.* T5 refer to the combination of Yacheng05-179 under S. *scitamineum* stress for 24, 48, or 120 h and "ROC"22 under sterile water stress after 24 h, respectively.

Table 8. Expression of resistance-related genes in sugarcane after *S. scitamineum* infection.

pathway	Gene	No. change	T2 vs. T1 No. ⇑/⇓	T3 vs. T1 No. ⇑/⇓	T4 vs. T1 No. ⇑/⇓	T6 vs. T5 No. ⇑/⇓	T7 vs. T5 No. ⇑/⇓	T8 vs. T5 No. ⇑/⇓
Plant hormone signal transduction	PYR/PYL	2	0/0	0/0	0/0	0/0	0/0	2/0
	PP2C	3	0/0	0/0	0/0	0/3	0/2	0/1
	SnRK2	3	0/0	1/0	1/0	0/0	0/1	0/0
	JAZ	7	3/0	3/0	1/0	2/0	5/0	3/0
	MYC2	1	0/0	0/0	0/0	1/0	1/0	0/0
Flavonoid biosynthesis	PAL	4	1/0	1/0	1/0	3/0	1/0	0/0
	C4H	4	1/0	1/0	1/0	1/0	2/0	2/0
	4CL	2	0/0	0/0	0/0	1/0	1/0	0/0
Plant-pathogen interaction	Glucanase	10	1/0	2/0	8/0	2/0	1/0	4/0
	Chitinase	26	0/3	4/5	15/3	11/0	7/0	3/0
	Catalase	1	0/0	1/0	1/0	0/0	1/0	1/0
Cell wall fortification pathways	Syntaxin	8	1/0	0/0	3/0	1/0	3/0	0/0
	HRGP	9	3/0	5/0	3/1	1/0	1/0	2/0
	CER1	1	0/0	0/0	1/0	0/0	0/0	0/0
Transcription factors	MYB	25	2/0	0/0	12/0	3/2	5/1	7/3
	WRKY	18	3/0	3/0	13/0	3/1	4/1	2/1
	ERF	18	6/0	2/1	2/1	8/1	8/0	9/0

Notes: T2 vs. T1, T3 vs. T1 and T4 vs. T1 refer to a combination of "ROC"22 under *S. scitamineum* stress for 24, 48, or 120 h and "ROC"22 under sterile water stress after 24 h, respectively. T6 vs. T5, T7 vs. T5 and T8 vs. T5 refer to the combination of Yacheng05-179 under *S. scitamineum* stress for 24 h, 48 h or 120 h and Yacheng05-179 under sterile water stress after 24 h, respectively.

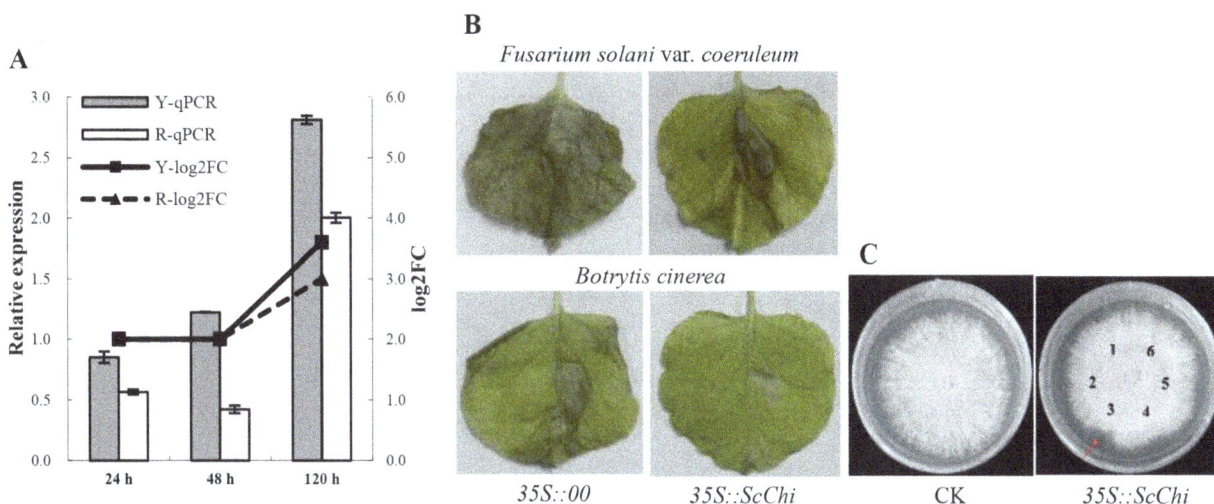

Figure 7. Analysis of the gene encoding sugarcane chitinase. (A) Transcript levels of *ScChi* during sugarcane-*Sporisorium scitamineum* interaction. The data of RT-qPCR was normalized to the *GAPDH* expression level. All data points (deduction its mock) are means ±SE (n = 3). Y, Yacheng05-179; R, "ROC"22. 24 h, 48 h and 120 h, sugarcane buds inoculation with *S. scitamineum* at 24 h, 48 h and 120 h, respectively; qPCR, detection results of real-time fluorescent quantitative PCR; $\log_2$FC, fold change of the differential expression of chitinase gene in the transcriptome. (B) The infection results of *Nicotiana benthamiana Fusarium solani* var. *coeruleum* and *Botrytis cinerea* by infiltrated with the *35S::ScChi*-containing *Agrobacterium* strain. The disease symptom was assessed 20 d after inoculation. (C) The antimicrobial action of chitinase (T_0 generation of ScChi transgenic *N. benthamiana*) on *Fusarium solani* var. *coeruleum*. CK, the control of normal culture on *Fusarium solani* var. *coeruleum*; *35S::ScChi*, the antimicrobial action of chitinase of T_0 generation of *ScChi* transgenic *N. benthamiana*; 1~3, chitinase from three different T_0 generation of *ScChi* transgenic *N. benthamiana*, respectively; 4, chitinase from T_0 generation of pCAMBIA 1301 transgenic *N. benthamiana*; 5, chitinase from untransgenic *N. benthamiana*; 6, 0.1 mol/L sodium acetate buffer (pH 5.0). Read arrow indicated the antibacterial effect.

alterations in plant appearance, physiology and biochemistry. Chief differences between resistant and susceptible plants can be found in the timing of host recognition of pathogen invasion, and the defense reaction speed and effectiveness. In susceptible plants, slow response and weaker defense signals allow pathogens to travel throughout and damage the plant [25]. How sugarcane and *S. scitamineum* interact at a molecular level is complex, so we investigated how two sugarcane cultivars confer resistance or susceptibility to *S. scitamineum* inoculation and whether this response is based on differential expression of genes involved with induction of resistance-associated metabolic pathways.

Plant Hormone Signal Transduction Pathways. Plant hormones are produced in response to environmental factors that regulate physiological reactions at low concentrations and include auxin, gibberellin acid (GA), cytokinin (CK), ABA, ethyne (ETH), SA, jasmonic acid (JA), brassinosteroid (BR) and polyamines. Plant hormones independently or collaboratively regulate plant growth, development and differentiation through cell division and elongation, differentiation of tissues and organs, sleep, seed germination, flowering and fruiting, aging and *in vitro* culture. Some plant hormones, such as SA, JA and ET which defend against pathogens have been well studied [30]. SA, JA and ET have been reported to form an orderly regulation network for plant-biotic stress interactions, improving plant tolerance to adverse environments. In particular, SA chiefly mediated acquired plant resistance, whereas JA and ET mediated induced systemic resistance (ISR) in response to biotic stresses [31]. ABA has been reported to affect plant responses to biotic stress mainly via interaction with other stress response pathways [32]. In the present study, many up-regulated genes were observed to be involved in plant hormone metabolism and signal transduction pathways, mainly ABA and JA pathways.

ABA is considered a negative regulatory factor in plant disease resistance, and its expression is associated with increased disease

sensitivity [32,33]. In the ABA signal transduction pathway, there are three core factors, ABA receptor PYR/PYL/RCAR protein (the most upstream regulator in the ABA signaling pathway), protein phosphatase (PP2C, negative regulatory factor) and SNF1-related protein kinase 2 (SnRK2, positive regulatory factor). These three factors form a double negative regulatory system to regulate ABA signal transduction and its downstream reactions. After the plant produces ABA induced by growth and development signals or environmental stimuli, ABA binds to PYR/PYL/RCAR protein, and interacts with PP2C to inhibit protein phosphatase activity as well as removes PP2C inhibition on SnRK2, thereby activating the ABA signal response [34]. Studies suggest that (Table 8), after *S. scitamineum* infection, TDFs of PYR/PYL, PP2C and SnRK2 are differentially expressed, indicating that the ABA signaling pathway is involved in the response of sugarcane to *S. scitamineum*. After 48 and 120 h of sugarcane and *S. scitamineum* interaction, in "ROC"22, sustained up-regulation of SnRK2 TDFs ($\log_2$FC = 1.43 and 1.53, respectively) was detected. In comparison, after 48 h of sugarcane and *S. scitamineum* interaction, in Yacheng05-179 one down-regulated SnRK2 TDF ($\log_2$FC = −1.10) was found. Thus, after the susceptible cultivar is infected by the pathogen, high expression of SnRK2 may activate the ABA signal, making it susceptible to infection or facilitating the reproduction and spread of the pathogen after infection.

When plants respond to environmental stress, signaling molecule JA acts the most rapidly, playing an important role in resistance reactions [23]. Studies indicate that after biotic and abiotic stresses, JA-related gene expression is up-regulated, causing JA accumulation [23]. MYC2 belongs to the myelocytomatosis protein family (MYCs), and is an MYC transcription factor with a role in regulating the JA response pathway and directly regulating downstream response genes. Jasmonate ZIM-Domain (JAZ) protein is a major inhibitory factor in the JA signaling pathway,

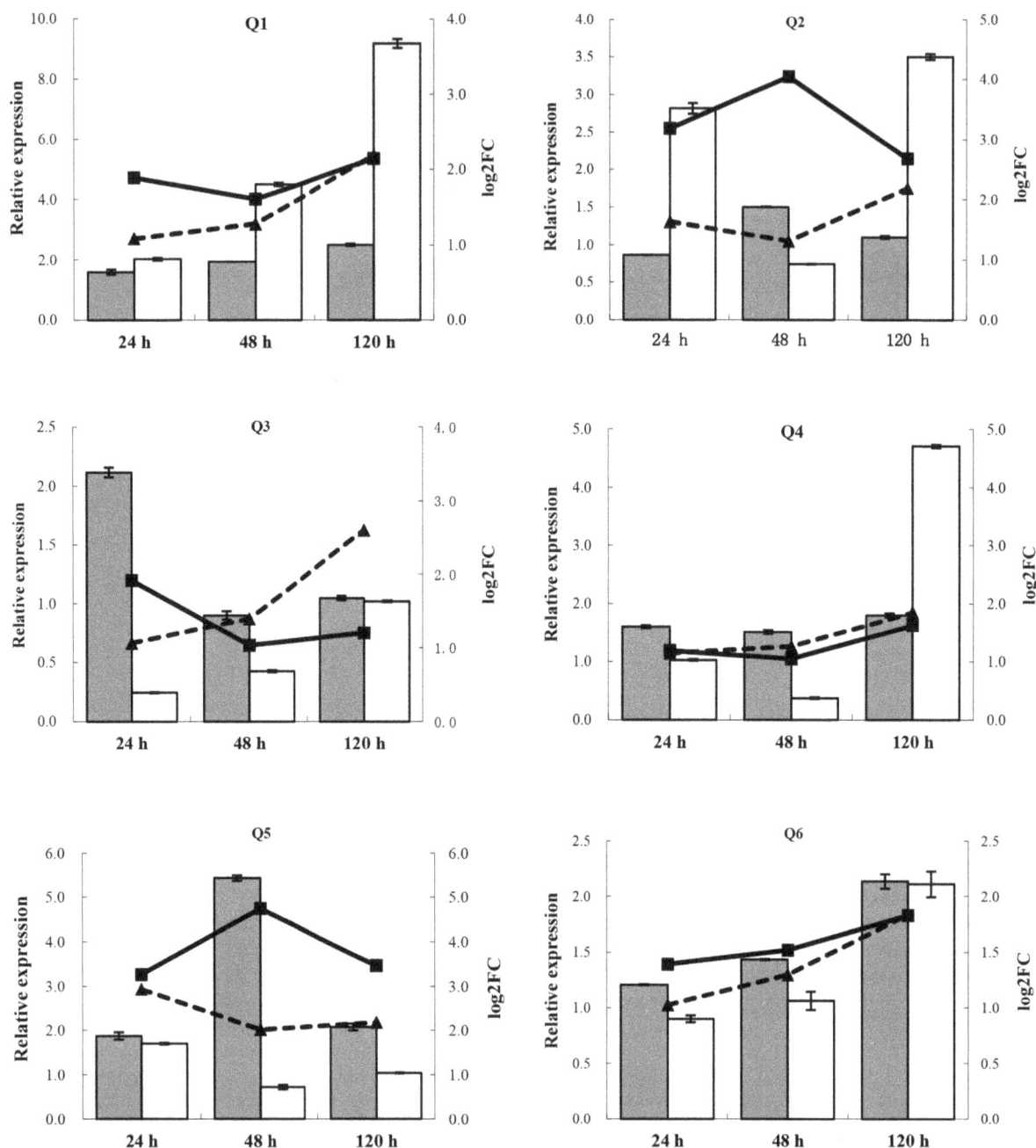

Figure 8. RT-qPCR validation of parts of differentially expressed genes identified by Illumina sequencing. Q1, metacaspase-1-like gene; Q2, ribonuclease 3-like gene; Q3, pathogenesis-related protein (PR-10) gene; Q4, sucrose transporter (SUT1) gene; Q5, vacuolar amino acid transporter 1-like gene; Q6, heat shock protein-like gene. Y, Yacheng05-179; R, "ROC"22; 24 h, 48 h and 120 h, sugarcane buds inoculation with *Sporisorium scitamineum* at 24 h, 48 h and 120 h, respectively; qPCR, detection results of real-time fluorescent quantitative PCR; log_2FC, fold change of the differential expression genes in the transcriptome.

able to bind to SCFcOI protein, leading to degradation mediated by 26S proteasome [23]. In *Arabidopsis thaliana* the COIl-JAZs-MYC2 complex function has been elucidated-when *A. thaliana* is not stimulated by JA type hormones, JAZ protein binds to MYC2, preventing MYC2 binding to downstream genes and JA response gene activation. When JA type hormones stimulate the plant, JA type signaling molecules activate the SCFCOIl protein complex, which competes with binding of JAZ to MYC2, releasing MYC2 proteins that bind to JA response genes, ultimately regulating plant

physiology [35]. As Table 8 shows, after *S. scitamineum* infection of sugarcane, JAZ and MYC2 genes were up-regulated suggesting that *S. scitamineum* can stimulate the JA biosynthesis and that the JA signaling pathway is involved in the response of sugarcane to *S. scitamineum*. In addition, as infection time prolonged, differentially expressed TDFs of JAZ and MYC2 increased as did their expression in Yacheng05-179 (more so than in "ROC"22), suggesting that in Yacheng05-179 JA signaling pathway activation in response to *S. scitamineum* is stronger.

Flavonoid Biosynthesis Pathway. Flavonoids are secondary metabolites widely present in plants with important roles in many biological processes (including response to biotic and abiotic stresses) [36,37]. The phenylpropanoid metabolic pathway is the key metabolic pathway leading to the flavonoid pathway. Under the catalysis of phenylalanine ammonia lyase (PAL), phenylpropanoid produces cinnamic acid; then after catalysis of cinnamate-4 hydroxylase (C4H), cinnamic acid produces 4-coumaric acid which then yields 4-coumarate CoA under the catalysis of 4-coumarate CoA ligase (4CL). Next, under the catalysis of chalcone synthase (CHS) and chalcone isomerase (CHI), 4-coumarate CoA and its derivatives enter the downstream flavonoid biosynthetic pathway [38]. PAL, C4H and 4CL are key enzymes in the phenylpropanoid metabolic pathway, with roles in plant responses to various biotic abiotic stresses including pathogen infection, mechanical damage and exogenous hormone stimuli [23].

Here, we observed that (Table 8) *S. scitamineum* induced upregulation of PAL, C4H, 4CL genes in both Yacheng05-179 and "ROC"22, yet there were more differentially expressed TDFs in Yacheng05-179 (PAL: sugar cane_unigene_BMK.40935, sugar cane_unigene_BMK.51492 and gi35076956; C4H: sugar cane_unigene_BMK.74288, sugar cane_unigene_BMK.65142 and gi35122896; 4CL: sugar cane_unigene_BMK.57158 and gi35030858) than in "ROC"22 (PAL: gi34918942; C4H: sugar cane_unigene_BMK.45497). These data show that resistance-associated genes involved in the flavonoid biosynthesis pathway in the resistant cultivar exceed those in the susceptible cultivar, and that sugarcane can synthesize polyphenolic compounds with antibacterial effects to defend itself from *S. scitamineum*.

Plant-pathogen Interaction Pathway. In long-term interactions with pathogens, plants form a series of defense mechanisms [39] and these include phytoalexin (PA) formation, hypersensitive response (HR) production, enzyme changes (such as peroxidase and polyphenol oxidase) and accumulation of pathogenesis-related proteins [40]. At 24, 48 and 120 h after *S. scitamineum* infection, in "ROC"22 proportions of differentially expressed genes involved in plant-pathogen interactions were affected (See Table 6). In Yacheng05-179, these proportions were also noted (Table 7), suggesting that the defense response occurs earlier than 48 h.

In addition, we observed that under *S. scitamineum* stress, pathogenesis-associated genes in sugarcane were expressed differentially (Table 8), including 10 glucanase, 26 chitinase genes and 1 catalase gene. Catalase is an important antioxidant enzyme in plants, functioning to clear and protect against metabolically produced H_2O_2. Studies suggest that catalase is needed for plant defense, stress response, and regulation of the cellular redox balance [41,42]. After 48 and 120 h of sugarcane and *S. scitamineum* interaction, transcription and expression of the catalase gene was induced. Chitinase and β-1,3-glucanase are two typical plant pathogenesis-associated proteins documented to act synergistically in defense against pathogenic fungi. Under normal condition, expression of chitinase and β-1,3-glucanase in plants is relatively low. After pathogenic fungal stress, β-1,3-glucanase and chitinase defense proteins accumulate in the cells. Currently, β-1,3-glucanase and chitinase have been detected in nearly 100 plant species, and β-1,3-glucanase and chitinase genes in many plants have been cloned [43]. Reports indicate that successful enhancement of plant resistance to diseases can be accomplished by introduction of exogenous β-1,3-glucanase and chitinase genes. In *O. sativa*, *Triticum aestivum*, *Nicotiana tabacum* and other species, trans-chitinase and/or trans-β-1,3-glucanase gene plants have been obtained [44,45]. Chitinases were encoded by a multi-gene family which have been reported to group into seven classes (Class I–VII) [46,47]. Thokoane and

Rutherford investigate differentially expressed genes after sugarcane exposure to *S. scitamineum* and sequence homology analysis revealed that chitinase protein family members were induced after *S. scitamineum* infection for 7 d [7]. RT-qPCR showed a high gene expression level of a sugarcane class IV chitinase gene ScChiB1 (EU914815.1) in the resistant cultivar than in the susceptible one during interaction with *Colletotrichum falcatum* [48]. Chitinases have been shown to inhibit the growth of chitin-containing fungi, both *in vitro* [49,50] and *in vivo* [51,52]. Our present study revealed that chitinases were triggered during *S. scitamineum* infection and data from inoculation experiments and the validation of *in vitro* antibacterial activity suggest a close relationship between the expression of *ScChi* and plant immunity. These data indicate that the sugarcane chitinase gene identified through RNA-Seq analysis is relevant to plant-pathogen interaction.

Cell Wall Fortification Pathway. The cell wall maintains plant cell morphology and participates in physiological activities such as extracellular signal identification [53]. Fungi obtain nutrition via saprophytes, parasitosis or symbiosis and *Ustilago maydis* in *Z. mays* and *Blumeria graminis f. sp. Hordei* in *Triticum aestivum* are documented to be biotrophic pathogens. In their genomes, genes encoding cell wall degradation enzymes are fewer than that in saprophytic fungi. Typically, these two fungi do not directly degrade plant cell walls; instead, they form haustoria in host epidermal cells to absorb nutrition [54,55]. Studies suggest that the cell wall is the first line of plant defense against pathogens [56,57]. After pathogen stress, with cell wall damage, disease signaling pathways are activated to initiate cell wall defense reactions. For example, cell walls surrounding invading pathogens produce and accumulate callose, phenolics and lignin, increasing the strength of the cell wall [58]. Genes encoding attachment protein receptors such as syntaxin in *Hordeum vulgare*, *N. tabacum* and *A. thaliana* participate in cell wall fortification, improving plant resistance [59]. We observed that (Table 8), after *S. scitamineum* inoculation, expression of genes involved in cell wall fortification were up-regulated in sugarcane, including 8 syntaxin genes and 9 hydroxyproline-rich glycoprotein (HRGP) genes. Also, 120 h after *S. scitamineum* inoculation, expression of waxy gene CER1 (gi36041011) was down-regulated ($\log_2FC = -1.63$) in "ROC"22, but remained unchanged in Yacheng05-179. Likely, increased expression of genes encoding proteins positively associated with the cell wall may enhance the resistance of sugarcane to *S. scitamineum*, and genes encoding proteins negatively associated with the cell wall may have the opposite effect. Thus, differential expression of these genes between resistant and susceptible sugarcane cultivars reflects cultivar resistance and susceptibility.

Resistance-associated Transcription Factors. Plant disease resistance involves coordinated expression of defense response genes. Transcriptional regulation commands expression of plant defense response genes, altering plant susceptibility or resistance and this is mediated by transcription factors [60]. When a plant is infected, signal transduction transcription factors in the plant are activated, and these then interact with the corresponding cis-acting elements via DNA-protein interaction, triggering expression of relevant defense response genes [61]. Studies suggest that transcription factors associated with plant disease resistance mainly include MYB transcription factor, the WRKY family in zinc finger proteins and ERF-type transcription factor [61].

MYB transcription factors can participate in plant systemic acquired resistance (SAR) and HR. Previously, over-expression of *A. thaliana* AtMYB30 in *A. thaliana* and *N. tabacum* were reported to result in HR to different pathogenic bacteria, and resistance to many bacterial diseases was enhanced [62]. *A. thaliana* MYB protein BOS1 can regulate plant resistance to *Botrytis cinerea*, *Alternaria alternata* and other necrotizing pathogens [63]. The WRKY family is a class of zinc finger proteins present in higher plants and all family members contain 1 to 2 WRKY (WRKYGQK) conserved domains. WRKY family proteins can bind the highly conserved element W-box in promoters of many plant defense response genes, offering important roles in plant defense responses [61]. Over-expression of the *O. sativa* WRKY45 gene can increase the resistance of *O. sativa* to *Pyricularia oryzae* Cav. [64]. *A. thaliana* WRKY11 and WRKY17 have negative regulatory roles in its basic resistance [61], and WRKY18, WRKY40 and WRKY60 can form complexes regulating plant disease resistance [65]. ERF proteins are members of the plant AP2/EREBP transcription factor superfamily and previously many ERF transcription factors were isolated from *A. thaliana*, *O. sativa*, *N. tabacum*, *T. aestivum* and other plants, and their roles in plant growth and development, as well as in response to biotic and abiotic stresses, have been documented [66]. Here, we found (Table 8) that in *S. officinarum* there were 25 MYB, 18 WRKY and 18 ERF differentially expressed genes. Compared to the susceptible cultivar, the number of activated transcription factors in the resistant cultivar was higher, and there were more up/down regulation of transcripts. These data suggest that the above-mentioned transcription factors actively regulate sugarcane resistance to *S. scitamineum*.

Conclusion

In summary, with RNA-Seq techniques, we performed transcriptome analysis on sugarcane genotypes at different resistance levels at different time points after *S. scitamineum* infection. Data indicate that as infection time was prolonged, activated differentially expressed genes in sugarcane increased. Differentially expressed genes induced by *S. scitamineum* inoculation in the resistant Yacheng05-179 cultivar and the susceptible "ROC"22 cultivar were similar. However, overall the expression time of resistance-associated genes in Yacheng05-179 (24–48 h) was earlier than that in "ROC"22 (48–120 h), and more transcript expressions were observed in the former, suggesting resistance specificity and early timing of these genes in non-affinity interactions between sugarcane and *S. scitamineum*. Data regarding potential functions of sugarcane TDFs in response to *S. scitamineum* will lay the foundation for future investigations into the role of these candidate genes in sugarcane-*S. scitamineum* interactions, and inform genomic studies on sugarcane smut resistance.

Supporting Information

File S1 Contains the following files: **Table S1.** Primers used for RT-qPCR analysis of differentially expressed genes. **Table S2.** Gene ontology classification of up⇧/down⇩ regulated genes in "ROC"22 after *S. scitamineum* inoculation. **Table S3.** Gene ontology classification of up⇧/down⇩ regulated genes in Yacheng05-179 after *S. scitamineum* inoculation.

Author Contributions

Conceived and designed the experiments: YQ YS LX. Performed the experiments: YQ YS QW JG. Analyzed the data: YQ YS JG QW LX. Contributed reagents/materials/analysis tools: YQ YS LX. Contributed to the writing of the manuscript: YQ YS LX. Revised and approved the final version of the paper: YQ LX.

References

1. Luthea J, Sattar A, Sandhu S (1940) Experiments on the control of smut of sugarcane (*Ustilago scitaminea* Syd.). Proc Natl Acad Sci India 12: 118–128.
2. Padmanaban P, Alexander K, Shanmugan N (1988) Effect of smut on growth and yield parameters of sugarcane. Indian Phytopathol 41: 367–369.
3. Wang BH (2007) Current status of sugarcane diseases and research progress in China. Sugar Crops China, 3: 48–51.
4. Chao C, Hoy J, Saxton A, Martin FA (1990) Heritability of resistance and repeatability of clone reactions to sugarcane smut in Louisiana. Phytopathology 80: 622–626.
5. Raboin L, Offmann B, Hoarau J, Costet L, Telismart H, et al. (2001) Undertaking genetic mapping of sugarcane smut resistance. Proc S Afr Sug Technol Ass 75: 94–98.
6. Xu LP, Chen RK (2005) Screening of RAPD markers linked to sugarcane smut resistant genes. Chin J Appl Environ Biol 10: 263–267.
7. Thokoane L, Rutherford R (2001) cDNA-AFLP differential display of sugarcane (*Saccharum* spp. hybrids) genes induced by challenge with the fungal pathogen *Ustilago scitaminea* (sugarcane smut). Proc S Afr Sug Technol Ass 75: 104–107.
8. Heinze B, Thokoane L, Williams N, Barnes JM, Rutherford RS (2001) The smut-sugarcane interaction as a model system for the integration of marker discovery and gene isolation. Proc S Afr Sug Technol Ass 75: 88–93.
9. Borrás-Hidalgo O, Thomma BP, Carmona E, Borroto CJ, Pujol M, et al. (2005) Identification of sugarcane genes induced in disease-resistant somaclones upon inoculation with *Ustilago scitaminea* or *Bipolaris sacchari*. Plant Physiol Biochem 43: 1115–1121.
10. Que YX, Lin JW, Song XX, Xu LP, Chen RK (2011) Differential gene expression in sugarcane in response to challenge by fungal pathogen *Ustilago scitaminea* revealed by cDNA-AFLP. Biomed Res Int 160934. Available: http://dx.doi:10.1155/2011/160934.
11. Wu QB, Xu LP, Guo JL, Su YC, Que YX (2013) Transcriptome profile analysis of sugarcane responses to *Sporisorium scitaminea* infection using Solexa sequencing technology. Biomed Res Int 298920. Available: http://dx.doi.org/10.1155/2013/298920.
12. Que YX, Yandg ZX, Xu LP, Chen RK (2009) Isolation and identification of differentially expressed genes in sugarcane infected by *Ustilago scitaminea*. Acta Agronom Sin 35: 452–458.
13. Su YC, Xu LP, Xue BT, Wu QB, Guo JL, et al. (2013) Molecular cloning and characterization of two pathogenesis-related β-1,3-glucanase genes *ScGluA1* and *ScGluD1* from sugarcane infected by *Sporisorium scitamineum*. Plant Cell Rep 32: 1503–1519.
14. Que YX, Xu LP, Lin JW, Ruan MH, Zhang MQ, et al. (2011) Differential protein expression in sugarcane during sugarcane-*Sporisorium scitamineum* interaction revealed by 2-DE and MALDI-TOF-TOF/MS. Comp Funct Genomics 989016. Available: http://www.hindawi.com/journals/ijg/2011/989016/.
15. Wu J, Zhang Y, Zhang H, Huang H, Folta M, et al. (2012) Whole genome wide expression profiles of *Vitis amurensis* grape responding to downy mildew by using Solexa sequencing technology. BMC Plant Biol 10: 234.
16. Ward J, Weber C (2011) Comparative RNA-seq for the investigation of resistance to Phytophthora root rot in the red raspberry 'Latham'. Proc X Int Rubus Ribes Symp 946. Available: http://www.actahort.org/books/946/946_7.htm.
17. Strauss T, van Poecke M, Strauss A, Römer P, Minsavage GV, et al. (2012) RNA-seq pinpoints a Xanthomonas TAL-effector activated resistance gene in a large-crop genome. Proc Natl Acad Sci USA 109: 19480–19485.
18. Li XL, Bai B, Wu J, Deng CY, Zhou B (2012) Application of second-generation sequencing techniques in transcriptome analysis of early interaction between *Oryza sativa* and *Magnaporthe grisea*. Hereditas 34: 102–112.
19. Moosawi-Jorf S, Mahin B (2007) In vitro detection of yeast-like and mycelial colonies of *Ustilago scitaminea* in tissue-cultured plantlets of sugarcane using polymerase chain reaction. J Appl Sci 7: 3768–3773.
20. Ye SJ, Tian DL, Xu JS (2013) Identification of genes associated with MVA pathway in Eucommia and selection of primers for fluorescent quantitative PCR. J Cent South Fores Univ 38: 50–56.
21. Que YX, Xu LP, Xu JS, Zhang JS, Zhang MQ, et al. (2009) Selection of internal reference gene in quantitative PCR analysis of sugarcane genes. Chinese J Trop Crop 30: 274–278.
22. Livak KJ, Schmittgen TD (2001) Analysis of relative gene expression data using real-time quantitative PCR and the $2^{-\Delta\Delta Ct}$ method. Methods 25: 402–408.
23. Chen QZ (2013) Identification of eucalyptus scorch virus in Fujian and transcriptome and proteome analyses of its interaction with eucalyptus. Dr. Phil. thesis, Fujian Agriculture and Forestry University. http://cdmd.cnki.com.cn/Article/CDMD-10389-1013327346.htm. Accessed 4 June 2013.
24. Li XY (2012) Solexa sequencing of bilberry fruit transcriptome library and expression analysis of genes associated with anthocyanin synthesis. Dr. Phil. thesis, Jilin Agricultural University. http://cdmd.cnki.com.cn/Article/CDMD-10193-1013126913.htm. Accessed 20 May 2013.
25. Li CW (2007) The cytology and transcriptome analysis of the interaction between tomato and powdery mildew. Dr. Phil. thesis, Chinese Academy of

Agricultural Sciences. http://epub.cnki.net/kns/brief/default_result.aspx. Accessed 10 April 2007.

26. Chen RK, Xu LP, Lin YQ, Deng ZH, Zhang MQ, et al. (2010) Modern sugarcane genetics and breeding. In: Chen RK, editor. Beijing: China Agriculture Press. 1–3.

27. Coram TE, Settles ML, Chen X (2008) Transcriptome analysis of high-temperature adult-plant resistance conditioned by Yr39 during the wheat–*Puccinia striiformis* f. sp. *tritici interaction*. Mol Plant Pathol 9: 479–493.

28. Bolton MD, Kolmer JA, Xu WW, Garvin DF (2008) Lr34-mediated leaf rust resistance in wheat: transcript profiling reveals a high energetic demand supported by transient recruitment of multiple metabolic pathways. Mol Plant Microbe Interact 21: 1515–1527.

29. Bozkurt TO, Mcgrann GR, Maccormack R, Boyd LA, Akkaya MS (2010) Cellular and transcriptional responses of wheat during compatible and incompatible race-specific interactions with *Puccinia striiformis* f. sp. *tritici*. Mol Plant Pathol 11: 625–640.

30. Bari R, Jones J D (2009) Role of plant hormones in plant defence responses. Plant Mol Biol 69: 473–488.

31. Zhou JX, Hu XW, Zhang HW, Huang RF (2008) The regulatory effect of ABA in biological response to stress. J Agri Biotechnol 16: 169–174.

32. Mauch-Mani B, Mauch F (2005) The role of abscisic acid in plant–pathogen interactions. Curr Opin Plant Biol 8: 409–414.

33. Fan J, Hill L, Crooks C, Doerner P, Lamb C (2009) Abscisic acid has a key role in modulating diverse plant-pathogen interactions. Plant Physiol 150: 1750–1761.

34. Hu S, Wang FZ, Liu ZN, Liu YP, Yu XL (2012) ABA signal transduction mediated by PYR/PYL/RCAR proteins in plants. Hereditas, 34: 560–572.

35. Shen Q, Lu X, Zhang L, Gao ED, Tang KX (2012) Progress in studies on the function of transcription factor MYC2 in plants. J Shanghai Jiaotong Univ (Agricultural Science) 30: 51–57.

36. Kimura M, Yamamoto YY, Seki M, Sakurai T, Sato M, et al. (2003) Identification of Arabidopsis genes regulated by high light-stress using cDNA microarray. Photochem Photobiol 77: 226–233.

37. Walia H, Wilson C, Condamine P, Liu X, Ismail AM, et al. (2005) Comparative transcriptional profiling of two contrasting rice genotypes under salinity stress during the vegetative growth stage. Plant Physiol 139: 822–835.

38. Boerjan W, Ralph J, Baucher M (2003) Lignin biosynthesis. Annu Rev Plant Biol 54: 519–546.

39. Jia JH, Wang B (2000) Advances in cloning of plant disease resistance genes. Prog Biotechnol 20: 21–26.

40. Hao LM, Wang LA, Ma LH, Yan ZF, Wang JB (2001) Progress in studies on the interaction between pathogenic fungi and the host plant. J Hebei Agri Sci 5: 73–78.

41. Su YC, Guo JL, Ling H, Chen SS, Wang SS, et al. (2014) Isolation of a novel peroxisomal catalase gene from sugarcane, which is responsive to biotic and abiotic stresses. PLos One 9: e84426.

42. Mittler R, Herr E H, Orvar B L, Camp WV, Willekens H, et al. (1999) Transgenic tobacco plants with reduced capability to detoxify reactive oxygen intermediates are hyperresponsive to pathogen infection. Proc Natl Acad Sci USA 96: 14165–14170.

43. Romero GO, Simmons C, Yaneshita M, Doan M, Thomas BR, et al. (1998) Characterization of rice endo-β-glucanase genes (Gns2–Gns14) defines a new subgroup within the gene family. Gene 223: 311–320.

44. Wang QW, Qu M, Zhang HL, Xu XL (2008) Cloning of bean chitinase gene *Bchi* and its expression in transgenic tobacco plants. Mol Plant Breeding 6: 53–58.

45. Gu LH, Zhang SZ, Yang BP, Cai WW, Huang DJ, et al. (2008) Introduction of chitinase and β-1, 3- glucanase genes into sugarcane. Mol Plant Breeding 6: 277–280.

46. Neuhaus JM, Fritig B, Linthorst HJM, Meins F, Mikkelsen JD, et al. (1996) A revised nomenclature for chitinase genes. Plant Mol Biol Rep 14: 102–104.

47. Singh A, Isaac-Kirubakaran S, Sakthivel N (2007) Heterologous expression of new antifungal chitinase from wheat. Protein Expre Purif 56: 100–109.

48. Rahul PR, Kumar VG, Sathyabhama M, Viswanathan R, Sundar AR, et al. (2013) Characterization and 3D structure prediction of chitinase induced in sugarcane during pathogenesis of *Colletotrichum falcatum*. J Plant Biochem Biot: 1–8.

49. Schlumbaum A, Mauch F, Vogeli U, Boller T (1986) Plant chitinases are potent inhibitors of fungal growth. Nature 324: 365–367.

50. Mauch F, Mauch-Mani B, Boller T (1988) Antifungal hydrolases in pea tissue II. Inhibition of fungal growth by combinations of chitinase and β-1, 3-glucanase. Plant Physiol 88: 936–942.

51. Maximova SN, Marelli JP, Young A, Pishak S, Verica JA, et al. (2006) Over-expression of a cacao class I chitinase gene in *Theobroma cacao* L. enhances resistance against the pathogen, *Colletotrichum gloeosporioides*. Planta 224: 740–749.

52. Xiao YH, Li XB, Yang XY, Luo M, Hou L, et al. (2007) Cloning and characterization of a balsam pear class I chitinase gene (*Mcchil1*) and its ectopic expression enhances fungal resistance in transgenic plants. Biosci Biotechnol Biochem 71: 1211–1219.

53. Keegstra K (2010) Plant cell walls. Plant Physiol 154: 483–486.

54. Spanu PD, Abbott JC, Amselem J, Burgis TA, Soanes DM, et al. (2010) Genome expansion and gene loss in powdery mildew fungi reveal tradeoffs in extreme parasitism. Science 330: 1543–1546.

55. Kämper J, Kahmann R, Bölker M, Ma LJ, Brefort T, et al. (2006) Insights from the genome of the biotrophic fungal plant pathogen *Ustilago maydis*. Nature 444: 97–101.

56. Cantu D, Vicente AR, Labavitch JM, Bennett AB, Powell AL (2008) Strangers in the matrix: plant cell walls and pathogen susceptibility. Trends Plant Sci 13: 610–617.

57. Underwood W, Somerville SC (2008) Focal accumulation of defences at sites of fungal pathogen attack. J Exp Bot 59: 3501–3508.

58. Fuchs H, Sacristán MD (1996) Identification of a gene in *Arabidopsis thaliana* controlling resistance to clubroot (*Plasmodiophora brassicae*) and characterization of the resistance response. Mol Plant Microbe Interact 9: 91–97.

59. Collins NC, Thordal-Christensen H, Lipka V, Bau S, Kombrink E, et al. (2003) SNARE-protein-mediated disease resistance at the plant cell wall. Nature 425: 973–977.

60. McGrath KC, Dombrecht B, Manners JM, Schenk PM, Edgar CI, et al. (2005) Repressor-and activator-type ethylene response factors functioning in jasmonate signaling and disease resistance identified via a genome-wide screen of *Arabidopsis* transcription factor gene expression. Plant Physiol 139: 949–959.

61. Luo HL, Chen YH (2011) Advances in transcription factors associated with plant resistance to diseases. J Trop Organisms 2: 83–88.

62. Daniel X, Lacomme C, Morel JB, Roby D (1999) A novel myb oncogene homologue in *Arabidopsis thaliana* related to hypersensitive cell death. The Plant Journal 20: 57–66.

63. Mengiste T, Chen X, Salmeron J, Dietrich R (2003) The BOTRYTIS SUSCEPTIBLE1 gene encodes an R2R3MYB transcription factor protein that is required for biotic and abiotic stress responses in *Arabidopsis*. The Plant Cell 15: 2551–2565.

64. Shimono M, Sugano S, Nakayama A, Jiang CJ, Ono K, et al. (2007) Rice WRKY45 plays a crucial role in benzothiadiazole-inducible blast resistance. The Plant Cell 19: 2064–2076.

65. Xu X, Chen C, Fan B, Chen Z (2006) Physical and functional interactions between pathogen-induced *Arabidopsis* WRKY18, WRKY40, and WRKY60 transcription factors. The Plant Cell 18: 1310–1326.

66. Xu Z S, Chen M, Li LC, Ma YZ (2008) Functions of the ERF transcription factor family in plants. Botany 86: 969–977.

3

Reflectance Variation within the In-Chlorophyll Centre Waveband for Robust Retrieval of Leaf Chlorophyll Content

Jing Zhang[1], Wenjiang Huang[2], Qifa Zhou[1]*

1 College of Life Sciences, Zhejiang University, Hangzhou, China, **2** Key Laboratory of Digital Earth Science, Institute of Remote Sensing and Digital Earth, Chinese Academy of Sciences, Beijing, China

Abstract

The in-chlorophyll centre waveband (ICCW) (640–680 nm) is the specific chlorophyll (Chl) absorption band, but the reflectance in this band has not been used as an optimal index for non-destructive determination of plant Chl content in recent decades. This study develops a new spectral index based solely on the ICCW for robust retrieval of leaf Chl content for the first time. A glasshouse experiment for solution-culture of one chlorophyll-deficient rice mutant and six wild types of rice genotypes was conducted, and the leaf reflectance (400–900 nm) was measured with a high spectral resolution (1 nm) spectrophotometer and the contents of chlorophyll a (Chl$a$), chlorophyll b (Chl$b$) and chlorophyll a+b (Chl$t$) of the rice leaves were determined. It was found that the reflectance curves from 640 nm to 674 nm and from 675 nm to 680 nm of the low-chlorophyll mutant leaf were drastically steeper than that of the wild types in the ICCW. The new index based on the reflectance variation within ICCW, the difference of the first derivative sum within the ICCW (DFDS_ICCW), was highly sensitive ($r = -0.77$, $n = 93$, $P < 0.01$) to Chl$t$ while the mean reflectance (R_ICCW) in the ICCW became insensitive ($r = -0.12$, $n = 93$, $P > 0.05$) to Chl$t$ when the leaf Chl$t$ was higher than 200 mg/m^2. The best equations of R-ICCW and DFDS_ICCW yielded an RMSE of 78.7, 32.9 and 107.3 mg/m^2, and an RMSE of 37.4, 16.0 and 45.3 mg/m^{-2}, respectively, for predicting Chl$a$, Chl$b$ and Chl$t$. The new index could rank in the top 10 for prediction of Chl$a$ and Chl$t$ as compared with the 55 existing indices. Additionally, most of the 55 existing Chl-related VIs performed robustly or strongly in simultaneous prediction of leaf Chl$a$, Chl$b$ and Chl$t$.

Editor: Jinxing Lin, Beijing Forestry University, China

Funding: This work was funded by the National Natural Science Foundation of China (grant numbers 41271363 and 41271412). QFZ received 41271363 and WJH received 41271412. The funders had no role in study design, data collection and analysis, decision to publish, or preparation of the manuscript.

Competing Interests: The authors have declared that no competing interests exist.

* Email: zqifa2002@yahoo.com

Introduction

Chlorophyll (Chl) a and Chl b are major constituents of the photosynthetic apparatus in higher plants. Chl a and Chl b are interconverted in the chlorophyll cycle [1]. Leaf Chl a concentration (Chl$a$) and Chl b concentration (Chl$b$) indicate a plant's photosynthetic capacity and health status, and determination of Chl$a$, Chl$b$ and ratios of Chl$a$ to Chl$b$ are also helpful for understanding the light acclimation mechanisms in higher plants [2]. Conventionally, leaf Chl$a$ and Chl$b$ are determined with a traditional wet extraction analysis based on measuring the extinction of the extract at the major red absorption maxima of Chl a (~664 nm) and b (~647 nm) in the in-chlorophyll centre waveband (640–680 nm), and by inserting these values into simultaneous equations [2,3]. In recent decades, there has been an increasing interest in non-destructively determining leaf and canopy Chl content by measuring leaf and canopy spectral reflectance. Particular efforts have been devoted to the development of robust algorithms for Chl$t$ determination from the leaf to canopy scale [4–10]. Contrastingly, studies conducted for determination of individual Chl$a$ or individual Chl$b$ with spectral vegetation indices (VIs) are much less frequent [4,6,11]. Reflec-

tance in the ICCW had been used for a long time as an indicator of chlorophyll content of leaves, but has not been used as an optimal index since Thomas and Gausman (1977) [12] found that reflectance near 675 nm became saturated at medium to high chlorophyll concentrations [6]. In recent decades, many studies have found that reflectance in the green and red-edge spectral regions was optimal for non-destructive estimation of leaf Chl content in a wide range of its variation [13–16]. The results of Féret et al. (2011) [17] showed that the reflectance in the red-edge and near infrared spectral regions simulated with the Prospect 5 radiative transfer model provided an accurate estimation of leaf Chl content. Recently, Main et al. (2011) [11] assessed the performance of 73 published VIs for leaf Chl estimation and also found that the indices using off-chlorophyll absorption centre wavebands (OCCW) performed better than those using ICCW. To our best knowledge, no VIs based solely on ICCW for Chl estimation have been developed since Thomas and Gausman (1977) [12] found the saturated reflection of plant leaves. Plant leaves have a reflectance minima around 675 nm, and there are substantial differences in reflectance among different wavelengths in the ICCW. Is the reflectance difference within the ICCW

Table 1. The existing vegetation indices used in this study.

Index	Formulation	Reference
log(1/R737)	log(1/R737)	Yoder, Pettigrew-Crosby (1995)
SIPI	(R800-R445)/(R800-R680)	Peñuelas et al. (1995)
Ratcart	R695/R760	Carter et al. (1996)
PSSRa	R800/R680	Blackburn (1998)
PSSRb	R800/R635	Blackburn (1998)
PSNDa	(R800-R675)/(R800+R675)	Blackburn (1998)
PSNDb	(R800-R650)/(R800+R650)	Blackburn (1998)
PSSRchla	R810/R676	Blackburn (1999)
PSRI	(R680-R500)/R750	Merzlyak et al. (1999)
SR705	R750/R705	Sims, Gamon (2002)
ND705	(R750-R705)/(R750+R705)	Sims, Gamon (2002)
mND705	(R750-R445)/(R700-R445)	Sims, Gamon (2002)
mSR705	(R750-R705)/(R750+R705-2×R445)	Sims, Gamon (2002)
Readone	R415/R695	Read et al. (2002)
RGRcan	(R612+R660)/(R510+R560)	Steddom et al. (2003)
NDVIcanste	(R760-R708)/(R760+R708)	Steddom et al. (2003)
Red edge Model	(R800/R700)-1	Gitelson et al. (2005)
Green Model	(R800/R550)-1	Gitelson et al. (2005)
OSAVI	1.16×(R800-R670)/(R800+R670+0.16)	Rondeaux et al. (1996)
CI $_{red\ edge}$	(R800/R700)-1	Gitelson et al. (2005)
EVI2	2.5×(R800-R660)/(1+R800+2.4×R660)	Jiang et al. (2008)
CARI	R700×(sqrt(a×670+R670+b)^2)/R670×(a^2+1)$^{0.5}$ a=(R700-R550)/150 b=R550-a×550	Kim et al. (1994)
Carter[A]	R695/R420	Carter (1994)
Carter2[A]	R695/R760	Carter (1994)
Carter3[A]	R605/R760	Carter (1994)
Carter4[A]	R710/R760	Carter (1994)
Carter5[A]	R695/R670	Carter (1994)
Carter6[A]	R550	Carter (1994)
DD	(R749-R720)-(R701-R672)	Le Maire et al. (2004)
Datt[A]	(R850-R710)/(R850-R680)	Datt (1999)
Datt2[A]	R850/R710	Datt (1999)
Datt4[A]	R672/(R550×R708)	Datt (1998)
Datt5[A]	R672/R550	Datt (1998)
Datt6[A]	R860/(R550×R708)	Datt (1998)
Gitelson2[A]	(R750-R800/R695-R740)-1	Gitelson et al. (2003)
Gitelson[A]	1/R700	Gitelson et al. (1999)
mNDVI	(R800-R680)/(R800+R680-2×R445)	Sims, Gamon (2002)
Maccioni[A]	(R780-R710)/(R780-R680)	Maccioni et al. (2001)
mSR	(R800-R445)/(R680-R445)	Sims, Gamon (2002)
SRPI	R430/R680	Peñuelas et al. (1995)
NDVI2[A]	(R750-R705)/(R750+R705)	Gitelson, Merzlyak (1994)
NPCI	(R680-R430)/(R680+R430)	Penuelas et al. (1994)
REP_LE[A]	700+40×(Rre-R700)/(R740-R700) Rre=(R670+R780)/2	Cho, Skidmore (2006)
REP_Li[A]	700+40×((R670+R780/2)/(R740-R700))	Guyot, Baret (1988)
SR1[A]	R750/R700	Gitelson, Merzlyak (1997)
SR2[A]	R752/R690	Gitelson, Merzlyak (1997)
SR3[A]	R750/R550	Gitelson, Merzlyak (1997)
SR4[A]	R700/R670	McMurtey et al. (1994)
SR5[A]	R675/R700	Chappelle et al. (1992)
SR6[A]	R750/R710	Zarco-Tejada, Miller (1999)

Table 1. Cont.

Index	Formulation	Reference
SR7[A]	R440/R690	Lichtenthaler et al. (1996)
Sum_Dr2[A]	sum of first derivative reflectance between R680 and R780	Filella, Penuelas (1994)
Vogelmann[A]	R740/R720	Vogelman et al. (1993)
Vogelmann2[A]	(R734-R747)/(R715+R726)	Vogelman et al. (1993)
SPAD reading	Based on the transmittance at 650 nm and 940 nm	Konica Minota, Japan

associated with the Chl content? This study has two objectives. The first is to examine the robustness of simultaneous estimation of Chl$a$, Chl$b$ and Chl$t$ with the existing Chl-related VIs and commercial chlorophyll meter readings by using a dataset of measured reflectance, Chl$a$, Chl$b$ and Chl$t$ of rice leaves of different genotypes including low-chlorophyll mutants (low in Chl content) at different stages. Second, we test if the reflectance difference within the ICCW is associated with the Chl content by using the constructed dataset and then solely using ICCW to develop a new VI simultaneously sensitive to Chl$a$, Chl$b$ and Chl$t$.

Materials and Methods

2.1. Plant materials and growth conditions

A pot experiment was conducted in a greenhouse with natural light (mean daily photosynthetically active radiation 130 μmol m^{-2} s^{-1} during the whole growth period) and controlled temperature (daily maximum 27.6°C, daily minimum 16.2°C during the rice growing period) and humidity (24.5–85.1% average daily relative humidity, RH, throughout the whole rice growing period) at Zhejiang University Experimental Farm, Hangzhou, China (30°14′ N, 120°10′ E). Six wild types of rice genotypes (IG1, IG23, IG24, DJ, NIP and ZH11) and one chlorophyll-deficient mutant (IG20) were solution-cultured according to the IRRI prescription [18], but the nitrogen level was designed as 1/5×40 mg l^{-1} (low N) and 40 mg l^{-1} (normal N), respectively, for two nitrogen treatments. The mutant 'IG 20' is an isogenic line of the recurrent parent "Zhefu 802" bred by China National Rice Research Institute. A completely random design with four replications was used. Each pot contained a 6.0-L nutrient solution and three seedlings. The nutrient solution was

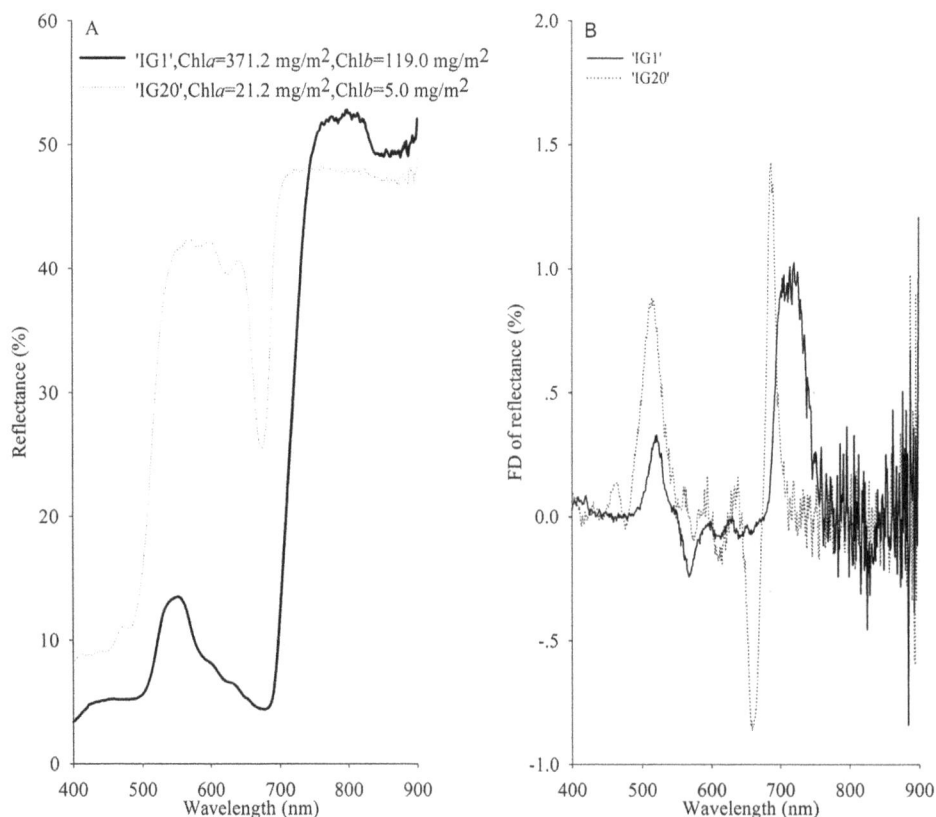

Figure 1. The reflectance curve (A) and the first derivative (FD) of reflectance curve (B) in the mutant (IG20) and wild type (IG1). Chl$a$ and Chl$b$ represent the leaf chlorophyll a content and chlorophyll b content, respectively.

Table 2. The best prediction equations of the existing vegetation indices.

Index	Prediction target	r	Prediction equation	R^2	RMSE (mg/m^2)	Rank
Log(1/R737)	Chla	0.34	$y = -34230x^2-110191x-88409$	0.25	73.8	a52
	Chlb	0.40	$y = -13672x^2-43899x-35148$	0.29	29.0	b47
	Chlt	0.37	$y = -47901x^2-154090x-123557$	0.28	99.0	t52
SIPI	Chla	−0.65	$y = 221.3x^{-6.194}$	0.78	59.5	a45
	Chlb	−0.51	$y = 63.261x^{-6.7392}$	0.50	29.6	b49
	Chlt	−0.62	$y = 288.46x^{-6.2375}$	0.76	84.1	t45
Ratcart	Chla	−0.83	$y = 577.68e^{-4.297x}$	0.94	37.8	a11
	Chlb	−0.70	$y = 2.4669x^{-2.057}$	0.77	15.8	b25
	Chlt	−0.82	$y = 768.46e^{-4.3833x}$	0.94	50.9	t19
PSSRa	Chla	0.81	$y = 4.3255x^{1.6601}$	0.90	50.4	a34
	Chlb	0.72	$y = 1.7069e^{0.327x}$	0.72	23.2	b36
	Chlt	0.81	$y = 5.2238x^{1.6927}$	0.89	68.1	t34
PSSRb	Chla	0.90	$y = 14.01x^{1.5063}$	0.93	41.2	a19
	Chlb	0.90	$y = 1.5702x^2+0.682x+0.6033$	0.84	13.6	b6
	Chlt	0.99	$y = 16.707x^{1.556}$	0.95	46.9	t10
PSNDa	Chla	0.74	$y = 1.3021e^{6.2414x}$	0.87	52.1	a37
	Chlb	0.61	$y = 0.1751e^{7.1639x}$	0.62	26.0	b43
	Chlt	0.72	$y = 1.5724e^{6.335x}$	0.86	72.5	t40
PSNDb	Chla	0.83	$y = 9.6049e^{4.182x}$	0.94	38.8	a15
	Chlb	0.73	$y = 717.58x^2-555.53x+77.434$	0.80	15.5	b19
	Chlt	0.83	$y = 11.591e^{4.2866x}$	0.95	49.0	t13
PSSRchla	Chla	0.81	$y = 3.9395x^{1.6948}$	0.90	50.4	a33
	Chlb	0.72	$y = 1.6415e^{0.3287x}$	0.72	23.2	b37
	Chlt	0.81	$y = 4.744x^{1.7285}$	0.89	68.0	t33
PSRI	Chla	−0.52	$y = 152.13e^{-13.23x}$	0.61	85.8	a55
	Chlb	−0.34	$y = 43.635e^{-12.77x}$	0.31	37.9	b55
	Chlt	−0.48	$y = 198.91e^{-13.064x}$	0.57	120.3	t55
SR705	Chla	0.91	$y = 23.775x^{2.5135}$	0.89	45.8	a26
	Chlb	0.88	$y = 19.518x^2-22.118x+8.0188$	0.81	15.2	b9
	Chlt	0.93	$y = 28.788x^{2.5989}$	0.91	54.2	t27
ND705	Chla	0.91	$y = 572.06x^{0.9776}$	0.94	37.5	a7
	Chlb	0.83	$y = 724.6x^2-161.79x+13.25$	0.80	15.3	b13
	Chlt	0.91	$y = 758.62x^{0.9945}$	0.93	51.6	t21
mND705	Chla	0.90	$y = 22.471x^{1.5336}$	0.89	47.9	a29
	Chlb	0.89	$y = 0.8471x^2+15.357x-14.561$	0.80	15.5	b21
	Chlt	0.92	$y = 27.138x^{1.5862}$	0.91	57.3	t29
mSR705	Chla	0.91	$y = 494.39x^{0.994}$	0.94	36.8	a6
	Chlb	0.83	$y = 517.31x^2-133.21x+13.164$	0.80	15.4	b15
	Chlt	0.91	$y = 654.1x^{1.0094}$	0.93	50.7	t18
Readone	Chla	0.88	$y = 1720.4x^{2.5357}$	0.85	54.9	a40
	Chlb	0.84	$y = 838.41x^{3.1792}$	0.73	20.9	b34
	Chlt	0.89	$y = 2403.7x^{2.619}$	0.87	69.2	t35
RGRcan	Chla	−0.68	$y = 6638.5e^{-5.523x}$	0.82	68.7	a51
	Chlb	−0.53	$y = 2736.3e^{-6.1144x}$	0.55	32.2	b54
	Chlt	−0.66	$y = 8855.4e^{-5.5601x}$	0.79	96.6	t51
NDVIcanste	Chla	0.91	$y = 609.94x^{0.925}$	0.94	36.6	a5
	Chlb	0.83	$y = 783.43x^2-128.31x+11.471$	0.80	15.5	b16
	Chlt	0.91	$y = 809.92x^{0.9412}$	0.93	50.3	t17
Red edge Model	Chla	0.91	$y = 117.36x^{0.821}$	0.95	35.5	a3
	Chlb	0.88	$y = 6.683x^2+11.629x+4.7987$	0.80	15.5	b17

Table 2. Cont.

Index	Prediction target	r	Prediction equation	R^2	RMSE (mg/m^2)	Rank
	Chlt	0.92	$y = 151.08x^{0.8386}$	0.95	45.6	t8
Green Model	Chla	0.91	$y = 118.66x^{1.0178}$	0.94	37.6	a9
	Chlb	0.93	$y = 4.6913x^2 + 28.539x - 2.6421$	0.87	12.5	b2
	Chlt	0.94	$y = 151.82x^{1.0515}$	0.96	41.2	t2
OSAVI	Chla	0.75	$y = 1.556e^{5.2403x}$	0.88	50.2	a31
	Chlb	0.62	$y = 0.2085e^{6.0466x}$	0.64	25.2	b40
	Chlt	0.73	$y = 1.8751e^{5.324\ x}$	0.87	69.7	t37
CI red edge	Chla	0.91	$y = 117.36x^{0.821}$	0.95	35.5	a4
	Chlb	0.88	$y = 6.683x^2 + 11.629x + 4.7987$	0.80	15.5	b18
	Chlt	0.92	$y = 151.08x^{0.8386}$	0.95	45.6	t9
EVI2	Chla	0.82	$y = 7.4037e^{1.9921x}$	0.93	41.2	a20
	Chlb	0.71	$y = 1.084e^{2.3895x}$	0.73	19.9	b32
	Chlt	0.81	$y = 8.9479e^{2.0371x}$	0.93	53.9	t26
CARI	Chla	−0.87	$y = 0.0159x^2 - 5.7648x + 540.52$	0.79	39.2	a16
	Chlb	−0.82	$y = 0.0141x^2 - 3.6719x + 247.15$	0.80	15.2	b10
	Chlt	−0.88	$y = 0.0299x^2 - 9.4367x + 787.67$	0.83	48.0	t11
Carter[A]	Chla	−0.87	$y = 1418.9e^{-0.839x}$	0.91	39.4	a17
	Chlb	−0.76	$y = 676.39x^{-3.0899}$	0.74	19.2	b31
	Chlt	−0.86	$y = 1941.8e^{-0.86x}$	0.92	49.8	t15
Carter2[A]	Chla	−0.83	$y = 577.68e^{-4.297x}$	0.94	37.8	a12
	Chlb	−0.70	$y = 2.4669x^{-2.057}$	0.77	15.8	b26
	Chlt	−0.82	$y = 768.46e^{-4.3833x}$	0.94	50.9	t20
Carter3[A]	Chla	−0.85	$y = 579.04e^{-4.3x}$	0.95	35.4	a2
	Chlb	−0.74	$y = 2.4748\ x^{-2.051}$	0.82	13.0	b4
	Chlt	−0.84	$y = 774.39e^{-4.4086}$	0.96	43.9	t5
Carter4[A]	Chla	−0.90	$y = 2561.5e^{-4.845x}$	0.92	38.2	a14
	Chlb	−0.81	$y = 593.69x^2 - 1001x + 423.27$	0.79	15.8	b24
	Chlt	−0.90	$y = 3589.7e^{-4.9847x}$	0.93	45.4	t7
Carter5[A]	Chla	−0.43	$y = -98.296x^2 + 341.52x34.281$	0.21	75.8	a53
	Chlb	−0.48	$y = -57.997x + 199.2$	0.23	30.2	b51
	Chlt	−0.45	$y = -99.659x^2 + 290.02x157.36$	0.22	102.6	t53
Carter6[A]	Chla	−0.88	$y = 1248.3e^{-0.102x}$	0.91	43.6	a23
	Chlb	−0.83	$y = 0.2785x^2 - 18.344x + 299.57$	0.86	12.9	b3
	Chlt	−0.89	$y = 1738.4e^{-0.106x}$	0.94	49.0	t12
DD	Chla	0.91	$y = 171.95e^{0.0753x}$	0.85	41.7	a22
	Chlb	0.83	$y = 0.1316x^2 + 4.8546x + 52.571$	0.80	15.5	b20
	Chlt	0.91	$y = 0.2558x^2 + 15.278x + 255.86$	0.87	42.2	t3
Datt[A]	Chla	0.90	$y = 18.526e^{4.5459x}$	0.90	44.5	a25
	Chlb	0.83	$y = 443.78x^2 - 139.88x + 14.677$	0.81	15.0	b8
	Chlt	0.91	$y = 22.272e^{4.6979x}$	0.92	51.9	t23
Datt2[A]	Chla	0.89	$y = 29.472x^{2.8339}$	0.83	57.3	a42
	Chlb	0.90	$y = 17.484x^2 + 14.947x - 30.522$	0.81	14.9	b7
	Chlt	0.92	$y = 35.395x^{2.9529}$	0.86	69.4	t36
Datt4[A]	Chla	0.69	$y = -237156x^2 + 25959x - 55.707$	0.48	61.4	a46
	Chlb	0.81	$y = 66027x^2 + 7841x - 38.216$	0.65	20.3	b33
	Chlt	0.75	$y = -171128x^2 + 33800x - 93.923$	0.56	77.3	t42
Datt5[A]	Chla	−0.27	$y = -5518x^2 + 3806.5x - 375.93$	0.44	63.9	a49
	Chlb	−0.09	$y = -2482x^2 + 1820.7x - 232.67$	0.46	25.3	b41
	Chlt	−0.23	$y = -8000x^2 + 5627.2x - 608.59$	0.46	85.3	t46

Table 2. Cont.

Index	Prediction target	r	Prediction equation	R^2	RMSE (mg/m^2)	Rank
Datt6[A]	Chla	0.86	$y = 2546.3x^{1.2194}$	0.85	54.5	a39
	Chlb	0.92	$y = -563.98x^2+748.31x-18.395$	0.86	13.0	b5
	Chlt	0.91	$y = 3709.6x^{1.2735}$	0.89	64.0	t32
Gitelson2[A]	Chla	−0.75	$y = 5.2141e^{-1.03x}$	0.63	83.9	a54
	Chlb	−0.74	$y = 0.4714e^{-1.3512x}$	0.59	31.4	b53
	Chlt	−0.77	$y = 5.7811e^{-1.0753x}$	0.66	109.9	t54
Gitelson[A]	Chla	0.88	$y = 50890x^{2.0381}$	0.89	50.3	a32
	Chlb	0.87	$y = 15333x^2-5.1781x-4.1117$	0.78	16.2	b28
	Chlt	0.90	$y = 80087x^{2.1079}$	0.91	60.4	t31
mNDVI	Chla	0.71	$y = 1.1004e^{5.2579x}$	0.84	56.5	a41
	Chlb	0.56	$y = 0.1657e^{5.8958x}$	0.58	28.5	b45
	Chlt	0.68	$y = 1.3561\ e^{5.3138x}$	0.82	80.0	t43
Maccioni[A]	Chla	0.90	$y = 468.03x^{1.1116}$	0.93	37.6	a10
	Chlb	0.81	$y = 524.36x^2-195.48x+19.343$	0.79	15.8	b23
	Chlt	0.90	$y = 22.432e^{4.6798x}$	0.93	44.9	t6
mSR	Chla	−0.32	$y = -0.0202x^2-3.6039x+105.02$	0.47	62.1	a48
	Chlb	−0.14	$y = -0.0073x^2-1.1513x+37.45$	0.31	28.5	b46
	Chlt	−0.28	$y = -0.0275x^2-4.7551x+142.47$	0.44	87.0	t47
NDVI2[A]	Chla	0.91	$y = 572.06x^{0.9776}$	0.94	37.5	a8
	Chlb	0.83	$y = 724.6x^2-161.79x+13.25$	0.80	15.3	b14
	Chlt	0.91	$y = 758.62x^{0.9945}$	0.93	51.6	t22
NPCI	Chla	−0.76	$y = 185.21e^{-5.54x}$	0.90	51.3	a35
	Chlb	−0.63	$y = 51.691e^{-6.487x}$	0.67	25.0	b39
	Chlt	−0.75	$y = 240.87e^{-5.6355x}$	0.89	70.6	t39
REP_LE[A]	Chla	0.73	$y = 2E-06e^{0.0261x}$	0.85	47.0	a27
	Chlb	0.59	$y = 8E-08e^{0.0288x}$	0.56	25.5	b42
	Chlt	0.71	$y = 2E-06e^{0.0263x}$	0.83	74.6	t41
REP_Li[A]	Chla	−0.62	$y = 7E+18x^{-5.741}$	0.76	57.8	a43
	Chlb	−0.49	$y = 3E+19x^{-6.1545}$	0.48	30.0	b50
	Chlt	−0.60	$y = 1E+19x^{-5.7714}$	0.74	95.2	t50
SR1[A]	Chla	0.91	$y = 20.424x^{2.0298}$	0.91	44.5	a24
	Chlb	0.88	$y = 7.965x^2-5.2604x+2.1228$	0.80	15.3	b11
	Chlt	0.92	$y = 24.674x^{2.0062}$	0.92	52.5	t25
SR2[A]	Chla	0.88	$y = 10.593x^{1.5411}$	0.94	41.6	a21
	Chlb	0.83	$y = 1.7565x^2-4.8351x+9.0096$	0.76	17.0	b29
	Chlt	0.89	$y = 12.789\ x^{1.5807}$	0.94	52.0	t24
SR3[A]	Chla	0.91	$y = 21.529x^{2.2107}$	0.85	51.5	a36
	Chlb	0.93	$y = 5.8111x^2+17.204x-25.878$	0.88	12.2	b1
	Chlt	0.94	$y = -15.607x^2+238.91x-238.84$	0.89	38.6	t1
SR4[A]	Chla	−0.06	$y = -114.37x^2+719.91x-871.51$	0.38	67.0	a50
	Chlb	−0.17	$y = -37.435x^2+226.79x-260.71$	0.28	29.3	b48
	Chlt	−0.10	$y = -151.81x^2+946.7x-1132.2$	0.37	92.5	t48
SR5[A]	Chla	−0.09	$y = -9120.2x^2+6433.1x-859.07$	0.48	61.7	a47
	Chlb	0.05	$y = -3456.2x^2+2521.7x-367.62$	0.41	26.4	b44
	Chlt	−0.05	$Y = -12576x^2+8954.8x-1226.7$	0.48	83.9	t44
SR6[A]	Chla	0.92	$y = 26.484x^{3.1156}$	0.87	47.7	a28
	Chlb	0.88	$y = 41.718x^2-58.098x+22.191$	0.80	15.3	b12
	Chlt	0.93	$y = 32.113x^{3.2249}$	0.90	57.0	t28
SR7[A]	Chla	0.90	$y = 429.79x^{2.2355}$	0.93	39.6	a18

Table 2. Cont.

Index	Prediction target	r	Prediction equation	R^2	RMSE (mg/m^2)	Rank
	Chlb	0.83	$y = 288.79x^2-165.83x+30.259$	0.75	17.1	b30
	Chlt	0.90	$y = 570.74x^{2.2932}$	0.94	49.3	t14
SRPI	Chla	0.78	$y = 4.2726e^{3.6528}$	0.90	52.4	a38
	Chlb	0.67	$y = 0.577e^{4.3543x}$	0.70	23.9	b38
	Chlt	0.77	$y = 5.141e^{3.7278x}$	0.90	70.4	t38
Sum_Dr2[A]	Chla	0.75	$y = 1E-05x^{4.4928}$	0.80	59.1	a44
	Chlb	0.59	$y = 0.1296e^{0.143x}$	0.53	30.4	b52
	Chlt	0.72	$y = 1E-05x^{4.532}$	0.78	94.5	t49
Vogelmann[A]	Chla	0.92	$y = 29.72x^{6.135}$	0.86	45.6	a30
	Chlb	0.87	$y = 313.02x^2-580.26x+274.16$	0.79	15.9	b27
	Chlt	0.93	$y = 36.19x^{6.3497}$	0.88	60.3	t30
Vogelmann2[A]	Chla	−0.91	$y = -10448x^2-3992.6x+21.527$	0.84	33.8	a1
	Chlb	−0.88	$y = 4359.1x^2-660.68x+5.9525$	0.79	15.6	b22
	Chlt	−0.93	$y = -6088.5x^2-4653.3x+27.479$	0.87	42.5	t4
SPAD	Chla	0.90	$y = 1.9176x^{1.3184}$	0.94	37.8	a13
	Chlb	0.82	$y = 2.9234e^{0.0794x}$	0.76	22.2	b35
	Chlt	0.90	$y = 2.2727 x^{1.3451}$	0.93	50.1	t16

replaced as the electric conductivity decreased to half of the original. The plants were transplanted on October 1, 2013.

2.2. Chlorophyll meter and spectral measurements

The second uppermost leaves of each treatment were measured in situ with a SPAD 502 model chlorophyll meter (Konica Minota Inc., Japan) around the midpoint at tillering, booting and heading. After the measurement of the chlorophyll meter, the leaves were immediately sampled and stored in an ice box, and transported to the lab for leaf reflectance measurements. The reflectance of the single leaf was measured with an integrating sphere (model LISR-3100, Shimadzu Scientific Instruments Inc., Japan) coupled to a UV-3600 UV-VIS-NIR spectrophotometer (Shimadzu Scientific Instruments Inc., Japan) in the wavelength range of 400–900 nm around the midpoint of each leaf. The spectral meter has a 1-nm resolution in the region of 400–900 nm.

2.3. Determination of leaf Chl contents

After spectral measurements, 15 leaf discs of 0.5 cm^2 from each leaf were sampled for determination of leaf Chl content. The Chl a and Chl b contents per unit area were measured spectrophotometrically using a solution of alcohol, acetone and water (4.5:4.5:1, V/V/V) as a solvent, employing the equations of Lichtenhaler and Wellburn (1983) [19]. The total Chl content was calculated as Chla plus Chlb. The leaves that appeared evidently desiccative were not used in this study. We measured a total of 108 leaves across tillering, booting and heading stages, which included 12 leaves of the mutant and 96 leaves of the wild types.

2.4. Data analysis

The scatterplots of the reflectance and the first derivative (FD) reflectance *vs* Chla, Chlb and Chlt were plotted, and the curves were visually analysed for extraction of spectral signatures of interest including shape, peak position, trough position and inflection point. FD was calculated with the following equation:

$$FD(\lambda) = R(\lambda+1) - R(\lambda) \qquad (1)$$

where FD(λ), R(λ) and R(λ+1) represent the first derivative reflectance at wavelength λ (nm), reflectance at λ and reflectance at λ+1, respectively.

The existing published Chl-related VIs selected in this study and their formulations were summarized in Table 1 [4–7,20–46]. Only leaf-scale indices were collected. Among the 55 indices, none were solely based on the ICCW, although 21 indices used the ICCW.

The sensitivity of the VIs to Chl contents were tested with the correlation coefficients between the VIs and the Chl content, and the correlation coefficients were computed with Excel 10.0 (Microsoft).

The relationship between the VIs and the Chl content (Chla or Chlb or Chlt) were fitted with linear, power, exponential, logarithmic and polynomial equations and the equation with the highest determination coefficients (R^2) was selected as the best equation. The root mean square error (RMSE) was computed for each best equation, and the predictive performance of the VIs was assessed by ranking the RMSE values in ascending order. The relationships were fitted with Excel.

Results

3.1. Rice leaf Chl content

All the leaves of both the normal N treatment and the low N treatment of the mutant 'IG 20' were yellow-green in color during the whole growth period. The leaves of the wild types were green in colour, although the low N treatments were shallower in leaf colour than the normal N treatments. The means and ranges of Chl content (mg/m^2) for the 96 leaf samples of the conventional genotypes as well as Chla/Chlb were 260.5 (148.7–378.5) for Chla, 81.8 (31.9–135.3) for Chlb, 342.3 (209.4–497.7) for Chlt and 3.76 (1.99–6.55) for Chla/Chlb. The values for the 12 leaf samples of the low-chlorophyll mutant were 52.2 (11.9–157.5) for Chla,

14.7 (0.2–40.5) for Chl*b*, 66.8 (16.9–198.0) for Chl*t* and 11.08 (1.05–114.35) for Chl*a*/Chl*b*. The leaves of the wild types had an evidently higher Chl*a*, Chl*b* and Chl*t* and a much lower ratio of Chl*a* to Chl*b* than the leaves of the mutant. As compared with the previous study [6] for constructing VIs for Chl*a*, Chl*b* and Chl*t* estimation, this study had a similar mean Chl content, a lower minimum Chl content, a lower maximum Chl content, and a significantly larger variation of ratios of Chl*a* to Chl*b*.

3.2. Leaf spectral reflectance signatures and construction of the new VI

As shown in Figure 1A, a profound difference in leaf spectral reflectance was observed between the conventional rice genotypes and the mutant. The reflectance curves from 640 nm to 674 nm

and from 675 nm to 680 nm of the mutant leaf of a low Chl content were drastically steeper than those of the wild types in the ICCW. For both the wild types and the mutant, the inflection point of the reflectance spectra in the ICCW was 645 nm, where the FD value of reflectance started to be positive (Figure 1B). Additionally, the reflectance trough around 620 nm became evident, and the green peak around 550 nm was broadened and deformed in the reflectance spectra of the mutant as compared with that of the wild types. The reflectance spectra of all the leaves of the mutant were visually similar in shape and reflection band positions.

Based on the spectral signatures in the ICCW we observed, we found that the reflectance variation within the ICCW was sensitive to the Chl content, and constructed a new VI—the difference of

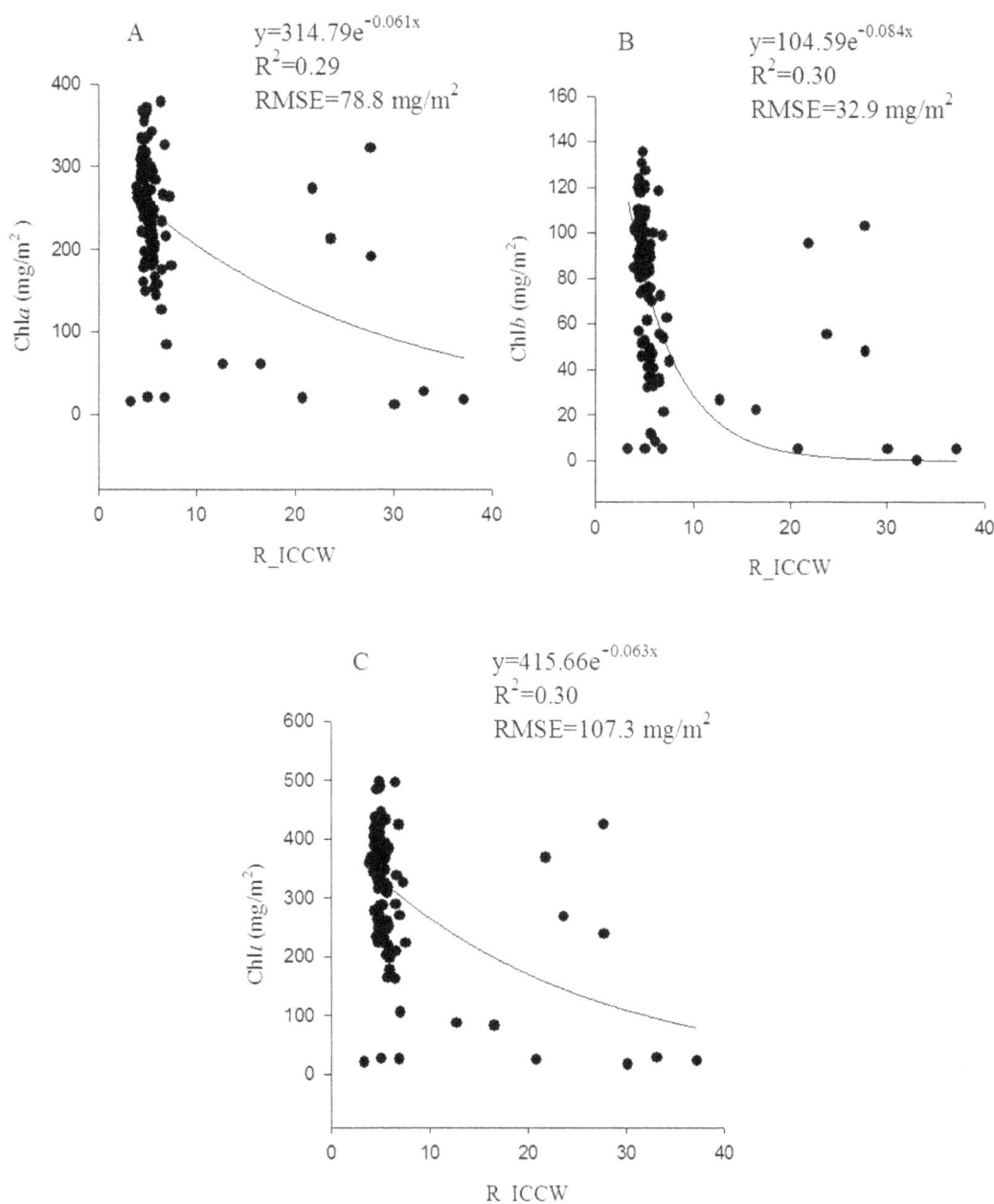

Figure 2. The best prediction models of R_ICCW for Chl*a* (A), Chl*b* (B) and Chl*t* (C).

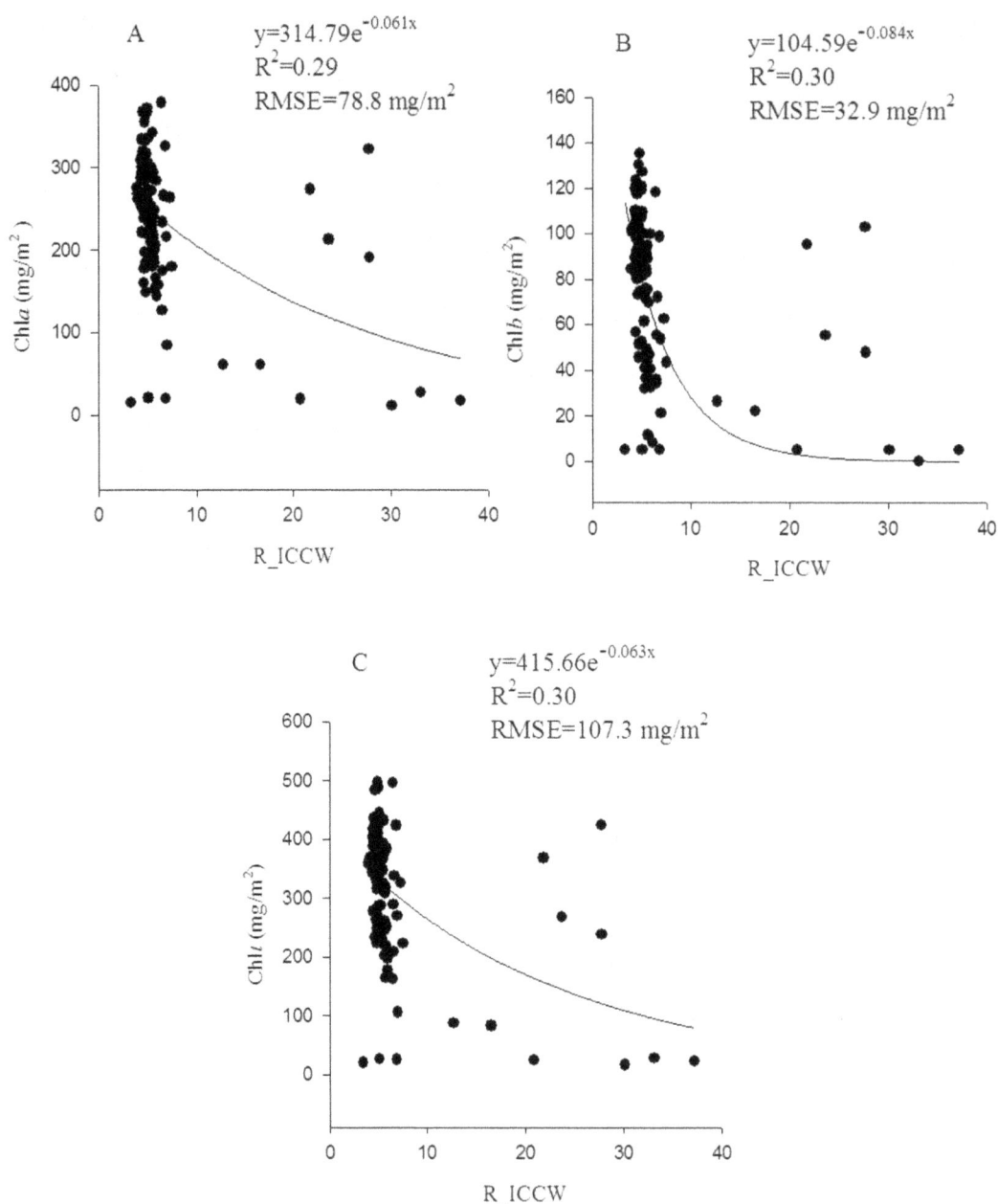

Figure 3. The best prediction models of DFDS_ICCW for Chl$a$ (A), Chl$b$ (B) and Chl$t$ (C).

first derivative sum within the ICCW (DFDS_ICCW)—for simultaneous retrieval of Chl$a$, Chl$b$ and Chl$t$:

$$DFDS\_ICCW = \text{sum of } FD_{675-680}\text{-sum of } FD_{640-674} \quad (2)$$

where the sum of $FD_{675-680}$ and the sum of $FD_{640-674}$ represent the sum of the first derivative reflectance between R675 and R680 and that between R640 and R674, respectively. R640, R674, R675 and R680 are the reflectance at 640 nm, 674 nm, 675 nm and 680 nm, respectively.

3.3. Sensitivity of the VIs to Chl$a$, Chl$b$ and Chl$t$

Of the 55 VIs tested (Table 2), 24 were robustly sensitive to the leaf Chl$t$ ($r2 \geq 0.81$, n = 108), 19 were strong ($0.49 \leq r2 < 0.81$, n = 108), 5 were moderate ($0.25 \leq R2 < 0.49$) and 5 were weak ($0.04 \leq R2 < 0.25$). Only 2 indices, SR4[A] and SR5[A], were insignificantly ($P > 0.05$, n = 108) related to the leaf Chl$a$, Chl$b$

and Chl$t$. Generally, the sensitivity of the indices to Chl$t$ was similar to that of Chl$a$, and the sensitivity of the indices to Chl$b$ was slightly lower than Chl$t$ or Chl$a$. The results showed that most of the tested indices were highly sensitive to Chl$a$, Chl$b$ and Chl$t$.

The mean reflectance in the ICCW (R-ICCW) was significantly ($P < 0.05$) related to Chl$a$, Chl$b$ and Chl$t$ with a low correlation strength, yielding an r (n = 108) of -0.45, -0.40 and -0.45, respectively. In contrast, the new VI, DFDS_ICCW, had an r (n = 108) of -0.86, -0.76 and -0.85 as correlated with Chl$a$, Chl$b$ and Chl$t$, respectively, indicating that this index was highly sensitive to Chl$t$, Chl$a$ and Chl$b$. When leaf Chl$t$ was higher than 200 mg/m^2, the r value was -0.77 (n = 93, $P < 0.01$) and -0.12 (n = 93, $P > 0.05$) respectively between DFDS-ICCW and Chl$t$ and between R_ICCW and Chl$t$. The results demonstrated that DFDS-ICCW was still highly sensitive, but R_ICCW became

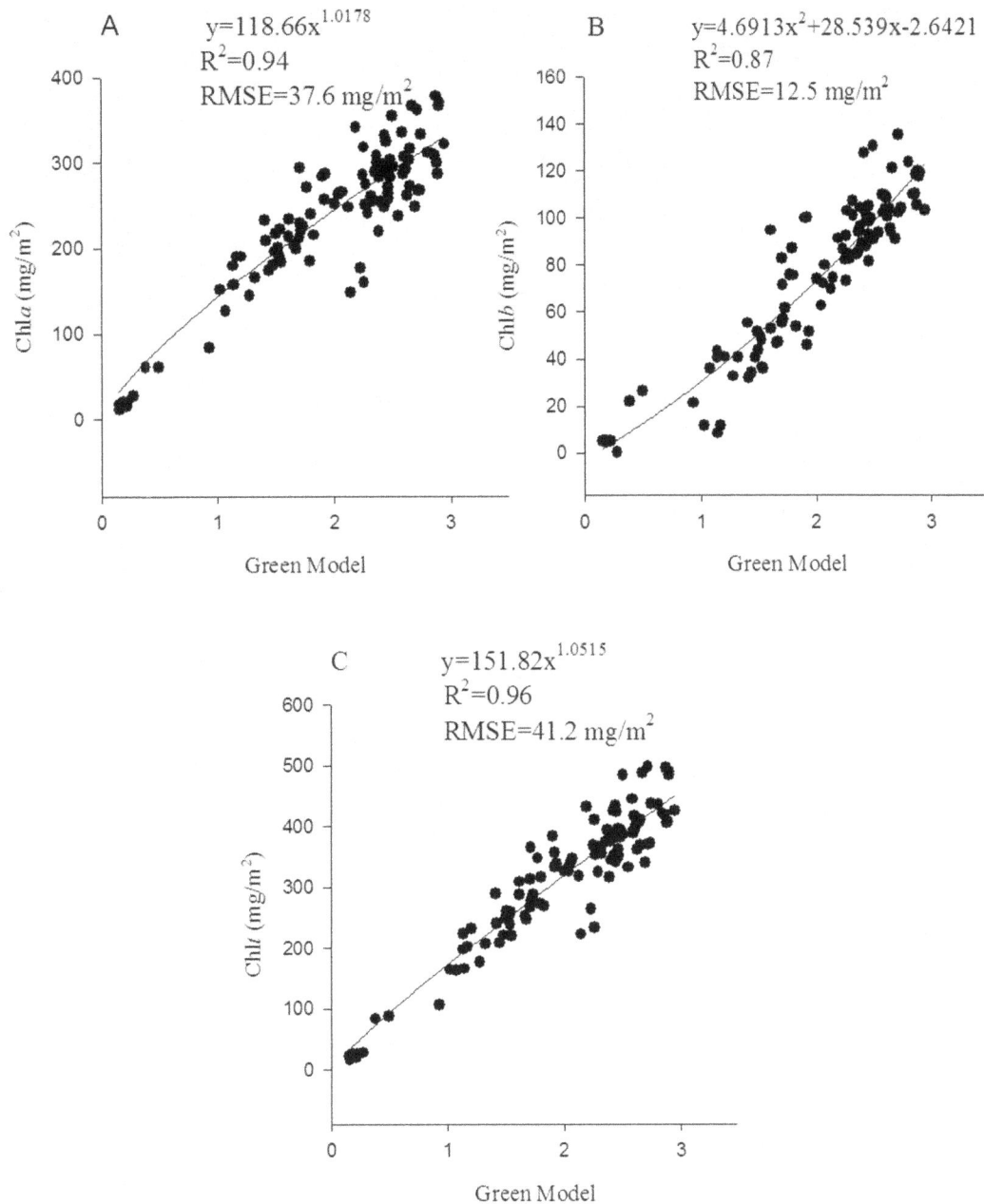

Figure 4. The best prediction models of Green Model for Chl*a* (A), Chl*b* (B) and Chl*t* (C).

insensitive to Chl*t* when Chl*t* was at medium and high levels. As shown in Figure 2C, the R-ICCW tended to be saturated when leaf Chl*t*>200 mg/m^2. Contrastingly, DFDS_ICCW decreased sensitively with the Chl*t* even when Chl*t* was higher than 200 mg/m^2 (Figure 3C). This result confirmed the saturated reflection of the leaves at medium to high Chl content.

3.4. Prediction of Chl*a*, Chl*b* and Chl*t* with the best-fit equations

The best equations of R-ICCW (Figure 2) and DFDS_ICCW (Figure 3) were all exponential equations. For R-ICCW, the exponential equations yielded an RMSE (mg/g^2) of 78.7 for Chl*a*, 32.9 for Chl*b* and 107.3 for Chl*t*. The DFDS_ICCW equations yielded an RMSE of 37.4 for Chl*a*, 16.0 for Chl*b* and 45.3 for

Chl*t*. The results indicated that DFDS_ICCW had a drastically higher prediction accuracy for Chl*a*, Chl*b* and Chl*t* than R_ICCW. The prediction accuracy of DFDS_ICCW was slightly lower for Chl*b* than Chl*a* or Chl*t*.

The prediction performance with a best prediction equation for all of the 55 existing indices are presented in Table 2. Interestingly, none of the best equations were linear; they were exponential, polynomial and power. The RMSE (mg/m^2) ranged from 33.8 to 85.8 for Chl*a*, from 12.2 to 37.9 for Chl*b* and from 38.6 to 120.3 for Chl*t*, which demonstrated that there was a large difference of prediction accuracy between the best index and the last index. However, the RMSE (mg/m^2) of the top 30 indices ranged from 33.8 to 49.6 for Chl*a*, from 12.2 to 17.1 for Chl*b* and from 38.6 to 60.3 for Chl*t*, indicating that the differences in the RMSE were not large in the top 30 indices. An index of high

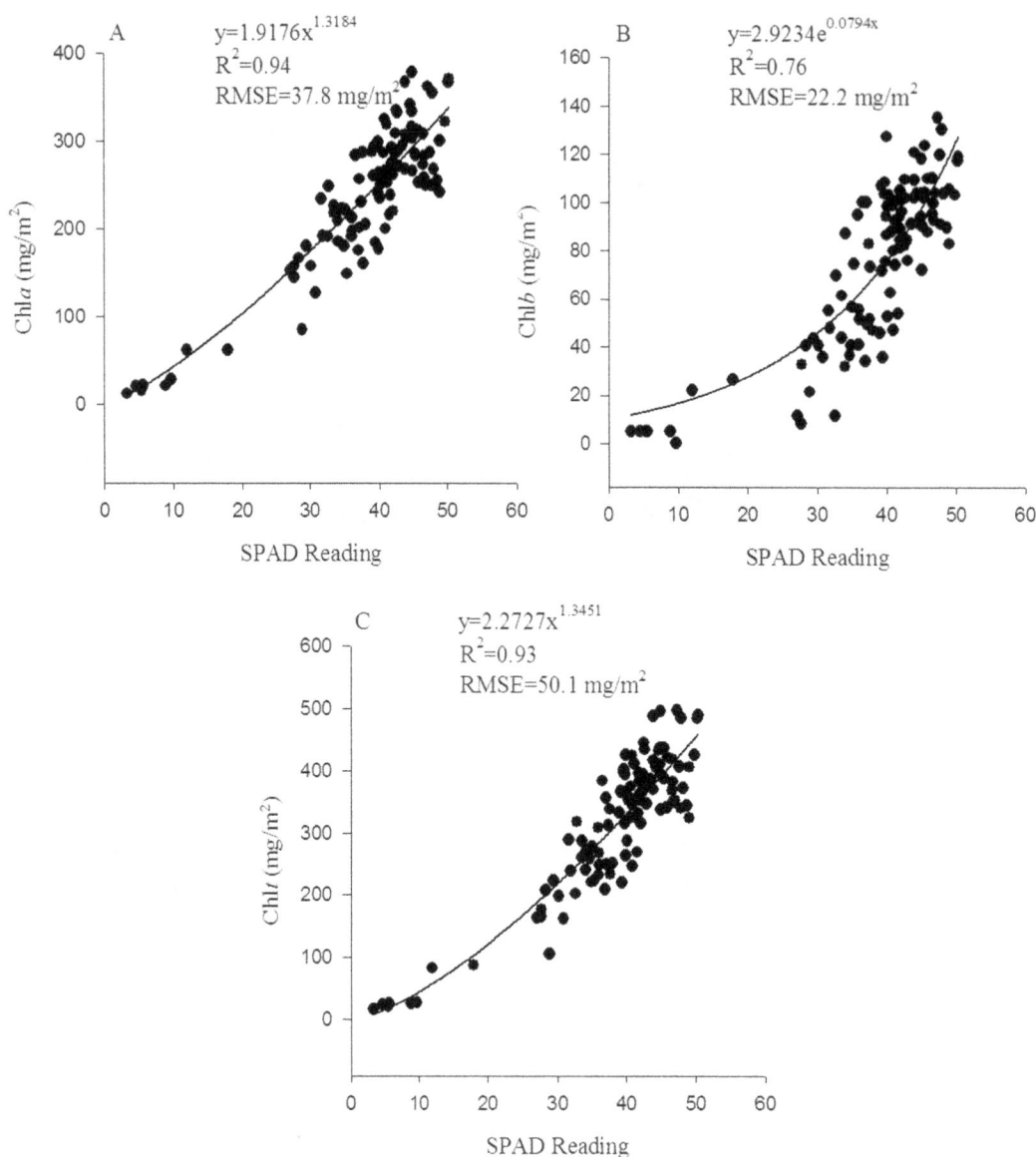

Figure 5. The best prediction models of SPAD readings for Chl*a* (A), Chl*b* (B) and Chl*t* (C).

predictive ability for Chl*t* (e.g. Green Model) generally also performed well for prediction of Chl*a* or Chl*b*, although the prediction accuracy for Chl*b* was generally and slightly lower than that for Chl*a* or Chl*t*, and an index of low predictive ability for Chl*t* (e.g. PSRI) was also weak for prediction of Chl*a* or Chl*b*. The SPAD reading ranked 13th, 35th and 16th among the 55 indices, respectively for prediction of Chl*a*, Chl*b* and Chl*t* with the polynomial equations, which indicated that it was also a strong index for predicting the leaf Chl contents.

The prediction results of the best VI, Green Model, together with the SPAD reading are also presented in Figure 4 and Figure 5, which confirm their high accuracy for prediction of Chl*a*, Chl*b* and Chl*t*.

The results in this study demonstrated that most of the existing indices could be used for simultaneous retrieval of Chl*a*, Chl*b* and Chl*t*.

As compared with the 55 indices, the prediction accuracy of DFDS_ICCW was similar to Datt2[A], ranking 7th for Chl*a*

prediction, similar to SR6[A] ranking 28th for Chl*b* prediction and similar to Carter4[A] ranking 7th for Chl*t* prediction. The results indicated that DFDS_ICCW could simultaneously and robustly predict Chl*a*, Chl*b* and Chl*t*.

Discussion

Most of the existing VIs as well as the SPAD reading were simultaneously and robustly or strongly related to Chl*a*, Chl*b* and Chl*t*, and achieved a high accuracy for Chl*a*, Chl*b* and Chl*t* prediction. As most of the indices were originally sought for prediction of Chl*t*, the results in this study suggested that the indices could be extended for simultaneous retrieval of Chl*a*, Chl*b* and Chl*t*. None of the best-fit equations for prediction of Chl*a*, Chl*b* and Chl*t* were linear equations; therefore, the ranking of the existing indices in this study was not in agreement with that of Main et al. (2011) [11], who used a linear equation for all indices. The VIs based on red edge (e.g. REP_LE[A] and REP_Li[A]) ranked high for leaf Chl prediction in the previous study, but ranked low in this study. The indices excluding the ICCW generally performed better than those including the ICCW in this study, which is consistent with the previous study [11]. Particularly, both the best index for Chl$a$ and Chl$t$, SR3^A, and the best index for Chl$b$, Vogelmann2^A, did not use ICCW. The simple ratio indices—SR4^A based on 670 nm in the ICCW and 700 nm and S5^A based on 675 nm in the ICCW and 700 nm—were the only indices that were insignificantly ($P>0.05$) related to the Chl contents. In contrast, another simple ratio index, SR3^A based on 550 nm and 750 nm in the OCCW, was the best index for prediction of Chl$a$ and Chl$t$. In the ICCW, the reflectance curves from 640 nm to 674 nm and from 675 nm to 680 nm of the mutant leaf of a low Chl content were drastically steeper than that of the wild types of medium to high Chl content. This spectral signature could enlighten us to use the reflectance variation within the ICCW for retrieval of plant Chl content, although further studies are needed for understanding the mechanisms causing this signature. The successful detection of the reflectance variation within the ICCW in this study could be attributed to the high spectral resolution (1 nm) of the spectral photometer, as the current widely-used spectral meter, the Field Spec spectro-radiometer (Analytical Spectral Devices, Boulder, CO, USA), has a spectral resolution of 3 nm in the red band.

Plant leaves tend to have saturated reflectance in the ICCW [6,12] when leaf Chl$t$ is medium to high, which has limited the use of this spectral region for non-destructive determination of leaf Chl. The results in this study also showed that the R-ICCW tended to be saturated when leaf Chl$t$ was higher than 200 mg/m^2. However, the new spectral index based on the reflectance variation within the ICCW decreased sensitively with the Chl$t$ even when Chl$t$ was greater than 200 mg/m^2. The new index could rank in the top 10 for prediction of Chl$a$ and Chl$t$ as compared with the 55 tested indices, and also achieved a promising accuracy for Chl$b$ prediction. Therefore, the results suggested that ICCW could also be used for development of robust

VIs for retrieval of plant Chl contents. Unlike the existing VIs, the new index is solely based on the specific Chl adsorption band. Therefore, the retrieval of Chl by using this index may not be confounded by non-Chl factors, e.g. other pigments and leaf structure. Further studies are needed for confirmation of this finding at different scales (e.g. canopy and region) and for different plant species.

Conclusions

Most of the 55 existing VIs could robustly or strongly and simultaneously predict Chl$a$, Chl$b$ and Chl$t$ in the rice leaves of a large variation of ratios of Chl$a$ to Chl$b$ in this study. It was found that the reflectance curves from 640 nm to 674 nm and from 675 nm to 680 nm of the mutant leaf were drastically steeper than those of the wild types in the ICCW, which implied that the reflectance variation within ICCW could be used for retrieval of Chl content. The new index based solely on the reflectance variation within the ICCW were simultaneously and strongly sensitive to Chl$a$, Chl$b$ and Chl$t$ and achieved a high accuracy for prediction of Chl$a$, Chl$b$ and Chl$t$. The results suggested that ICCW could also be of potential for development of robust VIs for retrieval of plant Chl content with non-destructive reflectance measurement approaches.

Acknowledgments

We thank Dr Wang Danying at the China National Rice Research Institute for providing the rice seeds. We also thank the reviewers and the editor for their valuable comments.

Author Contributions

Conceived and designed the experiments: QFZ JZ WJH. Performed the experiments: JZ QFZ. Analyzed the data: QFZ JZ. Contributed reagents/materials/analysis tools: JZ QFZ. Wrote the paper: QFZ WJH JZ.

References

1. Tanaka R, Tanaka A (2011) Chlorophyll cycle regulates the construction and destruction of the light-harvesting complexes. Biochimica et Biophysica Acta 1807: 968–976.
2. Porra RJ (2002) The chequered history of the development and use of simultaneous equations for the accurate determination of chlorophylls a and b. Photosynth Res 73: 149–156.
3. Arnon DI (1949) Copper enzymes in isolated chloroplasts. Polyphenoloxidase in *Beta vulgaris*. Plant Physiol 24: 1–15.
4. Chappelle EW, Kim MS, McMurtrey JE III (1992) Ratio analysis of reflectance spectra (RARS): an algorithm for the remote estimation of the concentrations of chlorophyll a, chlorophyll b and carotenoids in soybean leaves. Remote Sens Environ 39: 239–247
5. Blackburn GA (1998) Quantifying chlorophylls and carotenoids at leaf and canopy scales: an evaluation of some hyper-spectral approaches. Remote Sens Environ 66: 273–285.
6. Datt B (1998) Remote sensing of chlorophyll a, chlorophyll b, chlorophyll a + b and total carotenoid content in Eucalyptus leaves. Remote Sens Environ 66: 111–121.
7. Sims DA, Gamon JA (2002) Relationship between leaf pigment content and spectral reflectance across a wide range species, leaf structures and development stages. Remote Sens Environ 81: 337–354.
8. Gitelson AA, Gritz Y, Merzlyak MN (2003) Relationships between leaf chlorophyll content and spectral reflectance and algorithms for nondestructive chlorophyll assessment in higher plant leaves. J Plant Physiol 160: 271–282.
9. Gitelson AA, Vina A, Ciganda V, Rundquist DC, Arkebauer TJ (2005) Remote estimation of canopy chlorophyll content in crops. Geophys Res Lett 32: L08403. doi: 10.1029/2005 GL022688.
10. Schlemmer M, Gitelson A, Schepers J, Ferguson R, Peng Y, et al. (2013) Remote estimation of nitrogen and chlorophyll contents in maize at leaf and canopy levels. Int J Appl Earth Obs 25: 47–54.
11. Main R, Cho MA, Mathieu R, O'Kennedy RM, Ramoelo A, et al. (2011) An investigation into robust spectral indices for leaf chlorophyll estimation. ISPRS J Photogramm 66: 751–761.
12. Thomas JR, Gausman HW (1977) Leaf reflectance versus leaf chlorophyll and carotenoids concentration for eight crops. Agron J 63: 845–847.
13. Blackburn GA (2006) Hyperspectral remote sensing of plant pigments. J Exp Bot 58: 855–867.
14. Gitelson AA, Gritz Y, Merzlyak MN (2003) Relationships between leaf chlorophyll content and spectral reflectance and algorithms for non-destructive chlorophyll assessment in higher plant leaves. J Plant Physiol 160: 271–282.
15. Hatfield JL, Gitelson AA, Schepers JS, Walthall CL (2008) Application of spectral remote sensing for agronomic decisions. Agron J 100: S117–S131.
16. le Maire G, Francois C, Dufrene E (2004) Towards universal broad leaf chlorophyll indices using PROSPECT simulated database and hyperspectral reflectance measurements. Remote Sens Environ 89: 1–28.
17. Féret JB, Francois C, Gitelson AA, Barry KM, Panigada C, et al. (2011) Optimizing spectral indices and chemometric analysis of leaf chemical properties using radiative transfer modeling. Remote Sens Environ 115: 2742–2750.
18. He QX, Zhou QF, Sun XM (2005) Strikingly high content of grain protein in solution-cultured rice. J Sci Food Agr 85: 1197–1202.
19. Lichtenhaler HK, Wellburn AR (1983) Determination of total carotenoids and chlorophylls a and b of leaf extracts in different solvents. Biochem Soc T 11: 591–592.
20. Yoder BJ, Pettigrew-Crosby RE (1995) Predicting nitrogen and chlorophyll content and concentrations from reflectance spectra (400–2500 nm) at leaf and canopy scales. Remote Sens Environ 53: 199–211.
21. Peñuelas J, Baret F, Filella I (1995) Semiempirical indices to assess carotenoids chlorophyll-a ratio from leaf spectral reflectance. Photosynthetica 31: 221–230.
22. Carter GA, Cibula WG, Miller RL (1996) Narrow-band reflectance imagery compared with thermal imagery for early detection of plant stress. J Plant Physiol 148: 515–522.
23. Blackburn GA (1998) Quantifying chlorophylls and carotenoids at leaf and canopy scales: an evaluation of some hyper-spectral approaches. Remote Sens Environ 66: 273–285.
24. Blackburn GA (1999) Relationships between spectral reflectance and pigment concentrations in stacks of deciduous broadleaves. Remote Sens Environ 70: 224–237.
25. Merzlyak MN, Gitelson AA, Chivkunova OB, Rakitin VY (1999) Non-destructive optical detection of pigment changes during leaf senescence and fruit ripening. Physiol Plantarum 106: 135–141.

26. Sims DA, Gamon JA (2002) Relationship between leaf pigment content and spectral reflectance across a wide range species, leaf structures and development stages. Remote Sens Environ 81: 337–354.

27. Read JJ, Tarpley L, McKinion JM, Reddy KR (2002) Narrow-waveband reflectance ratios for remote estimation of nitrogen status in cotton. J Environ Qual 31: 1442–1452.

28. Steddom K, Heidel G, Jones D, Rush CM (2003) Remote detection of rhizomania in sugar beets. Phytopathology 93: 720–726.

29. Gitelson AA. Vina A, Ciganda V, Rundquist DC, Arkebauer TJ (2005) Remote estimation of canopy chlorophyll content in crops. Geophys Res Lett 32: L08403. doi: 10.1029/2005 GL022688.

30. Rondeaux G, Steven M, Baret F (1996) Optimization of soil adjusted vegetation indices. Remote Sens Environ 55: 95–107.

31. Jiang Z, Huete AR, Didan K, Miura T (2008) Development of a two-band enhanced vegetation index without a blue band. Remote Sens Environ 112: 3833–3845.

32. Kim MS, Daughtry CST, Chappelle EW, McMurtrey III JE, Walthall CL (1994) The use of high spectral resolution bands for estimating absorbed photosynthetically active radiation (Apar). In: Proc. Sixth Symposium on Physical Measurements and Signatures in Remote Sensing, Val D'Isere, France, January 17–21, pp. 299–306.

33. Gitelson AA, Buschmann C, Lichtenthaler HK (1999) Thechlorophyll fluorescence ratio F735/F700 as an accurate measure of the chlorophyll content in plants. Remote Sens Environ 69: 296–302.

34. Maccioni A, Agati G, Mazzinghi P (2001) New vegetation indices for remote measurement of chlorophylls based on leaf directional reflectance spectra. Journal of Photochemistry and Photobiology 61: 52–61.

35. Gitelson A, Merzlyak MN (1994) Quantitative estimation of chlorophyll-a using reflectance spectra: experiments with autumn chestnut and maple leaves. Journal of Photochemistry and Photobiology B: Biology 22: 247–252.

36. Peñuelas J, Gamon JA, Fredeen AL, Merino J, Field CB (1994) Reflectance indices associated with physiological changes in nitrogen and water limited sunflower leaves. Remote Sens Environ 48: 135–146.

37. Cho MA, Skidmore AK (2006) A new technique for extracting the red-edge position from hyperspectral data: the linear extrapolation method. Remote Sens Environ 101: 181–193.

38. Guyot G, Baret F (1988) Utilisation de la haute résolution spectrale pour suivrel'état des couverts végétaux. In: Guyenne TD, Hunt, JJ (Eds.), Proc. Fourth International Colloquium on Spectral Signatures of Objects in Remote Sensing, ESA SP-287, Assois, France, 18–22 January, pp. 279–286.

39. Gitelson AA, Merzlyak MN (1997) Remote estimation of chlorophyll content in higher plant leaves. Int J Remote Sens 18: 2691–2697.

40. McMurtey III JE, Chappelle EW, Kim MS, Meisinger JJ, Corp LA (1994) Distinguish nitrogen fertilization levels in field corns (Zea maysL.) with actively induced fluorescence and passive reflectance measurements. Remote Sens Environ 47: 36–44.

41. Zarco-Tejada PJ, Miller JR (1999) Land cover mapping at BOREAS using red-edge spectral parameters from CASI imagery. J Geophys Res 104: 27921–27933.

42. Lichtenthaler HK, Lang M, Sowinska M, Heisel F, Miehe JA (1996) Detection of vegetation stress via a new high resolution fluorescence imaging system. J Plant Physiol 148: 599–612.

43. Peñuelas J, Filella I, Lloret P, Munoz F, Vilajeliu M (1995) Reflectance assessment of mite effects on apple trees. Int J Remote Sens 16: 2727–2733.

44. Filella I, Peñuelas J (1994) The red-edge position and shape as indicators of plant chlorophyll content, biomass and hydric status. Int J Remote Sens 15: 1459–1470.

45. Vogelman JE, Rock BN, Moss DM (1993) Red-edge spectral measurements from sugar maple leaves. Int J Remote Sens 14: 1563–1575.

46. Jin XL, Diao WY, Xiao CH, Wang FY, Chen B, et al. (2013) Estimation of wheat agronomic parameters using new spectral indices. PLoS ONE 8: e72736. doi: 10.1371/journal.pone.0072736.

4

Infrageneric Phylogeny and Temporal Divergence of *Sorghum* (Andropogoneae, Poaceae) Based on Low-Copy Nuclear and Plastid Sequences

Qing Liu[1]*, Huan Liu[1,2], Jun Wen[3], Paul M. Peterson[3]*

1 Key Laboratory of Plant Resources Conservation and Sustainable Utilization, South China Botanical Garden, Chinese Academy of Sciences, Guangzhou, China, **2** University of Chinese Academy of Sciences, Beijing, China, **3** Department of Botany, National Museum of Natural History, Smithsonian Institution, Washington, D.C., United States of America

Abstract

The infrageneric phylogeny and temporal divergence of *Sorghum* were explored in the present study. Sequence data of two low-copy nuclear (LCN) genes, phosphoenolpyruvate carboxylase 4 (*Pepc4*) and granule-bound starch synthase I (*GBSSI*), from 79 accessions of *Sorghum* plus *Cleistachne sorghoides* together with those from outgroups were used for maximum likelihood (ML) and Bayesian inference (BI) analyses. Bayesian dating based on three plastid DNA markers (*ndhA* intron, *rpl32-trnL*, and *rps16* intron) was used to estimate the ages of major diversification events in *Sorghum*. The monophyly of *Sorghum* plus *Cleistachne sorghoides* (with the latter nested within *Sorghum*) was strongly supported by the *Pepc4* data using BI analysis, and the monophyly of *Sorghum* was strongly supported by *GBSSI* data using both ML and BI analyses. *Sorghum* was divided into three clades in the *Pepc4*, *GBSSI*, and plastid phylograms: the subg. *Sorghum* lineage; the subg. *Parasorghum* and *Stiposorghum* lineage; and the subg. *Chaetosorghum* and *Heterosorghum* lineage. Two LCN homoeologous loci of *Cleistachne sorghoides* were first discovered in the same accession. *Sorghum arundinaceum*, *S. bicolor*, *S. x drummondii*, *S. propinquum*, and *S. virgatum* were closely related to *S. x almum* in the *Pepc4*, *GBSSI*, and plastid phylograms, suggesting that they may be potential genome donors to *S. almum*. Multiple LCN and plastid allelic variants have been identified in *S. halepense* of subg. *Sorghum*. The crown ages of *Sorghum* plus *Cleistachne sorghoides* and subg. *Sorghum* are estimated to be 12.7 million years ago (Mya) and 8.6 Mya, respectively. Molecular results support the recognition of three distinct subgenera in *Sorghum*: subg. *Chaetosorghum* with two sections, each with a single species, subg. *Parasorghum* with 17 species, and subg. *Sorghum* with nine species and we also provide a new nomenclatural combination, *Sorghum sorghoides*.

Editor: Manoj Prasad, National Institute of Plant Genome Research, India

Funding: This work was supported by the National Natural Science Foundation of China (31270275, 31310103023), the Special Basic Research Foundation of Ministry of Science and Technology of the People's Republic of China (2013FY112100), the Key Project of Key Laboratory of Plant Resources Conservation and Sustainable Utilization, South China Botanical Garden, CAS (201212ZS), the 42nd Scientific Research Foundation for the Returned Overseas Chinese Scholars, State Education Ministry (2011-1139), and the Laboratories of Analytical Biology of the National Museum of Natural History, Smithsonian Institution. The funders had no role in study design, data collection and analysis, decision to publish, or preparation of the manuscript.

Competing Interests: The authors have declared that no competing interests exist.

* Email: liuqing@scib.ac.cn (QL); peterson@si.edu (PMP)

Introduction

Cultivated sorghum [*Sorghum bicolor* (L.) Moench] ranks fifth in both production and planted area of cereal crops worldwide, only behind wheat, rice, maize, and barley [1]. *Sorghum* Moench comprises 31 species exhibiting considerable morphological and ecological diversity [2–4] in global tropical, subtropical, and warm temperate regions [5]. The genus has panicles bearing short and dense racemes of paired spikelets (one sessile, the other pedicelled), whose sessile spikelets resemble the single sessile spikelets of *Cleistachne* Benth. These two genera were assigned to Sorghinae Clayton & Renvoize [6], one of the 11 subtribes of the tribe Andropogoneae Dumort. [7]. Previous studies of the genus using chloroplast DNA (cpDNA) and nuclear ribosomal DNA (nrDNA) internal transcribed spacer (ITS) sequences indicated that

Cleistachne was sister to or part of an unresolved polytomy within *Sorghum* [8–10]. The ambiguous relationship between *Sorghum* and *Cleistachne* is reflected by the absence of pedicelled spikelets and the unverified hypothesis for the allotetraploid origin of *Cleistachne sorghoides* Benth. [2,11]. Within Andropogoneae, *Sorghastrum* Nash has sometimes been considered as a subgenus in *Sorghum* due to its somatic chromosome number of 40 [2], or a distinct genus whose pedicelled spikelets are reduced to vestigial pedicels [12]. Therefore, the generic limits of *Sorghum* have long been a controversial issue that needs to be tested using highly informative molecular markers.

Five morphological subgenera are recognized in *Sorghum*: *Sorghum*, *Parasorghum*, *Stiposorghum*, *Chaetosorghum*, and *Heterosorghum* [2,3,8]. Subgenus *Sorghum* contains ten species (including the cultivated sorghum) that are distributed throughout

Africa, Asia, Europe, Australia, and the Americas [2,5]. The seven species of subg. *Parasorghum* occur in Africa, Asia, and northern Australia, and the ten species of subg. *Stiposorghum* occur in northern Australia and Asia. Subgenera *Chaetosorghum* and *Heterosorghum* are native to northern Australia and the Pacific Islands [3]. Culm nodes are glabrous or slightly pubescent in three subgenera: *Sorghum*, *Chaetosorghum*, and *Heterosorghum*, and bear a ring of hairs in subg. *Parasorghum* and *Stiposorghum* [2,13]. Subgenus *Sorghum* is characterized by the presence of well-developed pedicelled spikelets, while subg. *Chaetosorghum* and *Heterosorghum* are characterized by pedicelled spikelets which are reduced to glumes [2,3].

The five morphological subgenera of *Sorghum* are not shown to be concordant with molecular phylogenetic hypothesis [14–16]. The combined ITS1/*ndhF*/*Adh1* sequence data support a clade of *Sorghum* plus *Cleistachne sorghoides* that is divided into two lineages, one containing subg. *Sorghum*, *Chaetosorghum* and *Heterosorghum*, as well as *Cleistachne sorghoides*, and the other, subg. *Parasorghum* and *Stiposorghum* [14]. Uncertainty about relationships in *Sorghum* has led to the reclassification of three distinct genera: *Sarga* Ewart including species of subg. *Parasorghum* and *Stiposorghum*; *Sorghum* including *S. bicolor*, *S. halepense* (L.) Pers., and *S. nitidum* (Vahl) Pers.; and *Vacoparis* Spangler including species of sub. *Chaetosorghum* and *Heterosorghum* [15]. Ng'uni et al. [16] argued that this reclassification was unwarranted. Based on plastid and ITS sequence data, they found that *Sorghum* consisted of two lineages: one lineage containing species of subg. *Sorghum*, *Chaetosorghum* and *Heterosorghum*, and a second lineage containing species of subg. *Parasorghum* and *Stiposorghum*. More than 80% of samples were confined to Australia in previous molecular studies, which focused on resolving interspecific relationships in subg. *Sorghum*. Therefore, the molecular analysis based on a greater sampling of taxa throughout their geographic ranges is essential to explore the infrageneric relationships in *Sorghum*.

The species of *Sorghum* are an excellent group for understanding the evolutionary patterns in crop species and wild relatives since the genus contains a large tertiary gene pool (GP-3, a genetic entity developed by Harlan and De Wet [17] to deal with varying levels of interfertility among related taxa), and a relatively small secondary gene pool (GP-2) [9]. Members of primary gene pool (GP-1) from the same species (such as the cereal species) can interbreed freely. Members of GP-2 are closely related to members of GP-1, although there are some hybridization barriers between members of GP-1 and GP-2, which can occasionally produce fertile first-generation (F1) hybrids. Members of GP-3 are more distantly related to members of GP-1, while gene transfers between members of GP-1 and GP-3 are impossible without artificial disturbance measures [17]. Members of subg. *Sorghum* are found in GP-2, except for *S. bicolor*, which belongs to GP-1, while species of the other four subgenera are found in GP-3 [18]. Subgenus *Sorghum* is traditionally treated as two complexes: the *Arundinacea* complex, consisting of annual non-rhizomatous species such as *S. arundinaceum* (Desv.) Stapf, *S. bicolor*, *S. x drummondii* (Nees ex Steud.) Millsp. & Chase, and *S. virgatum* (Hack.) Stapf; and the *Halepensia* complex, consisting of perennial rhizomatous species such as *S. almum* Parodi, *S. halepense* (L.) Pers, *S. miliaceum* (Roxb.) Snowden, and *S. propinquum* (Kunth) Hitchc. [19]. Members of GP-3 contain wild genetic resources of important agronomic traits, e.g., drought tolerance and disease resistance. Nevertheless, the studies of interspecific relationships among GP-3 species has lagged behind due to small sampling, so a detailed understanding of relationships among GP-3 species is conducive for the exploitation of these valuable agronomic traits.

To date, 21.8% of grass species have been documented to have arisen as a result of hybridization events [20,21]. Plastid genes are commonly employed in phylogenetic reconstructions because they exist in high copy numbers in plant genomes and sequencing them often does not require cloning steps, and they are uniparentally (in most cases, maternally) inherited in angiosperms [22]. Low-copy nuclear (LCN) genes harbor the genetic information of bi-parental inheritance and often provide critical phylogenetic information for tracking evolution of plant lineages involving hybridization and allopolyploidization [23,24]. For these reasons, LCN gene data complementing plastid gene data are more effective in identifying allopolyploids and their genome donors. Several studies using this method have successfully resolved the backbone phylogenetic patterns of economically important crop genera, e.g., *Eleusine* Gaertn. [25], *Gossypium* L. [26], and *Hordeum* L. [27].

The middle Miocene-Pliocene interval of 1.8–17.6 million years ago (Mya) was a crucial period in the diversification of Poaceae [28]. The C_4 clades within the subfamily Panicoideae originated in the middle Miocene (ca. 14.0 Mya) in global tropical and subtropical regions. Subsequently, the ecological expansion of C_4 Panicoideae became associated with climate aridification and cooling through the late Miocene-Pliocene boundary (3.0–8.0 Mya) [29,30]. *Sorghum*, documented as an ecologically dominant member during the C_4 grassland expansion [28], is characterized by its modern geographic distribution spanning five continents [5,6,31]. Therefore, its ecological abundance in the late Tertiary, coupled with its wide geographic distribution in modern times, implies that *Sorghum* may have established conservative ecological traits during the early diversification process, i.e., *Sorghum* is a niche-conservative C_4 genus [32,33]. However, the paucity of accurate age estimations of major diversification events in *Sorghum* has impeded our understanding of whether temporal relationships existed between the diversification of *Sorghum* and palaeoclimatic fluctuations during the middle Miocene-Pliocene interval. Our study will shed some light on the impact of palaeoclimatic fluctuations on the diversification of niche-conservative C_4 grasses.

Here we explore the infrageneric phylogeny and temporal divergence of *Sorghum* by employing sequence data from two LCN and three plastid genes. The study aims to: (1) reconstruct infrageneric phylogenetic relationships in *Sorghum*; (2) investigate interspecific phylogenetic relationships among GP-3 species; and (3) estimate divergence times of major lineages in order to understand the impact of palaeoclimatic fluctuations on the diversification of *Sorghum*.

Materials and Methods

Plant Sampling and Sequencing

We sampled 79 accessions of 28 species in *Sorghum* [34–40], covering the morphological diversity and the geographic ranges of five subgenera (Table 1), plus the monotypic genus *Cleistachne*, together with seven species in six allied genera as outgroups [41,42]. Seeds were obtained from International Livestock Research Institute (ILRI), International Crops Research Institute for the Semi-Arid Tropics (IS), and United States Department of Agriculture (USDA). Leaf material was obtained from seedlings and dry herbarium specimens deposited at CANB, IBSC, K, and US (Table S1 [2,43–46]).

Two LCN genes, phosphoenolpyruvate carboxylase 4 (*Pepc4*) and granule-bound starch synthase I (*GBSSI*), were chosen for this study. The housekeeping *Pepc4* gene encodes PEPC enzyme responsible for the preliminary carbon assimilation in C_4 photosynthesis [47], whereas *GBSSI* gene encodes *GBSSI* enzyme

Table 1. Species of *Sorghum* included in the study. Chromosome numbers are based on the literature review.

Subgenus	Species	Longevity	Distribution	2n	References for Chromosome number
Sorghum					
	S. almum Parodi	Perennial	Americas, Australia, Asia	40	[34,35]
	S. arundinaceum (Desv.) Stapf	Annual	Africa, Asia, Australia, America	20	[11]
	S. bicolor (L.) Moench	Annual	Africa, Europe, Asia, Australia, America	20	[16,36]
	S. x *drummondii* (Nees ex Steud.) Millsp. & Chase	Annual	Africa, Asia, Australia, America	20	[11]
	S. halepense (L.) Pers.	Perennial	Mediterranean, Africa, Asia, Australia	40	[16,37]
	S. miliaceum (Roxb.) Snowden	Perennial	Asia, Africa, Australia	20	[38]
	S. propinquum (Kunth) Hitchc.	Perennial	Asia	20	[16]
	S. sudanense (Piper) Stapf	Annual	Africa, Asia, America, Europe	20	[39]
	S. virgatum (Hack.) Stapf	Annual	Africa, Asia	20	[34,40]
Parasorghum					
	S. grande Lazarides	Perennial	Australia	30/40	[3]
	S. leiocladum (Hack.) C.E. Hubb.	Perennial	Australia	20	[2,3,16]
	S. matarankense E.D. Garber & L.A. Snyder	Annual	Australia	10	[3,16]
	S. nitidum (Vahl) Pers.	Perennial	Asia, Australia	20/rarely 10	[2,3,16]
	S. purpureosericeum (Hochst. ex A. Rich.) Asch. & Schweinf.	Annual	Africa, Asia	10	[2,16]
	S. timorense (Kunth) Büse	Annual	Australia	10/rarely 20	[2,3,16]
	S. versicolor Andersson	Annual	Africa, Asia	10	[16]
Stiposorghum					
	S. amplum Lazarides	Annual	Africa, Australia	10/30	[3,36]
	S. angustum S.T. Blake	Annual	Australia	10	[3,16,36]
	S. brachypodum Lazarides	Annual	Australia	10	[3,16]
	S. bulbosum Lazarides	Annual	Australia	10	[3]
	S. ecarinatum Lazarides	Annual	Australia	10	[3,16]
	S. exstans Lazarides	Annual	Australia	10	[3,16]
	S. interjectum Lazarides	Perennial	Australia	30	[3,16]
	S. intrans F. Muell. ex Benth.	Annual	Australia	10	[2,3,16]
	S. plumosum (R.Br.) P. Beauv.	Perennial	Asia, Australia	10/20/30	[2,3]
	S. stipoideum (Ewart & Jean White) C.A. Gardner & C.E. Hubb.	Annual	Australia	10	[3,16]
Chaetosorghum					
	S. macrospermum E.D. Garber	Annual	Australia	40	[2,3]
Heterosorghum					
	S. laxiflorum F.M. Bailey	Annual	Australia, Asia	40	[2,3]

for amylose synthesis in plants and prokaryotes [48]. These two LCN genes have been used for accurate phylogenetic assessments in Poaceae [49,50]. They are predominantly low-copy in Poaceae, making it possible to establish orthology and track homoeologues arising by allopolyploidy [25,51]. Based on genome-wide researches on cereal crops, these two LCN genes appear to be on different chromosomes [48,52], thus each of the LCN markers can provide an independent phylogenetic estimation.

Genomic DNA extraction by means of DNeasy Plant Mini Kit (Qiagen, Valencia, CA, USA) was undertaken in accordance with the manufacturer's instructions. Two LCN markers were amplified using primers and protocols listed in Table 2 [53,54]. PCR products were purified by the PEG method [55]. Cycle sequencing reactions were conducted in 10 μL volumes containing 0.25 μL of BigDye v.3.1, 0.5 μL of primer, 1.75 μL of sequencing buffer (5×) and 1.0 μL of purified PCR product. For accessions that failed direct sequencing, the purified PCR products were cloned into pCR4-TOPO vectors and transformed into *Escherichia coli* TOP10 competent cells following the protocol of TOPO TA Cloning Kit (Invitrogen, Carlsbad, CA, USA). Transformed cells were plated and grown for 16 h on LB agar with X-Gal (Promega, Madison, WI, USA) and ampicillin (Sigma, St. Louis, MO, USA). We started with fewer colonies and picked more to ensure results, and eight to 24 colonies were selected from each individual via blue-white screening in order to assess allelic sequences and PCR errors [56,57]. Inserts were sequenced with primers T7 and T3 on the ABI PRISM 3730XL DNA Analyzer (Applied Biosystems, Forster City, CA, USA).

Cloned sequences of nuclear loci were initially aligned with MUSCLE v.3.8.31 [58] and adjusted in Se-Al v.2.0a11 (http:// tree.bio.ed.ac.uk/software/seal/). Subsequently, the corrected clones were assembled into individual-specific alignments that were analyzed separately using a maximum parsimony optimality criterion with the default parsimony settings in PAUP* v.4.0b10 [59]. The resulting trees were used to determine unique alleles present in each individual [56]. Alleles were recognized when one or more clones from a given individual were united by one or more characters [60]. After identifying all sequence clones for a given allele, the sequences were combined in a single project in Sequencher v.5.2.3 (Gene Codes Corp., Ann Arbor, Michigan, USA) and manually edited using a "majority-rule" criterion to form a final consensus allele sequence, and instances of PCR errors [56,57] were easily identified and never occurred in more than one sequence. Newly obtained consensus sequences of 62 *Pepc4* alleles and 76 *GBSSI* alleles were submitted to GenBank (http://ncbi.nlm.nih.gov/genbank; Table S1).

Three plastid markers (*ndhA* intron, *rpl32-trnL*, and *rps16* intron) were amplified and sequenced to estimate lineage ages in *Sorghum*. Primer sequences and amplification protocols for the plastid markers were listed in Table 2. PCR products were purified by the PEG method [55]. Cycle sequencing reactions were conducted in 10 μL volume and were run on an ABI PRISM 3730XL DNA Analyzer. Both strands were assembled in Sequencher v.5.2.3. Sequence alignment was initially performed using MUSCLE v.3.8.31 [58] in the multiple alignment routine

followed by manual adjustment in Se-Al v.2.0a11. The *Pepc4*, *GBSSI*, and combined plastid matrices were submitted to TreeBASE (http://purl.org/phylo/treebase/phylows/study/ TB2:S15625).

Phylogenetic analyses

Each data set was analyzed with maximum likelihood (ML) using GARLI v.0.96 [61], and Bayesian inference (BI) using MrBayes v.3.2.1 [62]. The substitution model for different data partitions was determined by the Akaike Information Criterion (AIC) implemented in Modeltest v.3.7 [63], and the best-fit model for each data set was listed in Table 3. ML topology was estimated using the best-fit model, and ML bootstrap support (MLBS) of internal nodes was determined by 1000 bootstrap replicates in GARLI v.0.96 with runs set for an unlimited number of generations, and automatic termination following 10,000 generations without a significant topology change (lnL increase of 0.01). The output file containing the best trees for bootstrap reweighted data was then read into PAUP* v.4.0b10 [59] where the majority-rule consensus tree was constructed to calculate bootstrap support values.

Bayesian inference (BI) analyses were conducted in MrBayes v.3.2.1 [62] using the best-fit model for *Pepc4* and *GBSSI* loci (Table 3). Each analysis consisted of two independent runs for 40 million generations; trees were sampled every 1000 generations, and the first 25% were discarded as burn-in. The majority-rule (50%) consensus trees were constructed after conservative exclusion of the first 10 million generations from each run as the burn-in, and the pooled trees (c. 60,000) were used to calculate the Bayesian posterior probabilities (PP) for internal nodes using the "sumt" command. The AWTY (Are We There Yet?) approach was used to explore the convergence of paired MCMC runs in BI analysis [64]. The stationarity of two runs was inspected by cumulative plots displaying the posterior probabilities of splits at selected increments over an MCMC run, and the convergence was

Table 2. Primer sequences and PCR protocols in the study.

Region	Location	Primers	Sequence (5′–3′)	PCR parameters	Reference
Pepc4	Chromosome 10	Pepc4-8F	GAT CGA CGC CAT CAC CAC	95°C/3 min; 16×(94°C/20 s; 65°C/40 s, −1°C/cycle; 72°C/90 s), 21×(94°C/20 s; 50°C/40 s; 72°C/90 s); 72°C/5 min	This study
		Pepc4-10R	GGA AGT TCT TGA TGT CCT TGT CG		This study
GBSSI	Chromosome 7	waxy-8F	ATC GTC AAC GGC ATG GAC GT	95°C/3 min; 16×(94°C/20 s; 65°C/40 s, −1°C/cycle; 72°C/90 s), 21×(94°C/20 s; 50°C/40 s; 72°C/90 s); 72°C/5 min	This study
		waxy-13R	GTT CTC CCA GTT CTT GGC AGG		This study
ndhA intron	Plastid	ndhA intron-1F	GCT GAC GCC AAA GAT TCC AC	95°C/3 min; 37×(94°C/40 sec; 51°C/40 S; 72°C/100 sec); 72°C/10 min	This study
		ndhA intron-1R	GTA CTA GCA ATA TCT CTA CG		This study
rpl32-trnL	Plastid	rpl32-F	CAGT TCC AAA AAA ACG TAC TTC	The same as above	[53]
		rpl32-trnL^(UAG)	CTG CTT CCT AAG AGC AGC GT		[53]
rps16 intron	Plastid	rps16-F2	AAA CGA TGT GGT AGA AAG CAA C	The same as above	[54]
		rps16-R2	ACA TCA ATT GCA ACG ATT CGA TA		[54]

Table 3. Sequence and tree statistics for LCN and plastid genes used in this study.

Region	N	Average sequence length	Aligned sequence length (SL)	GC%	VC	PIC	PIC/SL	Ti/Tv	Model
Pepc4	62	1056	1225	62.81%	415	249	20.3%	0.7428	GTR+I+G
GBSSI	76	1130	1501	54.66%	899	658	43.8%	0.7819	TIM+G
ndhA intron	62	1041	1131	33.30%	123	36	3.2%	1.1150	TVM+G
rpl32-trnL	62	730	807	28.45%	106	48	5.9%	0.6217	HKY+G
rps16 intron	62	872	920	34.26%	98	29	3.2%	0.4432	F81+I+G
Combined plastid	62	2223	2858	32.28%	327	113	4.0%	0.6870	TVM+I+G

GC = guanine and cytosine; N = number of sequences; VC = variable characters; PIC = parsimony informative characters; Ti/Tv = transition/transversion ratio.

visualized by comparative plots displaying posterior probabilities of all splits for paired MCMC runs.

The nuclear data were used to help determine bi-parental contributions, and multiple alleles were present for most polyploid taxa. Thus, the nuclear data cannot be combined with the plastid dataset, which provided the maternal phylogenetic framework. We rooted the *Pepc4* tree using species of *Apluda, Bothriochloa, Chrysopogon, Dichanthium* and *Sorghastrum* as outgroups and rooted the *GBSSI* tree using species of *Bothriochloa, Dichanthium, Microstegium* and *Sorghastrum* as outgroups [41,42] because clean *GBSSI* sequences of *Apluda* and *Chrysopogon* could not be isolated in the laboratory. The appropriate choice of outgroups was confirmed by phylogenetic proximity (the monophyletic ingroup being supported), genetic proximity (short branch length being observed) and base compositional similarity (ingroup-like GC%; Table 3) [65].

Molecular Dating

For molecular dating analyses using the plastid markers, a strict molecular clock model was rejected at a significance level of 0.05 (IL = 686.7024, d.f. = 60, $P = 0.025$) based on a likelihood ratio test [66]. A Bayesian relaxed clock model was implemented in BEAST v.1.7.4 [67] to estimate lineage ages in *Sorghum*. Three plastid markers were partitioned using BEAUti v.1.7.4 (within BEAST) with the best-fit model determined by Modeltest v.3.7 (Table 3).

The Andropogoneae crown age was estimated at 17.1±4.1 Mya [49] and within this confidence interval [68], although the most reliable fossils of subfamily Panicoideae were the petrified vegetative parts from the Richardo Formation in California [69] now dated to be approximately 12.5 Mya [70–72]. Because the lineages may have occurred earlier than the fossil record [73], the *Sorghum* stem age was set as a normal prior distribution (mean 17.1, SD 4.1). A Yule prior (Speciation: Yule Process) was employed. An uncorrelated lognormal distributed relaxed clock model was used, which permitted evolutionary rates to vary along branches according to lognormal distribution. Following optimal operator adjustment, as suggested by output diagnostics from preliminary BEAST runs, two independent MCMC runs were performed with 40 million generations, each run sampling every 1000 generations with the 25% of the samples discarded as burn-in. All parameters had a potential scale reduction factor [74] that was close to one, indicating that the posterior distribution had been adequately sampled. The convergence between two runs was checked using the "cumulative" and "compare" functions implemented in the AWTY [64]. A 50% majority rule consensus from the retained posterior trees (c. 60,000) of three runs were obtained using TreeAnnotator v.1.7.4 (within BEAST) with a PP limit of 0.5 and mean lineage heights.

Results

Phylogenetic analyses of *Pepc4* sequences

The aligned *Pepc4* matrix comprised 1225 characters, including partial exons 8 and 9, complete intron 9, at lengths of 841 bp, 190 bp, and 194 bp, respectively (Table 3). The *Pepc4* data provided a relatively high proportion of parsimony-informative characters (249 bp; 20.3%). The log likelihood scores of 56 substitution models ranged from 5883.8525 to 6165.2119, and Modeltest indicated that the best-fit model under AIC was GTR+I+G with base frequencies ($\pi_A = 0.19$, $\pi_C = 0.32$, $\pi_G = 0.31$, and $\pi_T = 0.18$), and substitution rates ($r_{AC} = 1.7$, $r_{AG} = 2.6$, $r_{AT} = 2.8$, $r_{CG} = 2.3$, $r_{CT} = 3.6$, and $r_{GT} = 1$). Within the Bayesian phylogenetic inference, two chains converged at similar topologies. The standard deviation of split frequencies reached values lower than 0.01 during analysis, and the stationarity was reached after 2.27 million generations (Figure S1). The ML and the BI analyses indicated an identical phylogenetic pattern for *Sorghum* plus *Cleistachne sorghoides*.

The monophyly of *Sorghum* plus *Cleistachne sorghoides* (with the latter nested within *Sorghum*) received strong support from the BI analysis (PP = 0.99). Three clades (designated as clades P-I, P-II, and P-III) were observed in the *Pepc4* phylogram with strong support (Figure 1). The *Pepc4* sequences from one accession of *Cleistachne sorghoides* fell into two divergent lineages [clade P-I and an independent branch with strong support (MLBP = 100%, PP = 1.00)], with clade P-I having A type sequence and the independent branch having B type sequences (putative homoeologues, a potential result caused by allotetraploidy, where each sequence type represents a different parental lineage). Clade P-I contained species of subg. *Sorghum*, *S. ecarinatum* Lazarides, and A-type sequence of *Cleistachne sorghoides* with strong support (MLBP = 100%, PP = 1.00). Clade P-II comprised subg. *Parasorghum* and *Stiposorghum* with strong or moderate support (MLBP = 88%, PP = 1.00). Clade P-III contained *S. laxiflorum* with strong support (MLBP = 95%, PP = 0.99). Clade P-I was sister to clade P-III (PP = 0.94), while clade P-II was sister to B-type sequences of *C. sorghoides* (PP = 0.58), and finally, the clade P-I+clade P-III was sister to the clade P-II and B-type sequences of *C. sorghoides* in the *Pepc4* phylogram (PP = 0.99) (Figure 1).

Phylogenetic analyses of *GBSSI* sequences

The aligned *GBSSI* matrix comprised 1501 characters, including partial exons 8 and 13, complete exons 9, 10, 11, and 12, introns 8, 9, 10, 11, and 12 at a length of 82 bp, 33 bp, 185 bp, 204 bp, 106 bp, 138 bp, 158 bp, 152 bp, 145 bp, 130 bp, and 168 bp, respectively (Table 3). The log likelihood scores of 56 substitution models ranged from 11947.3877 to 12361.0693, and Modeltest indicates that the best-fit model under AIC is TIM+G with base frequencies ($\pi_A = 0.23$, $\pi_C = 0.26$, $\pi_G = 0.28$, and $\pi_T = 0.23$) and substitution rates ($r_{AC} = 1.0$, $r_{AG} = 1.5$, $r_{AT} = 1.1$, $r_{CG} = 1.1$, $r_{CT} = 1.9$, and $r_{GT} = 1$). Within the Bayesian phylogenetic inference, two chains converged at similar topologies. The standard deviation of split frequencies reached values lower than 0.01 during analysis, and stationarity was reached after 1.09 million generations (Figure S2). The ML and the BI analyses generated an identical phylogenetic pattern for *Sorghum*.

The monophyly of *Sorghum* received strong support (MLBS = 100%, PP = 1.00) (Figure 2). Three clades (designated as clades G-I, G-II, and G-III) were recognized in the *GBSSI* phylogram with strong support. Clade G-I contained subg. *Sorghum* species, *S. leiocladum* (Hack.) C.E. Hubb., and *S. versicolor* Andersson with strong support (MLBP = 100%, PP = 1.00). Clade G-II comprised species of subg. *Parasorghum* and *Stiposorghum* with strong support (MLBP = 100%, PP = 1.00). Clade G-III consisted of *S. laxiflorum* and *S. macrospermum* with strong support (MLBP = 100%, PP = 1.00). Clade G-I was shown to be sister to clade G-II with weak support (MLBS = 0.61, PP = 0.71), and this group in turn, showed a strong association with clade G-III (MLBP = 100%, PP = 1.00) in the *GBSSI* phylogram (Figure 2).

Two (A- and B-type) homoeologous loci of *GBSSI* sequences were identified for two accessions of *Cleistachne sorghoides*, providing strong evidence for the presence of two divergent genomes. The A-type *GBSSI* sequences of *Cleistachne sorghoides* were characterized by three features: a large number of variations occurred in introns 8, 9, 11, and 12 (e.g., the strong support for A-type homoeologues of *C. sorghoides* and *Sorghastrum nutans* in

Figure 1); the A-type homoeologues of *C. sorghoides* being distantly related to B-type homoeologues of *C. sorghoides* (Figure 2); and 13 insertions (3–17 bp in length) distributed in introns 8, 9, 11, and 12, implying the likelihood of sequence divergence after the speciation event of *C. sorghoides*.

Divergence times

The combined plastid matrix of 62 accessions comprised 2858 characters, of which 113 were parsimony-informative (4.0%). The "cumulative" and "compare" results implemented in the AWTY showed that two runs had reached stationarity after 2.57 million generations (Figure S3). The BEAST analysis generated a well-supported tree (MLBP = 90%, PP = 0.99) for *Sorghum* plus *Cleistachne sorghoides* (Figure 3), which was identical to the topologies from ML and BI analyses. Three clades were recognized for *Sorghum* plus *Cleistachne sorghoides*. Clade II included *Cleistachne sorghoides* and subg. *Parasorghum* and *Stiposorghum* (lineage number 2), and clade I (i.e., subg. *Sorghum*) (lineage number 3) was sister to clade III (i.e., subg. *Chaetosorghum* and *Heterosorghum*). Here we discuss divergence times for the lineages of interest as shown in Table 4.

The uncorrelated-rates relaxed molecular clock suggests that the diversification of *Sorghum* plus *Cleistachne sorghoides* lineage occurred in the middle Miocene (12.7 Mya with 95% HPD of 5.5–16.7 Mya; lineage number 1 in Figure 3), which is the stem age for clade II (lineage number 2) and for clades I and III (lineage number 3). The crown age of clade II excluding *S. grande* was determined to be 10.5 (4.1–13.8) Mya in the late Miocene (lineage number 4), which is also the divergence time of clade II excluding *S. grande* and *Cleistachne sorghoides* (lineage number 5). The crown age of clade I was 10.5 (4.1–14.1) Mya in the late Miocene (lineage number 6), which is also the stem divergence time of clade III (lineage number 7) in Figure 3. Two lineages containing *S. bicolor* were estimated at 3.9 (0.3–4.3) Mya in the early Pliocene (the Africa-America-Asia-Europe lineage; lineage number 8) and 2.4 (0.0–3.4) Mya in the early Pliocene (the Africa-Asia lineage; lineage number 9), respectively (Table 4).

Discussion

Origin of *Cleistachne sorghoides*

Plastid, *Pepc4* and *GBSSI* data support the hypothesis for the allotetraploid origin of *Cleistachne sorghoides*. Based on the plastid data, *Cleistachne sorghoides* shared a common ancestor with clade II excluding *S. grande* (lineage number 4 in Figure 3), which may represent a source of the maternal parent for *C. sorghoides*. The plastid sequence similarity between *C. sorghoides* and clade II excluding *S. grande* also indicated that *C. sorghoides* became separated from the common ancestor in a relatively ancient time [10]. The *Pepc4* data provide evidence for this ancient allopolyploid origin because the conservative *Pepc4* gene evolved more slowly than non-housekeeping genes [75]. Two *Pepc4* homoeologous loci of *C. sorghoides* were isolated from the same accession, and this indicates the presence of two divergent genomes in *C. sorghoides*. The maternal lineage identified by the plastid tree was confirmed by the weak relationship between clade P-II and B-type homoeologues of *C. sorghoides* in the *Pepc4* phylogeny (Figure 1). The *GBSSI* tree was found to be complementary to the nrDNA ITS tree, in which *C. sorghoides* was deeply nested within the subg. *Parasorghum* and *Stiposorghum* lineage [8]. The authors inferred that the ITS sequences of *C. sorghoides* might have undergone complete homogenization towards the maternal parent, i.e. the subg. *Parasorghum* and *Stiposorghum* lineage. The B-type homoeologues of *Cleistachne sorghoides* showed no close relation-

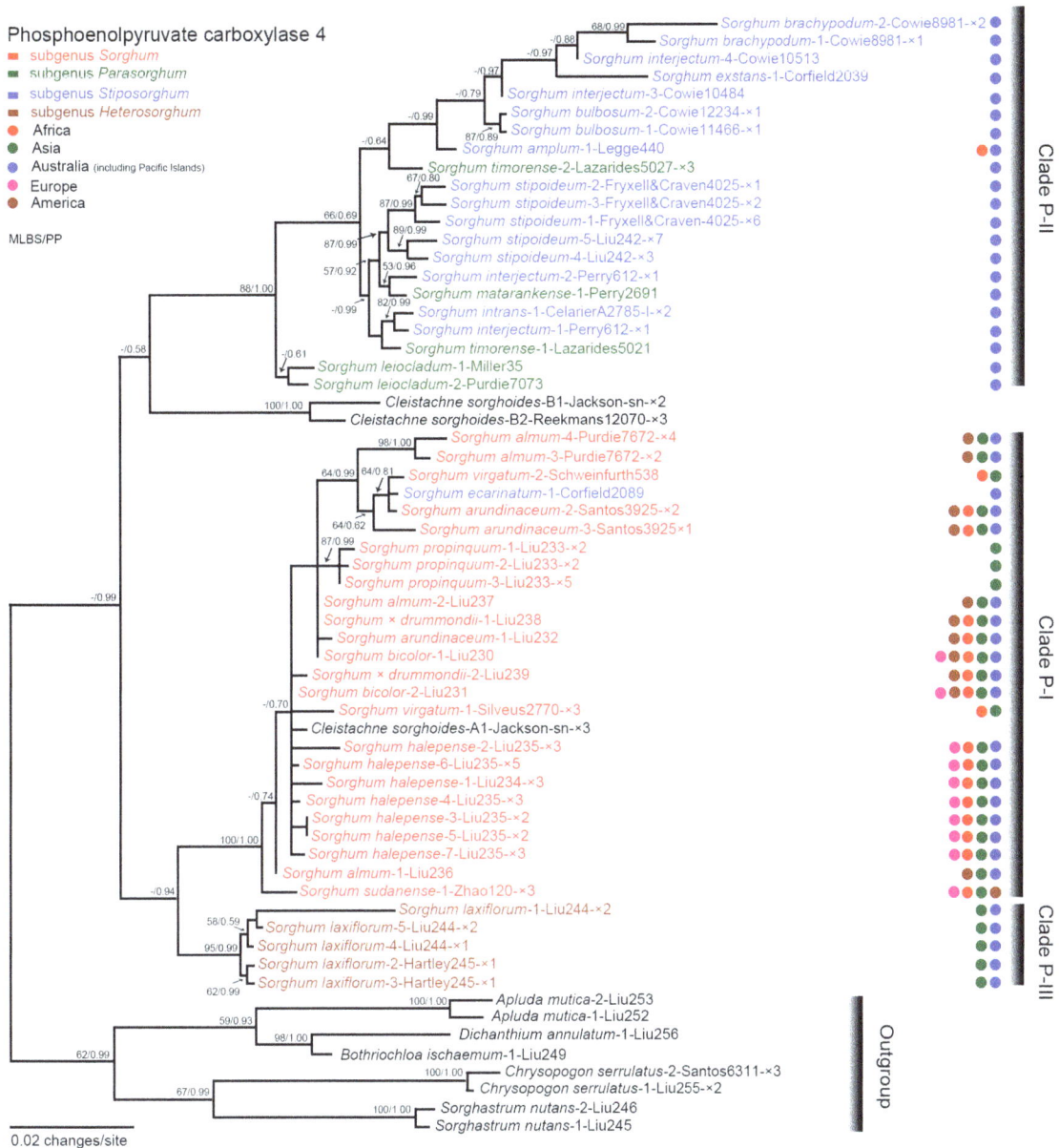

Figure 1. Maximum likelihood phylogeny of *Sorghum* inferred from nuclear *Pepc4* data. Numbers above branches are maximum likelihood bootstrap/Bayesian posterior probability (MLBS/PP). Taxon labels are in the format: *Sorghum brachypodum*-2-Cowie8981-×2 where *Sorghum brachypodum* indicates that the sequence belongs to the species *Sorghum brachypodum*; -2- = the second sequence listed in Table S1 for the species; Cowie8981 = specimen voucher information; -×2 indicates we recovered 2 clones for the sequence; and without any mark after specimen voucher information indicates the sequence is derived from PCR-direct sequencing. Coloured taxon labels and circles correspond to the listed subgenera and geographic ranges at the top left corner of the figure, respectively.

ship with any sampled species in the *GBSSI* tree (Figure 2), providing indirect evidence for the full divergence of B-type *GBSSI* homoeologues of *C. sorghoides* away from the maternal parent in *Sorghum* (clade II) in the *GBSSI* tree.

The paternal parent of *Cleistachne sorghoides* remains unresolved due to the incongruence between the two LCN trees. In the *Pepc4* tree, A-type homoeologue of *C. sorghoides* shared a common ancestor with clade P-I native to the Old World, while A-type *GBSSI* homoeologues of *C. sorghoides* showed a strong relationship with *Sorghastrum nutans* in the *GBSSI* tree. Considering its geographic range in North America, *Sorghastrum nutans* seems a much less likely candidate as the paternal parent

for *C. sorghoides* because geographically there is no opportunity for sexual contact with its potential maternal lineage.

To explain the paternal genome of *Cleistachne sorghoides*, it seems likely that *C. sorghoides* acquired the A-type *Pepc4* sequences via hybridization with the ancestor of subg. *Sorghum*, and subsequently the A-type *GBSSI* sequences of *C. sorghoides* experienced recombination (gene exchange) with species of the of African-American disjunct *Sorghastrum* [11]. A pre-requisite of this hypothesis is that East Africa and India would have been the geographic location of the recombination episode, perhaps in the fallow lands of Sudan, Uganda, Kenya, Congo, and India, where the native distribution of *C. sorghoides* is found [11]. Therefore, the recombination event of *C. sorghoides* placed its *GBSSI*

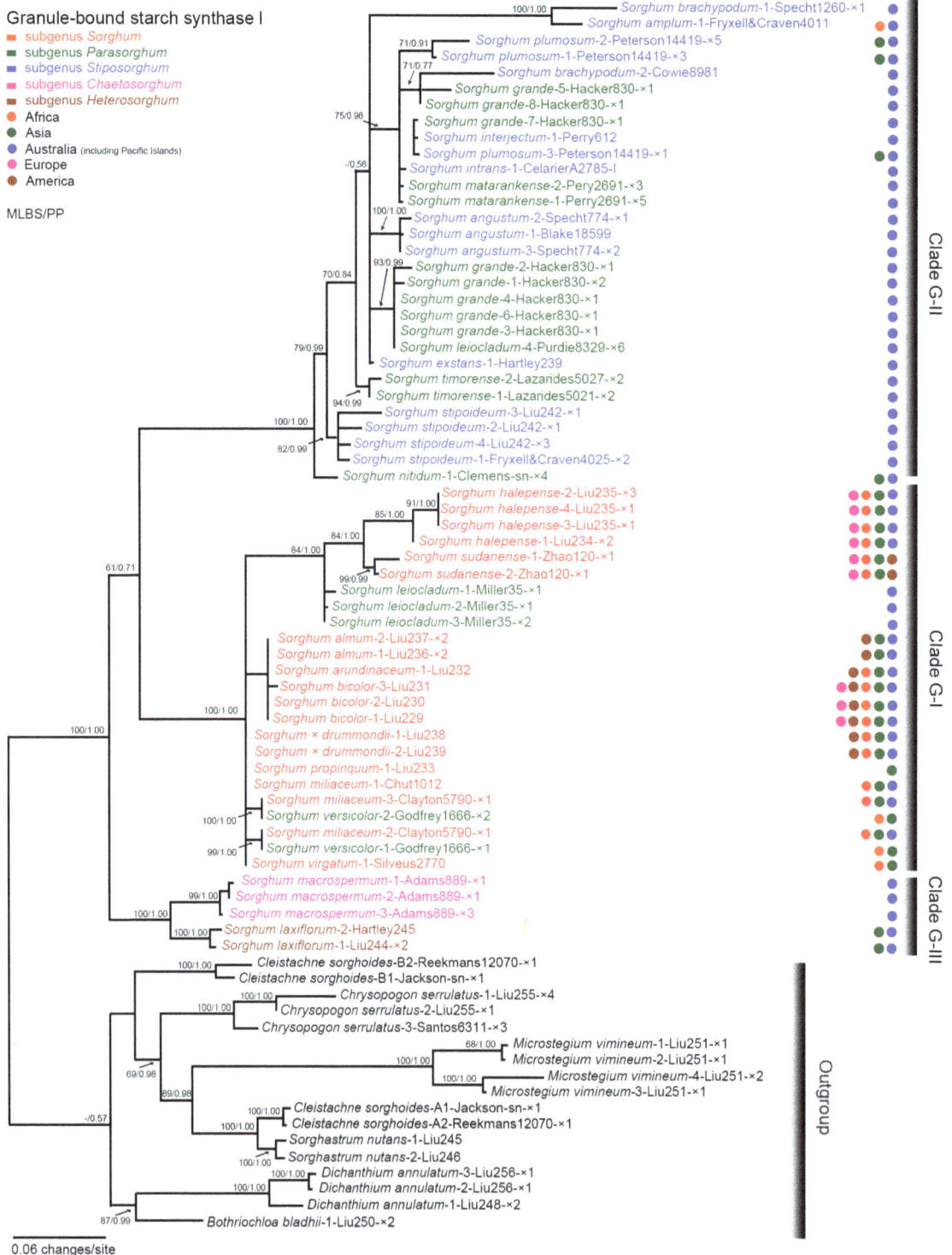

Figure 2. Maximum likelihood phylogeny of *Sorghum* inferred from nuclear *GBSSI* data. Numbers above branches are maximum likelihood bootstrap/Bayesian posterior probability (MLBS/PP). Taxon labels are in the format: *Sorghum matarankense*-2-Perry2691-×3 where *Sorghum matarankense* indicates that the sequence belongs to the species *Sorghum matarankense*; -2- = the second sequence listed in Table S1 for the species; Perry2691 = specimen voucher information; ×3 indicates we recovered 3 clones for the sequence; and without any mark after specimen voucher information indicates the sequence is derived from PCR-direct sequencing. Coloured taxon labels and circles correspond to the listed subgenera and geographic ranges at the top left corner of the figure, respectively.

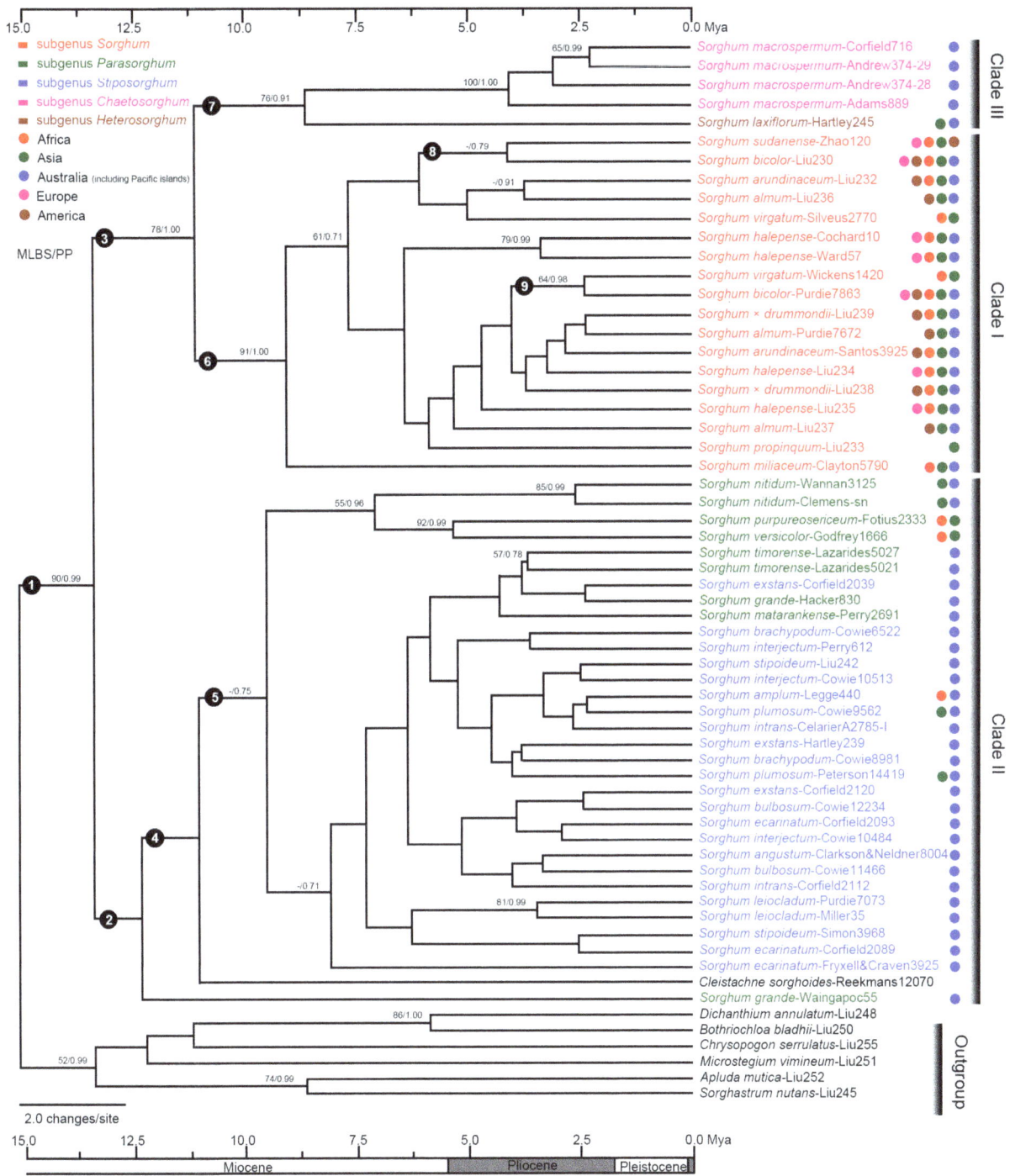

Figure 3. Chronogram of *Sorghum* and relatives based on three plastid sequences (*ndhA* intron, *rpl32-trnL*, and *rps16* intron) as inferred from BEAST. Numbers above the branches are maximum likelihood bootstrap/Bayesian posterior probability (MLBS/PP). Taxon labels are in the format: *Sorghum almum*-Liu236 where *Sorghum almum* indicates that the sequence belongs to the species *Sorghum almum*; -Liu236 = specimen voucher information. Coloured taxon labels and circles correspond to the listed subgenera and geographic ranges at the top left corner of the figure, respectively. Numbers 1–9 indicate the lineages of interest as shown in Table 4.

homoeologues near the outgroup location in the *GBSSI* phylogram. The LCN data indicate that *C. sorghoides* may have experienced a complex speciation process [2]. Based on support from *Pepc4*, combined plastid, and previous restriction site data [76], we chose to transfer *Cleistachne sorghoides* into *Sorghum* (Table 5).

Infrageneric phylogenetic relationships in *Sorghum*

The monophyly of *Sorghum* plus *Cleistachne sorghoides* is supported by *Pepc4* and plastid data, as well as the combined ITS1/*ndhF*/*Adh1* data [14], where *Sorghum* plus *Cleistachne sorghoides* are resolved into a distinct clade with 100% support. Nevertheless, the result contradicts the monophyly of *Sorghum*

Table 4. Posterior age distributions of lineages of interest in *Sorghum* plus *Cleistachne sorghoides*.

Lineage	N	Stem age (Mya)	Crown age (Mya)
Sorghum plus *Cleistachne sorghoides*	1	14.3 (5.6–18.0)	12.7 (5.5–16.7)
Clade II	2	12.7 (5.5–16.7)	11.7 (5.0–14.2)
Clades I+III	3	12.7 (5.5–16.7)	10.5 (4.1–14.1)
Clade II excluding *S. grande*	4	11.7 (5.0–14.2)	10.5 (4.1–13.8)
Clade II excluding *S. grande* and *Cleistachne sorghoides*	5	10.5 (4.1–13.8)	9.0 (3.3–11.5)
Clade I	6	10.5 (4.1–14.1)	8.6 (3.0–11.1)
Clade III	7	10.5 (4.1–14.1)	8.2 (2.3–11.1)
The *S. bicolor-S. sudanense* lineage (Africa, America, Asia, Europe)	8	5.8 (1.5–6.6)	3.9 (0.3–4.3)
The *S. bicolor-S. virgatum* lineage (Africa, Asia)	9	3.9 (0.1–4.0)	2.4 (0.0–3.4)

Lineage number (N) correspond to Figure 3; Lineage age is given by the mean age and the 95% highest posterior density (HPD) intervals in brackets; The age of each lineage is composed of the stem and the crown ages.

supported by *GBSSI* data. The absence of a definitive boundary for members of the subtribe Sorghinae has led others to suggest that the subtribe might have experienced rapid radiation [41]. The gene recombination event was inferred to explain the *GBSSI* sequence divergence of *C. sorghoides* from *Sorghum*, thus the unresolved phylogenetic position of the B-type *GBSSI* homoeologues of *C. sorghoides* in the *GBSSI* tree may indicate a complex phylogenetic history of the Sorghinae.

Three infrageneric lineages were supported by the LCN and the plastid data: the subg. *Sorghum* lineage; the subg. *Parasorghum* and *Stiposorghum* lineage; and the subg. *Chaetosorghum* and *Heterosorghum* lineage. The subg. *Chaetosorghum* and *Heterosorghum* lineage contained *S. macrospermum* and *S. laxiflorum*, respectively (Figures 2 and 3). These two species were easily distinguished from the remaining Australian native species of *Sorghum* in having glabrous culm nodes, reduced pedicelled spikelets, and a minute obtuse callus [2,3]. The two species possessed relatively smaller 2C DNA content (2.07 pg to 2.49 pg) than the remaining congeneric Australian species [3,36,77,78]. The close relationship between *S. macrospermum* and *S. laxiflorum* was also supported by nrDNA ITS [8,10] and the combined ITS1/*ndhF*/*Adh1* [9,14], On the basis of morphological, cytogenetic, and molecular sequence evidence, it is appropriate to recognize a distinct subg. *Chaetosorghum* comprising two sections: sect. *Chaetosorghum* (E.D. Garber) Ivanjuk. & Doronina (*S. macrospermum*) and sect. *Heterosorghum* (E.D. Garber) Ivanjuk. & Doronina (*S. laxiflorum*) (Table 5), although we could not get clean *Pepc4* sequences of *S. macrospermum* in the laboratory.

Most species of subg. *Parasorghum* and *Stiposorghum* were resolved into one well-supported lineage in the two LCN phylograms. The two subgenera were traditionally distinguished by length and shape of the callus on the sessile spikelet: *Parasorghum* was characterized by a short and blunt callus with an articulation joint, whereas *Stiposorghum* was characterized by a long and pointed callus with a linear joint [2,3]. However, doubts have recently been cast on the systematic value of the callus owing to the continuity of character states across the subgeneric boundary [14]. The subjective nature of determining callus morphology was also reflected by the molecular results because members of *Parasorghum* and *Stiposorghum* were aligned into a single lineage [7,8,40]. Since there were no well-defined taxonomic and genetic boundaries between these two subgenera,

the most practical solution is to combine them into a single subg. *Parasorghum* (Table 5).

Subgenus *Chaetosorghum* (including *S. macrospermum* and *S. laxiflorum*) appears closely related to subg. *Sorghum* with strong support (PP = 1.00) in the plastid tree (Figure 3); and such a relationship is consistent with nrDNA ITS [8], the combined ITS1/*ndhF*/*Adh1* [14], and *Pepc4* sequence data (Figure 1). Although the relationship between subg. *Chaetosorghum* and the clade G-I+clade G-II lineage received weak support (MLBS = 0.61, PP = 0.71) in the *GBSSI* tree, the placement of subg. *Chaetosorghum* in *Sorghum* is unequivocally supported by the sequence data [79].

Interspecific relationships within subg. *Sorghum* and GP-3 species

In the *Pepc4* phylogram, weak support (MPBS<50%, PP<0.5) was found for *S. bicolor* (Australian and Mexican accessions) and its immediate wild relatives, i.e., *S. almum*, *S. arundinaceum*, *S. x drummondii*, *S. propinquum*, and *S. virgatum* (Figure 1). The five species formed a strongly supported clade G-I (Figure 2). Based on the short branch lengths within clade P-I and clade G-I, the ease to hybrid formation between *S. bicolor* and certain members of subg. *Sorghum* [80], and their similar karyotypes [81], it is reasonable to infer that the ancestors of *S. bicolor* may be members of subg. *Sorghum* [82]. It was suggested that *S. almum* was a recent fertile hybrid between *S. bicolor* and *S. halepense* [80], but *S. arundinaceum*, *S. bicolor*, *S. x drummondii*, *S. propinquum*, and *S. virgatum* appear closely related to *S. almum* in *Pepc4*, *GBSSI*, and plastid phylograms, suggesting that they may be potential genome donors to *S. almum* [16].

Sorghum bicolor is an annual diploid species native to Africa [13]. Four main hypotheses have been proposed to explain its early evolutionary history: (1) annual *S. arundinaceum* was assumed to be the wild progenitor of *S. bicolor* based on a cytological study [11]; (2) *S. bicolor* was thought to be an interspecific hybrid and a descendant of two diploid species ($2n = 10$) [83]; (3) *S. bicolor* may have arisen by chromosome doubling from one diploid ancestor ($2n = 10$) [84]; or (4) *S. bicolor* may share a common ancestor with sugarcane and maize through an ancient polyploidization event [85]. The first hypothesis is supported by our study, where *S. arundinaceum* is confirmed to have a close relationship with *S. bicolor*, and this is seen in our LCN trees. Being an ancient forest-savanna species native to tropical Africa [86], *Sorghum arundinaceum* extends eastwards to

Table 5. A proposed new subgeneric classification of *Sorghum* Moench (subtribe Sorghinae Clayton & Renvoize, tribe Andropogoneae Dumort.) based on plastid and nuclear DNA data (*not examined in this study).

---

subg. ***Chaetosorghum*** E.D. Garber

  sect. ***Chaetosorghum*** (E.D. Garber) Ivanjuk. & Doronina

    S. macrospermum E.D. Garber

  sect. ***Heterosorghum*** (E.D. Garber) Ivanjuk. & Doronina

    S. laxiflorum F.M. Bailey

subg. ***Parasorghum*** (Snowden) E.D. Garber [syn: subg. *Stiposorghum* E.D. Garber]

    S. amplum Lazarides, *S. angustum* S.T. Blake, *S. bulbosum* Lazarides, *S. brachypodum* Lazarides, *S. ecarinatum* Lazarides, *S. exstans* Lazarides, *S. grande* Lazarides, *S. interjectum* Lazarides, *S. intrans* F. Muell. ex Benth., *S. leiocladum* (Hack.) C.E. Hubb., *S. matarankense* E.D. Garber & L.A. Snyder, *S. nitidum* (Vahl) Pers., *S. plumosum* (R.Br.) P. Beauv., *S. purpureosericeum* (Hochst. ex A. Rich.) Asch. & Schweinf., *S. stipoideum* (Ewart & Jean White) C.A. Gardner & C.E. Hubb., *S. timorense* (Kunth) Büse, *S. versicolor* Andersson

subg. ***Sorghum***

    S. almum Parodi, *S. arundinaceum* (Desv.) Stapf, *S. bicolor* (L.) Moench, *S.* x *drummondii* (Nees ex Steud.) Millsp. & Chase, *S. halepense* (L.) Pers., *S. miliaceum* (Roxb.) Snowden, *S. propinquum* (Kunth) Hitchc., *S. sudanense* (Piper) Stapf, *S. virgatum* (Hack.) Stapf

Incertae sedis

    S. burmahicum* Raizada, **S. controversum* (Steud.) Snowden, **S.* x *derzhavinii* Tzvelev, *S. sorghoides* (Benth.) Q. Liu & P.M. Peterson,S. trichocladum* (Rupr. ex Hack.) Kuntze

---

India, Australia, and is introduced to tropical America [5,11]. It is possible that the cultivated sorghum originated from *S. arundinaceum* native to forest-savanna in the sub-Saharan belt at the north of the equator before it colonized regions from the Atlantic to the Indian Oceans.

The separation of *S. sudanense* (Sudan grass) from *S.* x *drummondii* is supported by our study. The two species are distributed from Sudan to Egypt in East Africa [13] and naturalized in China and the Americas [39]. The relationship between these two species was incongruent based on the two LCN gene phylograms. The *Pepc4* sequences suggest that *S. sudanense* is sister to the lineage containing *S.* x *drummondii* and the remainder of subg. *Sorghum* with strong support (MLBS = 100%, PP = 1.00, Figure 1), it appears that *S. sudanense* is genetically distant from *S.* x *drummondii*. While in the *GBSSI* phylogram, the two species are nested within a strongly supported clade G-I (MLBS = 100%, PP = 1.00, Figure 2). An interpretation of the incongruent pattern might be that *S. sudanense* was a consequence of sympatric speciation among different East African populations of *S.* x *drummondii* occurring abundant genetic variation [87]. *Sorghum sudanense* has obovate caryopses with smooth surfaces whereas *S.* x *drummondii* has obovate or elliptic caryopses with striate surfaces (H. Liu et al., unpublished data). Perhaps caryopses with different surface sculptures are the phenotypic consequence of adaptation to different microhabitats [88,89]. Recognition of the two taxa at the specific level, as opposed to merging them as varieties [13] is compatible with our results.

The genome origin of *S. halepense* has been debated for years. It was believed that *S. halepense* experienced homoeologous chromosome transpositions [90] from potential progenitors *S. bicolor* and *S. propinquum* [91,92]. Some workers proposed that *S. halepense* was a segmental allotetraploid hybrid between *S. arundinaceum* and *S. propinquum* [12,80]. If so, the maternal parents of *S. halepense* may have come from members of subg. *Sorghum*, since *S. halepense* is deeply nested within lineage number 6 (Figure 3). Furthermore, the plastid data supports *S. arundinaceum* and *S.* x *drummondii* as potential progenitors of *S. halepense*. An alternative hypothesis is that *S. halepense* is an interspecific hybrid and a descendant of *S. bicolor* and *S. virgatum*

[93]. However, the *Pepc4* and *GBSSI* data contradict this hypothesis since no corresponding loci were isolated from *S. halepense*. In *GBSSI* tree, four sequences of *S. halepense* formed a lineage (MLBS = 85%, PP = 1.00), which was sister to the *S. sudanense* lineage. These results are consistent with the hypothesis that *S. halepense* arose via homoeologous chromosome transpositions from members of subg. *Sorghum*. *Sorghum halepense* exhibits disomic inheritance [38,83], allowing the independent assortment of DNA segments between progenitors resulting in a complex evolutionary pattern [94]. This assumption is substantiated in allozyme studies, where high-frequency alleles found in *S. halepense* were not detected in *S. bicolor* or *S. propinquum*, providing further evidence for the absence of alleles from progenitors of *S. halepense* [95].

Based on *GBSSI* and plastid data, *Sorghum nitidum* is nested within the subg. *Parasorghum* and *Stiposorghum* lineage. *Sorghum nitidum* is distributed in southeast Asia, the Pacific Islands, and northern Australia [2], and exhibits significant morphological variation. The species is characterized by a hairy ring around the nodes, awnless or awned lemmas in sessile spikelets, and relatively small chromosomes [81]. Based on ITS and *ndhF* analyses, *S. nitidum* is embedded in subg. *Sorghum* [16]. However, the genome size of *S. nitidum* (2.20 pg) resembles that of members of subg. *Parasorghum* and *Stiposorghum* (0.64 pg–2.30 pg) rather than that of subg. *Sorghum* (0.26 pg–0.42 pg) [36]. Our study supports a close relationship between *S. nitidum* and the subg. *Parasorghum* and *Stiposorghum* lineage [2,9].

Palaeoclimatic hypothesis for lineage divergence in *Sorghum*

It is recognized that the evolution of organisms is profoundly influenced by past tectonic activities and climate changes [30,96]. Two *Sorghum* major lineages (lineage numbers 2 and 3) diverged from a common ancestor at 12.7 (95% HPD: 5.5–16.7) Mya (Figure 3) in the middle Miocene-Pliocene interval marked by aridification, which induced C_4 grassland emergences in Africa [28,97]. The Eastern branch of East Africa Rift has continuously uplifted since the early Miocene [98,99], and the increasingly arid climate of tropical and subtropical Africa was caused by the topographic barrier of the eastern branch Rift to moist maritime air from the Indian Ocean [100,101]. The resultant formation of new ecological niches [28] presumably catalyzed the diversification of *Sorghum* (e.g., lineage numbers 8 and 9 in Figure 3) in Africa at a time when significant faunal turnover was observed, e.g., leaf-mining flies [102], savanna-inhabiting crickets [103], prairie-adapted rodents [104], and grass-feeding mammals [105].

The northern Australian endemic species of *Sorghum* (mostly in lineage number 5, Figure 3) diverged by 9.0 (HPD: 3.3–11.5) Mya around the late Miocene/Pliocene boundary, when the monsoonal palaeoclimate was characterized by south-eastward dry trade winds in winter and north-westward moist flow in summer [106–108]. The Australian endemic species [e.g., *S. intrans*, *S. leiocladum*, *S. matarankense* E.D. Garber & L.A. Snyder, and *S. timorense* (Kunth) Büse] are geographically restricted to rocky hills, coastal dunes, and seasonally flooded swamps in northern Australia [3,5] where the local vegetation was affected by the lowering seas, leading to the dominance of monsoonal savannas [109]. Meanwhile, the highly dissected tropical areas became even more scattered in northern Australia causing complex topography in the monsoonal savannas. Therefore, it is reasonable to hypothesize that the dominance of monsoonal savanna in the late Miocene contributed to the high level of endemism of *Sorghum* in Australia.

Taxonomy

Traditionally, *Cleistachne* has been separated from *Sorghum* because it has only single spikelets whose pedicels are thought to represent raceme peduncles, whereas *Sorghum* has sessile and pedicelled spikelets, although the sessile spikelets can be much reduced [6,11]. Our study and that of early workers agree that *Cleistachne* is allied with *Sorghum* [6,11,110]; we thus propose the new combination as below.

Sorghum sorghoides (Benth.) Q. Liu & P.M. Peterson, **comb. nov.** Basionym: *Cleistachne sorghoides* Benth., Hooker's Icon. Pl. 14: t. 1379. 1882.

We also propose a new subgeneric classification of *Sorghum* (Table 5). Within *Sorghum* we recognized three subgenera: *Chaetosorghum*, *Parasorghum*, and *Sorghum*; and chose to retain two sections within *Chaetosorghum*: *Chaetosorghum* and *Heterosorghum*. Alternatively, based on our molecular results, one could use the new generic name *Sarga* to represent species in subg. *Parasorghum*, *Sorghum* for species in subg. *Sorghum*, *Vacoparis* for species in *Chaetosorghum* and retain *Cleistachne*. Perhaps with a greater number of molecular markers, the apparent hybrid origin of *S. sorghoides* and phylogenetic position of *S. burmahicum* Raizada, *S. controversum* (Steud.) Snowden, *S. derzhavinii* Tzvelev, and *S. trichocladum* (Rupr. ex Hack.) Kuntze (all incertae sedis in our classification) will be elucidated.

Conclusions

The monophyly of *Sorghum* plus *Cleistachne sorghoides* is supported by the *Pepc4* and the plastid data, and we provide a new combination, *Sorghum sorghoides*. Molecular results support the allotetraploid origin of *S. sorghoides*. Based on combined plastid data, members of subg. *Parasorghum* may represent the maternal parents, while the paternal parents of *S. sorghoides* remained unresolved because of incongruence between the *Pepc4* and the *GBSSI* phylograms. *Sorghum macrospermum* is sister to *S. laxiflorum*, forming a distinct clade, which we refer to as subg. *Chaetosorghum* with two sections *Chaetosorghum* (*S. macrospermum*) and *Heterosorghum* (*S. laxiflorum*). Most of members of the

two subgenera *Parasorghum* and *Stiposorghum* are resolved into one well-supported lineage by the two LCN phylograms. Therefore, we choose to recognize a single subg. *Parasorghum*, and place *Stiposorghum* in synonymy. The two LCN gene trees and the combined plastid tree are consistent with the hypothesis that *S. halepense* originated via homoeologous chromosome transpositions. During the middle Miocene-Pliocene interval, the formation of new ecological niches in tropical and subtropical Africa presumably catalysed the diversification of *Sorghum* in Africa. Furthermore, it seems reasonable to infer that the dominance of monsoonal savanna in the late Miocene contributed to the high level of endemism of *Sorghum* in Australia. Molecular results support the recognition of three distinct subgenera in *Sorghum*: subg. *Chaetosorghum* with two sections each containing a single species, subg. *Parasorghum* with 17 species, and subg. *Sorghum* with nine species.

Supporting Information

Figure S1 Results of the exploration of *Pepc4* MCMC convergence using the AWTY (Are We There Yet?) approach. (a) Cumulative plot of the posterior probabilities of 20 splits at selected increments over one of two MCMC runs. (b) Comparative plot of posterior probabilities of all splits for paired MCMC runs.

Figure S2 Results of the exploration of *GBSSI* MCMC convergence using the AWTY (Are We There Yet?) approach. (a) Cumulative plot of the posterior probabilities of 20 splits at selected increments over one of two MCMC runs. (b) Comparative plot of posterior probabilities of all splits for paired MCMC runs.

Figure S3 Results of the exploration of three plastid sequences (*ndhA* intron, *rpl32-trnL* and *rps16* intron) MCMC convergence using the AWTY (Are We There Yet?) approach. (a) Cumulative plot of the posterior probabilities of 20 splits at selected increments over one of two MCMC runs. (b) Comparative plot of posterior probabilities of all splits for paired MCMC runs.

Table S1 Taxon name, chromosome number, source, and GenBank accession numbers of *Pepc4*, *GBSSI*, and three plastid (*ndhA* intron, *rpl32-trnL*, and *rps16* intron) sequences used in the study.

Acknowledgments

We thank ILRI-Addis Ababa, IS-Andhra Pradesh, and USDA-Beltsville Germplasm System for seeds, and six anonymous reviewers for their constructive comments that improved the manuscript.

Author Contributions

Conceived and designed the experiments: QL PMP. Performed the experiments: QL HL. Analyzed the data: QL HL. Contributed reagents/materials/analysis tools: QL HL JW PMP. Contributed to the writing of the manuscript: QL HL JW PMP. Obtained necessary plant material: QL HL PMP.

References

1. FAO (Food and Agriculture Organization of the United Nations) (2011) FAOSTAT Database. FAO, Rome, Italy. Available: http://faostat.fao.org. Accessed 30 September 2011.

2. Garber ED (1950) Cytotaxonomic studies in the genus *Sorghum*. Univ Calif Publ Bot 23: 283–361.

3. Lazarides M, Hacker JB, Andrew MH (1991) Taxonomy, cytology and ecology of indigenous Australian sorghums (*Sorghum* Moench: Andropogoneae: Poaceae). Aust Syst Bot 4: 591–635.

4. Clayton WD, Vorontsova MS, Harman KT, Williamson H (2006 onwards). GrassBase –The online world grass flora. Available: http://www.kew.org/data/grasses-db.html. Accessed 8 November 2006.

5. Liu H, Liu Q (2014) Geographical distribution of *Sorghum* Moench (Poaceae). J Trop Subtrop Bot 22: 1–11.

6. Clayton WD, Renvoize SA (1986) Genera graminum: grasses of the world. Kew Bull Addit Ser 13: 320–375.

7. Soreng RJ, Davidse G, Peterson PM, Zuloaga FO, Judziewicz EJ, et al. (2014) A world-wide phylogenetic classification of Poaceae (Gramineae): cão (草), capim, çayır, çimen, darbha, ghaas, ghas, gish, gramas, graminius, gräser, grasses, gyokh, he-ben-ke, hullu, kasa, kusa, nyasi, pastos, pillu, pullu, zlaki, etc. Available: http://www.tropicos.org/projectwebportal.aspx?pagename = ClassificationNWG &projectid = 10. Accessed 13 January 2014.

8. Dillon SL, Lawrence PK, Henry RJ (2001) The use of ribosomal ITS to determine phylogenetic relationships within *Sorghum*. Plant Syst Evol 230: 97–110.

9. Dillon SL, Lawrence PK, Henry RJ, Ross L, Price HJ, et al. (2004) *Sorghum laxiflorum* and *S. macrospermum*, the Australian native species most closely related to the cultivated *S. bicolor* based on ITS1 and *ndhF* sequence analysis of 25 *Sorghum* species. Plant Syst Evol 249: 233–246.

10. Sun Y, Skinner DZ, Liang GH, Hulbert SH (1994) Phylogenetic analysis of *Sorghum* and related taxa using internal transcribed spacers of nuclear ribosomal DNA. Theor Appl Genet 89: 26–32.

11. Clayton WD, Renvoize SA (1982) Gramineae (Part 3). In: Polhill RM, editor. Flora of Tropical East Africa. Rotterdam: August Aimé Balkema. pp. 320–734.

12. Celarier RP (1958) Cytotaxonomy of the Andropogoneae. III. Subtribe Sorgheae, genus *Sorghum*. Cytologia 23: 395–418.

13. De Wet JMJ (1978) Systematics and evolution of *Sorghum* sect. *Sorghum* (Gramineae). Am J Bot 65: 477–484.

14. Dillon SL, Lawrence PK, Henry RJ, Price HJ (2007) *Sorghum* resolved as a distinct genus based on combined ITS1, *ndhF* and *Adh1* analyses. Plant Syst Evol 268: 29–43.

15. Spangler RE (2003) Taxonomy of *Sarga*, *Sorghum* and *Vacoparis* (Poaceae: Andropogoneae). Aust Syst Bot 16: 279–299.

16. Ng'uni D, Geleta M, Fatih M, Bryngelsson T (2010) Phylogenetic analysis of the genus *Sorghum* based on combined sequence data from cpDNA regions and ITS generate well-supported trees with two major lineages. Ann Bot 105: 471–480.

17. Harlan JR, De Wet JMJ (1971) Toward a rational classification of cultivated plants. Taxon 20: 509–517.

18. Stenhouse JW, Prasada Rao KE, Gopal Reddy V, Appa Rao KD (1997) Sorghum. In: Fuccillo D, Sears L, Stapleton P, editors. Biodiversity in Trust: Conservation and Use of Plant Genetic Resources in CGIAR Centers. Cambridge: Cambridge University Press. pp. 292–308.

19. Snowden JD (1955) The wild fodder sorghums of the section *Eu-sorghum*. J Linn Soc Lond 55: 191–260.

20. Knobloch IW (1968) A check list of crosses in the Gramineae. New York: Stechert- Hafner Service Agency.

21. Knobloch IW (1972) Intergeneric hybridization in flowering plants. Taxon 21: 97–103.

22. Ness RW, Graham SW, Barrett SCH (2011) Reconciling gene and genome duplication events: using multiple nuclear gene families to infer the phylogeny of the aquatic plant family Pontederiaceae. Mol Biol Evol 28: 3009–3018.

23. Zhang N, Zeng LP, Shan HY, Ma H (2012) Highly conserved low-copy nuclear genes as effective markers for phylogenetic analyses in angiosperms. New Phytol 195: 923–937.

24. Zimmer EA, Wen J (2012) Using nuclear gene data for plant phylogenetics: progress and prospects. Mol Phylogenet Evol 65: 774–785.

25. Liu Q, Triplett JK, Wen J, Peterson PM (2011) Allotetraploid origin and divergence in *Eleusine* (Chloridoideae, Poaceae): evidence from low-copy nuclear gene phylogenies and a plastid gene chronogram. Ann Bot 108: 1287–1298.

26. Cronn R, Wendel JF (2004) Cryptic trysts, genomic mergers, and plant speciation. New Phytol 161: 133–142.

27. Brassac J, Jakob SS, Blattner FR (2012) Progenitor-derivative relationships of *Hordeum* polyploids (Poaceae, Triticeae) inferred from sequences of *TOPO6*, a nuclear low-copy gene region. PLoS ONE 7: e33808.

28. Edwards EJ, Osborne CP, Strömberg CAE, Smith SA, C4 Grasses Consortium (2010) The origins of C4 grasslands: integrating evolutionary and ecosystem science. Science 328: 587–591.

29. Cerling TE, Harris JM, Macfadden BJ, Leakey MG, Quade J, et al. (1997) Global vegetation change through the Miocene/Pliocene boundary. Nature 389: 153–158.

30. Strömberg CAE (2005) Decoupled taxonomic radiation and ecological expansion of open-habitat grasses in the Cenozoic of North America. Proc Natl Acad Sci USA 102: 11980–11984.

31. Hartley W (1958) Studies on the origin, evolution, and distribution of the Gramineae. I. The tribe Andropogoneae. Aust J Bot 6: 116–128.

32. Keng YL (1939) The gross morphology of Andropogoneae. Sinensia 10: 274–343.

33. Liu Q, Peterson PM, Ge XJ (2011) Phylogenetic signals in the realized climate niches of Chinese grasses (Poaceae). Plant Ecol 212: 1733–1746.

34. Li N (2009) Cytology and seed biology of *Sorghum halepense* and its three related species. M.S. Thesis. Jinhua: Zhejiang Normal University.

35. Martin JH (1959) Sorghum and pearl millet. In: Happert H, Rudorf W, editors. Handbuch der Pflanzenzüchtung, 2nd edition, vol. 2. Berlin: Paul Parey. pp. 565–587.

36. Price HJ, Dillon SL, Hodnett G, Rooney WL, Ross L, et al. (2005) Genome evolution in the genus *Sorghum* (Poaceae). Ann Bot 95: 219–227.

37. De Wet JMJ, Huckabay JP (1967) The origin of *Sorghum bicolor*. II. Distribution and domestication. Evoluton 21: 787–802.

38. Reddi VR (1970) Chromosome association in one induced and five natural tetraploids of *Sorghum*. Genetica 41: 321–333.

39. Chen SL, Phillips SM (2006) Sorghum Moench. In: Wu ZY, Raven PH, editors. Flora of China, vol. 22. Beijing: Science Press and St. Louis: Missouri Botanical Garden Press. pp. 602–604.

40. Lu QS (2006) Sorghum. In: Dong YC, Liu X, editors. Crops and Their Wild Relatives in China: Food Crops. Beijing: China Agriculture Press. pp. 360–405.

41. Mathews S, Spangler RE, Mason-Gamer RJ, Kellogg EA (2002) Phylogeny of Andropogoneae inferred from phytochrome B, *GBSSI*, and *ndhF*. Int J Plant Sci 163: 441–450.

42. Spangler RE, Zaitchik B, Russo E, Kellogg E (1999) Andropogoneae evolution and generic limits in *Sorghum* (Poaceae) using *ndhF* sequences. Syst Bot 24: 267–281.

43. Nadeem Ahsan SM, Vahidy AA, Ali SI (1994) Chromosome numbers and incidence of polyploidy in Panicoideae (Poaceae) from Pakistan. Ann Mo Bot Gard 81: 775–783.

44. Baltisberger M, Kocyan A (2010) IAPT/IOPB chromosome data 9. Taxon 59: 1298–1302.

45. Celarier RP (1956) Cytotaxonomy of the Andropogoneae. I. Subtribes Dimeriinae and Saccharinae. Cytologia 21: 272–291.

46. Vahidy AA, Davidse A, Shigenobu Y (1987) Chromosome counts of Missouri Asteraceae and Poaceae. Ann Mo Bot Gard 74: 432–433.

47. Lepiniec L, Vidal J, Chollet R, Gadal P, Crétin C (1994) Phosphoenolpyruvate carboxylase: structure, regulation and evolution. Plant Sci 99: 111–124.

48. Mason-Gamer RJ, Weil CF, Kellogg EA (1998) Granule-bound starch synthase: structure, function, and phylogenetic utility. Mol Biol Evol 15: 1658–1673.

49. Christin PA, Besnard G, Samaritani E, Duvall MR, Hodkinson TR, et al. (2008) Oligocene CO2 decline promoted C4 photosynthesis in grasses. Curr Biol 18: 37–43.

50. Mahelka V, Kopecký D (2010) Gene capture from across the grass family in the allohexaploid *Elymus repens* (L.) Gould (Poaceae, Triticeae) as evidenced by ITS, *GBSSI*, and molecular cytogenetics. Mol Biol Evol 27: 1370–1390.

51. Fortuné PM, Schierenbeck K, Ainouche A, Jacquemin J, Wendel JF, et al. (2007) Evolutionary dynamics of *Waxy* and the origin of hexaploid *Spartina* species. Mol Phylogenet Evol 43: 1040–1055.

52. Paterson AH, Bowers JE, Bruggmann R, Dubchak I, Grimwood J, et al. (2009) The *Sorghum bicolor* genome and the diversification of grasses. Nature 457: 551–556.

53. Shaw J, Lickey EB, Schilling EE, Small RL (2007) Comparison of whole chloroplast genome sequences to choose noncoding regions for phylogenetic studies in Angiosperms: the tortoise and the hare III. Am J Bot 94: 275–288.

54. Peterson PM, Romaschenko K, Johnson G (2010) A classification of the Chloridoideae (Poaceae) based on multi-gene phylogenetic trees. Mol Phylogenet Evol 55: 580–598.

55. Hiraishi A, Kamagata Y, Nakamura K (1995) Polymerase chain reaction amplification and restriction fragment length polymorphism analysis of 16S rRNA genes from methanogens. J Ferment Bioeng 79: 523–529.

56. Li FW, Pryer KM, Windham MD (2012) *Gaga*, a new fern genus segregated from *Cheilanthes* (Pteridaceae). Syst Bot 37: 845–860.

57. Rothfels CJ, Schuettpelz E (2013) Accelerated rate of molecular evolution for vittarioid ferns is strong and not driven by selection. Syst Biol 63: 31–54.

58. Edgar RC (2004) MUSCLE: multiple sequence alignment with high accuracy and high throughput. Nucleic Acids Res 32: 1792–1797.

59. Swofford DL (2003) PAUP*. Phylogenetic analysis using parsimony (* and other methods), ver. 4.0b10. Sunderland: Sinauer Associates.

60. Grusz AL, Windham MD, Pryer KM (2009) Deciphering the origins of apomictic polyploids in the *Cheilanthes yavapensis* complex (Pteridaceae). Am J Bot 96: 1636–1645.

61. Zwickl DJ (2006). Genetic algorithm approaches for the phylogenetic analysis of large biological sequence datasets under the maximum likelihood criterion. Ph.D. Thesis. Austin: University of Texas at Austin.

62. Ronquist F, Teslenko M, van der Mark P, Ayres DL, Darling A, et al. (2012) MrBayes 3.2: efficient Bayesian phylogenetic inference and model choice across a large model space. Syst Biol 61: 539–542.

63. Posada D, Crandall KA (1998) Modeltest: testing the model of DNA substitution. Bioinformatics 14: 817–818.

64. Nylander JAA, Wilgenbusch JC, Warren DL, Swofford DL (2008) AWTY (are we there yet?): a system for graphical exploration of MCMC convergence in Bayesian phylogenetics. Bioinformatics 24: 581–583.

65. Rota-Stabelli O, Telford MJ (2008) A multi criterion approach for the selection of optimal outgroups in phylogeny: recovering some support for Mandibulata over Myriochelata using mitogenomics. Mol Phylogenet Evol 48: 103–111.

66. Felsenstein J (1981) Evolutionary trees from DNA sequences: a maximum likelihood approach. J Mol Evol 17: 368–376.

67. Drummond AJ, Rambaut A (2007) BEAST: Bayesian evolutionary analysis by sampling trees. BMC Evol Biol 7: 214.

68. Vicentini A, Barber JC, Aliscioni SS, Giussani LM, Kellogg EA (2008) The age of the grasses and clusters of origins of C_4 photosynthesis. Glob Change Biol 14: 2963–2977.

69. Nambudiri EMV, Tidwell WD, Smith BN, Hebbert NP (1978) A C_4 plant from the Pliocene. Nature 276: 816–817.

70. Jacobs BF, Kingston JD, Jacobs LL (1999) The origin of grass-dominated ecosystems. Ann Mo Bot Gard 86: 590–643.

71. Kellogg EA (2000) Molecular and morphological evolution in the Andropogoneae. In: Jacobs SWL, Everett J, editors. Grasses: Systematics and Evolution. Collingwood: Commonwealth Scientific and Industrial Research Organization Publishing. pp. 149–158.

72. Whistler DP, Burbank DW (1992) Miocene biostratigraphy and biochronology of the Dove Spring Formation. Mojave Desert, California, and characterization of the Clarendonian mammal age (late Miocene) in California. Geol Soc Am Bull 104: 644–658.

73. Ho SYW, Phillips MJ (2009) Accounting for calibration uncertainty in phylogenetic estimation of evolutionary divergence times. Syst Biol 58: 367–380.

74. Gelman A, Rubin DB (1992) Inference from iterative simulation using multiple sequences. Statist Sci 7: 457–511.

75. Hata S, Izui K, Kouchi H (1998) Expression of a soybean nodule-enhanced phosphoenolpyruvate carboxylase gene that shows striking similarity to another gene for a house-keeping isoform. Plant J 13: 267–273.

76. Duvall MR, Doebley JF (1990) Restriction site variation in the chloroplast genome of Sorghum (Poaceae). Syst Bot 15: 472–480.

77. Wu TP (1990) Sorghum macrospermum and its relationship to the cultivated species S. bicolor. Cytologia (Tokyo) 55: 141–151.

78. Wu TP (1993) Cytological and morphological relationships between Sorghum laxiflorum and S. bicolor. J Hered 84: 484–489.

79. Liao F, Liu Y, Yang XL, Huang GM, Niu CJ (2009) Molecular phylogenetic relationships among species in the genus Sorghum based on partial Adh1 gene. Hereditas 31: 523–530.

80. Doggett J (1970) Sorghum. Longmans: Green and Company.

81. Gu MH, Ma HT, Liang GH (1984) Karyotype analysis of seven species in the genus Sorghum. J Hered 75: 196–202.

82. Van Oosterhout SAM (1992) The biosystems and ethnobotany of Sorghum bicolor in Zimbabwe. Ph.D. Thesis. Harare: University of Zimbabwe.

83. Tang H, Liang GH (1988) The genomic relationship between cultivated sorghum [Sorghum bicolor (L.) Moench] and Johnsongrass [S. halepense (L.) Pers.]: a re-evaluation. Theor Appl Genet 76: 277–284.

84. Swigoňová Z, Lai JS, Ma JX, Ramakrishna W, Llaca V, et al. (2004) Close split of sorghum and maize genome progenitors. Genome Res 14: 1916–1923.

85. Paterson AH, Bowers JE, Chapman BA (2004) Ancient polyploidization predating divergence of the cereals, and its consequences for comparative genomics. Proc Natl Acad Sci USA 26: 9903–9908.

86. House LR (1985) A guide to sorghum breeding. Andhra Pradesh: International Crops Research Institute for the Semi-Arid Tropics.

87. Bolnick DI, Fitzpatrick BM (2007) Sympatric speciation: models and empirical evidence. Annu Rev Ecol Evol Syst 38: 459–487.

88. Jiang B, Peterson P M, Liu Q (2011) Caryopsis micromorphology of Eleusine Gaertn. (Poaceae) and its systematic implications. J Trop Subtrop Bot 19: 195–204.

89. Zhang Y, Hu XY, Liu YX, Liu Q. 2014. Caryopsis micromorphological survey of Themeda (Poaceae) and allied spathaceous genera in the Andropogoneae. Turk J Bot 38: 1206–1212.

90. Udall JA, Quijada PA, Osborn TC (2005) Detection of chromosomal rearrangements derived from homologous recombination in four mapping populations of Brassica napus L. Genetics 169: 967–979.

91. Doggett H (1976) Sorghum. In: Simmonds NW, editor. Evolution of Crop Plants. London: Longman Scientific and Technical. 112–117.

92. Paterson AH, Schertz KF, Lin YR, Liu C, Chang YL (1995) The weediness of wild plants: molecular analysis of genes influencing dispersal and persistence of Johnsongrass, Sorghum halepense (L.) Pers. Proc Natl Acad Sci USA 92: pp. 6127–6131.

93. Bhatti AG, Endrizzi JE, Reeves RG (1960) Origin of Johnsongrass. J Hered 51: 107–110.

94. Gaut BS, Doebley JF (1997) DNA sequence evidence for the segmental allotetraploid origin of maize. Proc Natl Acad Sci USA 94: 6809–6814.

95. Morden CW, Doebley J, Schertz KF (1990) Allozyme variation among the spontaneous species of Sorghum section Sorghum (Poaceae). Theor Appl Genet 80: 296–304.

96. Linder HP, Rudall PJ (2005) Evolutionary history of Poales. Annu Rev Ecol Evol Syst 36: 107–124.

97. Zachos J, Pagani M, Sloan L, Thomas E, Billups K (2001) Trends, rhythms, and aberrations in global climate 65 Ma to present. Science 292: 686–693.

98. Guiraud R, Bosworth W, Thierry J, Delaplanque A (2005) Phanerozoic geological evolution of Northern and Central Africa: an overview. J Afr Earth Sci 43: 83–143.

99. Lærdal T, Talbot MR (2002) Basin neotectonics of Lakes Edward and George, East African Rift. Palaeogeogr Palaeoclimatol Palaeoecol 187: 213–232.

100. Sepulchre P, Ramstein G, Fluteau F, Schuster M, Tiercelin JJ, et al. (2006) Tectonic uplift and East Africa aridification. Science 313: 1419–1423.

101. Swezey CS (2009) Cenozoic stratigraphy of the Sahara, Northern Africa. J Afr Earth Sci 53: 89–121.

102. Winkler IS, Mitter C, Scheffer SJ (2009) Repeated climate-linked host shifts have promoted diversification in a temperate clade of leaf-mining flies. Proc Natl Acad Sci USA 106: 18103–18108.

103. Voje KL, Hemp C, Flagstad Ø, Saetre GP, Stenseth NC (2009) Climatic change as an engine for speciation in flightless Orthoptera species inhabiting African mountains. Mol Ecol 18: 93–108.

104. Finarelli J, Badgley C (2010) Diversity dynamics of Miocene mammals in relation to the history of tectonism and climate. Proc R Soc Lond B Biol Sci 277: 2721–2726.

105. Janis CM, Damuth J, Theodor JM (2000) Miocene ungulates and terrestrial primary productivity: where have all the browsers gone? Proc Natl Acad Sci USA 97: 7899–7904.

106. Wheeler MC, McBride JL (2005) Australian-Indonesian monsoon. In: Lau WKM, Waliser DE, editors. Intraseasonal Variability in the Atmosphere-Ocean Climate System. Heidelberg: Springer-Praxis. pp. 125–173.

107. Martin HA (2006) Cenozoic climatic change and the development of the arid vegetation in Australia. J Arid Environ 66: 533–563.

108. Russell-Smith J, Needham S, Brock J (1995) The physical environment. In: Press T, Lea D, Webb A, Graham A, editors. Kakadu: Natural and Cultural Heritage Management. Darwin: Australian Nature Conservation Agency. pp. 94–126.

109. Fujita MK, McGuire JA, Donnellan SC, Moritz C (2010) Diversification and persistence at the arid-monsoonal interface: Australia-wide biogeography of the Bynoe's gecko (Heteronotia binoei; Gekkonidae). Evolution 64: 2293–2314.

110. Phillips S (1995) Poaceae (Gramineae). In: Hedberg I, Edwards S, editors. Flora of Ethiopia and Eritrea. Addis Ababa: Addis Ababa University and Uppsala: Uppsala University.

5

HPLC-MS/MS Analyses Show That the Near-Starchless *aps1* and *pgm* Leaves Accumulate Wild Type Levels of ADPglucose: Further Evidence for the Occurrence of Important ADPglucose Biosynthetic Pathway(s) Alternative to the pPGI-pPGM-AGP Pathway

Abdellatif Bahaji[1,9], Edurne Baroja-Fernández[1,9], Ángela María Sánchez-López[1,9], Francisco José Muñoz[1], Jun Li[1], Goizeder Almagro[1], Manuel Montero[1], Pablo Pujol[2], Regina Galarza[2], Kentaro Kaneko[3], Kazusato Oikawa[3], Kaede Wada[3], Toshiaki Mitsui[3], Javier Pozueta-Romero[1]*

1 Instituto de Agrobiotecnología, Universidad Pública de Navarra/Consejo Superior de Investigaciones Científicas/Gobierno de Navarra, Mutiloabeti, Nafarroa, Spain, 2 Servicio de Apoyo a la Investigación, Universidad Pública de Navarra, Campus de Arrosadia, Iruña, Nafarroa, Spain, 3 Department of Applied Biological Chemistry, Niigata University, Niigata, Japan

Abstract

In leaves, it is widely assumed that starch is the end-product of a metabolic pathway exclusively taking place in the chloroplast that (a) involves plastidic phosphoglucomutase (pPGM), ADPglucose (ADPG) pyrophosphorylase (AGP) and starch synthase (SS), and (b) is linked to the Calvin-Benson cycle by means of the plastidic phosphoglucose isomerase (pPGI). This view also implies that AGP is the sole enzyme producing the starch precursor molecule, ADPG. However, mounting evidence has been compiled pointing to the occurrence of important sources, other than the pPGI-pPGM-AGP pathway, of ADPG. To further explore this possibility, in this work two independent laboratories have carried out HPLC-MS/MS analyses of ADPG content in leaves of the near-starchless *pgm* and *aps1* mutants impaired in pPGM and AGP, respectively, and in leaves of double *aps1/pgm* mutants grown under two different culture conditions. We also measured the ADPG content in wild type (WT) and *aps1* leaves expressing in the plastid two different ADPG cleaving enzymes, and in *aps1* leaves expressing in the plastid GlgC, a bacterial AGP. Furthermore, we measured the ADPG content in *ss3/ss4/aps1* mutants impaired in starch granule initiation and chloroplastic ADPG synthesis. We found that, irrespective of their starch contents, *pgm* and *aps1* leaves, WT and *aps1* leaves expressing in the plastid ADPG cleaving enzymes, and *aps1* leaves expressing in the plastid GlgC accumulate WT ADPG content. In clear contrast, *ss3/ss4/aps1* leaves accumulated ca. 300 fold-more ADPG than WT leaves. The overall data showed that, in Arabidopsis leaves, (a) there are important ADPG biosynthetic pathways, other than the pPGI-pPGM-AGP pathway, (b) pPGM and AGP are not major determinants of intracellular ADPG content, and (c) the contribution of the chloroplastic ADPG pool to the total ADPG pool is low.

Editor: Frederik Börnke, Leibniz-Institute for Vegetable and Ornamental Plants, Germany

Funding: This research was partially supported by the grants [BIO2010-18239] from the Comisión Interministerial de Ciencia y Tecnología and Fondo Europeo de Desarrollo Regional (Spain) and [IIM010491.RI1] from the Government of Navarra, and by Iden Biotechnology. This research was also supported by Scientific Research on Innovative Areas [22114507] and Grants-in-Aid for Scientific Research (B) [22380186] from the Ministry of Education, Culture, Sports, Science and Technology, Japan. The funders had no role in study design, data collection and analysis, decision to publish, or preparation of the manuscript.

Competing Interests: This research was partially supported by the grants BIO2010-18239 from the Comisión Interministerial de Ciencia y Tecnología and Fondo Europeo de Desarrollo Regional (Spain), IIM010491.RI1 from the Government of Navarra, and by the Scientific Research on Innovative Areas [22114507] and Grants-in-Aid for Scientific Research (B) [22380186] from the Ministry of Education, Culture, Sports, Science and Technology, Japan. This research was also supported by Iden Biotechnology S.L. A-M.S.-L. acknowledges a predoctoral fellowship from the Spanish Ministry of Science and Innovation.

* Email: javier.pozueta@unavarra.es

9 These authors contributed equally to this work.

Introduction

Starch is a branched homopolysaccharide of α-1,4-linked glucose subunits with α-1,6-linked glucose at the branched points. Synthesized by starch synthases (SSs) using ADPglucose (ADPG)

as the sugar donor molecule, this polyglucan accumulates as predominant storage carbohydrate in most plants. In leaves, up to 50% of the photosynthetically fixed carbon is retained within the chloroplasts of mesophyll cells during the day to synthesize starch [1,2], which is then remobilized during the subsequent night to support non-photosynthetic metabolism and growth by continued export of carbon to the rest of the plant. Due to the diurnal rise and fall cycle of its levels, foliar starch is termed "transitory starch".

It is widely assumed that the whole starch biosynthetic process occurring in mesophyll cells of leaves resides exclusively in the chloroplast [3–5]. According to this classical view of starch biosynthesis, starch is considered the end-product of a metabolic pathway that is linked to the Calvin-Benson cycle by means of the plastidic phosphoglucose isomerase (pPGI). This enzyme catalyzes the conversion of fructose-6-phosphate from the Calvin-Benson cycle into glucose-6-phosphate (G6P), which is then converted into glucose-1-phosphate (G1P) by the plastidic phosphoglucomutase (pPGM). ADPG pyrophosphorylase (AGP) then converts G1P and ATP into inorganic pyrophosphate and ADPG necessary for starch biosynthesis (**Figure 1A**). These three enzymatic steps are reversible, but the last step is rendered irreversible upon hydrolytic breakdown of PPi by plastidial alkaline pyrophosphatase.

The classic view of transitory starch biosynthesis also implies that AGP is the sole source of ADPG, and functions as the major regulatory step in the starch biosynthetic process [3–7]. Plant AGPs are heterotetrameric enzymes comprising two types of homologous but distinct subunits, the small (APS) and the large (APL) subunits [9,10]. In Arabidopsis, six genes encode proteins with homology to AGP. Two of these genes (*APS1* and *APS2*) code for small subunits, and four (*APL1-APL4*) encode large subunits [9–11]. *APS2* is in a process of pseudogenization [12] since its expression level is two orders of magnitude lower than that of *APS1* [10], and its product lacks activity due to the absence of essential amino acids involved in the catalysis and/or in the binding of G1P and 3-phosphoglycerate [9]. Whereas APS1, APL1 and APL2 are catalytically active, APL3 and APL4 have lost their catalytic properties during evolution [13]. In Arabidopsis, the large subunits are highly unstable in the absence of small subunits [14]. Therefore, *APS1* null mutants lack not only APS1, but also the large subunits, which results in a total lack of AGP activity [13,15].

In *Arabidopsis,* genetic evidence showing that transitory starch biosynthesis occurs solely by the pPGI-pPGM-AGP pathway has been obtained from the characterization of mutants impaired in pPGI [16,17], pPGM [18,19] and AGP [13,14,20]. Despite the monumental amount of data apparently supporting the classic interpretation of transitory starch biosynthesis in mesophyll cells involving the pPGI-pPGM-AGP pathway (**Figure 1A**), mounting evidence has exposed inconsistencies that previews the occurrence of important pathway(s) of transitory starch biosynthesis wherein (a) pPGI plays a minor role in the connection of the Calvin-Benson cycle with the starch biosynthetic pathway, (b) a sizable pool of ADPG linked to starch biosynthesis is produced in the cytosol by enzymes such as sucrose synthase (SuSy) [27–32], (c) cytosolic ADPG is transported to the chloroplast by the action of a yet to be identified ADPG translocator [33], and (d) pPGM and AGP play important roles in the scavenging of glucose units derived from starch breakdown occurring during starch biosynthesis and during the biogenesis of the starch granule [27,28,32,34]. According to this interpretation of transitory starch biosynthesis (schematically illustrated in **Figure 1B**), starch accumulation in leaves is the result of the balance between *de novo* starch synthesis from ADPG entering the chloroplast and breakdown, and the efficiency by which starch breakdown products are recycled back to starch by means of pPGM and AGP. Thus, according to this interpretation of transitory starch biosynthesis, starch is actively synthesized in pPGM and AGP mutants, but its accumulation is prevented due to the blockage of the mechanism of scavenging of glucose units

derived from the starch breakdown [15,34]. The occurrence of starch turnover during illumination is not surprising since pulse-chase and starch-preloading experiments using isolated chloroplasts [35,36], intact leaves [37,38], or cultured photosynthetic cells [39] have shown that chloroplasts can synthesize and mobilize starch simultaneously. Furthermore, recent metabolic flux analyses carried out using illuminated Arabidopsis plants cultured in $^{13}CO_2$-enriched environment revealed rapid labelling of maltose, the main starch degradation product [40]. Also, leaves of *sex1-1* mutants impaired in β-amylolytic starch breakdown accumulated 3–4 fold more starch than WT leaves when plants were cultured under continuous light conditions [41]. Moreover, simultaneous synthesis and breakdown of glycogen has been shown to widely occur in animals [42–44] and in bacteria [45–48]. In this respect we must emphasize that many bacterial species co-express glycogen biosynthetic and breakdown genes in a single transcriptional unit, which guarantees simultaneous synthesis and breakdown of glycogen [49–53] (for a review see [54]).

The possible occurrence of sources, other than the pPGI-pPGM-AGP pathway, of ADPG linked to starch biosynthesis has been a matter of debate for more than 20 years [3–5,28–30, 32,34,55–60]. In attempting to solve this controversy, we recently carried out HPLC analyses of ADPG content in leaves of the near-starchless *adg1-1* and *aps1* Arabidopsis mutants impaired in AGP [15]. We also measured the ADPG content in the leaves of both wild type (WT) and *aps1* plants ectopically expressing the *Escherichia coli* ADPG hydrolase EcASPP [61] either in the cytosol or the chloroplast [15]. We reasoned that if leaves produce starch from ADPG exclusively synthesized in the plastid, plastidial expression of EcASPP competing with SS for ADPG, but not cytosolic EcASPP expression, should lead to reduction of both starch and ADPG content. Conversely, if ADPG linked to starch biosynthesis occurs both in the plastid and in the cytosol, but mainly accumulates in the cytosol, plants expressing EcASPP in the cytosol should accumulate reduced levels of both ADPG and starch content, whereas plants expressing EcASPP in the plastid should accumulate normal ADPG but reduced starch. We also measured the starch and ADPG contents in leaves of *aps1* mutants expressing in the chloroplast the *E. coli* AGP (GlgC) [15]. We found that *adg1-1* and *aps1* leaves accumulate nearly WT ADPG contents, the estimated values of ca. 0.3–0.4 nmol ADPG/g fresh weight (FW) being comparable to those reported by Szecowka et al. [40], Barratt et al. [58] and Crumpton-Taylor et al. (Table S3 in [62]) for WT leaves using HPLC-MS/MS, and those reported by Chen and Thelen [63] using HPLC. These values, however, were 5–10 fold lower than those reported for WT leaves by Lunn et al. [64], Ragel et al. [65], Martins et al. [66] and Crumpton-Taylor et al. (Table 1 in [62]) using HPLC-MS/MS. We also found that *aps1* leaves expressing GlgC in the plastid accumulate WT levels of both starch and ADPG [15]. As expected, expression of EcASPP in the chloroplast resulted in the reduction of starch content [15]. Noteworthy, this reduction in starch content was not accompanied by a significant reduction in the intracellular levels of ADPG. Moreover, plants expressing EcASPP in the cytosol accumulated reduced levels of both starch and ADPG [15]. The overall data thus provided strong evidence that (a) there occur important source(s) other than AGP, of ADPG linked to starch biosynthesis, (b) AGP is a major determinant of starch accumulation but not of intracellular ADPG content in Arabidopsis, (c) most of ADPG has an extraplastidial localization in WT leaves and (d) cytosolic ADPG is linked to starch biosynthesis. The occurrence of an important pool of cytosolic ADPG is not surprising since leaf cells possess cytosolic ADPG metabolizing enzymes such as ADPG phosphorylase (ADPGP) [67] and glucan

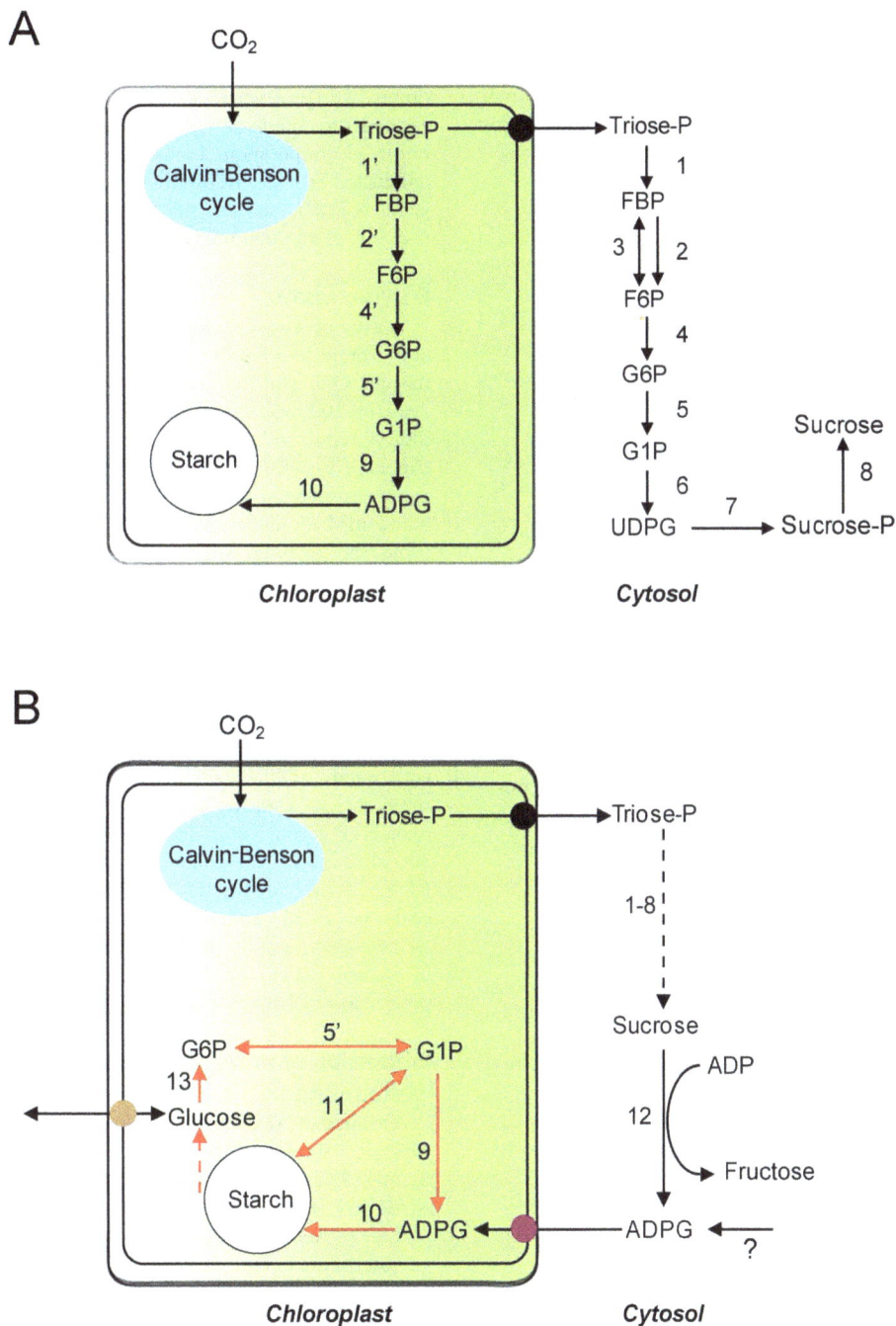

Figure 1. Suggested models of starch biosynthesis in leaves. (A) The classic model of starch biosynthesis according to which (a) the starch biosynthetic process takes place exclusively in the chloroplast, segregated from the sucrose biosynthetic process taking place in the cytosol, and (b) AGP exclusively catalyzes the synthesis of ADPG. (B) Suggested ''additional/alternative'' model of starch biosynthesis wherein (a) ADPG is produced in the cytosol by enzyme(s) such as SuSy and then is transported to the chloroplast by the action of an ADPG translocator, and (b) pPGM and AGP play an important role in the scavenging of glucose units derived from starch breakdown. Starch to glucose conversion would involve the coordinated actions of amylases, isoamylase and disproportionating enzyme [21–23]. According to this interpretation of transitory starch biosynthesis starch accumulation in leaves is the result of the balance between *de novo* starch synthesis from ADPG entering the chloroplast and breakdown, and the efficiency by which starch breakdown products are recycled back to starch by means of pPGM and AGP. Thus, this view predicts that the recovery towards starch biosynthesis of the glucose units derived from the starch breakdown will be deficient in pPGM and AGP mutants, resulting in a parallel decline of starch accumulation and enhancement of soluble sugars content since starch breakdown derived products (especially glucose) will leak out the chloroplast through the very active glucose translocator [24]. The enzyme activities involved are numbered as follows: 1, 1', fructose-1, 6-bisphosphate aldolase; 2, 2', fructose 1,6-bisphosphatase; 3, PPi:fructose-6-phosphate phosphotransferase; 4, 4', PGI; 5, 5', PGM; 6, UDPG pyrophosphorylase; 7, sucrose phosphate synthase; 8, sucrose-phosphate-phosphatase; 9, AGP; 10, SS; 11, starch phosphorylase; 12, SuSy; 13, plastidial hexokinase [25,26]. FBP: fructose bis-phosphate; UDPG: UDP-glucose.

synthase [68]. Likewise, low steady state concentration of ADPG in the plastid should be expected since (a) this nucleotide-sugar spontaneously hydrolyzes to AMP and glucose1,2-monophosphate under conditions of alkaline pH and high Mg^{2+} concentration occurring during starch biosynthesis in the illuminated chloroplast [56,69], and (b) SSs rapidly remove ADPG from the stroma to produce starch.

Our previous HPLC analyses of ADPG content in Arabidopsis leaves have been recently questioned by Stitt and Zeeman [5] and Ragel et al. [65], who used HPLC-MS/MS to measure ADPG content in Arabidopsis leaves. These authors reported that *aps1* leaves and leaves of *pgm* plants impaired in pPGM accumulate far lower levels of ADPG than WT leaves. However, values of ADPG content in *pgm* and *aps1* leaves were not shown [5] or not clearly presented [65]. Subcellular localization and determination of ADPG content is critically important to understand starch metabolism and its regulation and connection with other metabolic pathways. Thus, to further investigate the possible occurrence of important ADPG biosynthetic pathway(s) alternative to pPGI-pPGM-AGP, and to test the validity of our previous HPLC-based results and conclusions on ADPG content and subcellular localization, in this work two independent laboratories have carried out HPLC-MS/MS based analysis of ADPG content in leaves of the near-starchless *pgm*, *aps1* and double *pgm/aps1* mutants cultured under two different conditions. We also measured the ADPG content in WT and *aps1* leaves ectopically expressing in the plastid two different ADPG cleaving enzymes, and in *aps1* leaves expressing GlgC in the plastid. Furthermore, we measured the ADPG content in *ss3/ss4/aps1* mutants impaired in starch granule initiation and chloroplastic ADPG synthesis. We found that *pgm*, *aps1* and *pgm/aps1* leaves, and leaves with reduced starch content as a consequence of the ectopic expression of ADPG breakdown enzymes in the plastid accumulate nearly WT ADPG content. Furthermore, we found that *aps1* leaves ectopically expressing GlgC in the plastid accumulate WT starch and ADPG contents. In clear contrast, *ss3/ss4/aps1* leaves accumulated ca. 300 fold-more ADPG than WT leaves. The overall data showed that, in Arabidopsis leaves, (a) there are important ADPG sources other than the pPGI-pPGM-AGP pathway, (b) pPGM and AGP are not major determinants of intracellular ADPG content, and (c) the contribution of the chloroplastic ADPG pool to the total ADPG pool is low.

Materials and Methods

Plants, bacterial strains and plant transformation

The work was carried out using WT *Arabidopsis* (ecotype Columbia), the *aps1::T-DNA* mutant (SALK_040155) [13], the double *aps1::T-DNA/pgm::T-DNA* mutant, the double *ss3::T-DNA/ss4::T-DNA* mutant [70], the *pgm::T-DNA* mutant (GABI_094D07), the triple *ss3::T-DNA/ss4::T-DNA/aps1::T-DNA* mutant as well as WT, *aps1* and *ss3/ss4* plants transformed with either *35S-TP-P541-glgC* [15], *35S-TP-P541-EcASPP* [15,27], *35S-TP-P541-AtADPGP* or *35S-TP-P541-AtADPGP-GFP* (this work, see below). Plants were grown in pots either on soil or solid MS medium at ambient CO_2 in growth chambers at 20°C under a 16 h light (90 μmol photons sec^{-1} m^{-2})/8 h dark regime.

Triple *ss3/ss4/aps1* and double *aps1/pgm* mutants were obtained by crossing and selecting mutants from segregating the F2 populations by PCR on genomic DNA, using the oligonucleotide primers listed in **Table S1 and Table S2**, respectively. Different *35S-TP-P541-AtADPGP*, *35S-TP-P541-EcASPP* and *35S-TP-P541-AtADPGP-GFP* plasmid constructs conferring resistance to

either kanamycin or hygromycin were produced as illustrated in **Figure S1**. Constructs conferring resistance to hygromycin were used to transform *ss3/ss4* plants. Plasmid constructs were electroporated and propagated in *E. coli* TOP 10. Transfer of the plasmid construct to *Agrobacterium tumefaciens* EHA105 cells was carried out by electroporation. Transformation of *Arabidopsis* plants were conducted as described by Clough and Bent [71]. Transgenic plants were selected on the adequate (kanamycin- or hygromicin-containing) selection medium.

Enzyme assays

Leaves of 4-weeks old plants were harvested, freeze-clamped and ground to a fine powder in liquid nitrogen with a pestle and mortar. One g of the frozen powder was resuspended at 4°C in 5 ml of 100 mM HEPES (pH 7.5), 2 mM EDTA and 5 mM dithiothreitol, and desalted by ultrafiltration on Centricon YM-10 (Amicon, Bedford, MA). The proteins retained in the filter then were resuspended in 100 mM HEPES (pH 7.5), 2 mM EDTA and 5 mM dithiothreitol. ADPG hydrolytic activity was assayed using the two-step spectrophotometric determination of G1P described by Rodríguez-López et al. [72]. ADPGP activity was assayed at 37°C in the direction of ADPG breakdown in two steps: (1) ADPGP reaction, and (2) measurement of G1P. In step one, the ADPGP assay mixture contained 50 mM HEPES (pH 7.0), 1 mM ADPG, 2 mM Pi, 1 mM $MgCl_2$, 1 mM dithiothreitol and the leaf extract in a total volume of 50 μl. The reaction was initiated by adding the leaf extract to the assay mixture. All assays were run with minus ADPG blanks. After 3 min at 37°C, reactions were stopped by boiling the assay reaction mixture for 2 min. In step two, G1P formed was determined spectrophotometrically in a 300 μl mixture containing 50 mM HEPES (pH 7.0), 1 mM EDTA, 2 mM $MgCl_2$, 15 mM KCl, 0.6 mM NAD^+, 1 unit (U) each of PGM and G6P dehydrogenase from *Leuconostoc mesenteroides*, and 30 μl of the step-one reaction. We define 1 U of enzyme activity as the amount of enzyme that catalyzes the production of 1 μmol of product per min.

Production of polyclonal antisera against AtADPGP and western blot analyses

A complete AtADPGP encoding cDNA from the Arabidopsis Biological Center at Ohio State University [73] was cloned into the pDEST17 expression vector (Invitrogen) to create pDEST17-AtADPGP (**Figure S2**). BL21 C43 (DE3) cells transformed with pDEST-AtADPGP were grown in 100 ml of liquid LB medium to an absorbance at 600 nm of 0.5 and then 1 mM IPTG was added. After 5 h, cells were centrifuged at 6,000 g for 10 min. The pelleted bacteria were resuspended in 6 ml of His-bind binding buffer (Novagen), sonicated and centrifuged at 10,000 g for 10 min. The supernatant thus obtained was subjected to His-bind chromatography (Novagen). The eluted His-tagged AtADPGP was then rapidly desalted by ultrafiltration on Centricon YM-10 (Amicon, Bedford, MA).

The purified recombinant AtADPGP was electrophoretically separated by 12% SDS-PAGE and stained with Coomassie Blue. A ca. 38 kDa protein band was eluted and utilized to produce polyclonal antisera by immunizing rabbits.

For immunoblot analyses, samples were separated on 10% SDS-PAGE, transferred to nitrocellulose filters, and immunodecorated by using the antisera raised against either AtADPGP or EcASPP [27] as primary antibody, and a goat anti-rabbit IgG alkaline phosphatase conjugate (Sigma) as secondary antibody.

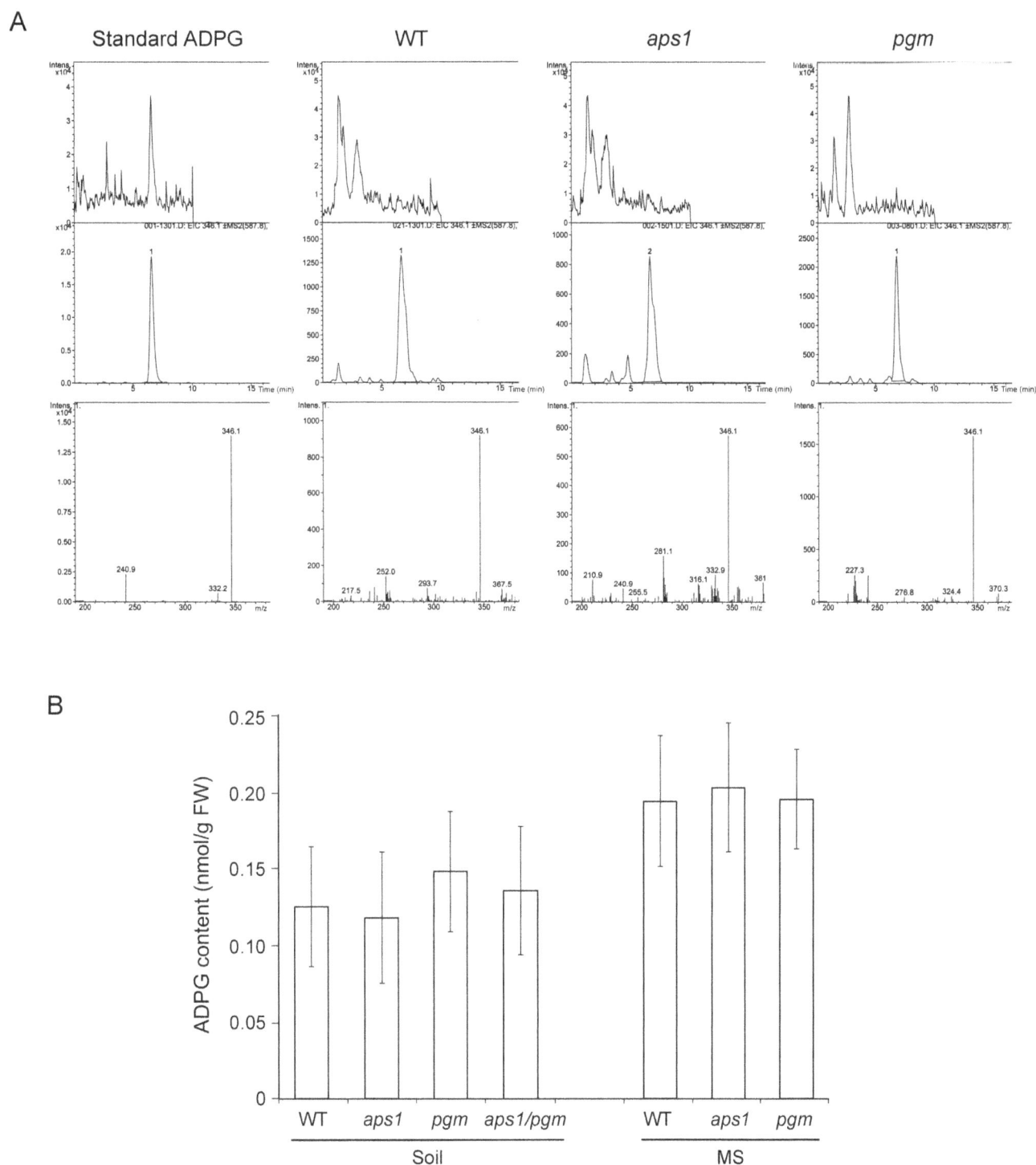

Figure 2. *aps1* and *pgm* leaves accumulate WT ADPG content. (A) HPLC-MS/MS detection of ADPG in WT, *aps1* and *pgm* leaves. Upper panels: Total ion chromatograms (TIC) of extracts from the indicated plants in which the selected fragmentation parent ion was 587.8 m/z. Middle panels: Extracted ion chromatograms (EIC) in which the selected ion for fragmentation of the parent ion was 346.1 m/z. Lower panels: Mass spectra (MS2) obtained from fragmentation of parent ion. ADPG was measured using an Agilent 1100 HPLC fitted with a Xbridge C18 column (100×3.0 mm I.D. particle size 3.5 μm) coupled to a MSD-Trap spectrometer (Agilent) (see Materials and Methods for further details). (B) ADPG content in WT, *aps1, pgm* and *aps1/pgm* leaves. Plants were simultaneously grown either in soil or solid MS. Leaves from 4-weeks old WT, *aps1, pgm* and *aps1/pgm* plants were simultaneously harvested after 10 h of illumination. ADPG was simultaneouly extracted from leaves of WT, *aps1, pgm* and *aps1/pgm* plants and content was simultaneously measured by HPLC-MS/MS as described in Materials and Methods. Note that, consistent with [15], leaves of *aps1, pgm* and *aps1/pgm* plants accumulated WT ADPG content. Values represent the mean ±SD of determinations on three independent samples.

Assay of ADPG content by HPLC-MS/MS

Fully expanded source leaves of 4-weeks old plants were harvested at the indicated illumination period, freeze-clamped and ground to a fine powder in liquid nitrogen with a pestle and mortar. ADPG was then immediately extracted as described by Lunn et al. [64]. Aliquots (50–100 mg FW) of the frozen powdered leaves were transferred to pre-cooled tubes and quenched by adding 250 μL of ice-cold $CHCl_3/CH_3OH$ (3:7, v/v). The frozen mixture was warmed to $-20°C$ with vigorous shaking, and incubated at $-20°C$ for 2 h. ADPG was extracted from the $CHCl_3$ phase by adding 200 μL of water and warming to 4°C with repeated shaking. After centrifugation at 420 g for 4 min, the upper, aqueous-CH_3OH phase was transferred to a new tube, and kept at 4°C. The lower, $CHCl_3$ phase was re-extracted with 200 μL of cold water, centrifuged, and the second aqueous-CH_3OH extract was added to the first. The combined aqueous-CH_3OH extract was freeze-dried using a lyophilizer and re-dissolved in 250 μL of water. High molecular-mass components were removed from the samples by ultrafiltration on vivaspin 500 centrifugal concentrator (Sartorius) at 2,300 g for 2–3 h, 20°C. Recovery experiments were carried out by the addition of known amounts of ADPG disodium salt (Sigma-Aldrich A0627) standards to the frozen tissue slurry immediately after addition of the cold $CHCl_3/CH_3OH$. As described in the "Results and Discussion" section, we found endogenous levels of ADPG of ca. $0.13±0.03$ nmol ADPG/g FW (equivalent to 3.1 pmol per 100 μL of extract) in leaves of plants cultured on soil. We thus added 5, 10 or 20 pmol of authenticated ADPG standard to the 50–100 mg samples of the frozen plant material (containing 6.5–13 pmol of ADPG) before extraction. Recoveries of the added 5, 10, and 20 pmol of ADPG were $94±5.3$, $93±4.9$ and $89±6.6$, respectively, demonstrating that even the smallest amounts of ADPG were quantitatively recovered.

ADPG content in leaves of plants cultured on soil was measured in the Research Support Service at the Public University of Navarra using an Agilent 1100 HPLC fitted with a Xbridge C18 column (100×3.0 mm I.D. particle size 3.5 μm) coupled to a MSD-Trap spectrometer (Agilent). The column was equilibrated with a mixture of 99% solution A (15 mM acetic acid and 10 mM triethylamine, pH 4.95) and 1% solution B (methanol) for 7 min before each sample run. The extracts were eluted with a multi-step gradient as follows: 0–4 min, 99% A; 4–25 min, 99–10% A; 25–30 min, 10–10% A; 30–35 min, 10–95% A. ADPG peak detection in the HPLC elute was made after entering directly into the MSD-Trap, which was operated in a multiple reaction monitoring mode, with an electrospray ionization source in negative ionization mode. For ADPG measurement the parent and product ions selected were 587.8 m/z and 346.1 m/z, respectively, and the fragmentation amplitude was 2.0. ADPG was quantified by comparison of the integrated MSD-Trap signal peak area with a calibration curve obtained using ADPG disodium salt as standard.

ADPG of leaves of plants cultured on solid MS was extracted as described above and measured in Niigata University using liquid chromatography-mass spectrometer consisting of LaChrom Elite-HPLC system with L-2130 pump (Hitachi) and LTQ Orbitrap XL (ThermoFisher Scientific) controlled by Xcalibur 2.0 software as described previously [74]. Reversed-phase ion-pair chromatography separation was carried out on a Hypersil GOLD column (50×2.1 mm, 5 μm particle size, ThermoFisher Scientific). An aliquot of sample (10 μl) was loaded onto the Hypersil GOLD column equilibrated with solvent A at flow rate of 150 μl min^{-1}. Solvent A was 97:3 water:methanol with 10 mM tributylamine and 15 mM acetic acid; solvent B was methanol. The gradient is: 0–2.5 min, 100% A; 2.5–5 min, 100–80% A; 5–7.5 min, 80% A;

7.5–13 min, 80–45% A; 13–15.5 min, 45–5% A; 15.5–18.5 min, 5% A; 18.5–19 min, 5–100% A; 19–25 min, 100% A. Other liquid chromatography parameters are autosampler temperature 4°C, injection volume 10 μl, and column temperature 30°C. The MS data was acquired full scans from 450–1000 m/z at 1 Hz and 30,000 resolution in negative ion mode using only Orbitrap.

Confocal microscopy

Subcellular localization of AtADPGP-GFP was performed using D-Eclipse C1 confocal microscope (NIKON, Japan) equipped with standard Ar 488 laser excitation, BA515/30 filter for green emission, BA650LP filter for red emission and transmitted light detector for bright field images.

Starch measurement

Starch was measured by using an amyloglucosydase–based test kit (Boehringer Manheim).

Results and Discussion

pgm and aps1 leaves accumulate WT ADPG content

We conducted HPLC-MS/MS analyses of ADPG content in pgm and aps1 mutants cultured on soil and solid MS medium conditions (see Materials and Methods for further details). As shown in **Figure 2**, these analyses revealed that ADPG contents in leaves of WT plants cultured on soil and solid MS medium after 10 h of illumination were $0.13±0.03$ and $0.19±0.05$ nmol ADPG/g FW, respectively. These values were comparable to those of previous HPLC and HPLC-MS/MS analyses of ADPG content [15,31,58,63]. Most importantly, these analyses also revealed that, irrespective of the culture conditions, ADPG contents in pgm and aps1 leaves were comparable to those of WT leaves (**Figure 2B**). Moreover, leaves of WT, pgm and aps1 cultured on soil accumulated ca. 0.015 nmol ADPG/g FW in the end of the dark period (not shown).

Leaves impaired in pPGM and AGP accumulate 0.5%–3% of the WT starch [4,14,15,17,75]. The occurrence of low starch content in pgm leaves has been ascribed to marginally low import to the chloroplast of cytosolic G1P and subsequent AGP-mediated conversion into ADPG [8,57,65,75], whereas the occurrence of reduced starch content in aps1 leaves has been ascribed to residual AGP activity from the large subunits [5,65]. The latter explanation is highly unlikely, since the AGP large subunits are unstable in the absence of APS1 in Arabidopsis [14,15]. Therefore, aps1 null mutants lack not only the small APS1 AGP subunits, but also the large AGP subunits [14,15], which results in total lack of AGP activity [13,15].

We reasoned that if the above interpretations for the occurrence of reduced starch content in aps1 and pgm leaves were correct, and if the pPGI-pPGM-AGP pathway is the sole source of ADPG in leaves, leaves of double aps1/pgm mutants should not accumulate any starch and ADPG at all. To test this hypothesis we crossed aps1 and pgm mutants as indicated in Materials and Methods. The resulting aps1/pgm mutants were cultured on soil and the leaf ADPG and starch contents were measured. As shown in **Figure 2B**, this analyses revealed that, similar to pgm and aps1 leaves, aps1/pgm leaves accumulated nearly WT ADPG content (**Figure 2B**). Furthermore, leaves of the double aps1/pgm mutant accumulated ca. 1.5% of the WT starch content.

The overall data (a) showed that pPGM and AGP are not major determinants of intracellular ADPG content, and (b) were consistent with the occurrence in Arabidopsis leaves of important ADPG source(s) other than the pPGI-pPGM-AGP pathway.

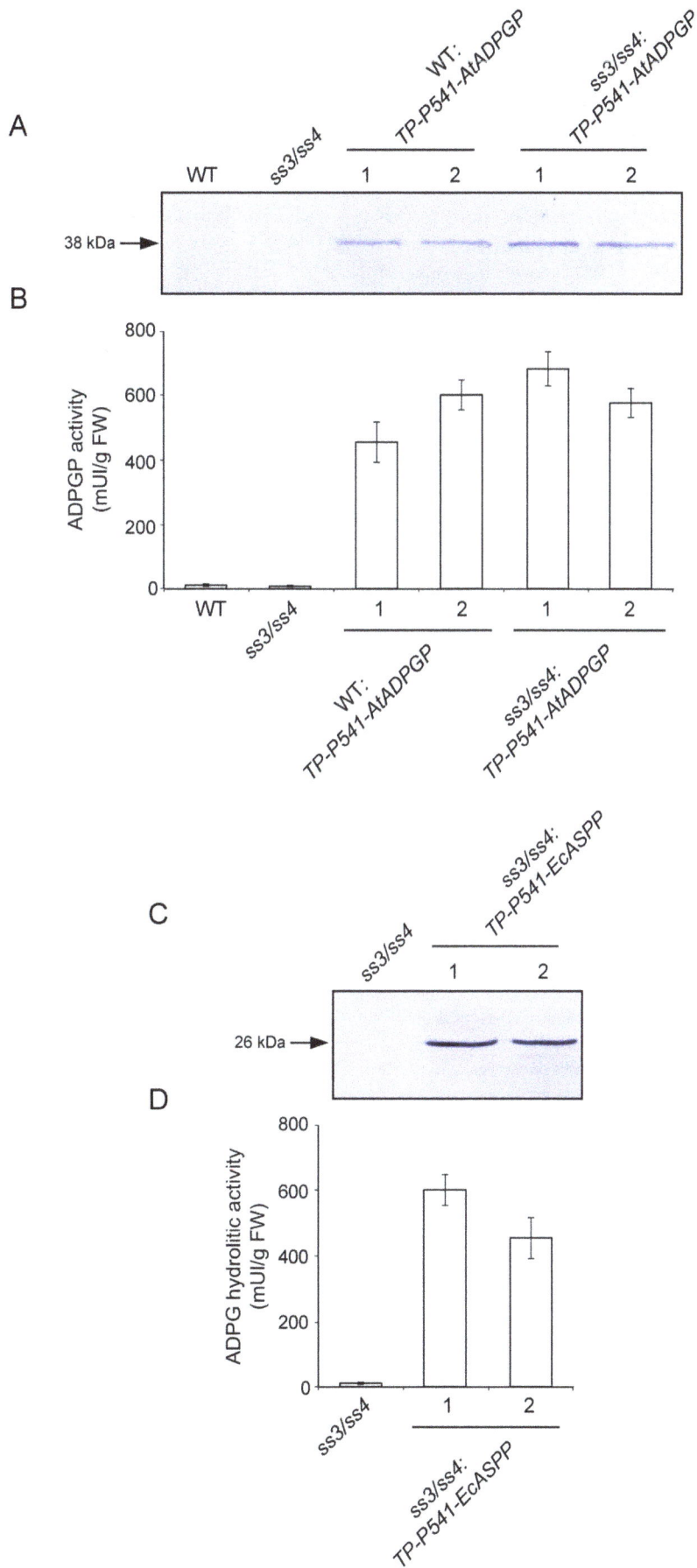

Figure 3. Production of WT and *ss3/ss4* plants expressing AtADPGP or EcASPP in the plastid. (A) Western blot of AtADPGP in leaves of WT and *ss3/ss4* plants, and leaves of two independent lines each of *TP-P541-AtADPGP* expressing WT plants and *TP-P541-AtADPGP* expressing *ss3/ss4* plants. (B) ADPGP activity in leaves of WT and *ss3/ss4* plants, and leaves of *TP-P541-AtADPGP* expressing WT plants and *TP-P541-AtADPGP* expressing *ss3/ss4* plants. (C) Western blot of EcASPP in *ss3/ss4* leaves, and leaves of two independent lines of *TP-P541-EcASPP* expressing *ss3/ss4* plants. (D) ASPP activity in *ss3/ss4* leaves and leaves of two independent *TP-P541-EcASPP* expressing *ss3/ss4* plants. In "A" and "C", the gels were loaded with 20 µg per lane of protein and AtADPGP and EcASPP were immunodecorated by using antisera specifically raised against AtADPGP and EcASPP.

The contribution of the chloroplastic ADPG pool to the total ADPG pool is low in WT and *aps1* leaves

Our previous HPLC based ADPG content measurement analyses revealed that *aps1* leaves expressing in the plastid either GlgC (*TP-P541-glgC* expressing *aps1* plants) or EcASPP (*TP-P541-EcASPP* expressing *aps1* plants) accumulate WT ADPG content [15]. *TP-P541-glgC* expressing *aps1* leaves accumulated WT starch content, whereas *TP-P541-EcASPP* expressing *aps1* leaves accumulated ca. 50% less starch than *aps1* leaves [15]. The overall data thus provided evidence that (a) AGP is not a major determinant of intracellular ADPG levels, and (b) the contribution of plastidic ADPG to the total ADPG pool is very low in *Arabidopsis* leaves [15]. To further test the validity of these conclusions, we measured by HPLC-MS/MS the ADPG content in leaves of two independent *TP-P541-glgC* expressing *aps1* lines and two independent *TP-P541-EcASPP* expressing *aps1* and WT plants. We must emphasize that our previous studies showed that leaves of *TP-P541-EcASPP* expressing plants accumulate as much as ca. 50% of the starch accumulated by the leaves of the parental plants [15,27]. This moderate reduction in the starch content exerted by the ectopic expression of EcASPP in the chloroplast may be ascribed to the relatively low affinity of EcASPP for ADPG (K_m being 160 µM [61]) combined with very low concentration of ADPG in the chloroplast. Therefore, to further reduce the plastidic ADPG pool we also produced WT plants expressing *Arabidopsis thaliana* ADPG phosphorylase (AtADPGP) in the chloroplast (*TP-P541-AtADPGP* expressing WT plants) (**Figure 3A,B**). ADPGP (E.C. 2.7.7.36) is a widely distributed cytosolic enzyme exhibiting high affinity for ADPG (K_m for ADPG being 7 µM [67]) that catalyzes the phosphorolytic breakdown of ADPG into ADP and G1P [67,76,77]. Fluorescence distribution pattern in TP-P541-AtADPGP-GFP expressing cells confirmed the exclusive plastidial localization of AtADPGP in *TP-P541-AtADPGP* expressing cells (**Figure S3**).

To test whether EcASPP and AtADPGP are active in the chloroplast, we also produced *ss3/ss4* plants ectopically expressing in the plastid either EcASPP or AtADPGP (**Figure 3A–D**). *ss3/ss4* leaves display a high ADPG content phenotype as a consequence of impairments in starch granule initiation and synthesis [65], a phenotype that can be partially reverted by the introduction of the *aps1* mutation [65]. Furthermore, *ss3/ss4* plants display a severe dwarf phenotype due to accumulation of high ADPG content in the chloroplast [65]. We reasoned that if EcASPP and AtADPGP are active in the chloroplast *TP-P541-EcASPP* and *TP-P541-AtADPGP* expressing *ss3/ss4* plants should display a WT growth phenotype and their leaves should accumulate less ADPG than *ss3/ss4* leaves. Consistent with this presumption *TP-P541-EcASPP* and *TP-P541-AtADPGP* expressing *ss3/ss4* plants exhibited a nearly WT growth phenotype (**Figure 4A**). Moreover, the ectopic expression in the plastid of EcASPP and AtADPGP resulted in a 5-fold decrease of ADPG content in *ss3/ss4* leaves (**Figure 4B**). Leaves of *TP-P541-EcASPP* and *TP-P541-AtADPGP* expressing *ss3/ss4* plants still exhibited ca. 300-400 fold more ADPG than WT leaves. This high ADPG content was comparable to that of *ss3/ss4/aps1* leaves

impaired in both starch granule initiation and chloroplastic ADPG synthesis (**Figure 4B**). The overall data thus showed that (a) both EcASPP and AtADPGP are active in the chloroplast, and (b) plastidic expression of EcASPP and AtADPGP can be utilized as a trait to reduce the plastidic ADPG pool.

As shown in **Figure 5A**, starch contents in leaves of plants from two independent *TP-P541-AtADPGP* and *TP-P541-EcASPP* expressing WT lines were ca. 3% and 50% of that of WT leaves, respectively, which further confirms that active ADPGP and EcASPP have access to the plastidic pool of ADPG. The fact that the expression of AtADPGP (whose K_m for ADPG is 7 µM) in the chloroplast resulted in a strong (ca. 97%) reduction of the starch content, whereas expression of EcASPP (whose K_m for ADPG is 160 µM) in the chloroplast resulted in a moderate (ca. 50%) reduction of starch content [15,27] (**Figure 5A**) points to the occurrence of very low concentration of ADPG in the chloroplast. In line with this presumption, despite the considerable reduction of starch content exerted by the ectopic expression of EcASPP and AtADPGP in the chloroplast, *TP-P541-AtADPGP* and *TP-P541-EcASPP* expressing WT leaves accumulated nearly WT ADPG content when plants were cultured on soil and MS medium (**Figure 5B, Figure S4**).

aps1 leaves accumulate 1–2% of the WT starch content, a phenotype that is reverted to WT by the ectopic expression of GlgC in the plastid [15] (**Figure 5A**). Furthermore, *TP-P541-EcASPP* expressing *aps1* leaves accumulate 40–50% of the starch accumulated by *aps1* leaves [15] (**Figure 5A**). Despite the considerable enhancement and reduction of starch content exerted by the ectopic expression of GlgC and EcASPP in the chloroplast of *asp1* leaves, respectively, leaves of plants from two independent *TP-P541-glgC* expressing *aps1* lines, and leaves of plants from two independent *TP-P541-EcASPP* expressing *aps1* lines accumulated nearly WT ADPG content in two different culture conditions (**Figure 5B, Figure S4**).

The overall results thus provided strong evidence that (a) most of ADPG accumulates outside the chloroplast, and (b) the contribution of the chloroplastic ADPG pool to the total ADPG pool is low in WT and *aps1* leaves.

Additional concluding remarks

HPLC-MS/MS studies carried out in this work by two independent laboratories showed that *aps1*, *pgm* and *aps1/pgm* leaves accumulate WT ADPG content (**Figure 2**), which provides strong evidence that leaves possess important ADPG sources other than the pPGI-pPGM-AGP pathway. As expected, leaves of plants ectopically expressing in the chloroplast either ADPG synthesis or breakdown enzymes accumulated higher and lower starch levels, respectively, than their parental lines (**Figure 5A**). However, these changes in starch content were not accompanied by concomitant changes in ADPG content (**Figure 5B**). This and the fact that the expression of ADPG cleaving enzymes in the cytosol results in reducing levels of both ADPG and starch [15,27] strongly indicate that (i) most of ADPG accumulates outside the chloroplast, and (ii) a sizable pool of ADPG occurring in the cytosol is linked to starch biosynthesis. That *ss3/ss4/aps1* leaves impaired in both starch granule initiation and AGP-mediated chloroplastic ADPG syn-

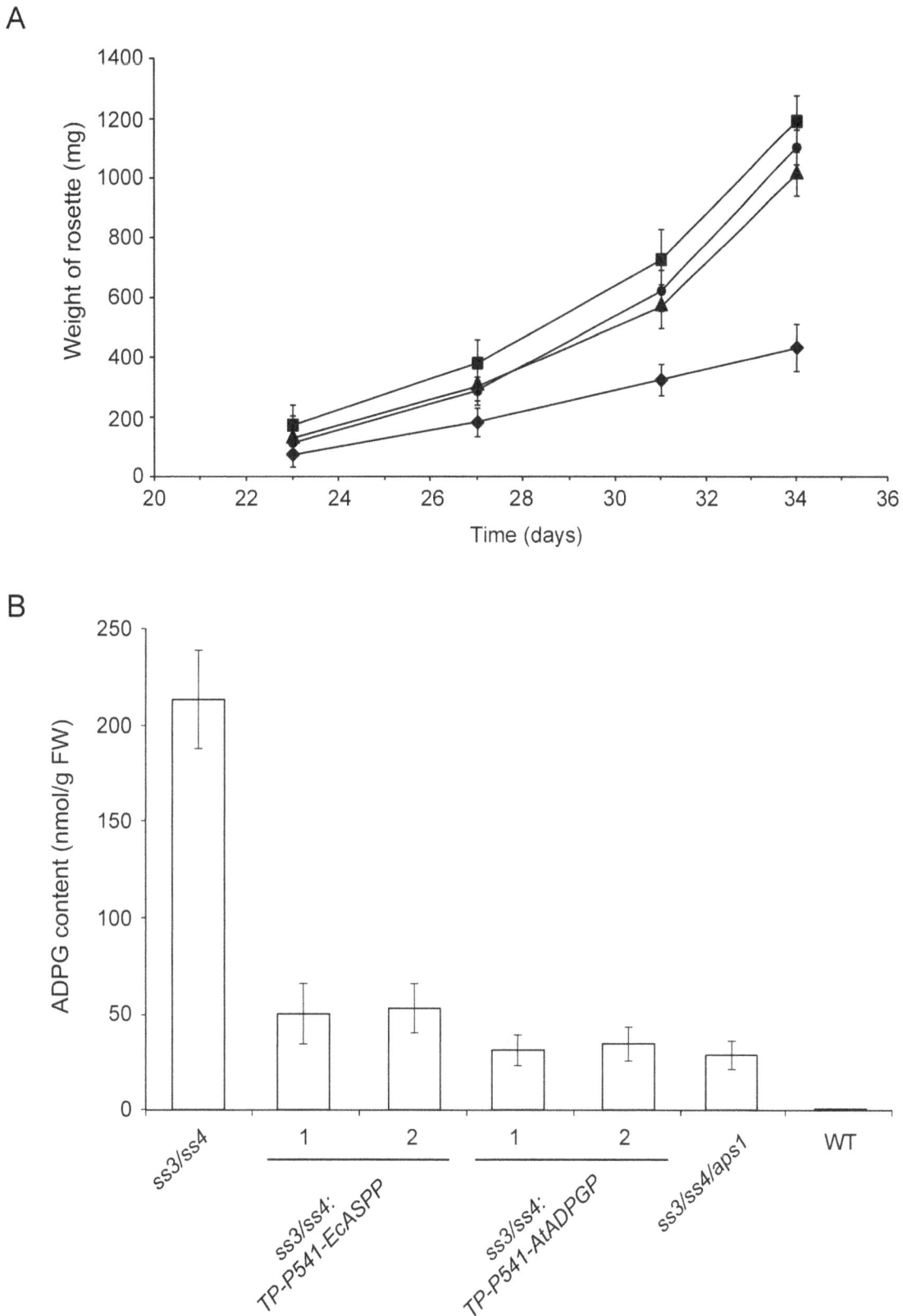

Figure 4. Ectopic expression of EcASPP and AtADPGP in the plastid restores the WT growth and partially reverts the ADPG excess phenotype of *ss3/ss4* plants. (A) Time-course of fresh weight of rosettes of WT (■) and *ss3/ss4* (◆) plants, and rosettes of one representative line each of *TP-P541-AtADPGP* expressing *ss3/ss4* plants and *TP-P541-EcASPP* expressing *ss3/ss4* plants (● and ▲, respectively). Plants were grown under long-day conditions (16 h light/8 h dark, 20°C) and at an irradiance of 90 μmol photons sec^{-1} m^{-2}. Values represent the mean ±SD of determinations on five independent plants. (B) ADPG content in WT, *aps1* and *ss3/ss4* leaves, and leaves of plants of two independent *TP-P541-EcASPP-* and *TP-P541-AtADPGP-* expressing *ss3/ss4* lines. Leaves were harvested after 10 h of illumination. Values represent the mean ±SD of determinations on three independent samples.

A

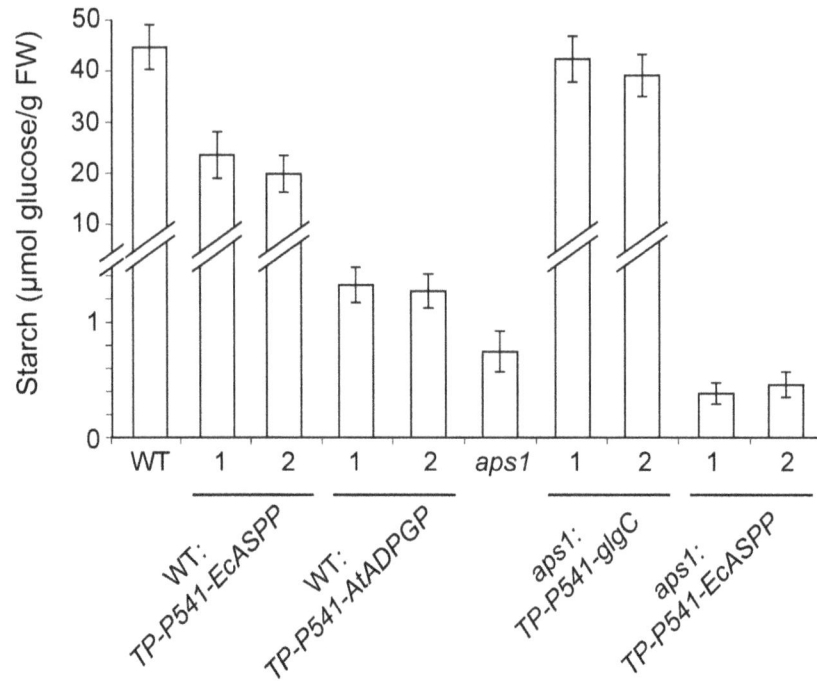

B

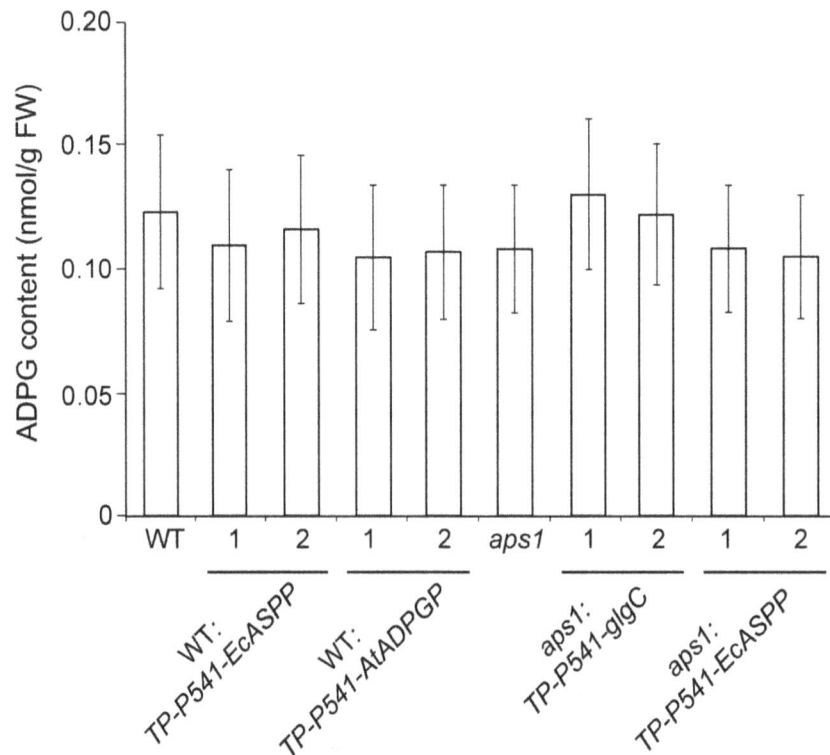

Figure 5. The contribution of the chloroplastic ADPG pool to the total ADPG pool is low in WT and *aps1* leaves. (A) Starch and (B) ADPG contents in leaves of WT and *aps1* plants, and leaves of two independent lines each of *TP-P541-AtADPGP* expressing WT plants, *TP-P541-EcASPP* expressing WT plants, *TP-P541-glgC* expressing *aps1* plants, and *TP-P541-EcASPP* expressing *aps1* plants. Plants were simultaneously grown and leavesf from 4-weeks old plants were simultaneously harvested after 10 h of illumination. ADPG was simultaneouly extracted and measured by HPLC-

MS/MS as described in Materials and Methods. Note that, consistent with [15], leaves of *TP-P541-EcASPP* expressing WT plants, *TP-P541-glgC* expressing *aps1* plants, and *TP-P541-EcASPP* expressing *aps1* plants accumulated WT ADPG content. Values represent the mean ±SD of determinations on three independent samples.Values represent the mean ±SD of determinations on three independent samples.

thesis accumulate ca. 300-fold more ADPG than WT leaves (**Figure 4**) further supports the idea that ADPG can be synthesized outside the chloroplast.

Using HPLC-MS/MS we have recently shown that leaves of *pgi1* mutants impaired in pPGI accumulate WT ADPG content [78], which further reinforces the idea that leaves possess important ADPG sources other than the pPGI-pPGM-AGP pathway. These studies also showed that the low starch content phenotype of *pgi1* mutants is largely the consequence of combined factors including reduction of photosynthetic activity, rather than the lack of pPGI-mediated flow between the Calvin-Benson cycle and the pPGM-AGP starch biosynthetic pathway [78]. Moreover, our studies showed that *pgi1* leaves of plants exposed for few hours to the action of microbial volatiles can accumulate up to 15-fold more starch than WT leaves [79]. This, and the facts that (a) chloroplasts can incorporate extraplastidial ADPG by means of a yet to be identified transporter and convert it into starch [33], (b) a sizable pool of ADPG accumulates outside the chloroplast [15,27,34] (this work), (c) cytosolic enzymes such as SuSy can produce ADPG [30,69,80–84], and (d) SuSy expresses in the mesophyll cells [85,86] prompted us to propose the model of transitory starch metabolism similar to that illustrated in **Figure 1B**.

The suggested starch biosynthetic model illustrated in **Figure 1B** involves simultaneous synthesis and breakdown of starch, and the pPGM and AGP-mediated scavenging of the starch breakdown products, thus making up a starch futile cycle. In this respect we must emphasize that many phylogenetically distant bacteria arrange all glycogen synthetic and breakdown genes in a single transcriptional unit, which guarantees simultaneous expression of glycogen synthesis and breakdown enzymes, and scavenging of glycogen breakdown products [49–53]. The resulting glycogen futile cycling would entail advantages such as dissipation of excess energy, sensitive regulation and rapid channeling of metabolic intermediates toward various metabolic pathways in response to biochemical needs [47,48,87,88]. Also, many phylogenetically distant bacteria possess various important sources, other than AGP, of ADPG linked to glycogen biosynthesis [89–93]. Since glycogen may play relevant roles in the survival of bacteria to sporadic periods of famine, and because the metabolism of this polyglucan is highly interconnected with multiple and important cellular processes [94,95], it is conceivable that both glycogen futile cycling and redundancy of ADPG sources were selected during bacterial evolution to guarantee the production of glycogen and its connection with other metabolic processes in response to physiological needs imposed by the environment and lifestyle [54]. Like in bacteria, starch is believed to act as a major integrator of the plant metabolic status that accumulates to cope with temporary starvation imposed by the environment [96]. Starch futile cycling may thus entail advantages such as rapid metabolic channeling toward various pathways (such as biosynthesis of fatty acids, OPPP, sulfolipid) [97,98] in response to physiological and biochemical needs. It is thus conceivable that, similar to bacteria, both redundancy of ADPG sources and starch futile cycling have been selected during plant evolution to warrant starch production and rapid connection of starch metabolism with other metabolic pathways.

Supporting Information

Figure S1   Stages to produce the *35S-TP-P541-AtADPGP, 35S-TP-P541-Ec-ASPP* and *35S-TP-P541-AtADPGP-GFP* plasmid constructs used to transform *Arabidopsis* plants. For AtADPGP constructs, a complete *AtADPGP* cDNA was obtained from the ABRC cDNA collection (C105280). *TP-P541-AtADPGP* were generated by cloning *AtADPGP* in the plasmid pSK TP-P541 which was used as template to generate *35S-TP-P541-AtADPGP* and *35S-TP-P541-AtADPGP* plasmid constructs using the forward 5′-GGGGACAAGTTTGTACAA-AAAAGCAGGCTTAATGACGTCACCGAGCCAT-3′ and the reverse 5′- GGGGACCACTTTGTACAAGAAAGCTGGGTA-TCAAGT-AAGGCTAACTTCCCGC-3′ primers and the Gateway technology (Invitrogen, http://www.invitrogen.com). To produce the *35S-TP-P541-AtADPGP-GFP* plasmid construct the reverse primer 5′-GGGGACCACTTTGTACAAGAAAGC-TGGGTAAG-TAAGGCTAACTTCCCGCATAAC-3′ was used to remove the stop codon from *AtADPGP*. DNA sequences of all constructs were confirmed by sequencing.

Figure S2   Stages to produce the pDEST17-AtADPGP plasmid construct used to transform *E. coli* cells.

Figure S3   Plastidial localization of AtADPGP-GFP in leaves of *TP-P541-AtADPGP-GFP* expressing WT plants. The upper panels show that AtADPGP-GFP fluorescence has a plastidial localization in leaf epidermal and mesophyll cells (bar 10 μm). In the middle panel note that AtADPGP-GFP was present among the grana in the central part of the chloroplast, as well as in the grana-free peripheral part of the chloroplast of mesophyll cells (bar 10 μm). The lower panels show a detailed view of plastid stromules in leaf epidermal cells (bar 2 μm). Note that GFP fluorescence labelled long stroma-filled tubular extensions corresponding to plastid stromules.

Figure S4   ADPG content in leaves of WT plants, and leaves of *TP-P541-AtADPGP* expressing WT plants (line #1), *TP-P541-EcASPP* expressing WT plants (line #1), and *TP-P541-EcASPP* expressing *aps1* plants (line #1) cultured on solid MS medium. Leaves were harvested after 10 h of illumination. Values represent the mean ±SD of determinations on three independent samples.

Table S1   Primers used for the identification of the triple *ss3/ss4/aps1* mutant plants.

Table S2   Primers used for the identification of the double *aps1/pgm* mutant plants.

Acknowledgments

We thank Dr. Ángel Mérida (Instituto de Bioquímica Vegetal y Fotosíntesis, Sevilla, Spain) who very kindly provided us the *ss3/ss4* seeds. We are thankfull to María Teresa Sesma and Maite Hidalgo (Institute of Agrobiotechnology, Navarra, Spain) for technical support.

Author Contributions

Conceived and designed the experiments: AB EB-F AMS-L TM JP-R. Performed the experiments: AB EB-F AMS-L FJM JL GA MM PP RG

KK KO KW. Analyzed the data: AB EB-F AMS-L FJM PP KW TM JP-R. Contributed reagents/materials/analysis tools: TM JP-R. Contributed to the writing of the manuscript: AB EB-F AMS-L TM JP-R.

References

1. Stitt M, Quick WP (1989) Photosynthetic carbon partitioning: its regulation and possibilities for manipulation. Physiol Plantarum 77: 633–641.
2. Rao M, Terry N (1995) Leaf phosphate status, photosynthesis, and carbon partitioning in sugar beet. Plant Physiol 107: 195–202.
3. Neuhaus HE, Häusler RE, Sonnewald U (2005) No need to shift the paradigm on the metabolic pathway to transitory starch in leaves. Trends Plant Sci 10: 154–156.
4. Streb S, Egli B, Eicke S, Zeeman SC (2009) The debate on the pathway of starch synthesis: a closer look at low-starch mutants lacking plastidial phosphoglucomutase supports the chloroplast-localised pathway. Plant Physiol 151: 1769–1772.
5. Stitt M, Zeeman SC (2012) Starch turnover: pathways, regulation and role in growth. Curr Opin Plant Biol 15: 1–11.
6. Kleczkowsk LA (1999) A phosphoglycerate to inorganic phosphate ratio is the major factor in controlling starch levels in chloroplasts via ADP-glucose pyrophosphorylase regulation. FEBS Lett 448: 153–156.
7. Kleczkowski LA (2000) Is leaf ADP-glucose pyrophosphorylase an allosteric enzyme? Biochim Biophys Acta 1476: 103–108.
8. Streb S, Zeeman SC (2012) Starch metabolism in Arabidopsis. The Arabidopsis book 10:e0160. DOI: 10.1199/tab.0160.
9. Crevillén P, Ballicora MA, Mérida A, Preiss J, Romero JM (2003) The different large subunit isoforms of *Arabidopsis thaliana* ADP-glucose pyrophosphorylase confer distinct kinetic and regulatory properties to the heterotetrameric enzyme. J Biol Chem 278: 28508–28515.
10. Crevillén P, Ventriglia T, Pinto F, Orea A, Mérida A, et al. (2005) Differential pattern of expression and sugar regulation of *Arabidopsis thaliana* ADP-glucose pyrophosphorylase-encoding genes. J Biol Chem 280: 8143–8149.
11. Sokolov LN, Déjardin A, Kleczkowski LA (1998) Sugars and light/dark exposure trigger differential regulation of ADP-glucose pyrophosphorylase genes in *Arabidopsis thaliana* (thale cress). Biochem J 336: 681–687.
12. Zhang J (2003) Evolution by gene duplication: an update. Trends Ecol Evol 18: 292-298.
13. Ventriglia T, Kuhn ML, Ruiz MT, Ribeiro-Pedro M, Valverde F, et al. (2008) Two Arabidopsis ADP-glucose pyrophosphorylase large subunits (APL1 and APL2) are catalytic. Plant Physiol 148: 65–76.
14. Wang SM, Lue WL, Yu T-S, Long JH, Wang CN, et al. (1998) Characterization of *ADG1*, an Arabidopsis locus encoding for ADPG pyrophosphorylase small subunit, demonstrates that the presence of the small subunit is required for large subunit stability. Plant J 13: 63–70.
15. Bahaji A, Li J, Ovecka M, Ezquer I, Muñoz FJ, et al. (2011) *Arabidopsis thaliana* mutants lacking ADP-glucose pyrphosphorylase accumulate starch and wild-type ADP-glucose content: further evidence for the occurrence of important sources, other than ADP-glucose pyrophosphorylase, of ADP-glucose linked to leaf starch biosynthesis. Plant Cell Physiol 52: 1162–1176.
16. Yu T-S, Lue W-L, Wang S-M, Chen J (2000) Mutation of Arabidopsis plastid phosphoglucose isomerase affects leaf starch synthesis and floral initiation. Plant Physiol 123: 319–325.
17. Kunz HH, Häusler RE, Fettke J, Herbst K, Niewiadomski P, et al. (2010) The role of plastidial glucose-6-phosphate/phosphate translocators in vegetative tissues of Arabidopsis thaliana mutants impaired in starch biosynthesis. Plant Biol (Stuttg) 12 Suppl 1: 115–128.
18. Caspar T, Huber SC, Somerville C (1985) Alterations in growth, photosynthesis, and respiration in a starchless mutant of *Arabidopsis thaliana* (L.) deficient in chloroplast phosphoglucomutase activity. Plant Physiol 79: 11–17.
19. Kofler H, Häusler RE, Schulz B, Gröner F, Flügge U-I, et al. (2000) Molecular characterization of a new mutant allele of the plastid phosphoglucomutase in *Arabidopsis*, and complementation of the mutant with the wild-type cDNA. Mol Gen Genet 262: 978–986.
20. Lin T-P, Caspar T, Somerville C, Preiss J (1988) Isolation and characterization of a starchless mutant of *Arabidopsis thaliana* (L.) Heynh lacking ADPglucose pyrophosphorylase activity. Plant Physiol 86: 1131–1135.
21. Asatsuma S, Sawada C, Itoh K, Okito M, Kitajima A, et al. (2005) Involvement of α-amylase I-1 in starch degradation in rice chloroplasts. Plant Cell Physiol 46: 858–869.
22. Fulton DC, Stettler M, Mettler T, Vaughan CK, Li J, et al. (2008) β-amylase4, a noncatalytic protein required for starch breakdown, acts upstream of three active β-amylases in *Arabidopsis* chloroplasts. Plant Cell 20: 1040–1058.
23. Streb S, Eicke S, Zeeman SC (2012) The simultaneous abolition of three starch hydrolases blocks transient starch breakdown in Arabidopsis. J Biol Chem 287: 41745–41756.
24. Cho MH, Lim H, Shin DH, Jeon JS, Bhoo SH, et al. (2011) Role of the plastidic glucose translocator in the export of starch degradation products from the chloroplasts in *Arabidopsis thaliana*. New Phytol 109: 101–112.
25. Olsson T, Thelander M, Ronne H (2003) A novel type of chloroplast stromal hexokinase is the major glucose-phosphorylating enzyme in the moss *Physcomitrella patens*. J Biol Chem 278: 44439–44447.
26. Giese JO, Herbers K, Hoffmann M, Klösgen RB, Sonnewald U (2005) Isolation and functional characterization of a novel plastidic hexokinase from *Nicotiana tabacum*. FEBS Lett 579: 827–831.
27. Baroja-Fernández E, Muñoz FJ, Zandueta-Criado A, Moran-Zorzano MT, Viale AM, et al. (2004) Most of ADP-glucose linked to starch biosynthesis occurs outside the chloroplast in source leaves. Proc Natl Acad Sci USA 101: 13080–13085.
28. Baroja-Fernández E, Muñoz FJ, Pozueta-Romero J (2005) A response to Neuhaus, et al. Trends Plant Sci 10: 156–158.
29. Baroja-Fernández E, Muñoz FJ, Bahaji A, Almagro G, Pozueta-Romero J (2012a) Reply to Smith, et al.: No pressing biological evidence to challenge the current paradigm on starch and cellulose biosynthesis involving sucrose synthase activity. Proc Natl Acad Sci USA 109:E777.
30. Baroja-Fernández E, Muñoz FJ, Li J, Bahaji A, Almagro G, et al. (2012b) Sucrose synthase activity in the *sus1/sus2/sus3/sus4* Arabidopsis mutant is sufficient to support normal cellulose and starch production. Proc Natl Acad Sci USA 109: 321–326.
31. Muñoz FJ, Baroja-Fernández E, Morán-Zorzano MT, Viale AM, Etxeberria E, et al. (2005) Sucrose synthase controls the intracellular levels of ADPglucose linked to transitory starch biosynthesis in source leaves. Plant Cell Physiol 46: 1366–1376.
32. Muñoz FJ, Morán-Zorzano MT, Alonso-Casajús N, Baroja-Fernández E, Etxeberria E, et al. (2006) New enzymes, new pathways and an alternative view on starch biosynthesis in both photosynthetic and heterotrophic tissues of plants. Biocatal Biotransformation 24: 63–76.
33. Pozueta-Romero J, Ardila F, Akazawa T (1991) ADPglucose transport by adenylate translocator in chloroplasts is linked to starch biosynthesis. Plant Physiol 97: 1565–1572.
34. Bahaji A, Li J, Sánchez-López AM, Baroja-Fernández E, Muñoz FJ, et al. (2014) Starch biosynthesis, its regulation and biotechnological approaches to improve crop yields. Biotechnol Adv 32: 87–106.
35. Stitt M, Heldt HW (1981) Simultaneous synthesis and degradation of starch in spinach chloroplasts in the light. Biochim Biophys Acta 638: 1–11.
36. Fox TC, Geiger DR (1984) Effects of decreased net carbon exchange on carbohydrate metabolism in sugar beet source leaves. Plant Physiol 76: 763–768.
37. Scott P, Kruger NJ (1995) Influence of elevated fructose-2,6-bisphosphate levels on starch mobilization in transgenic tobacco leaves in the dark. Plant Physiol 108: 1569–1577.
38. Walters RG, Ibrahim DG, Horton P, Kruger NJ (2004) A mutant of *Arabidopsis* lacking the triose-phosphate/phosphate translocator reveals metabolic regulation of starch breakdown in the light. Plant Physiol 135: 891–906.
39. Lozovaya VV, Zabotina OA, Widholm JM (1996) Synthesis and turnover of cell-wall polysaccharides and starch in photosynthetic soybean suspension cultures. Plant Physiol 111: 921–929.
40. Szecowka M, Heise R, Tohge T, Nunes-Nesi A, Vosloh D, et al. (2013) Metabolic fluxes in an illuminated *Arabidopsis* rosette. Plant Cell 25: 694–714.
41. Caspar T, Lin TP, Kakefuda G, Benbow L, Preiss J, et al. (1991) Mutants of *Arabidopsis* with altered regulation of starch degradation. Plant Physiol. 95: 1181–1188.
42. David M, Petit WA, Laughlin MR, Shulman RG, King JE, et al. (1990) Simultaneous synthesis and degradation of rat liver glycogen. J Clin Invest 86: 612–617.
43. Massillon D, Bollen M, De Wulf H, Overloop K, Vanstapel F, et al. (1995) Demonstration of a glycogen/glucose 1-phosphate cycle in hepatocytes from fasted rats. J Biol Chem 270: 19351–19356.
44. Bollen M, Keppens S, Stalmans W (1998) Specific features of glycogen metabolism in the liver. Biochem J 336: 19–31.
45. Lehmann M, Wöber G (1976) Accumulation, mobilization and turn-over of glycogen in the blue-green bacterium *Anacystis nidulans*. Arch Microbiol 111: 93–97.
46. Gaudet G, Forano E, Dauphin G, Delort A-M (1992) Futile cycling of glycogen in *Fibrobacter succinogenes* as shown by in situ ^{1}H-NMR and ^{13}C-NMR investigation. Eur J Biochem 207: 155–162.
47. Belanger AE, Hatfull GF (1999) Exponential-phase glycogen recycling is essential for growth of *Mycobacterium smegmatis*. J Bacteriol 181: 6670–6678.
48. Guedon E, Desvaux M, Petitdemange H (2000) Kinetic analysis of *Clostridium cellulolyticum* carbohydrate metabolism: importance of glucose 1-phosphate and glucose 6-phosphate branch points for distribution of carbon fluxes inside and outside cells as revealed by steady-state continuous culture. J Bacteriol 182: 2010–2017.
49. Kiel JAKW, Boels JM, Geldman G, Venema G (1994) Glycogen in *Bacillus subtilis*: molecular characterization of an operon encoding enzymes involved in glycogen biosynthesis and degradation. Mol Microbiol 11: 203–218.
50. Ugalde JE, Lepek V, Uttaro A, Estrella J, Iglesias A, et al. (1998) Gene organization and transcription analysis of the *Agrobacterium tumefaciens*

glycogen (glg) operon: two transcripts for the single phosphoglucomutase gene. J Bacteriol 180: 6557–6564.

51. Marroquí S, Zorreguieta A, Santamaría C, Temprano F, Soberón M, et al. (2001) Enhanced symbiotic performance by *Rhizobium tropici* glycogen synthase mutants. J Bacteriol 183: 854–864.

52. Lepek VC, D'Antuono AL, Tomatis PE, Ugalde JE, Giambiagi S, et al. (2002) Analysis of *Mesorhizobium loti* glycogen operon: effect of phosphoglucomutase (pgm) and glycogen synthase (glgA) null mutants on nodulation of *Lotus tenuis*. Mol Plant Microbe Interact 15: 368–375.

53. Montero M, Almagro G, Eydallin G, Viale AM, Muñoz FJ, et al. (2011) *Escherichia coli* glycogen genes are organized in a single *glgBXCAP* transcriptional unit possessing an alternative suboperonic promoter within *glgC* that directs *glgAP* expression. Biochem J 433: 107–118.

54. Wilson WA, Roach PJ, Montero M, Baroja-Fernández E, Muñoz FJ, et al. (2010) Regulation of glycogen metabolism in yeast and bacteria. FEMS Microbiol Rev 34: 952–985.

55. Okita TW (1992) Is there an alternative pathway for starch synthesis? Plant Physiol 100: 560–564.

56. Baroja-Fernández E, Muñoz FJ, Akazawa T, Pozueta-Romero J (2001) Reappraisal of the currently prevailing model of starch biosynthesis in photosynthetic tissues: A proposal involving the cytosolic production of ADP-glucose by sucrose synthase and occurrence of cyclic turnover of starch in chloroplast. Plant Cell Physiol 42: 1311–1320.

57. Zeeman SC, Smith SM, Smith AM (2007) The diurnal metabolism of leaf starch. Biochem J 401: 13–28.

58. Barratt DHP, Derbyshire P, Findlay K, Pike M, Wellner N, et al. (2009) Normal growth of *Arabidopsis* requires cytosolic invertase but not sucrose synthase. Proc Natl Acad Sci USA 106: 13124–13129.

59. Geigenberger P (2011) Regulation of starch biosynthesis in response to a fluctuating environment. Plant Physiol 155: 1566–1577.

60. Smith AM, Kruger NJ, Lunn JE (2012) Source of nucleotides for starch and cellulose synthesis. Proc Natl Acad Sci USA 109: E776.

61. Moreno-Bruna B, Baroja-Fernández E, Muñoz FJ, Bastarrica-Berasategui A, Zandueta-Criado A, et al. (2001) Adenosine diphosphate sugar pyrophosphatase prevents glycogen biosynthesis in *Escherichia coli*. Proc Natl Acad Sci USA 98: 8128–8132.

62. Crumpton-Taylor M, Pike M, Lu K-J, Hylton CM, Feil R, et al. (2013) Starch synthase 4 is essential for coordination of starch granule formation with chloroplas division during Arabidopsis leaf expansion. New Phytol 200: 1064–1075.

63. Chen M, Thelen JJ (2011) Plastid uridine salvage activity is required for photoassimilate allocation and partitioning in *Arabidopsis*. Plant Cell 23: 2991–3006.

64. Lunn JE, Feil R, Hendriks JHM, Gibon Y, Morcuende R, et al. (2006) Sugar-induced increases in trehalose 6-phosphate are correlated with redox activation of ADPglucose pyrophosphorylase and higher rates of starch synthesis in *Arabidopsis thaliana*. Biochem J 397: 139–148.

65. Ragel P, Streb S, Feil R, Sahrawy M, Annunziata MG, et al. (2013) Loss of starch granule initiation has a deleterious effect on the growth of *Arabidopsis thaliana* plants due to accumulation of ADP-glucose. Plant Physiol 163: 75–85.

66. Martins MCM, Hejazi M, Fettke J, Steup M, Feil R, et al. (2013) Feedback inhibition of starch degradation in Arabidopsis leaves mediated by trehalose 6-phosphate. Plant Physiol 163: 1142–1163.

67. McCoy JG, Arabshahi A, Bitto E, Bingman CA, Ruzicka FJ, et al. (2006) Structure and mechanism of an ADP-glucose phosphorylase from *Arabidopsis thaliana*. Biochemistry 45: 3145–3162.

68. Tacke M, Yang Y, Steup M (1991) Multiplicity of soluble glucan-synthase activity in spinach leaves: enzyme pattern and intracellular location. Planta 185: 220–226.

69. Zervosen A, Römer U, Elling L (1998) Application of recombinant sucrose synthase-large scale synthesis of ADP-glucose. J Mol Catalysis B: Enzymatic 5: 25–28.

70. Szydlowski N, Ragel P, Raynaud S, Roldán I, Montero M, et al. (2009) Starch granule initiation in Arabidopsis requires the presence of either class IV or class III starch synthase. Plant Cell 21: 2443–2457.

71. Clough SJ, Bent AF (1998) Floral dip: a simplified method for *Agrobacterium*-mediated transformation of *Arabidopsis thaliana*. Plant J 16: 735–743.

72. Rodríguez-López M, Baroja-Fernández E, Zandueta-Criado A, Pozueta-Romero J (2000) Adenosine diphosphate glucose pyrophosphatase: a plastidial phosphodiesterase that prevents starch biosynthesis. Proc Natl Acad Sci USA 97: 8705–8710.

73. Yamada K, Lim J, Dale JM, Chen H, Shinn P, et al. (2003) Empirical analysis of transcriptional activity in the Arabidopsis genome. Science 302: 842–846.

74. Kaneko K, Inomata T, Masui T, Koshu T, Umezawa Y, et al. (2014) Nucleotide pyrophosphatase/phosphodiesterase 1 exerts a negative effect on starch accumulation and growth in rice seedlings under high temperature and CO2 concentration conditions. Plant Cell Physiol 55: 320–332.

75. Fettke J, Malinova I, Albrecht T, Hejazi M, Steup M (2011) Glucose 1-phosphate transport into protoplasts and chloroplasts from leaves of *Arabidopsis thaliana*. Plant Physiol 155: 1723–1734.

76. Dankert M, Gonçalves IRJ, Recondo E (1964) Adenosine diphosphate glucose:orthophosphate adenylyltransferase in wheat germ. Biochim Biophys Acta 81: 78–85.

77. Murata T (1977) Partial purification and some properties of ADP-glucose phosphorylase from potato tubers. Agric Biol Chem 41: 1995–2002.

78. Bahaji A, Sánchez-López AM, Li J, Baroja-Fernández E, Muñoz FJ, et al. (2013) The Calvin-Benson cycle is not directly linked to transitory starch biosynthesis by means of phosphoglucose isomerase in plants exposed to microbial volatiles. XIII Spain-Portugal Congress on Plant Physiology (24–27 July 2013, Lisbon, Portugal).

79. Bahaji A, Sánchez-López AM, Muñoz FJ, Baroja-Fernández E, Li J, et al. (2014) Reevaluating the involvement of plastidic phosphoglucoseisomerase in starch biosynthesis in mesophyll cells. XII Plant Molecular Biology Meeting, Cartagena, Spain.

80. Murata T, Sugiyama T, Minamikawa T, Akazawa T (1966) Enzymic mechanism of starch synthesis in ripening rice grains. Mechanism of the sucrose-starch conversion. Arch Biochem Biophys 113: 34–44.

81. Delmer DP (1970) The purification and properties of sucrose synthase from etiolated *Phaseolus aureus* seedlings. J Biol Chem 247: 3822–3828.

82. Silvius JE, Snyder FW (1979) Comparative enzymic studies of sucrose metabolism in the taproots and fibrous roots of *Beta vulgaris* L. Plant Physiol 64: 1070–1073.

83. Tanase K, Yamaki S (2000) Purification and characterization of two sucrose synthase isoforms from japanese pear fruit. Plant Cell Physiol 41: 408–414.

84. Baroja-Fernández E, Muñoz FJ, Saikusa T, Rodríguez-López M, Akazawa T, et al. (2003) Sucrose synthase catalyzes the *de novo* production of ADPglucose linked to starch biosynthesis in heterotrophic tissues of plants. Plant Cell Physiol 44: 500–509.

85. Fu H, Kim SY, Park WD (1995) A potato *sus3* sucrose synthase gene contains a context-dependent 3'element and a leader intron with both positive and negative tissue-specific effects. Plant Cell 7: 1395–1403.

86. Wang A-Y, Kao M-H, Yang W-H, Sayion Y, Liu L-F, et al. (1999) Differentially and developmentally regulated expression of three rice sucrose synthase genes. Plant Cell Physiol 40: 800–807.

87. Newsholme EA, Challiss RAJ, Crabtree B (1984) Substrate cycles: their role in improving sensitivity in metabolic control. Trends Biochem Sci 9: 277–280.

88. Neijssel OM, Buurman ET, Texeira de Matos MJ (1990) The role of futile cycles in the energetics of bacterial growth. Biochim Biophys Acta 1018: 252–255.

89. Martin MC, Scheneider D, Bruton CJ, Chater KF, Hardison C (1997) A *glgC* gene essential only for the first two spatially distinct phases of glycogen synthesis in *Streptomyces coelicolor*. J Bacteriol 179: 7784–7789.

90. Eydallin G, Morán-Zorzano MT, Muñoz FJ, Baroja-Fernández E, Montero M, et al. (2007) An *Escherichia coli* mutant producing a truncated inactive form of GlgC synthesizes glycogen: further evidences for the occurrence of various important sources of ADPglucose in enterobacteria. FEBS Lett 581: 4417–4422.

91. Morán-Zorzano MT, Alonso-Casajús N, Muñoz FJ, Viale AM, Baroja-Fernández E, et al. (2007) Occurrence of more than one important source of ADPglucose linked to glycogen biosynthesis in *Escherichia coli* and *Salmonella enterica*. FEBS Lett 581: 4423–4429.

92. Sambou T, Dinadayala P, Stadthagen G, Barilone N, Bordat Y, et al. (2008) Capsular glucan and intracellular glycogen of *Mycobacterium tuberculosis*: biosynthesis and impact on the persistence in mice. Mol Microbiol 70: 762–774.

93. Guerra LT, Xu Y, Bennette N, McNeely K, Bryant DA, et al. (2013) Natural osmolytes are much less effective substrates than glycogen for catabolic energy production in the marine cyanobacterium *Synechococcus* sp. strain PCC 7002. J Biotechnol 166: 65–75.

94. Eydallin G, Viale AM, Morán-Zorzano MT, Muñoz FJ, Montero M, et al. (2007) Genome-wide screening of genes affecting glycogen metabolism in *Escherichia coli*. FEBS Lett 581: 2947–2953.

95. Montero M, Eydallin G, Viale AM, Almagro G, Muñoz FJ, et al. (2009) *Escherichia coli* glycogen metabolism is controlled by the PhoP-PhoQ regulatory system at submillimolar environmental Mg2+ concentrations, and is highly interconnected with a wide variety of cellular processes. Biochem J 424: 129–141.

96. Sulpice R, Pyl E-T, Ishihara H, Trenkamp S, Steinfath M, et al. (2009) Starch as a major integrator in the regulation of plant growth. Proc Natl Acad Sci USA 106: 10348–10353.

97. Zeeman SC, Thorneycroft D, Schupp N, Chapple A, Weck M, et al. (2004) Plastidial α-glucan phosphorylase is not required for starch degradation in *Arabidopsis* leaves but has a role in the tolerance of abiotic stress. Plant Physiol 135: 849–858.

98. Okazaki Y, Shimojima M, Sawada Y, Toyooka K, Narisawa T, et al. (2009) A chloroplastic UDP-glucose pyrophosphorylase from *Arabidopsis* is the committed enzyme for the first step of sulfolipid biosynthesis. Plant Cell 21: 892–909.

6

Phylogeny in Defining Model Plants for Lignocellulosic Ethanol Production: A Comparative Study of *Brachypodium distachyon*, Wheat, Maize, and *Miscanthus* x *giganteus* Leaf and Stem Biomass

Till Meineke, Chithra Manisseri, Christian A. Voigt*

Phytopathology & Biochemistry, Biocenter Klein Flottbek, University of Hamburg, Hamburg, Germany

Abstract

The production of ethanol from pretreated plant biomass during fermentation is a strategy to mitigate climate change by substituting fossil fuels. However, biomass conversion is mainly limited by the recalcitrant nature of the plant cell wall. To overcome recalcitrance, the optimization of the plant cell wall for subsequent processing is a promising approach. Based on their phylogenetic proximity to existing and emerging energy crops, model plants have been proposed to study bioenergy-related cell wall biochemistry. One example is *Brachypodium distachyon*, which has been considered as a general model plant for cell wall analysis in grasses. To test whether relative phylogenetic proximity would be sufficient to qualify as a model plant not only for cell wall composition but also for the complete process leading to bioethanol production, we compared the processing of leaf and stem biomass from the C_3 grasses *B. distachyon* and *Triticum aestivum* (wheat) with the C_4 grasses *Zea mays* (maize) and *Miscanthus* x *giganteus*, a perennial energy crop. Lambda scanning with a confocal laser-scanning microscope allowed a rapid qualitative analysis of biomass saccharification. A maximum of 108–117 mg ethanol·g^{-1} dry biomass was yielded from thermo-chemically and enzymatically pretreated stem biomass of the tested plant species. Principal component analysis revealed that a relatively strong correlation between similarities in lignocellulosic ethanol production and phylogenetic relation was only given for stem and leaf biomass of the two tested C_4 grasses. Our results suggest that suitability of *B. distachyon* as a model plant for biomass conversion of energy crops has to be specifically tested based on applied processing parameters and biomass tissue type.

Editor: Samuel P. Hazen, University of Massachusetts Amherst, United States of America

Funding: The study was financially supported by the German Federal Ministry of Education and Research (grant no. FKZ 0315521A to CAV and CM) and the Dr. Elisabeth Appuhn Stiftung (to TM). The funders had no role in study design, data collection and analysis, decision to publish, or preparation of the manuscript.

Competing Interests: The authors have declared that no competing interests exist.

* Email: christian.voigt@uni-hamburg.de

Introduction

The increasing global energy demand leads to an elevated consumption of fossil energy resources, which is not only associated with the observed climate change [1] but also a with a growing threat for sensitive ecosystems due to expanded explorations for fossil energy sources [2–4]. One strategy to face this challenge is the substitution of fossil by renewable energy sources. In the sector of transportation, liquid fossil fuels have a predominant role. Here, second generation biofuels from lignocellulosic feedstock have the potential for an extended substitution in the near future [5]. A main source of lignocellulosic feedstock could be stover and straw from field crops, which would mainly derive from *Zea mays* (maize) and *Triticum aestivum* (wheat) in temperate climates, where these two crops are predominantly cultivated in agriculture according to the 2012 FAO statistics of crop production [6]. However, to meet the expected demand on lignocellulosic feedstock and to improve a sustainable production of biomass, an extended cultivation of energy crops is a promising solution [7]. In terms of sustainability, perennial C_4 grasses are preferred low-input energy crops because they can recycle their nutrients and store them in their rhizomes in the non-growing season, which results in a minimized fertilizer input. In addition, they have a reduced requirement for herbicide application because of fast growth, providing habitats for animals, and reducing soil erosion with their extended, permanent root systems [8–11]. Among the group of perennial C_4 grasses, *Miscanthus* x *giganteus* not only meets the mentioned criteria for sustainable biomass production [12] but also proved to be the most productive energy crop under temperate climate conditions [13–15]. Hence, the calculated lignocellulosic ethanol yield per ha cropland for *M.* x *giganteus* was higher than in other designated energy crops including *Z. mays* and *Panicum virgatum* (switchgrass) [14].

However, the efficiency of lignocellulosic ethanol production is restricted by the recalcitrance of the cell wall, which results in a relatively low saccharification. Main factors that determine biomass saccharification are cellulose crystallinity [16–18] but also lignin and hemicellulose content [19,20]. Apart from modifying and adapting different biomass pretreatment methods, the optimization of the cell wall for improved saccharification is a strategy to overcome biomass recalcitrance. Whereas different genetic tools and their application have already been established for major crops like *T. aestivum* and *Z.mays*, none of these tools are available for *M. x giganteus*. Therefore, the wild grass *Brachypodium distachyon* with its relatively small and sequenced diploid genome, efficient transformation, easy cultivation, and fast generation cycle has been proposed as a suitable model to test modified or engineered cell walls for altered saccharification. However, whereas genomic and cell wall composition studies already revealed the suitability of *B. distachyon* as a model for other grasses [21–25], comparative analyses of the complete process starting from biomass pretreatment via saccharification efficiency to fermentation leading to ethanol production have not been performed.

Our study aimed to test whether phylogenetic relation could be used as a marker for similarities in the processing of biomass for ethanol production with a special focus on *B. distachyon* and its qualities as a model plant for energy crops regarding ethanol production via fermentation. In case phylogenetic proximity would correlate with the process of biomass fermentation, we expected that results from *B. distachyon* would reveal highest similarity to *T. aestivum*, which are both C_3 grasses, and not to the C_4 grasses *Z. mays* and *M. x giganteus*. We distinguished between leaf and stem biomass because we identified a specific requirement to establish a model plant for stem biomass. In general, biomass from perennial grasses like *M. x giganteus* mainly derives from stems because it is typically harvested in early spring when most leaves have fallen off and the water content is low. We cultivated all plants and performed all experiments under the same conditions to ensure that observed differences or similarities would not be based on the experimental setup but on cell wall characteristics.

Our results revealed that phylogenetic relation cannot be regarded as a general marker for similarities in the processing of biomass for bioethanol production because other factors like biomass pretreatment as well as tissue type of the biomass seem to have an additional influence.

Materials and Methods

Plant material

B. distachyon (inbred line Bd21 [26]), *T. aestivum* (cultivar Nandu, Lochow-Petkus, Bergen-Wohlde, Germany), *M. x giganteus*, and *Z. mays* (inbred line A188 [27]) were all cultivated in two parts of soil (Einheitserdewerk Uetersen, Germany, ED 73+10% sand) and one part of sand under greenhouse conditions with an additional light supply to provide 16 h light if required. Plant material that reached its final developmental stage due to complete, natural senescence with subsequent drying for 2 weeks without irrigation was harvested from *B. distachyon* after 4 month, from *T. aestivum* and *Z. mays* after 6 month, and from *M. x giganteus* after 1 year. Plant material was manually harvested and separated into stem biomass, including all nodes, and leaf biomass, comprising blades and sheaths. Spikes and spikelets from *B. distachyon* and *T. aestivum* as well as cobs from *Z. mays* were not used in this study. After harvest, leaf and stem biomass was dried at 50°C for 2 days. Biomass was homogenized with a mill fitted with a 1.5 mm mesh screen for its use in pretreatment and fermentation experiments and with a 0.2 mm mesh screen for compositional analysis.

Preparation of alcohol insoluble residue

Milled plant biomass was processed as described by Arora et al. [28] with the modifications as described by Ellinger et al. [29] to receive alcohol insoluble residue (AIR) and destarched AIR, which was used for lignin, cellulose, and hemicellulose quantification.

Lignin quantification using acetyl bromide

The lignin content of destarched AIR was determined with a modified acetyl bromide method [28]. Lignin contents were determined with the absorbance at 280 nm and calculated with an extinction coefficient of 17.747 $g^{-1} \cdot L \cdot cm^{-1}$ for maize leaf and stem, 17.377 $g^{-1} \cdot L \cdot cm^{-1}$ for *B. distachyon* leaf and stem, 19.808 $g^{-1} \cdot L \cdot cm^{-1}$ for wheat leaf and 17.542 $g^{-1} \cdot L \cdot cm^{-1}$ for wheat stem [30]. For *M. x giganteus* leaf and stem, a specific extinction coefficient of 17.78 $g^{-1} \cdot L \cdot cm^{-1}$ was used for calculation [31].

Cellulose and hemicellulose quantification using phenol/sulfuric acid

Destarched AIR biomass was used for quantification of cellulose and hemicellulose following a modified protocol of DuBois et al. [32]. For determination of the total cell wall sugar amount, 5 mg of AIR samples were mixed with 50 μL of 72% (w/w) sulfuric acid and incubated at 30°C for 1 h. After cooling down, these solutions were diluted with water to 4% (w/w) sulfuric acid and again incubated at 120°C for 1 h. Supernatants obtained after centrifugation at 13,000× *g* for 5 min were diluted 1:400 with water. 0.5 mL of these dilutions were mixed with 0.3 mL 5% phenol and 1.8 mL sulfuric acid and incubated at room temperature for 20 min. Absorbance of these solutions were measured at 480 nm. For determination of the hemicellulose amount, 5 mg of AIR samples were treated with trifluoro acetic acid (TFA, 2 M) at 120°C for 1 h. This treatment hydrolyzes non-cellulosic polysaccharides but does not affect cellulose [33]. TFA hydrolyzates were completely dried using a speedvac concentrator (Savant, USA) at 32°C under vacuum. Monosaccharides liberated after TFA hydrolysis were redissolved in 1 mL of water. After centrifugation at 13,000× *g* for 5 min, 0.5 mL of 1:200-diluted supernatants were mixed with 0.3 mL 5% phenol and 1.8 mL sulfuric acid and incubated at room temperature for 20 min. Absorbance of these solutions were measured at 480 nm. For the preparation of standards, 5–25 μg glucose and xylose in 0.5 mL total volume were mixed with 0.3 mL 5% phenol and 1.8 mL sulfuric acid and incubated at room temperature for 20 min. Amounts of hemicellulose were quantified with xylose as standard. Cellulose amounts were quantified by subtraction of the amount of hemicellulose from the calculated values from total cell wall sugar amount using the glucose standard.

Monosaccharide composition by high-performance anion exchange chromatography with pulsed amperometric detection (HPAEC-PAD)

The non-cellulosic monosaccharide composition of leaf and stem cell walls before and after pretreatment and fermentation was analyzed by HPAEC-PAD on an ICS-5000 system (Dionex, USA) equipped with electrochemical detector and a CarboPac PA 20 column (3×150 mm, Dionex, USA), according to the description in Ellinger et al. [29].

Pretreatment and fermentation of plant biomass

7.5 g milled leaf and stem biomass were mixed with 42.5 mL 1.75% (v/v) sulfuric acid and autoclaved for 1 h at 120°C in 300 mL fermentation reactors (DASGIP, Germany). 20 mL of

$10\times$ YP Broth (200 g peptone and 100 g yeast extract per liter) were added to the fermentation reactors and volume was adjusted to 200 mL with water. The $4\times$ parallel fermenter (DASGIP) was programmed to pH 5.0 at $30°C$ with 600–900 rpm. Samples (2×2 mL) were taken when these values were reached. The fermentation broth was inoculated with 1 mL of an overnight grown *Saccharomyces cerevisiae* culture (strain MaV203, Life Technologies, Germany). Samples (2×2 mL) were taken every 24 h. In experiments with additional enzymatic pretreatment, the thermo-chemical pretreated biomass (200 mL as described above) was mixed with 1.875 mL of the enzyme mixture Accellerase 1500 (Genencor, Netherlands) according to the manufactures instructions. The fermentation broth was stirred for 24 h at $55°C$. After samples were taken (2×2 mL), fermenters were inoculated with *S. cerevisiae* (as described above), which generally only ferments hexoses. The fermentation process was monitored and controlled with the DASGIP Control 4.0 software. Samples were centrifuged at 13,000 rpm for 5 min. Supernatants were filtered through 0.44 µm PTFE-filters. Glucose as well as ethanol concentrations were analyzed as described by Bonn and Bobleter [34] on an ICS-5000 system (Dionex) with a HPX 87H column (Bio-Rad, USA, mobile phase 0.005 M H_2SO_4, flow rate 0.6 mL·min^{-1}, column temperature: $50°C$, refractive index detector). Residues were used for lambda scanning in laser confocal fluorescence microscopy and for cell wall analysis HPAEC-PAD.

Lambda scanning using confocal laser-scanning microscopy

Milled, untreated biomass, diluted acid pretreated biomass, diluted acid and enzyme pretreated biomass as well as fermented biomass after 96 h of fermentation (both pretreatments) were analyzed with a LSM 780 confocal laser-scanning microscope (Zeiss, Germany). For the rapid qualitative analysis of biomass saccharification, confocal lambda scanning was applied with a 405 nm diode laser and a 561 nm diode-pumped solid-state (DPSS) laser for excitation. Emission spectra were acquired with a meta-detector in the range of 411–691 nm [35]. Lambda scans were processed with integral functions of the ZEN 2010 (Zeiss) operating software. For each scan, three regions of the cell wall were manually selected and the relative intensities for each wavelength were measured.

Phylogenetic analysis

DNA sequences from the second intergenic spacer (ITS2) of nuclear ribosomal DNA were aligned using the Clustal Omega [36] multiple sequence alignment tool provided by the EMBL European Bioinformatics Institute (http://www.ebi.ac.uk/tools/msa/clustalo/). Manual corrections of initial alignment followed guidelines of Kelchner [37]. PHYLIP interleaved alignment output format of the Clustal Omega online tool was used to generate a phylogenetic tree using the online phylogenetic tree drawing application Phylodendron (version 0.8 d; http://iubio.bio.indiana.edu/treeapp/). ITS2 sequence information previously used to generate phylogenetic trees was obtained from GenBank: *M.* x *giganteus* [AJ426562], *Z. mays* [AF019811], *Sorghum bicolor* (sorghum) [SBU04789], *Saccharum officinarum* (sugarcane) [AY116284], *T. aestivum* [FM998919], *Hordeum vulgare* (barley) [AF438194], *B. distachyon* [AF303399], *Oryza sativa* (rice) [JN402189], *Secale cereale* (rye) [AF303400], *Arabidopsis thaliana* [AJ232900], *Solanum lycopersicum* (tomato) [AJ300201], *Glycine max* (soybean) [AJ011337].

Statistical analysis

Descriptive statistics including the mean and the standard error of the mean (SE) along Tukey range test for multiple comparison procedure in conjunction with an ANOVA were used to determine significant differences. $P<0.05$ was considered significant.

Principal component analysis (PCA)

Separate two-factor PCA and associated PC loadings for factor 1 and 2 for leaf and stem biomass with thermo-chemical pretreatment and combined thermo-chemical and subsequent enzymatic pretreatment were calculated with the statistical data analysis software STATISTICA (version 9, Statsoft, Germany). In analogy to constant-sum normalization [38], PCA parameters of a specific plant species and biomass pretreatment were introduced as relative amounts ($g·g^{-1}$ dry biomass) and set to the sum of 1 to achieve an internal normalization between the different datasets. PCA parameters used for calculation were relative amounts of lignin, cellulose, and hemicellulose, relative amounts of glucose deriving from the hemicellulose fraction, relative amounts of hydrolyzed glucose from biomass, and relative amounts of produced ethanol.

Results and Discussion

Biomass composition

To test the application of *B. distachyon* as model plant for biomass conversion, we compared cell wall composition, saccharification, and lignocellulosic ethanol production of *B. distachyon* leaf and stem biomass with biomass of the crop *T. aestivum*, which is a C_3 grass like *B. distachyon*, and the C_4 grasses *Z. mays* and *M.* x *giganteus*. We generated a phylogenetic tree of the plant species used in this study and related crop species based on the alignment of the ITS2 nuclear ribosomal DNA spacer region (Fig. 1). As expected from previous phylogenetic analyses of the respective plant species, the C_4 grasses *M.* x *giganteus*, *Z. mays*, *S. biocolor*, and *S. officinarum* formed a group that separated from the group of the C_3 grasses *B. distachyon*, *T. aestivum*, *H. vulgare*, *S. cereale*, and *O. sativa*. Within these two groups, phylogenetic distances resembled those of previous studies where *M.* x *giganteus* revealed higher phylogenetic relation to *S. biocolor* and *S. officinarum* than to *Z. mays* [39], and *B. distachyon* a higher phylogenetic relation to the small grain cereals *T. aestivum*, *H. vulgare*, and *S. cereale* than to *O. sativa* [40]. The dicotyledonous model plant *A. thaliana* as well as the dicotyledonous crops *S. lycopersicum* and *G. max* outgrouped due to the relatively high phylogenetic distance to the monocotyledonous grasses [41].

In our study, we started to determined the cellulose, hemicellulose, and lignin content of the harvested, naturally senesced biomass after an additional drying process (Fig. 2). In *Z. mays* stem biomass, we determined a relative cellulose content of 40%, which was only slightly higher than previously reported [42] whereas the hemicellulose content was slightly decreased, 29% compared to 33% [42]. In *M.* x *giganteus*, the cellulose content of stem biomass, which ranged from 42% to 45% in previous reports [31,42], was reduced (36%) whereas we observed an increase in the hemicellulose content, 31% compared to the reported 27% [42] (Fig. 2B,C). The lignin content of 22% in *Z. mays* stem biomass (Fig. 2A) resembled the value that was previously reported (21%) [43]. In that study, the lignin content was also determine by the acetyl bromide method whereas amounts of lignin were reported to be lower in those studies where Klason lignin was determined [44,45]. This contributes to the observation that values for the lignin content deriving from different methods are

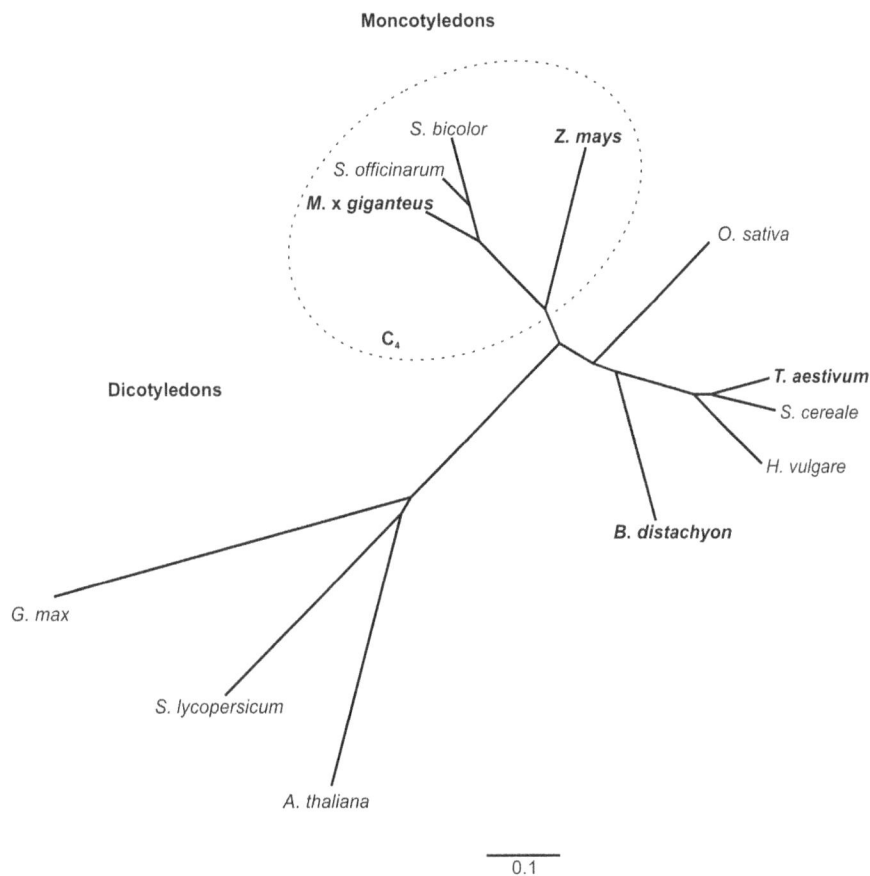

Figure 1. Phylogenetic relation of tested plant species. Phylogenetic tree based on the nucleotide alignment of the ITS2 region of ribosomal DNA of the C_4 grasses *Z. mays* (maize), *S. officinarum* (sugarcane), *S. bicolor* (sorghum), and *M.* x *giganteus*, the C_3 grasses *O. sativa* (rice), *T. aestivum* (wheat), *H. vulgare* (barley), *S. cereale* (rye), and *B. distachyon*, and the C_3 dicots *G. max* (soybean), *S. lycopersicum* (tomato), and *A. thaliana*, which were used as an outgroup. Bold: plant species used in this study, dotted line: plant species with a C_4 carbon fixation.

not directly comparable [30]. In addition, the lignin content in *Z. mays* stem biomass was strongly dependent on the tested inbred line and growing conditions [44,45]. Whereas we used biomass from the *Z. mays* inbred line A188 grown under greenhouse conditions, other studies mainly refer to biomass from different, field grown *Z. maize* cultivars.

In *M.* x *giganteus* stem biomass, the lignin content was reduced compared to a previous report from field grown *M.* x *giganteus* (24% compared to 27% [31]) (Fig. 2A). Whether reduction of lignin content in *M.* x *giganteus* comparing greenhouse and field grown plants would indicate general differences in modifying cell wall composition in response to different growing conditions in this C_4 grass, could be tested in comparative growth experiments. Regarding general cell wall composition of the analyzed C_3 grasses, we observed a higher lignin content in *B. distachyon* stem biomass than previously reported, 17% compared to a reported content of 14% [43], but no differences in cellulose and hemicellulose content [43]. In *T. aestivum*, the cellulose content with 42% was in the range of 35% to 48% as previously reported [42,46]. Also the hemicellulose content with 28% was in the range as indicated in previous studies (23% [42] to 28% [46]). The lignin content (21%) was also similar to previously reported amounts [30,42] (Fig. 2).

Whereas it was possible to compare our data of the cell wall composition of stem biomass with existing data in literature for all tested plant species, comparable data of the cell wall composition

were not available for leaf biomass. Therefore, our data from leaf biomass would represent a new reference for the cell wall composition of senesced, dry leaf biomass.

Comparing the content of the cell wall components cellulose and hemicellulose between the different plant species, we determined a significant difference in cellulose and hemicellulose content in leaf biomass only between the model grass *B. distachyon* and *T. aestivum* (Fig. 2B,C). Based on the phylogenetic relation between *B. distachyon* and *T. aestivum* and the other tested C_4 grasses (Fig. 1), we would have anticipated differences in the cell wall composition between the two groups, C_3 and C_4 grasses, rather than within a group. Therefore, these differences in cellulose and hemicellulose content were the first indication that a phylogenetic relation might not directly refer to relatively high similarities in cell wall composition. Regarding lignin, highest amounts were found in biomass from *M.* x *giganteus* (Fig. 2A). The lignin content of *M.* x *giganteus* leaf biomass was significantly higher than in the two C_3 grasses *B. distachyon* and *T. aestivum*. For stem biomass, differences in the lignin content resembled the phylogenetic relation only between the C_4 grasses and *B. distachyon* as we determined significantly higher lignin content in *M.* x *giganteus* and *Z. mays* than in *B. distachyon*, whereas the lignin content of *T. aestivum* was not significantly different from the C_4 grasses. A lignin content of 24% in *M.* x *giganteus* stem biomass, which was 40% higher than in *B. distachyon* stem biomass, marked the highest value in all tested samples (Fig. 2A).

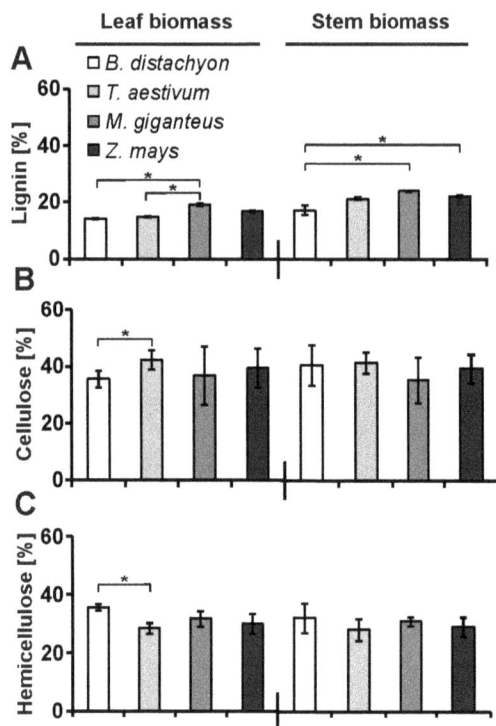

Figure 2. Cell wall composition of leaf and stem biomass. Milled leaf and stem samples from senesced, dry plants were used for the determination of the three major cell wall polymers (**A**) lignin, (**B**) cellulose, and (**C**) hemicelluloses. Values represent the relative amount of each polymer. *$P<0.05$ by Tukey's test. Error bars represent SE, and $n = 3$.

Apart from cellulose crystallinity that directly effects enzymatic saccharification [16–18], lignin is considered a major cell wall component, which reduces saccharification efficiency of biomass [19,20]. Therefore, modification or reduction of the lignin content might improve saccharification efficiency, especially in *M. x giganteus* with its relatively high lignin content. The reduction of lignin biosynthesis by RNA interference, which in *S. officinarum* [47] and *Populus tomentosa* (Chinese white poplar) [48] correlated with reduced recalcitrance of the cell wall, might represent a promising approach. However, strategies that affect lignin biosynthesis should be accompanied by experiments that test putative alterations in plant defense to invading pathogens because lignin has a critical role in establishing disease resistance [49].

Non-cellulosic monosaccharide composition

High-performance anion exchange chromatography with pulsed amperometric detection (HPAEC-PAD) facilitated a detailed analysis of the non-cellulosic monosaccharide composition of the dried leaf and stem biomass. As expected from secondary grass cell walls [25], xylose, glucose, and arabinose were the predominant monosaccharides (Fig. 3). After destarching and before biomass pretreatment, the relative monosaccharide content of all four tested plant species was similar (Fig. 3A). Slight aberrations were only determined in the glucose and xylose content in *T. aestivum* leaf biomass and in the arabinose and xylose content in *B. distachyon* stem biomass. However, these slight aberrations did not influence the overall observation that the ratio of arabinose, glucose, and xylose resembled previous cell wall studies of these plant species [25,31,42,50]. Differences in cell wall

composition can generally derive from differences in plant cultivation. In this regard, it has been shown that cultivation conditions could influence the process of maturation in *M. x giganteus* and *Z. mays* [51–53]. Hence, the composition of the cell wall may not be directly comparable between different studies. This supports the approach in our study where we compare cell wall composition of plant biomass that reached its final development stage from different plant species grown under the same condition, which contributes to a higher comparability of datasets between different plant species.

We then analyzed the non-cellulosic monosaccharide composition after application of two different methods of biomass pretreatment, i) thermo-chemical treatment with diluted sulfuric acid (1.75% (v/v)) and ii) with an additional application of an commercially available enzyme mixture (Accellerase 1500) after a previous diluted sulfuric acid treatment. The combination of a thermo-chemical pretreatment using diluted acid with a subsequent enzyme application is regarded as one of the most promising strategies to overcome biomass recalcitrance and release of fermentable carbohydrates [54].

In the leaf biomass of all analyzed plant species, we observed a shift in the glucose to xylose mol%-ratio from 1/7 before treatment to nearly 1/1 ratio after sulfuric acid and combined sulfuric acid and enzymatic treatment (Fig. 3), which was even stronger for *B. distachyon* leaf biomass after the combined pretreatment. Here, the glucose to xylose mol%-ratio shifted to almost 2/1 (Fig. 3C). For stem biomass, we observed a similar increased recalcitrance of glucose-containing hemicelluloses compared to xylose-containing, which was again most prominent for *B. distachyon*, where the glucose to xylose mol %-ratio shifted from about 1/7 before treatment (Fig. 3A) to 1/1 after sulfuric acid pretreatment (Fig. 3B) and to almost 2/1 after combined acid and enzyme pretreatment (Fig. 3C). In addition to the differences in the non-cellulosic monosaccharide composition observed for *B. distachyon*, we found additional significant differences especially for the arabinose content of *M. x giganteus* leaf and stem biomass, which was highest after both, sulfuric acid and combined sulfuric acid and enzymatic treatment compared to the other plant species, and for *Z. mays* stem biomass where the galacturonic acid content was highest after the two types of biomass treatment (Fig. 3B,C). These results revealed that the applied pretreatment methods favored the release of xylose rather than glucose from hemicelluloses. Our results confirmed previous studies where xylan, the main hemicellulose and main source of xylose, was better hydrolyzed during thermo-chemical biomass pretreatment than glucomannan, which mainly consists of glucose residues [55–57]. Because the relative non-cellulosic glucose content of almost all *B. distachyon* biomass samples after pretreatment was significantly higher than in all other analyzed species, which correlated with a significantly lower xylose content (Fig. 3A,B), a relatively efficient hydrolysis of xylan can be expected in *B. distachyon*.

Rapid qualitative analysis of biomass saccharification by confocal lambda scanning

To test the application of a rapid qualitative analysis of biomass composition and saccharification after different methods of biomass pretreatment, we applied a lambda scan via confocal laser-scanning microscopy to the different tissue types of biomass deriving from all four analyzed plant species. The qualitative mapping of the plant cell walls is based on its auto-fluorescence during lambda scanning. The excitation of the plant biomass with the 405 nm diode laser induced the emission of green fluorescence, which mainly derives from bound ferulic acid in cell walls [58,59] and carbohydrates; especially from cellulose [60,61] as the

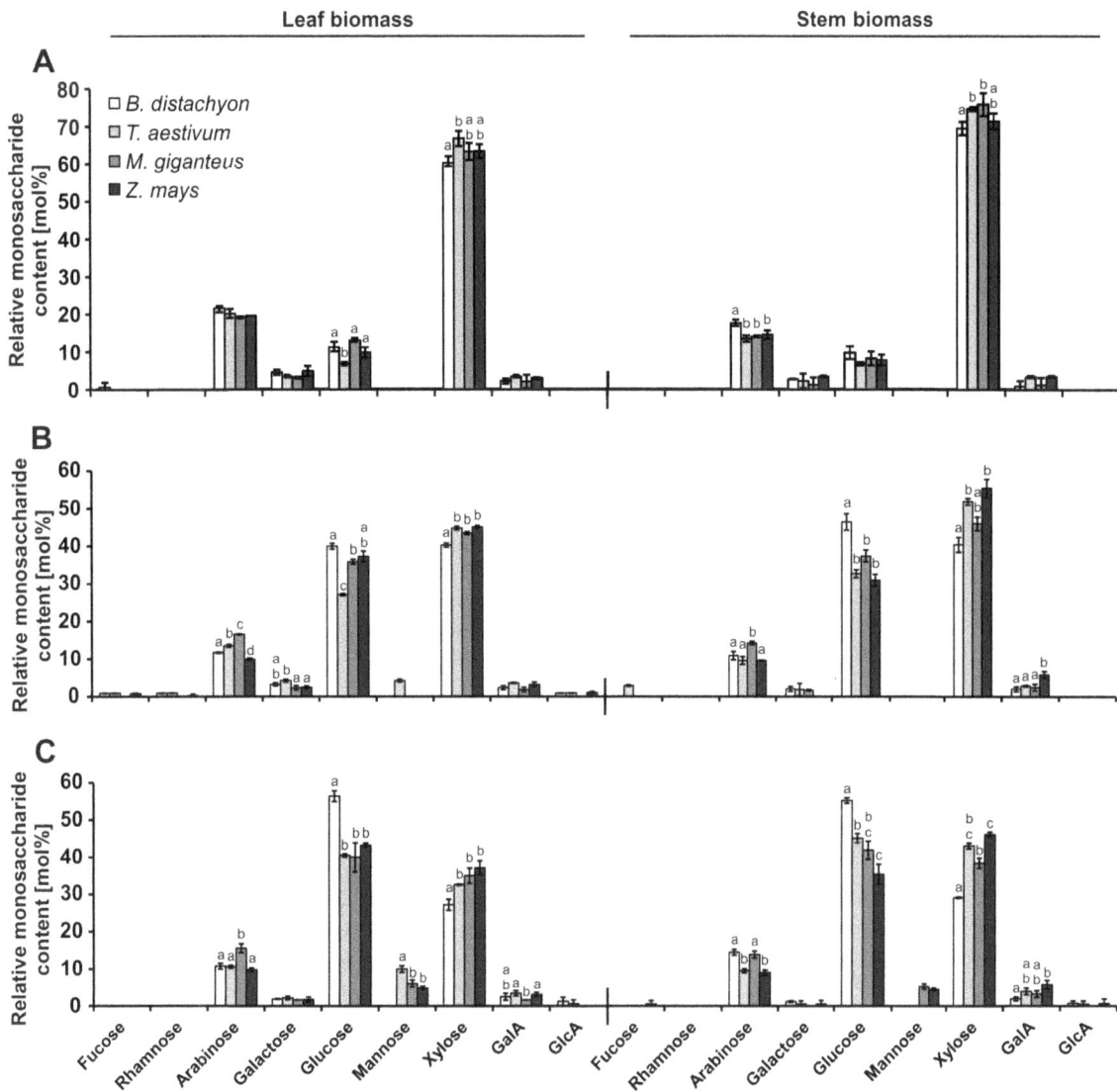

Figure 3. Relative non-cellulosic monosaccharide composition of leaf and stem biomass. Cell wall extracts from senesced, dry, and milled leaf and stem biomass samples were used. Non-cellulosic monosaccharide composition determined by HPAEC-PAD from (**A**) untreated biomass, (**B**) biomass autoclaved in diluted sulfuric acid (1.75% (v/v)), and (**C**) biomass autoclaved in diluted sulfuric acid with subsequent enzymatic hydrolysis (Accellerase 1500 enzyme mixture). Letters a, b, c, and d indicate groups with $P<0.05$ by Tukey's test. Error bars represent SE, and $n = 3$. GalA: galacturonic acid, GlcA: glucuronic acid.

most abundant polymer (Fig. 2). In this regard, a decrease in green fluorescence during pretreatment procedures as well as subsequent fermentation would not only indicate the hydrolysis of cellulose but also a reduction of cross-linked polysaccharides. The interdependency of these two processes is reflected by the fact that ester-linked ferulic acids at arabinose side chains of arabinoxylan are involved in the cross-linking of these cell wall polysaccharides and in forming ferulate-polysaccharide-lignin complexes that support cell wall cross-linking [62]. Therefore, the emitted green fluorescence can be seen as a marker for the recalcitrance of cell wall material to saccharification [63]. The parallel excitation of the plant cell wall material with 561 nm DPSS laser induced the emission of red-shifted fluorescence from *p*-hydroxyphenyl, guiacyl, and syringyl families of lignins with an accumulated wavelength of about 600 nm [35]. Manual selection of regions used for spectral analysis facilitated a precise identification of cell wall parts in the heterogeneous cell wall suspension,

which minimized the interference of the measurement by background auto-fluorescence of the fermentation broth (Fig. 4). As expected from biomass without pretreatment, the leaf and stem cell walls of all plant species revealed a strong green auto-fluorescence peaking at about 500 nm, which indicated a relatively high content of cellulose and other carbohydrates as well as ferulic acid in cell walls (Fig. 4). The absolute strength of the green fluorescence overlaid the red fluorescence deriving from excited lignin, which was detected in parallel and indicated by a small shoulder formation at 600 nm mainly visible in leaf biomass samples (Fig. 4). After thermo-chemical pretreatment with diluted sulfuric acid, the predominant auto-fluorescence of the leaf cell walls from *B. distachyon*, *T. aestivum*, and *Z. mays* as well as *T. aestivum* stem cell walls was still green and peaked at about 500 nm. Hence, the efficiency of saccharification of this biomass was slightly lower compared to stem cell walls from *B. distachyon*, *Z. mays*, and especially *M.* x *giganteus*, where also leaf cell walls revealed a second peak at about 600 nm after spectral analysis (Fig. 5). This peak revealed a higher relative content of lignin due to hydrolysis of carbohydrates. Interestingly, the recalcitrance of leaf biomass was slightly higher than of stem biomass after thermo-chemical pretreatment. Based on previous reports [64,65], we expected a higher recalcitrance of stem biomass, which would be mainly caused by a higher lignin content of stem biomass compared to leaf biomass (Fig. 2A). Therefore, other factors that determine biomass recalcitrance have to be considered. In this regard, the slightly higher hemicellulose content of leaf biomass samples than of stem biomass samples (Fig. 2C) might be involved as the content of hemicellulose affects cell wall digestibility due to a putatively higher cross-linking [62]. The combined biomass pretreatment with an initial diluted sulfuric acid and a subsequent enzymatic application resulted in a clearly improved saccharifica-tion of the cell walls. Auto-fluorescence from lignin components peaked at about 600 nm for almost all cell wall samples except the leaf and stem samples from *M.* x *giganteus* and the stem samples from *B. distachyon*, where a second, weaker but distinct 500 nm peak was present (Fig. 6), which indicated a relatively high carbohydrate content of the cell walls. Whereas increased recalcitrance of *M.* x *giganteus* biomass correlated with a relatively high lignin content in leaf and stem samples (Fig. 2A), we could not attribute a high lignin content to the relatively high recalcitrance of *B. distachyon* stem biomass because it was lowest in this measurement (Fig. 2A). Also the hemicellulose content, which can affect cell wall digestibility [62], was not significantly different from *T. aestivum* or *Z. mays* stem biomass that did not reveal an increased recalcitrance. This may indicate that cellulose crystallinity and/or ferulic acid-mediated cross-linking could be altered in *B. distachyon* stem cell walls. To analyze whether the observed recalcitrance of the leaf and stem cell walls from *M.* x *giganteus* and the stem cell walls from *B. distachyon* would be conserved during fermentation, we performed a lambda scan of sulfuric acid and enzymatic pretreated cell walls 72 h after start of fermentation. Whereas the degree of saccharification for the *B. distachyon* stem sample was comparable to the *B. distachyon* leaf sample and all *T. aestivum* and *Z. mays* samples, the *M.* x *giganteus* samples were more recalcitrant to carbohydrate hydrolysis compared to all other samples because spectral analysis still revealed a relatively strong 500 nm peak (Fig. S1). In general, the effectiveness of saccharification is not only dependent on the lignin content of the biomass but also on the level of hemicelluloses and cellulose. Comparing the cellulose und hemicelluloses level in our study with those reported by Xu et. al. [66], the dried *M.* x *giganteus* biomass that we used in this study revealed most similarities to those *Miscanthus* accessions that gave the lowest total sugar yield after diluted sulfuric acid biomass pretreatment with subsequent enzymatic hydrolysis. This might explain the relatively high degree of recalcitrance of *M.* x *giganteus* cell walls as indicated by the relatively strong emission of green fluorescence in confocal lambda scanning (Fig. 6).

Determination of glucose content and ethanol production during fermentation

The quantification of the glucose content via HPLC in the pre-fermentation broth after biomass pretreatment resembled the observations that we previously made during lambda scanning. The degree of saccharification, which was indicated by a shift of the cell wall's auto-fluorescence from green to red (Figs. 4 and 6), correlated with the amount of released glucose; a low efficiency of saccharification and a low amount of released glucose after diluted sulfuric acid pretreatment (Figs. 5 and 7A), and an efficient saccharification with a high glucose concentration of the pre-

fermentation broth after combined sulfuric acid and enzymatic biomass pretreatment (Figs. 6 and 7C). We also observed a correlation between the results of the spectral cell wall analysis and the HPLC analysis of glucose deriving from biomass that was only pretreated with diluted sulfuric acid. Those samples that showed the strongest shift of the auto-fluorescence spectrum from green to red also revealed the highest amount of released glucose. This was most prominent for stem biomass where *M.* x *giganteus* and *Z. mays* showed the most distinct green to red shift in the spectral cell wall analysis (Fig. 5), which resembled the three-fold and two-fold higher amount of hydrolyzed glucose compared to *B. distachyon* and *T. aestivum* (Fig. 5A; Tab. S1). Consequently, we detected a three-fold higher ethanol production for *M.* x *giganteus* and a two-fold higher ethanol production for *Z. mays* stem biomass compared to *B. distachyon* and *T. aestivum* (Fig. 7B; Tab. S1). Unlike stem biomass after diluted sulfuric acid treatment, we detected glucose in supernatants of the fermentation broth from leaf biomass even 48 h and 72 h after the start of fermentation (Fig. 7A). This indicated that the fermentation could be inhibited by compounds or cell wall components that might be only found in the leaf but not the stem where we did not detected an increased inhibition of fermentation. In general, furans formed by carbohy-drate hydrolysis and phenolic monomers, which derive from lignin degradation, are known inhibitors of biomass fermentation with *S. cerevisiae* [67]. Apart from biochemical approaches to reduce fermentation inhibitors [68] or the generation of *S. cerevisiae* strains with improved resistance to inhibitors [69], different process technologies like continuous fermentation have been discussed to reduce the problem of inhibition [70]. The increase in the released glucose amount of sulfuric acid pretreated leaf biomass from *T. aestivum* between 0 h and 48 h after the start of fermentation (Fig. 7A) could indicate an ongoing saccharification process. Under these pretreatment conditions, the stem biomass of the C_4 grasses *M.* x *giganteus* and *Z. mays* revealed the best ethanol production performance (Fig. 5), which was mainly based in relatively high glucose yield of 11% and 8% of the total biomass, respectively (Tab. S1). The differences in saccharification efficien-cy and ethanol production, which we observed with biomass that was only pretreated with diluted sulfuric acid, were almost equalized when an additional enzymatic pretreatment followed the initial acid pretreatment. On average, results from leaf biomass revealed a four- to seven fold higher biomass to glucose conversion efficiency due to enzymatic hydrolysis for the C_3 grasses and a three- to five-fold higher conversion efficiency for the C_4 grasses (Fig. 7C; Tab. S1). The increased amount of produced ethanol (three-fold for *B. distachyon*, ten-fold for *T. aestivum*, and almost six-fold for *M.* x *giganteus* and *Z. mays*) resembled the improved saccharification efficiency (Tab. S1). Interestingly, we did not detect glucose in the supernatant of the fermentation broth from leaf biomass with combined acid and enzymatic pretreatment (Fig. 7C) as we did after single acid treatment (Fig. 7A). This not only indicated a more efficient hexose release due to the pretreatment before fermentation but probably also less putative fermentation inhibitors. Hence, a comparative compound analysis of the differentially pretreated leaf biomass might support the identification of fermentation inhibitors, especially from *T. aestivum* leaf biomass, where our data suggest a higher level of inhibitors than in the other three plant species. Our observation that the enzyme mixture may reduce the concentration of putative fermentation inhibitors would support their identification. Putative fermentation inhibitors might be tested for the susceptibility to degradation by the enzyme mixture used for saccharification. Comparing leaf and stem biomass after combined acid and enzymatic pretreatment, we did not determine major differences

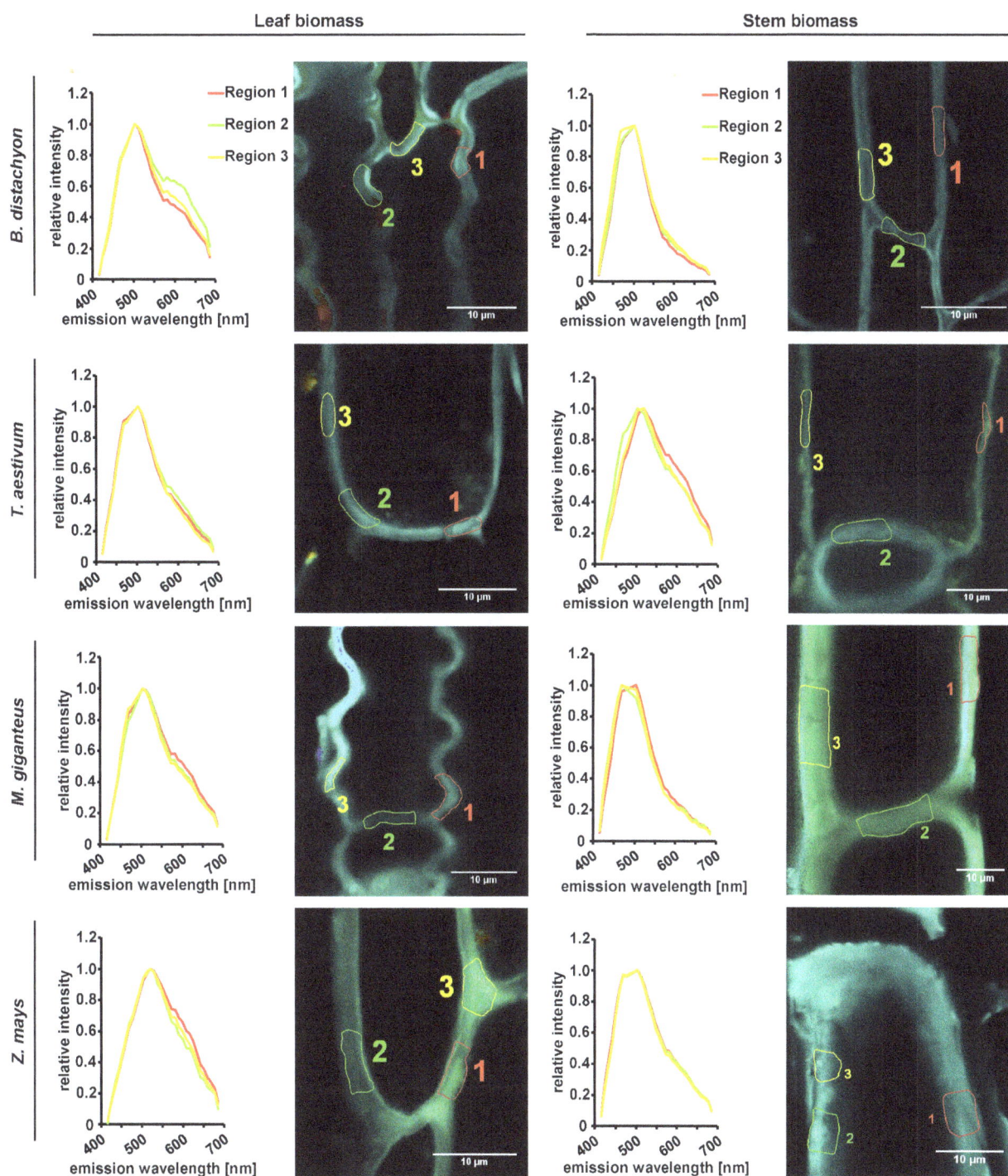

Figure 4. Rapid qualitative analysis of carbohydrate/lignin ratio in plant cell walls of untreated biomass by confocal lambda scanning. Cell wall particles from senesced, dry, and milled leaf and stem biomass samples were used for lambda scanning with a confocal laser-scanning microscope. Three defined cell wall regions in each leaf and stem sample were manually selected for measurement of emission spectra. Green fluorescence emitted from ferulic acid, cellulose, and additional carbohydrates, red fluorescence indicative for high lignin content. Micrographs are representative for each sample after evaluating at least five independent replicates.

in saccharification or ethanol production (Fig. 7C,D). Within stem biomass samples, only *B. distachyon* stem biomass showed an increased amount of released glucose, which was significantly higher than in *Z. mays*, but did not result in a significantly higher ethanol production (Fig. 7D). The amounts of produced ethanol

from *M.* x *giganteus* of 105–110 mg·g^{-1} dry biomass that we determined for diluted acid and enzymatically pretreated biomass (Tab. S1) were in the range of ethanol amounts (100–110 mg·g^{-1} dry biomass) that were yielded in previous *M.* x *giganteus* studies where different methods of biomass pretreatment were applied,

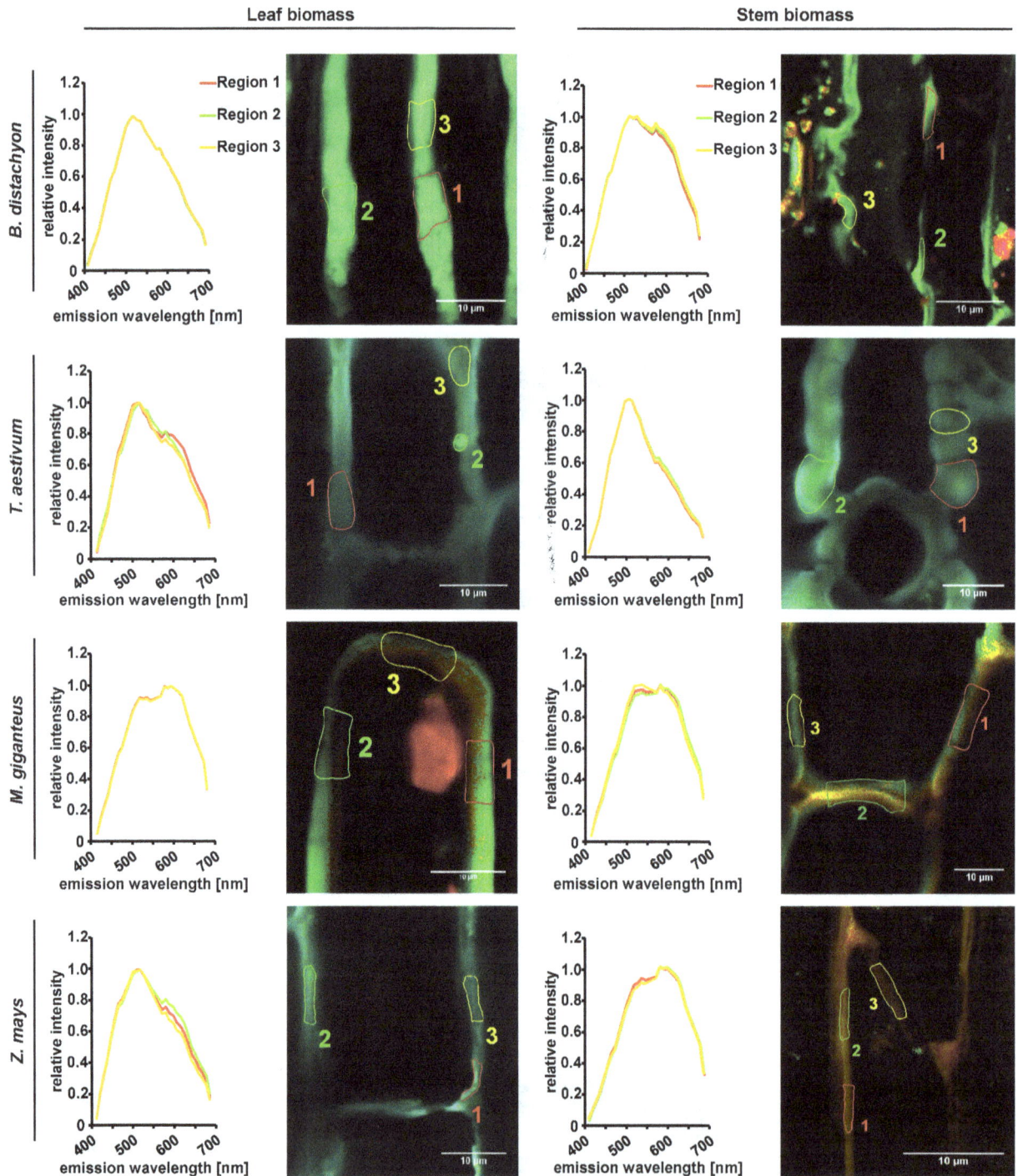

Figure 5. Rapid qualitative analysis of carbohydrate/lignin ratio in plant cell walls of thermo-chemical pretreated biomass by confocal lambda scanning. Cell wall particles from senesced, dry, and milled leaf and stem biomass samples were autoclaved in diluted sulfuric acid (1.75% (v/v)) before used for lambda scanning with a confocal laser-scanning microscope. Three defined cell wall regions in each leaf and stem sample were manually selected for measurement of emission spectra. Green fluorescence emitted from ferulic acid, cellulose, and additional carbohydrates, red fluorescence indicative for high lignin content. Micrographs are representative for each sample after evaluating at least five independent replicates.

like the usage of ammonia fiber expansion or the ethanol organosolv process [71–73]. For *Z. mays* biomass, our ethanol yields after diluted acid and enzymatic pretreatment of 106–109 mg ethanol·g^{-1} dry biomass outcompeted the ethanol yields of 20–100 mg·g^{-1} dry biomass as reported in previous studies with alternative pretreatment methods [74–77]. Only for *T. aestivum*, we yielded less ethanol than in a comparable study with a dilute acid and enzymatic pretreatment [46]. The higher ethanol

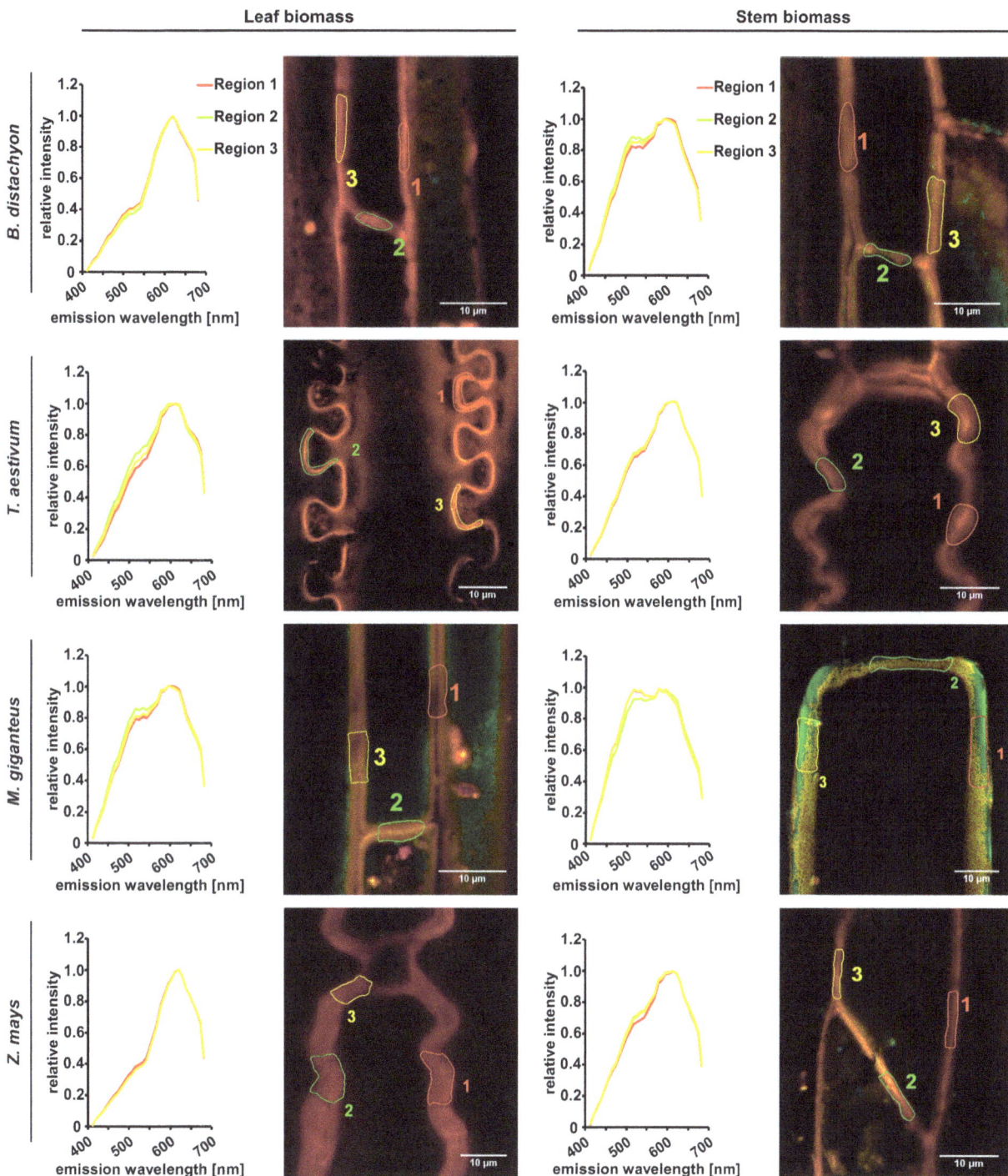

Figure 6. Rapid qualitative analysis of carbohydrate/lignin ratio in plant cell walls of thermo-chemical pretreated and hydrolyzed biomass by confocal lambda scanning. Cell wall particles from senesced, dry, and milled leaf and stem biomass samples were autoclaved in diluted sulfuric acid (1.75% (v/v)) with a subsequent enzymatic hydrolysis (Accellerase 1500 enzyme mixture) before used for lambda scanning with a confocal laser-scanning microscope. Three defined cell wall regions in each leaf and stem sample were manually selected for measurement of emission spectra. Green fluorescence emitted from ferulic acid, cellulose, and additional carbohydrates, red fluorescence indicative for high lignin content. Micrographs are representative for each sample after evaluating at least five independent replicates.

yield in that study could be explained by the usage of a recombinant *Escherichia coli* strain and overliming of the biomass hydrolyzates, which seems to be more effective in *T. aestivum* biomass fermentation than the usage of *S. cerevisiae* without overliming.

Principal component analysis

To evaluate and summarize the potential of *B. distachyon* as model plant for the production of ethanol by fermentation of senesced and dried leaf and stem biomass, we performed separated two-factored principal component analyses (PCA) of stem and leaf

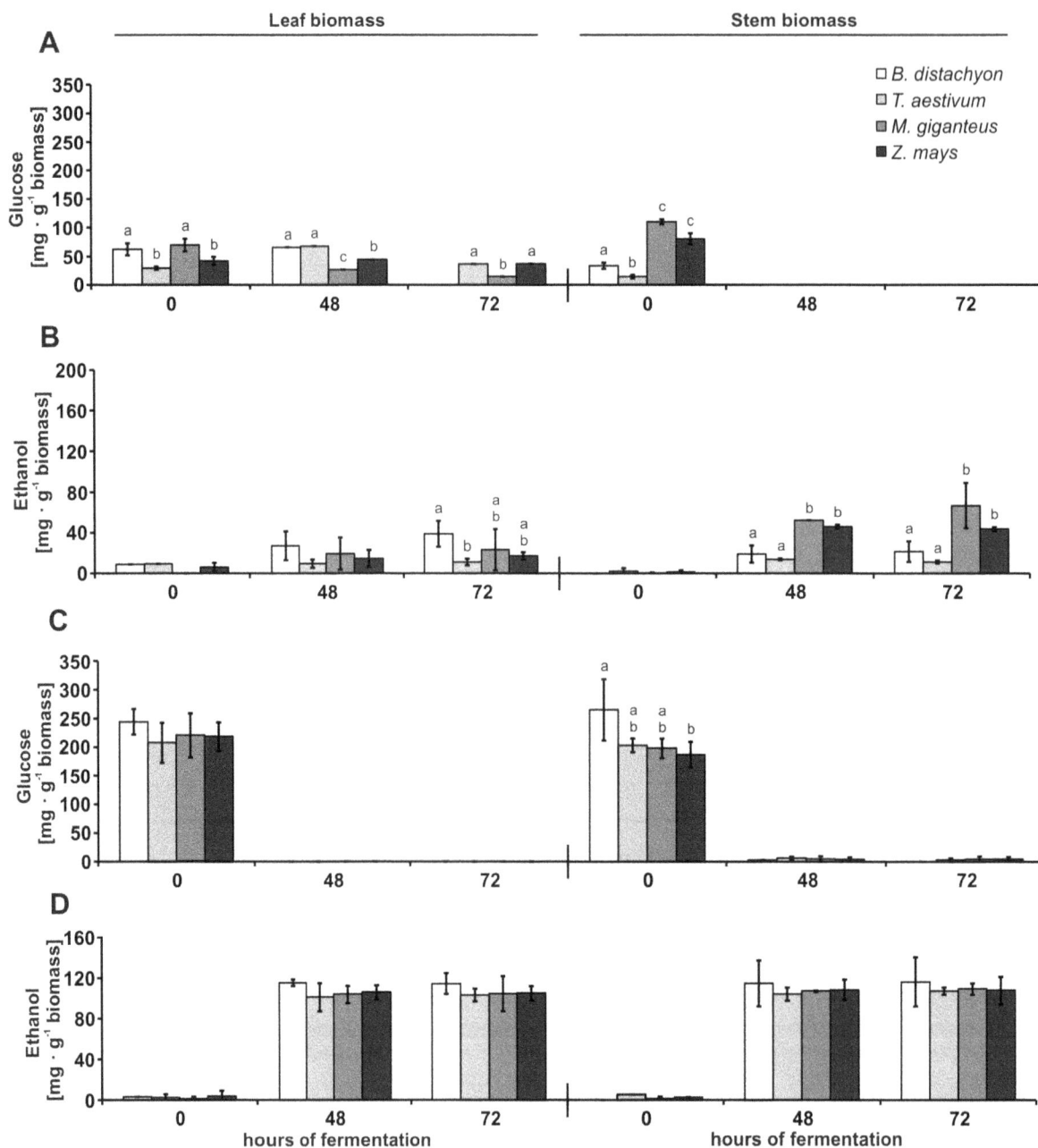

Figure 7. Glucose and ethanol concentration during fermentation of leaf and stem biomass. Glucose and ethanol concentrations in supernatants of the leaf and stem fermentation broth were determined by HPLC using a refractive index detector. Samples taken at the start of fermentation with *S. cerevisiae* and after 48 h and 72 h. (**A**) Glucose concentration and (**B**) ethanol concentration of biomass autoclaved in diluted sulfuric acid (1.75% (v/v)); (**C**) glucose concentration and (**D**) ethanol concentration of biomass autoclaved in diluted sulfuric acid with subsequent enzymatic hydrolysis (Accellerase 1500 enzyme mixture). Letters a, b, and c indicate groups with $P<0.05$ by Tukey's test. Error bars represent SE, and $n = 3$.

biomass with a thermo-chemical diluted sulfuric acid pretreatment and the combined pretreatment with diluted sulfuric acid followed by enzymatic hydrolysis. The calculation of the PCA and factor analysis was based on six different parameters that comprised cell wall composition, saccharification efficiency, and ethanol production from pretreated biomass of *B. distachyon*, *T. aestivum*, *M.* x *giganteus*, and *Z. mays*. A parallel analysis of multiple parameters would allow a more precise prediction of possible similarities than an evaluation of a single parameter or simple combinations as shown in *Z. mays* [78]. The PCA revealed that *B. distachyon* is

most suitable as model for the process of lignocellulosic ethanol production from stem biomass of the tested *T. aestivum* line after thermo-chemical biomass pretreatment. Previously, *B. distachyon* was also described as a suitable model to study modifications of arabinoxylan structure and biosynthesis that could be applied on monocot energy crops for saccharification improvements [79]. However, based on our PCA results, *B. distachyon* would neither qualify as a good model for the process of lignocellulosic ethanol production from stem biomass of the tested crop lines after combined thermo-chemical and enzymatic biomass pretreatment

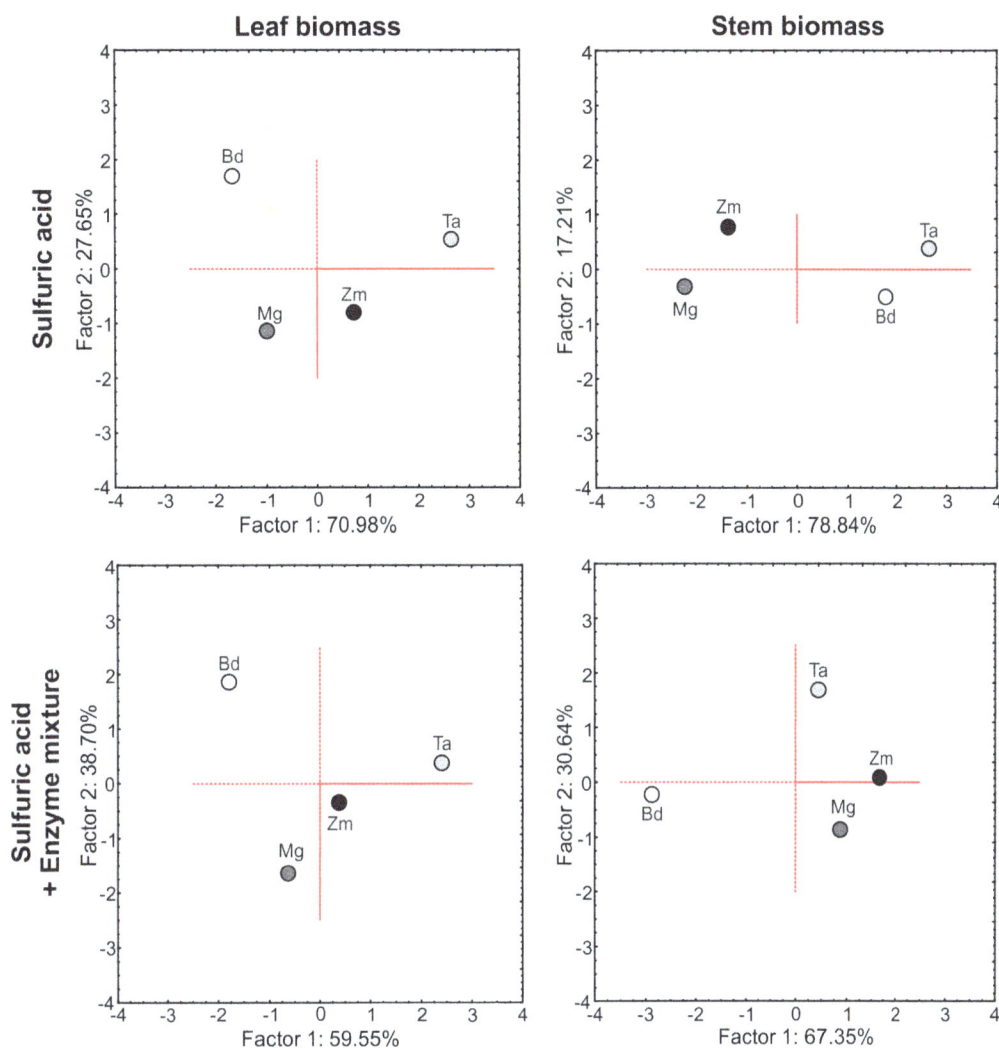

Figure 8. Principal components analysis of leaf and stem biomass after different pretreatment methods. Two-factor principal component analysis on the basis of six parameters defining efficiency of lignocellulosic ethanol production. Bd: *B. distachyon*, Mg: *M. x giganteus*, Ta: *T. aestivum*, Zm: *Z. mays*.

nor as a good model for lignocellulosic ethanol production from leaf biomass independent of the applied method for biomass pretreatment (Fig. 8). A constant, direct correlation between phylogenetic relation and similarity in lignocellulosic ethanol production was only found for the C_4 grasses *M. x giganteus* and *Z. mays*, which was independent of the applied biomass pretreatment method and type of biomass (Fig. 8). In contrast to its phylogenetic relation, *T. aestivum* revealed a higher similarity in lignocellulosic ethanol production to the C_4 grasses *M. x giganteus* and *Z. mays* than to the C_3 model grass *B. distachyon* especially in the two scenarios with a combined thermo-chemical and enzymatic biomass pretreatment (Fig. 8). This may reflect previous observations where even within a plant species, *Z. mays*, strong differences in cell wall composition and saccharification were reported from different cultivars [45]. Therefore, it has to be generally considered that phylogenetic relation of plant species does not directly indicate a good suitability as model for lignocellulosic ethanol production. The PCA analysis also showed that the degree of similarity in lignocellulosic ethanol production can be altered by the application of different biomass pretreatment methods. This was most obvious for stem biomass of *T. aestivum*

where the combined acidic and enzymatic pretreatment increased similarity in ethanol production to *Z. mays* and *M. x giganteus* but broke the relatively high similarity to *B. distachyon* compared to a single acidic biomass pretreatment (Fig. 8). For stem biomass, we also determined a difference of the principal component (PC) loadings associated with the PCA after different types of biomass pretreatment. Whereas the variable "cellulose amount" contributed most to the separation of factor 1 and "lignin amount" to factor 2 after single acidic biomass pretreatment, the PC loading pattern changed to "ethanol production" for factor 1 and "cellulose amount" for factor 2 after combined acidic and enzymatic pretreatment (Tab. S2). For leaf biomass, "cellulose amount" was the variable with the highest contribution to the separation of factor 1 and "ethanol production" of factor 2, which was independent of the applied biomass pretreatment method (Tab. S2). Except for stem biomass after single acidic pretreatment, the variable "glucose amount from hemicellulose fraction" revealed a constantly low contribution to the separation of the two factors associated with the PCA.

Our PCA results suggest that predictions for the suitability of model plant for lignocellulosic ethanol production can strongly differ among species, tissue type and applied pretreatment method. Similar observations were previously made for possible predictions of cell wall digestibility in *P. virgatum* and *Andropogon gerardii* (big bluestem) [65].

In general, PCA is a method that can provide a fast an easy overview of the suitability of a plant species to be considered as model for another plant species. This kind of analysis can easily be extended with additional plant species and parameters, or plant species parameters can be modified, replaced or reduced if the focus of comparative analysis is on different traits than saccharification and ethanol production from leaf and stem biomass as in our study of the four plant species.

Conclusions

The suitability of *B. distachyon* as model plant for lignocellulosic ethanol production from leaf and stem biomass was analyzed in comparative study with the two major crops from temperate climate, *T. aestivum* and *Z. mays*, and the emerging energy crop *M.* x *giganteus*. The model qualities of *B. distachyon* were best for stem biomass from *T. aestivum* after thermo-chemical biomass pretreatment whereas model qualities for leaf biomass were only limited for all tested crop lines. The ethanol yield from stem biomass ranging from 108 mg·g^{-1} dry biomass for *T. aestivum* to 117 mg·g^{-1} dry biomass for *B. distachyon* after combined thermo-chemical and enzymatic pretreatment was generally higher than from leaf biomass for all tested plant species. The results revealed that phylogenetic relation does not automatically reflect suitability as model for biomass conversion as the C$_3$ grass *B. distachyon* offered promising qualities as model plant for biomass from the phylogenetically related crop *T. aestivum* only in one out of four approaches.

Supporting Information

Figure S1 Rapid qualitative analysis of carbohydrate/ lignin ratio in plant cell walls of thermo-chemical

pretreated and hydrolyzed biomass after fermentation by confocal lambda scanning. Cell wall particles from senesced, dry, and milled leaf and stem biomass samples were autoclaved in diluted sulfuric acid (1.75% (v/v)) with a subsequent enzymatic hydrolysis (Accellerase 1500 enzyme mixture). After 72 h of fermentation, remaining cell wall particles were used for lambda scanning with a confocal laser-scanning microscope. Three defined cell wall regions in each leaf and stem sample were manually selected for measurement of emission spectra. Green fluorescence emitted from ferulic acid, cellulose, and additional carbohydrates, red fluorescence indicative for high lignin content. Micrographs are representative for each sample after evaluating at least five independent replicates.

Table S1 Overview of glucose saccharification and ethanol production of leaf and stem biomass with different pretreatment methods. Biomass pretreatment: acid, biomass autoclaved in diluted sulfuric acid (1.75% (v/v)); acid+enzymes, biomass autoclaved in diluted sulfuric acid (1.75% (v/v) with subsequent enzymatic hydrolysis (Accellerase 1500 enzyme mixture). Values in brackets represent SE, and $n = 3$.

Table S2 Principal component (PC) loadings associated with the PC analysis. PC loadings indicate importance of each variable in accounting for the variability of factor 1 and 2. Biomass pretreatment: sulfuric acid, biomass autoclaved in diluted sulfuric acid (1.75% (v/v)); sulfuric acid+enzyme mixture, biomass autoclaved in diluted sulfuric acid (1.75% (v/v) with subsequent enzymatic hydrolysis (Accellerase 1500 enzyme mixture).

Author Contributions

Conceived and designed the experiments: CAV CM TM. Performed the experiments: CM TM. Analyzed the data: CAV CM TM. Contributed to the writing of the manuscript: CAV.

References

1. Ash C, Culotta E, Fahrenkamp-Uppenbrink J, Malakoff D, Smith J, et al. (2013) Natural systems in changing climates. Once and future climate change. Introduction. Science 341: 472–473.
2. Finer M, Jenkins CN, Pimm SL, Keane B, Ross C (2008) Oil and gas projects in the Western Amazon: threats to wilderness, biodiversity, and indigenous peoples. PLoS One 3: e2932.
3. Jernelov A (2010) The threats from oil spills: now, then, and in the future. AMBIO 39: 353–366.
4. Sojinu OS, Wang JZ, Sonibare OO, Zeng EY (2010) Polycyclic aromatic hydrocarbons in sediments and soils from oil exploration areas of the Niger Delta, Nigeria. J Hazard Mater 174: 641–647.
5. Sims RE, Mabee W, Saddler JN, Taylor M (2010) An overview of second generation biofuel technologies. Bioresour Technol 101: 1570–1580.
6. FAOSTAT Gateway (2014) Statistics Division of the Food and Agricultural Organization of the United Nations. Database: FAOSTAT. Available: http://faostat3.fao.org/faostat-gateway/go/to/home/E. Accessed 20 March 2014.
7. Somerville C, Youngs H, Taylor C, Davis SC, Long SP (2010) Feedstocks for lignocellulosic biofuels. Science 329: 790–792.
8. Beale CV, Long SP (1997) Seasonal dynamics of nutrient accumulation and partitioning in the perennial C4-grasses *Miscanthus* x *giganteus* and *Spartina cynosuroides*. Biomass Bioenerg 12: 419–428.
9. Hohenstein WG, Wright LL (1994) Biomass energy production in the United States: an overview. Biomass Bioenerg 6: 161–173.
10. Hughes JK, Lloyd AJ, Huntingford C, Finch JW, Harding RJ (2010) The impact of extensive planting of *Miscanthus* as an energy crop on future CO$_2$ atmospheric concentrations. Glob Chang Biol Bioenerg 2: 79–88.
11. Roth AM, Sample DW, Ribic CA, Paine L, Undersander DJ, et al. (2005) Grassland bird response to harvesting switchgrass as a biomass energy crop. Biomass Bioenerg 28: 490–498.
12. van der Weijde T, Alvim Kamei CL, Torres AF, Vermerris W, Dolstra O, et al. (2013) The potential of C$_4$ grasses for cellulosic biofuel production. Front Plant Sci 4: 107.
13. Dohleman FG, Long SP (2009) More productive than maize in the Midwest: How does *Miscanthus* do it? Plant Physiol 150: 2104–2115.
14. Heaton EA, Dohleman FG, Long SP (2008) Meeting US biofuel goals with less land: the potential of *Miscanthus*. Glob Chang Biol 14: 2000–2014.
15. van Hulle S, van Waes C, de Vliegher A, Baert J, Muylle H. (2012) Comparison of dry matter yield of lignocellulosic perennial energy crops in a long-term Belgian field experiment. In: Golinski P, Warda M, Stypinski P, editors. Grassland - a European Resource?. Zurich: European Grassland Federation. pp. 499–501.
16. Hall M, Bansal P, Lee JH, Realff MJ, Bommarius AS (2010) Cellulose crystallinity–a key predictor of the enzymatic hydrolysis rate. FEBS J 277: 1571–1582.
17. Mansfield SD, Mooney C, Saddler JN (1999) Substrate and enzyme characteristics that limit cellulose hydrolysis. Biotechnol Prog 15: 804–816.
18. Puri VP (1984) Effect of crystallinity and degree of polymerization of cellulose on enzymatic saccharification. Biotechnol Bioeng 26: 1219–1222.
19. Chen F, Dixon RA (2007) Lignin modification improves fermentable sugar yields for biofuel production. Nat Biotechnol 25: 759–761.
20. Yoshida M, Liu Y, Uchida S, Kawarada K, Ukagami Y, et al. (2008) Effects of cellulose crystallinity, hemicellulose, and lignin on the enzymatic hydrolysis of *Miscanthus sinensis* to monosaccharides. Biosci Biotechnol Biochem 72: 805–810.
21. Alves SC, Worland B, Thole V, Snape JW, Bevan MW, et al. (2009) A protocol for Agrobacterium-mediated transformation of *Brachypodium distachyon* community standard line Bd21. Nat Protoc 4: 638–649.
22. Bevan MW, Garvin DF, Vogel JP (2010) *Brachypodium distachyon* genomics for sustainable food and fuel production. Curr Opin Biotechnol 21: 211–217.
23. Draper J, Mur LAJ, Jenkins G, Ghosh-Biswas GC, Bablak P, et al. (2001) *Brachypodium distachyon*. A new model system for functional genomics in grasses. Plant Physiol 127: 1539–1555.
24. International Brachypodium I (2010) Genome sequencing and analysis of the model grass *Brachypodium distachyon*. Nature 463: 763–768.

25. Gomez LD, Bristow JK, Statham ER, McQueen-Mason SJ (2008) Analysis of saccharification in *Brachypodium distachyon* stems under mild conditions of hydrolysis. Biotechnol Biofuels 1: 15.

26. Vogel JP, Gu YQ, Twigg P, Lazo GR, Laudencia-Chingcuanco D, et al. (2006) EST sequencing and phylogenetic analysis of the model grass *Brachypodium distachyon*. Theor Appl Genet 113: 186–195.

27. Green CE, Philips RL (1975) Plant regeneration from tissue cultures of maize. Crop Sci 15: 471–421.

28. Arora R, Manisseri C, Li C, Ong MD, Scheller HV, et al. (2010) Monitoring and analyzing process streams towards understanding ionic liquid pretreatment of switchgrass (*Panicum virgatum* L.). Bioenergy Res 3: 134–145.

29. Ellinger D, Naumann M, Falter C, Zwikowics C, Jamrow T, et al. (2013) Elevated early callose deposition results in complete penetration resistance to powdery mildew in Arabidopsis. Plant Physiol 161: 1433–1444.

30. Fukushima RS, Hatfield RD (2004) Comparison of the acetyl bromide spectrophotometric method with other analytical lignin methods for determining lignin concentration in forage samples. J Agric Food Chem 52: 3713–3720.

31. Lygin AV, Upton J, Dohleman FG, Juvik J, Zabotina OA, et al. (2011) Composition of cell wall phenolics and polysaccharides of the potential bioenergy crop –*Miscanthus*. Glob Chang Biol Bioenerg 3: 333–345.

32. DuBois M, Gilles DE, Hamilton JK, Rebers PA, Smith F (1956) Colorimetric method for the determination of sugars and related substances. Anal Chem 28: 350–356.

33. Fanta GF, Abbott TP, Herman AI, Burr RC, Doane WM (1984) Hydrolysis of wheat straw hemicellulose with trifluoroacetic acid. Fermentation of xylose with *Pachysolen tannophilus*. Biotechnol Bioeng 26: 1122–1125.

34. Bonn G, Bobleter O (1984) HPLC-analyses of plant biomass hydrolysis and fermentation solutions. Chromatographia 18: 445–448.

35. Singh S, Simmons BA, Vogel KP (2009) Visualization of biomass solubilization and cellulose regeneration during ionic liquid pretreatment of switchgrass. Biotechnol Bioeng 104: 68–75.

36. Sievers F, Wilm A, Dineen D, Gibson TJ, Karplus K, et al. (2011) Fast, scalable generation of high-quality protein multiple sequence alignments using Clustal Omega. Mol Syst Biol 7: 539.

37. Kelchner S (2000) The evolution of non-coding chloroplast DNA and its application in plant systematics. Ann Missouri Bot Gard 87: 182–198.

38. Worley B, Powers R (2013) Multivariate analysis in metabolomics. Curr Metabolomics 1: 92–107.

39. Swaminathan K, Alabady MS, Varala K, De Paoli E, Ho I, et al. (2010) Genomic and small RNA sequencing of *Miscanthus x giganteus* shows the utility of sorghum as a reference genome sequence for Andropogoneae grasses. Genome Biol 11: R12.

40. Peng Z, Lu T, Li L, Liu X, Gao Z, et al. (2010) Genome-wide characterization of the biggest grass, bamboo, based on 10,608 putative full-length cDNA sequences. BMC Plant Biol 10: 116.

41. Wolfe KH, Gouy M, Yang YW, Sharp PM, Li WH (1989) Date of the monocot-dicot divergence estimated from chloroplast DNA sequence data. Proc Natl Acad Sci U S A 86: 6201–6205.

42. Pauly M, Keegstra K (2008) Cell-wall carbohydrates and their modification as a resource for biofuels. Plant J 54: 559–568.

43. Rancour DM, Marita JM, Hatfield RD (2012) Cell wall composition throughout development for the model grass *Brachypodium distachyon*. Front Plant Sci 3: 266.

44. Jung HG, Casler MD (2006) Maize stem tissues Crop Sci 46: 1793–1800.

45. Méchin V, Argillier O, Menanteau V, Barrière Y, Mila I, et al. (2000) Relationship of cell wall composition to *in vitro* cell wall digestibility of maize inbred line stems. J Sci Food Agr 80: 574–580.

46. Saha BC, Iten LB, Cotta MA, Wu V (2005) Dilute acid pretreatment, enzymatic saccharification and fermentation of wheat straw to ethanol. Process Biochem 40: 3693–7000.

47. Jung JH, Fouad WM, Vermerris W, Gallo M, Altpeter F (2012) RNAi suppression of lignin biosynthesis in sugarcane reduces recalcitrance for biofuel production from lignocellulosic biomass. Plant Biotechnol J 10: 1067–1076.

48. Wang H, Xue Y, Chen Y, Li R, Wei J (2012) Lignin modification improves the biofuel production potential in transgenic *Populus tomentosa*. Ind Crop Prod 37: 170–177.

49. Vance CP, Kirk TK, Sherwood RT (1980) Lignification as a mechanism of disease resistance. Ann Rev Phytopathol 18: 259–288.

50. Gaillard BDE (1965) Comparison of the hemicelluloses from plants belonging to two different plant families. Phytochemistry 4: 631–634.

51. Abedon BG, Hatfield RD, Tracy WF (2006) Cell wall composition in juvenile and adult leaves of maize (*Zea mays* L.). J Agric Food Chem 54: 3896–3900.

52. Le Ngoc Huyen T, Remond C, Dheilly RM, Chabbert B (2010) Effect of harvesting date on the composition and saccharification of *Miscanthus x giganteus*. Bioresour Technol 101: 8224–8231.

53. Morrison TA, Jung HG, Buxton DR, Hatfield RD (1998) Cell-wall composition of maize internodes of varying maturity. Crop Sci 38: 455–460.

54. Pu Y, Hu F, Huang F, Davison BH, Ragauskas AJ (2013) Assessing the molecular structure basis for biomass recalcitrance during dilute acid and hydrothermal pretreatments. Biotechnol Biofuels 6: 15.

55. Kumar L, Chandra R, Chung PA, Saddler J (2010) Can the same steam pretreatment conditions be used for most softwoods to achieve good, enzymatic hydrolysis and sugar yields? Bioresour Technol 101: 7827–7833.

56. Kumar L, Chandra R, Saddler J (2011) Influence of steam pretreatment severity on post-treatments used to enhance the enzymatic hydrolysis of pretreated softwoods at low enzyme loadings. Biotechnol Bioeng 108: 2300–2311.

57. Marzialetti T, Olarte MBV, Sievers C, Hoskins TJC, Agrawal PK, et al. (2008) Dilute acid hydrolysis of Loblolly pine: A comprehensive approach. Ind Eng Chem Res 47: 7131–7140.

58. Harris PJ, Hartley RD (1976) Detection of bound ferulic acid in cell walls of the Gramineae by ultraviolet fluorescence microscopy. Nature 259: 508–510.

59. Fincher GB (1976) Ferulic acid in barley cell walls: a fluorescenec study. J Inst Brew 82: 347–349.

60. Pöhlker C, Huffman JA, Pöschl U (2012) Autofluorescence of atmospheric bioaerosols – fluorescent biomolecules and potential interferences. Atmos Meas Tech 5: 37–71.

61. Roshchina VV (2012) Vital autofluorescence: application to the study of plant living cells. Int J Spectrosc 2012: ID 124672.

62. de O. Buanafina MM (2009) Feruloylation in grasses: current and future perspectives. Mol Plant 2: 861–872.

63. Himmel ME, Ding SY, Johnson DK, Adney WS, Nimlos MR, et al. (2007) Biomass recalcitrance: engineering plants and enzymes for biofuels production. Science 315: 804–807.

64. Hu Z, Foston M, Ragauskas AJ (2011) Comparative studies on hydrothermal pretreatment and enzymatic saccharification of leaves and internodes of alamo switchgrass. Bioresour Technol 102: 7224–7228.

65. Jung H-JG, Vogel KP (1992) Lignification of switchgrass (*Panicum virgatum*) and big bluestem (*Andropogon gerardii*) plant parts during maturation and its effect on fibre degradability. J Sci Food Agr 59: 169–176.

66. Xu N, Zhang W, Ren S, Liu F, Zhao C, et al. (2012) Hemicelluloses negatively affect lignocellulose crystallinity for high biomass digestibility under NaOH and H_2SO_4 pretreatments in *Miscanthus*. Biotechnol Biofuels 5: 58.

67. Klinke HB, Thomsen AB, Ahring BK (2004) Inhibition of ethanol-producing yeast and bacteria by degradation products produced during pre-treatment of biomass. Appl Microbiol Biotechnol 66: 10–26.

68. Weil JR, Dien B, Bothast R, Hendrickson R, Mosier NS, et al. (2002) Removal of fermentation inhibitors formed during pretreatment of biomass by polymeric adsorbents. Ind Eng Chem Res 41: 6132–6138.

69. Larsson S, Cassland P, Jonsson LJ (2001) Development of a *Saccharomyces cerevisiae* strain with enhanced resistance to phenolic fermentation inhibitors in lignocellulose hydrolysates by heterologous expression of laccase. Appl Environ Microbiol 67: 1163–1170.

70. Palmqvist E, Hahn-Hägerdal B (2000) Fermentation of lignocellulosic hydrolysates. I: inhibition and detoxification. Bioresour Technol 74: 17–24.

71. Brosse N, Sannigrahi P, Ragauskas AJ (2009) Pretreatment of *Miscanthus x giganteus* using the ethanol organosolv process for ethanol production. Ind Eng Chem Res 48: 8328–8334.

72. Murnen HK, Balan V, Chundawat SPS, Bals B, Sousa LC, et al. (2008) Optimization of ammonia fiber expansion (AFEX) pretreatment and enzymatic hydrolysis of *Miscanthus x giganteus* to fermentable sugars. Biotechnol Progr 23: 846–850.

73. Sørensen A, Teller PJ, Hilstrom T, Ahring BK (2008) Hydrolysis of Miscanthus for bioethanol production using dilute acid presoaking combined with wet explosion pre-treatment and enzymatic treatment. Bioresour Technol 99: 6602–6607.

74. Kim S, Holtzapple MT (2005) Lime pretreatment and enzymatic hydrolysis of corn stover. Bioresour Technol 96: 1994–2006.

75. Ohgren K, Rudolf A, Galbe M, Zacchi G (2006) Fuel ethanol production from steam-pretreated corn stover using SSF at higher dry matter content. Biomass Bioenerg 30: 863–869.

76. Varga E, Klinke HB, Reczey K, Thomsen AB (2004) High solid simultaneous saccharification and fermentation of wet oxidized corn stover to ethanol. Biotechnol Bioeng 88: 567–574.

77. Wang XJ, Feng H, Li ZY (2012) Ethanol production from corn stover pretreated by electrolyzed water and a two-step pretreatment method. Chin Sci Bull 57: 1796–1802.

78. Jung H-JG, Buxton DR (1994) Forage quality variation among maize inbreds: relationships of cell-wall composition and in-vitro degradability for stem internodes. J Sci Food Agr 66: 313–322.

79. Kulkami AR, Pattathil S, Hahn MG, York WS, O'Neill MA (2012) Comparison of arabinoxylan structure in bioenergy and model grasses. Ind Biotechnol 8: 222–229.

7

Modeling Spatial Patterns of Soil Respiration in Maize Fields from Vegetation and Soil Property Factors with the Use of Remote Sensing and Geographical Information System

Ni Huang[1], Li Wang[1]*, Yiqiang Guo[2], Pengyu Hao[1], Zheng Niu[1]

1 The State Key Laboratory of Remote Sensing Science, Institute of Remote Sensing and Digital Earth, Chinese Academy of Sciences, Beijing, China, 2 Land Consolidation and Rehabilitation Center, Ministry of Land and Resources, Beijing, China

Abstract

To examine the method for estimating the spatial patterns of soil respiration (R_s) in agricultural ecosystems using remote sensing and geographical information system (GIS), R_s rates were measured at 53 sites during the peak growing season of maize in three counties in North China. Through Pearson's correlation analysis, leaf area index (LAI), canopy chlorophyll content, aboveground biomass, soil organic carbon (SOC) content, and soil total nitrogen content were selected as the factors that affected spatial variability in R_s during the peak growing season of maize. The use of a structural equation modeling approach revealed that only LAI and SOC content directly affected R_s. Meanwhile, other factors indirectly affected R_s through LAI and SOC content. When three greenness vegetation indices were extracted from an optical image of an environmental and disaster mitigation satellite in China, enhanced vegetation index (EVI) showed the best correlation with LAI and was thus used as a proxy for LAI to estimate R_s at the regional scale. The spatial distribution of SOC content was obtained by extrapolating the SOC content at the plot scale based on the kriging interpolation method in GIS. When data were pooled for 38 plots, a first-order exponential analysis indicated that approximately 73% of the spatial variability in R_s during the peak growing season of maize can be explained by EVI and SOC content. Further test analysis based on independent data from 15 plots showed that the simple exponential model had acceptable accuracy in estimating the spatial patterns of R_s in maize fields on the basis of remotely sensed EVI and GIS-interpolated SOC content, with R^2 of 0.69 and root-mean-square error of 0.51 μmol CO_2 m^{-2} s^{-1}. The conclusions from this study provide valuable information for estimates of R_s during the peak growing season of maize in three counties in North China.

Editor: Ben Bond-Lamberty, DOE Pacific Northwest National Laboratory, United States of America

Funding: This work was supported by the National Natural Science Foundation of China (41301498), the Public Service Sectors (Ministry of Land and Resources) Special Fund Research (201311127), the Special Foundation for Young Scientists of the State Laboratory of Remote Sensing Science (13RC-07), and the Major State Basic Research Development Program of China (2013CB733405). The funders had no role in study design, data collection and analysis, decision to publish, or preparation of the manuscript.

Competing Interests: The authors have declared that no competing interests exist.

* Email: wangli@radi.ac.cn

Introduction

Soil CO_2 efflux from terrestrial ecosystems to the atmosphere has been considered the second largest global carbon flux and is a vital component of ecosystem respiration [1]. In recent decades, significant progress has been made in identifying the biophysical factors that influence soil respiration (R_s) to predict soil CO_2 emission accurately in time and space [2–4].

The majority of R_s arises from root and microbial tissue. Therefore, understanding the spatial and temporal changes of these sources will facilitate the modeling of R_s. However, the large spatial and temporal heterogeneity of root and microbial activity within the landscape and the covariation of potentially important factors (i.e., temperature and water content) pose great challenges to the development of mechanistically based models that account for spatial and temporal variability in R_s [2]. Thus, many different

statistical models of R_s have been developed on the basis of data collected from different ecosystems [5]. Numerous studies have established R_s models based on soil temperature, soil moisture, or both [6,7]. Aside from soil temperature and moisture, plant productivity proxies [e.g., leaf area index (LAI), canopy chlorophyll content (Chl$_{canopy}$), and plant biomass] [8–10] and soil properties [e.g., soil organic carbon (SOC) content, soil total nitrogen (STN) content, and soil C and N ratio (soil C/N)] [11,12] also potentially influence R_s and are often included in models of R_s. However, most of the factors that affect variations in R_s tend to be derived through field measurements [13]. Furthermore, direct observation of these variables across long time spans or large spatial scales is expensive because of the required manpower and material resources. A simple method to derive data related to variations in R_s is necessary to facilitate the determination of the spatial and temporal distribution of R_s.

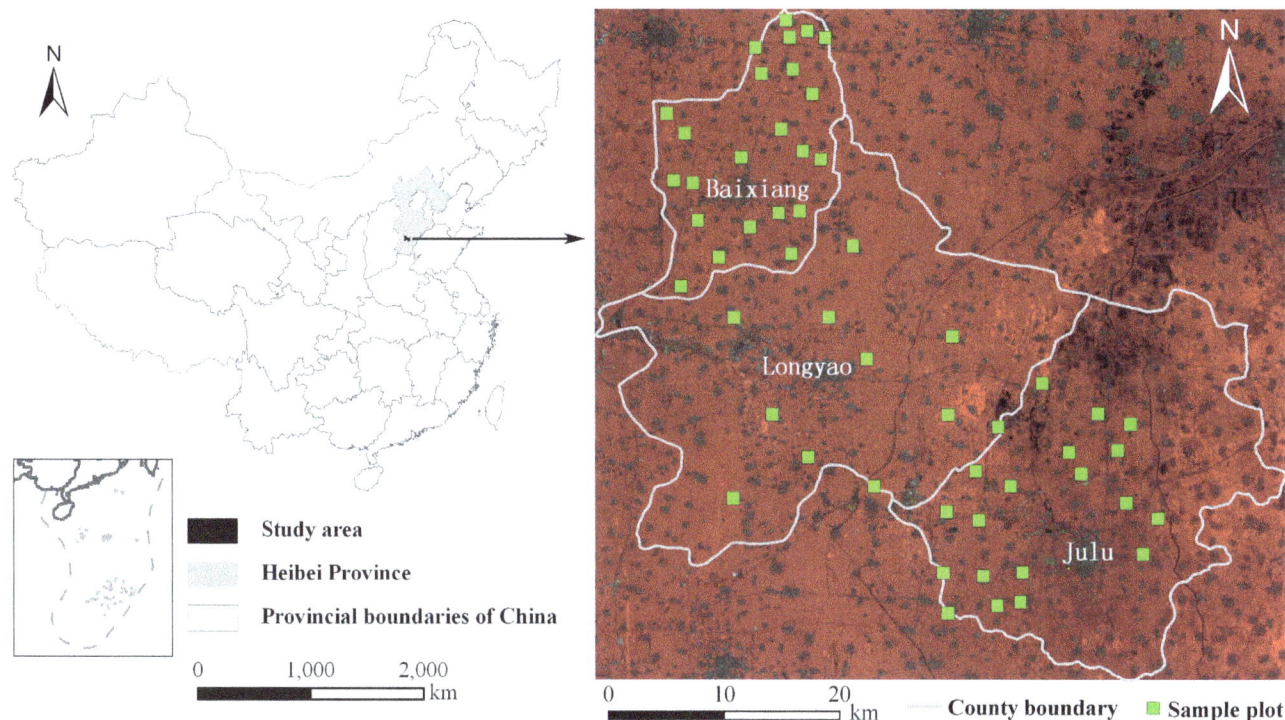

Figure 1. Spatial location of the sample plots for field experiments in three counties in North China. The box in the bottom left corner of Figure 1 shows the South China Sea islands.

Remote sensing and geographical information system (GIS) provide powerful tools for data acquisition, spatial analysis, and graphical display [14–16]. In the field of global change research, significant advances have been made in the development and application of remote sensing and GIS. These advances include land cover and land-use changes [17,18], environmental vulnerability and risk assessment [19,20], ecological restoration and management [21–23], and terrestrial ecosystem carbon cycle [24–26]. However, applying the data derived from remote sensing and GIS into R_s modeling remains controversial, especially for remote sensing data, because remotely sensed data in principle are independent measurements of site properties, not functionally important variables (e.g., soil temperature, soil moisture, and plant growth variables) that control R_s [3,27,28]. On the basis of statistical analysis of field experiments, previous studies found that remotely sensed vegetation indices (VIs) correlate with R_s in crop sites that lack drought stress [10] and can be used to model the spatial patterns of R_s during the peak growing season of alpine grasslands in the Tibetan Plateau [26]. However, few studies explore the potential of remote sensing and GIS data for estimating the spatial patterns of R_s in agricultural land, which may be affected by more complex factors than natural grasslands because of the influence of human activity. Although modern agriculture has successfully increased food production, the processes involved have profoundly affected the global carbon cycle through tillage, drainage and conversion of natural to agricultural ecosystems [29,30]. Therefore, a simple method should be identified to study the spatial characteristics of R_s in agricultural ecosystems.

This study aims to examine a potential new approach for estimating the spatial patterns of R_s during the peak growing season of maize by using remote sensing and GIS technology in Baixiang, Longyao and Julu Counties, which are typical agricul-

tural areas in the north plain of China. Studying the spatial characteristics of soil CO_2 efflux in maize fields will contribute to eco-agricultural development.

Materials and Methods

Ethics Statement

No specific permissions were required for the 53 sample plots in this study. We confirmed that the field studies did not involve endangered or protected species, and the specific location of the sample plots was provided in the manuscript (Fig. 1).

Study Site

The study site is situated within three counties (Baixiang, Longyao and Julu) in Southern Hebei Province of North China (Fig. 1). The total area of the study site is 1.64×10^3 km². This area is located in the North China Plain with a flat open terrain, single landform type, and a mean elevation of 30 m above sea level. Calcareous alluvial soil with high capacity to retain water and fertilizer is the main soil type in the study area. The study site is suitable for farming, and maize is the main crop. The climate is continental monsoon with four distinct seasons and adequate light and heat resources. Long records of meteorological data near the study site (http://cdc.cma.gov.cn) indicate that the mean annual temperature is 13.5°C with the coldest temperatures in January and the hottest in July. The mean annual precipitation is 502.8 mm, but precipitation is distributed unevenly in the four seasons with the greatest precipitation occurring in summer (362.5 mm). Therefore, drought influences agricultural development, and agriculture mainly involves irrigation in this study site.

Fifty-three sample plots located in the maize fields were identified within the study site (Fig. 1). The distance between any two sample plots was larger than 2 km. Each sample plot (greater than 100 m×100 m) has a large maize area, flat terrain, and maize under uniform growing conditions. All measurements were performed from August 11, 2013 to August 20, 2013, which corresponded to the tassel stage and peak growing period of maize. During the 10 days of field measurements, continuous measurements were performed, except on August 12 because of a minor precipitation event. Therefore, all field measurements required 9 days.

Field measurements

Soil respiration measurements. In each sample plot, R_s was measured by using a soil respiration chamber (LI-6400-09; LiCor, Lincoln, Nebraska, USA) connected to a portable photosynthesis system (LI-6400; LiCor, Lincoln, Nebraska, USA). The soil respiration chamber was mounted on a PVC soil collar that was sharpened at the bottom. Each PVC collar (5 cm long, 11 cm inside diameter) was inserted 2 cm to 3 cm into the ground and was installed at least 24 h prior to performing any measurements. To reduce the difference in root biomass, soil collars were placed in three locations on the basis of their distance to the maize plant: near a maize plant, inter-plant, and inter-row. Two collars were placed in each of the three positions for each R_s measurement. At least three to four consecutive measurements on each collar were performed to prevent any systematic error in the R_s estimates. An average R_s value was used for each collar, and the average value from six collars was used to represent the R_s value at plot level. Each R_s measurement was conducted between 09:00 h and 15:00 h (local time) because fluxes measured during this time interval are usually representative of the daily mean flux.

Soil temperature and soil moisture measurements. After the soil respiration measurement on a PVC soil collar in each plot, soil temperature and soil moisture were measured in this collar to minimize sample difference. Soil temperature was measured at a 10 cm depth (T_{s10}) by using a ground thermometer. Volumetric soil moisture at a depth of 0 cm to 20 cm (SM_{20}) was determined by using a portable time domain reflectometry probe (HydroSense, Campbell, USA). Thus, six soil temperature and moisture measurements were performed in each plot. The average value was used to represent soil temperature or soil moisture at the plot level.

Maize biophysical parameter measurements. LAI was measured by using an LAI-2000 (LI-COR Inc., Lincoln, Nebraska). In each plot, six representative positions were selected for LAI measurement, and in every position, two repeated measurements were performed. Leaf chlorophyll content (Chl_{leaf}) was determined by using a portable chlorophyll meter (SPAD-502, New Jersey, USA). Fully expanded leaves, which depended on the height of the maize plant, were randomly selected from three locations that corresponded to the upper, middle, and lower parts

of the maize plant. For each leaf location, 10 SPAD values were randomly collected. The vertical leaf area distribution in maize canopy was analyzed by measuring the area of each green leaf from the bottom to the top of eight randomly distributed maize plants with the use of an area meter (LI-3100, LI-COR, Lincoln, Nebraska). The area-weighted mean SPAD reading was used to derive Chl_{leaf}. However, the SPAD reading was in arbitrary units rather than in actual amounts of chlorophyll per unit area of the leaf tissue. A transform relationship exists between the SPAD readings and the actual chlorophyll content in maize [31]. To convert the SPAD readings to chlorophyll content per unit leaf area ($\mu g\ cm^{-2}$), this study used the transform relationship ($Chl_{leaf} = 0.95 \times SPAD\ reading - 3.25$) derived by Wu et al. [32] in maize plots, and the same SPAD meter was employed in this study. Chl_{canopy} was then determined by using the following equation:

$$Chl_{canopy} = Chl_{leaf} \times GLAI \qquad (1)$$

where Chl_{canopy} is the canopy chlorophyll content ($g\ m^{-2}$), Chl_{leaf} is the leaf chlorophyll content of maize ($g\ m^{-2}$), and GLAI represents the green leaf area per unit ground area.

In each sample plot, three representative maize plants were harvested for aboveground biomass (AGB) measurement. These fresh maize plants were sealed in a plastic bag and immediately transported to a nearby laboratory for subsequent analysis. Thereafter, fresh samples were oven dried at 65°C until the mass of the sample reached a constant weight. The AGB in each plot can be derived by multiplying the average dry weight per plant (g plant^{-1}) and the average plant density of maize (plants m^{-2}).

Soil property measurements. Soil within the six PVC collars in each plot was destructively sampled after measuring R_s, soil temperature and soil moisture. Soil was sampled to a depth of approximately 20 cm by a cylindrical soil driller (4 cm diameter, 20 cm height), in which fine root biomass and microbial activity are the highest [33,34]. These collected soil samples were sealed in plastic bags and stored at room temperature while being transported to the laboratory. Six collected soil samples in each plot were uniformly mixed to form a composite sample for laboratory analysis. The composite sample was air-dried in the laboratory to a constant weight for soil chemical analyses. The air-dried soil samples were ground to pass through a 0.2 mm sieve after any visible plant tissues and debris were manually removed. The SOC content was estimated by using the standard Mebius method [35]. The STN content was analyzed by using the Kjeldahl digestion procedure [36]. In this study, soil C/N was calculated by the ratio of SOC and STN contents.

Table 1. Calculation for vegetation indices [a].

Vegetation index	Formula	Reference
Normalized difference vegetation index	$NDVI = \dfrac{R_{Nir} - R_{Red}}{R_{Nir} + R_{Red}}$	Rouse et al. [47], Gamon et al. [48]
Modified soil adjusted vegetation index	$MSAVI = \dfrac{2R_{Nir} + 1 - \sqrt{(2R_{Nir}+1)^2 - 8(R_{Nir} - R_{Red})}}{2}$	Qi et al. [49]
Enhanced vegetation index	$EVI = 2.5 \times \dfrac{R_{Nir} - R_{Red}}{1 + R_{Nir} + 6 \times R_{Red} - 7.5 \times R_{Blue}}$	Huete et al. [50]

[a] R_{Blue}, R_{Red}, and R_{Nir} are reflectance of blue, red, and NIR band in the HJ-1A CCD optical image, respectively.

Table 2. Pearson's correlation among soil respiration and factors affecting soil respiration in maize fields during the peak growing season in three counties in North China.

	R_s	T_{s10}	SWC_{20}	Chl_{canopy}	LAI	AGB	SOC content	STN content	Soil C/N
R_s	1.00	-0.27	-0.18	0.54***	0.75***	0.59***	0.76***	0.59***	-0.23
T_{s10}		1.00	0.18	-0.15	-0.28	-0.27	-0.49**	-0.66***	0.51**
SWC_{20}			1.00	-0.17	-0.05	-0.07	0.16	-0.00	0.06
Chl_{canopy}				1.00	0.83***	0.81***	0.26	0.20	-0.18
LAI					1.00	0.76***	0.44**	0.38*	-0.28
AGB						1.00	0.45**	0.34*	-0.15
SOC content							1.00	0.78***	-0.29
STN content								1.00	-0.79***
Soil C/N									1.00

R_s is the daily mean soil respiration rate ($\mu mol\ CO_2\ m^{-2}\ s^{-1}$), T_{s10} is the soil temperature at 10 cm depth (°C), SWC_{20} is the soil water content at 0 cm to 20 cm depth ($m^3\ m^{-3}$), Chl_{canopy} is the canopy chlorophyll content ($g\ m^{-2}$), LAI is the leaf area index, AGB is the aboveground biomass ($kg\ m^{-2}$), SOC content is the soil organic carbon content ($g\ kg^{-1}$), STN content is the soil total nitrogen content ($g\ kg^{-1}$), and soil C/N is the soil C: N ratio. Significance levels:
*p<0.05,
**p<0.01,
***p<0.001.

Spatial data acquisition

Maize classification data. This study aimed to derive the spatial distribution of R_s in maize fields based on the field measurements at the plot scale. Maize classification data is necessary to spatially extrapolate R_s at the plot scale to the whole study area. Multi-temporal normalized difference vegetation index (NDVI) data collected over the growing season were used to classify maize at the study site [37–39]. Clouds are common occurrences in the study area during the growing season. Thus, obtaining a time sequence of cloud-free scenes is difficult. Two types of satellite data were used to establish the time-series NDVI data. One was the Operational Land Imager (OLI) image of Landsat 8, and the other was the small constellation for environmental and disaster mitigation (HJ-1A and B) charge coupled device (CCD) image [40–42]. Five scenes of OLI images acquired on May 3, 2013, May 19, 2013, July 6, 2013, October 10, 2013, and October 26, 2013 were downloaded from the U.S. Geological Survey (http://earthexplorer.usgs.gov/). Three HJ-1A and B CCD optical images acquired on June 6, 2013, August 17, 2013, and September 15, 2013 were downloaded from the China Center for Resource Satellite Data and Applications (http://www.cresda.com). The two types of remote sensing images exhibit same spatial resolution (30 m). The 30 m spatial resolution is appropriate for classifying maize patterns in the study area given the relatively large field in the region, which could spatially corresponded to five or more 30 m pixels. The strong relationship of the NDVI with biophysical vegetation characteristics, such as LAI and green biomass [43,44], enables the discrimination of land cover types on the basis of their unique phenological responses. Before land-use classification, pre-processing (i.e., radiometric calibration, atmospheric correction and geometric correction) of OLI images and HJ-1A and B CCD optical images was accomplished by using the Environment for Visualizing Images (ENVI) software (Version 4.7, Research Systems Inc., Boulder, Colorado, USA) [45,46]. This process ensured the consistency between the two types of remote sensing data and the seasonality of the NDVI time series. The maximum likelihood classification method, integrated in the ENVI software, was applied to the eight-date NDVI time series that spanned one maize growing season of the study site.

Spectral vegetation index for vegetation biophysical parameter estimation. Three greenness indices, namely, NDVI, enhanced vegetation index (EVI), and modified soil adjusted vegetation index (MSAVI), were derived from the HJ-1A CCD optical image acquired on August 17, 2013 (Table 1) for vegetation biophysical parameter estimation. Previous studies reported that greenness VIs offer important and convenient measures for vegetation biophysical parameters, such as LAI and Chl_{canopy} [51–54]. Meanwhile, LAI and Chl_{canopy} are also found to be good indicators of plant canopy photosynthesis [55–57] and are used in the modeling of R_s [58]. To obtain the spatial patterns of vegetation biophysical parameters in maize fields, the spatial distribution of vegetation biophysical parameters over the whole study area was overlapped with the maize classification data.

Quantifying the spatial pattern of SOC content. Statistics and geostatistics have been widely applied to quantify the spatial distribution patterns of SOC at a regional scale [59–61]. Based on the theory of regionalized variables, geostatistics provides advanced tools to quantify the spatial features of soil parameters and to conduct spatial interpolation [62,63]. In this study, geostatistical analyses were performed by using the geostatistical analyst module of ArcGIS software (Version 9.3, 2008) to quantify the spatial pattern of SOC content. To obtain the spatial pattern of the SOC content in the maize fields, the spatial distribution of

Table 3. Spatial characteristics of soil respiration (R_s, µmol CO_2 m^{-2} s^{-1}), soil temperature at 10 cm depth (T_{s10}, °C), soil water content at 0 cm to 20 cm depth (SWC_{20}, m^3 m^{-3}), canopy chlorophyll content (Chl_{canopy}, g m^{-2}), leaf area index (LAI), aboveground biomass (AGB, kg m^{-2}), soil organic carbon content (SOC content, g kg^{-1}), soil total nitrogen content (STN content, g kg^{-1}) and soil C: N ratio (soil C/N) in maize fields during the peak growing season in three counties in North China.

Variables	Mean	Maximum	Minimum	CV (%)
R_s	5.43	7.33	2.64	15.45
T_{s10}	28.32	30.93	25.78	4.73
SWC_{20}	27.54	33.27	19.54	12.48
Chl_{canopy}	0.18	0.21	0.16	6.54
LAI	3.75	4.53	2.81	8.64
AGB	0.94	1.89	0.44	31.93
SOC content	11.86	17.26	6.40	16.71
STN content	1.25	1.78	0.53	24.47
Soil C/N	9.82	14.38	7.07	18.53

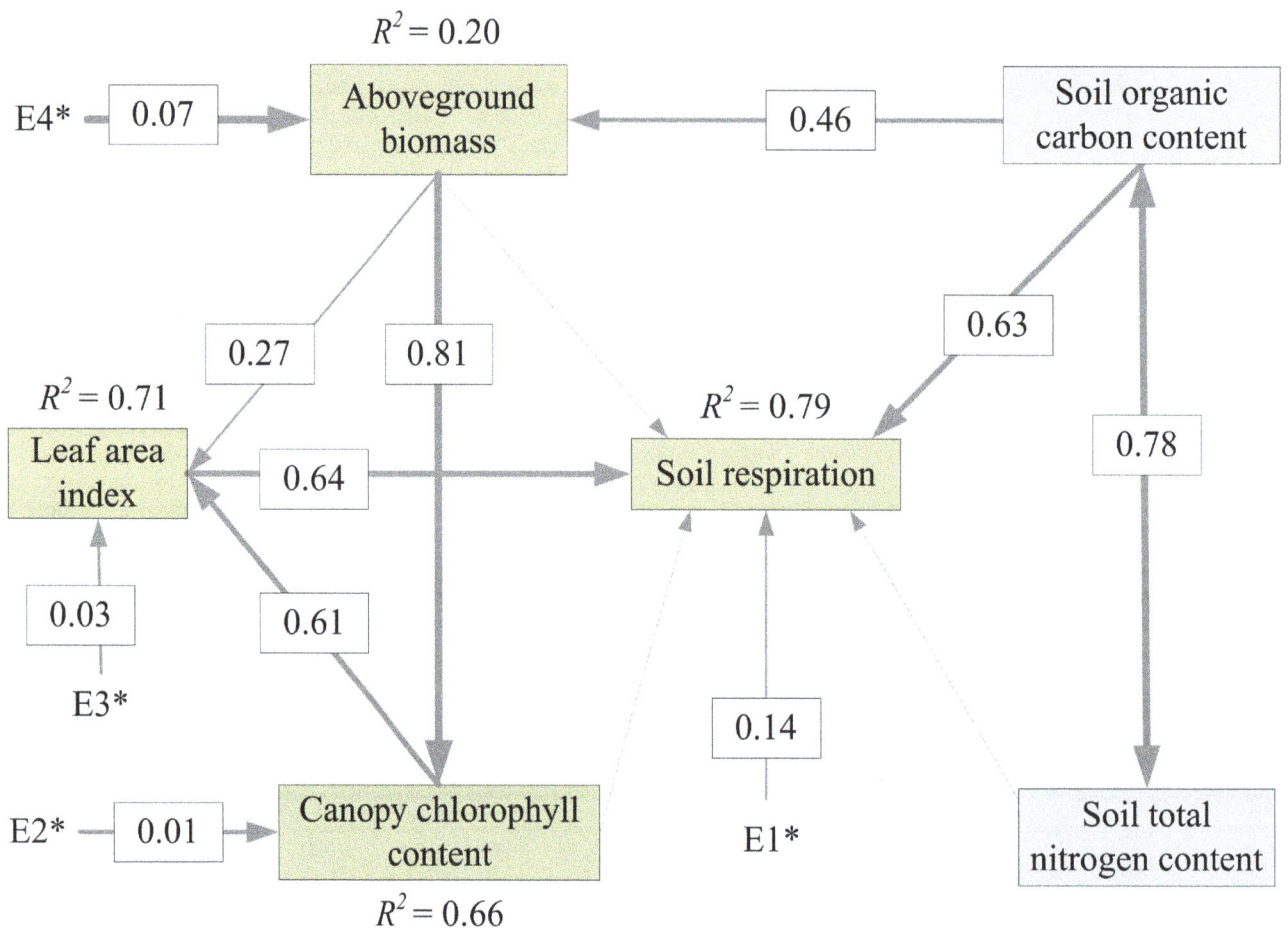

Figure 2. Final structural equation modeling (SEM) for soil respiration. Non-significant paths are shown in dashed line. The thickness of the solid arrows reflects the magnitude of the standardized SEM coefficients. Standardized coefficients are listed on each significant path. * represents error terms for observed variables, among them, E1, E2, E3, and E4 represent measurement errors for soil respiration, canopy chlorophyll content, leaf area index, and aboveground biomass, respectively.

Table 4. Total, direct, and indirect effects in the structural equation modeling.

Variable	Direct effect	Indirect effect	Total
Soil respiration			
Aboveground biomass	−0.10ns	0.46	0.36
Soil organic carbon content	0.63	0.16	0.79
Soil total nitrogen content	−0.09ns	0.30	0. 21
Leaf area index	0.64	-	0.64
Canopy chlorophyll content	−0.04ns	0.39	0.35
Aboveground biomass			
Soil organic carbon content	0.46	-	0.46
Soil total nitrogen content	−0.01ns	-	−0.01ns
Leaf area index			
Aboveground biomass	0.27	0.50	0.77
Soil organic carbon content	-	0.35	0.35
Soil total nitrogen content	-	−0.01ns	−0.01ns
Canopy chlorophyll content	0.61	-	0.61
Canopy chlorophyll content			
Aboveground biomass	0.81	-	0.81
Soil organic carbon content	-	0.37	0.37
Soil total nitrogen content	-	−0.01ns	−0.01ns

These effects were calculated using standardized path coefficients. Non-significant effects are indicated by "ns".

the SOC content over the whole study area was overlapped with the maize classification data.

Modeling spatial patterns of soil respiration

Identifying factors affecting spatial variability of soil respiration. The variables that explain the spatial variability of R_s are as follows: (1) soil properties, measured by SOC content, STN content and soil C/N; (2) environmental factors, encompassing T_{s10} and SM_{20}, and (3) plant photosynthesis proxy factors, including AGB, LAI and Chl_{canopy}. Pearson's correlation requires variables to be normally distributed and mutually independent. Each variable was tested for normal distribution by using the Shapiro–Wilk normality test and for randomness by the runs test of the Statistical Package for the Social Sciences (SPSS, Chicago, Illinois, USA). The results of the statistical analysis showed that each of these measured variables followed a normal distribution (Shapiro-Wilk, p>0.05) and showed randomness (runs test, p> 0.05). Thus, Pearson's correlation analysis, as implemented in the SPSS software, was used to screen important variables that influence R_s. Five variables with statistically significant correlation (p<0.05) with R_s, namely, SOC content, STN content, LAI, AGB, and Chl_{canopy}, were screened out (Table 2). However, these variables were cross-correlated [64–66] and included both direct and indirect effects. To solve this problem, structural equation modeling (SEM) was used to evaluate explicitly the causal relationships among these interacting variables [67–69] and to divide the total effects of variables on R_s into direct and indirect effects. On the basis of the theoretical knowledge on the major factors that influence spatial patterns of R_s at regional scales [8,13,26], we developed an SEM model to relate R_s to SOC content, STN content, LAI, AGB, and Chl_{canopy}. This SEM model was used to identify the direct effect factors for R_s estimation. The SEM model was fitted by using AMOS 18.0 for Windows [70]. After using the SEM, the fit indices, namely, comparative fit

index = 0.984 and goodness-of-fit index = 0.946. Thus, the theoretical model showed a good fit with the sample data.

Quantifying the spatial patterns of soil respiration in maize fields. In this study, the direct effect factors of R_s identified by SEM were used to estimate R_s. The spatial distribution data of these direct effect factors were first obtained on the basis of remote sensing or GIS to quantify the spatial patterns of R_s in maize fields. A simple exponential model that used the proxy data was then employed to estimate the spatial pattern of R_s during the peak growing season of maize. The accuracy of this method was examined by separating the observed data into two datasets through a random generator. One dataset consisted of 38 sample plots for analysis, whereas the other consisted of 15 for testing the accuracy of the R_s estimation.

Result

Spatial characteristics of soil respiration

Based on field-measured data at 38 plots, the daily mean R_s of maize during the peak growing season was 5.43 µmol CO_2 m^{-2} s^{-1} with a range of 2.64 µmol CO_2 m^{-2} s^{-1} to 7.33 µmol CO_2 m^{-2} s^{-1} and a coefficient of variation (CV) of 15.45% (Table 3). The spatial variability of soil temperature at 10 cm depth (T_{s10}) was relatively small at the study site with a CV of 4.73% and was far less than the spatial variation in soil water content at 0 cm to 20 cm depth (SWC_{20}). The AGB of maize showed greater spatial variability (CV = 31.93%) than LAI (CV = 8.64%) and Chl_{canopy} (CV = 6.54%).

Mean SOC content, STN content, and soil C/N at 0 cm to 20 cm depth in maize fields of the study site were 11.86 g kg^{-1} (ranged from 6.40 g kg^{-1} to 17.26 g kg^{-1}), 1.25 g kg^{-1} (ranged from 0.53 g kg^{-1} to 1.78 g kg^{-1}), and 9.82 (ranged from 7.07 to 14.38), respectively. Their CVs were not similar with the STN content which showed greater spatial variability than the SOC content and soil C/N.

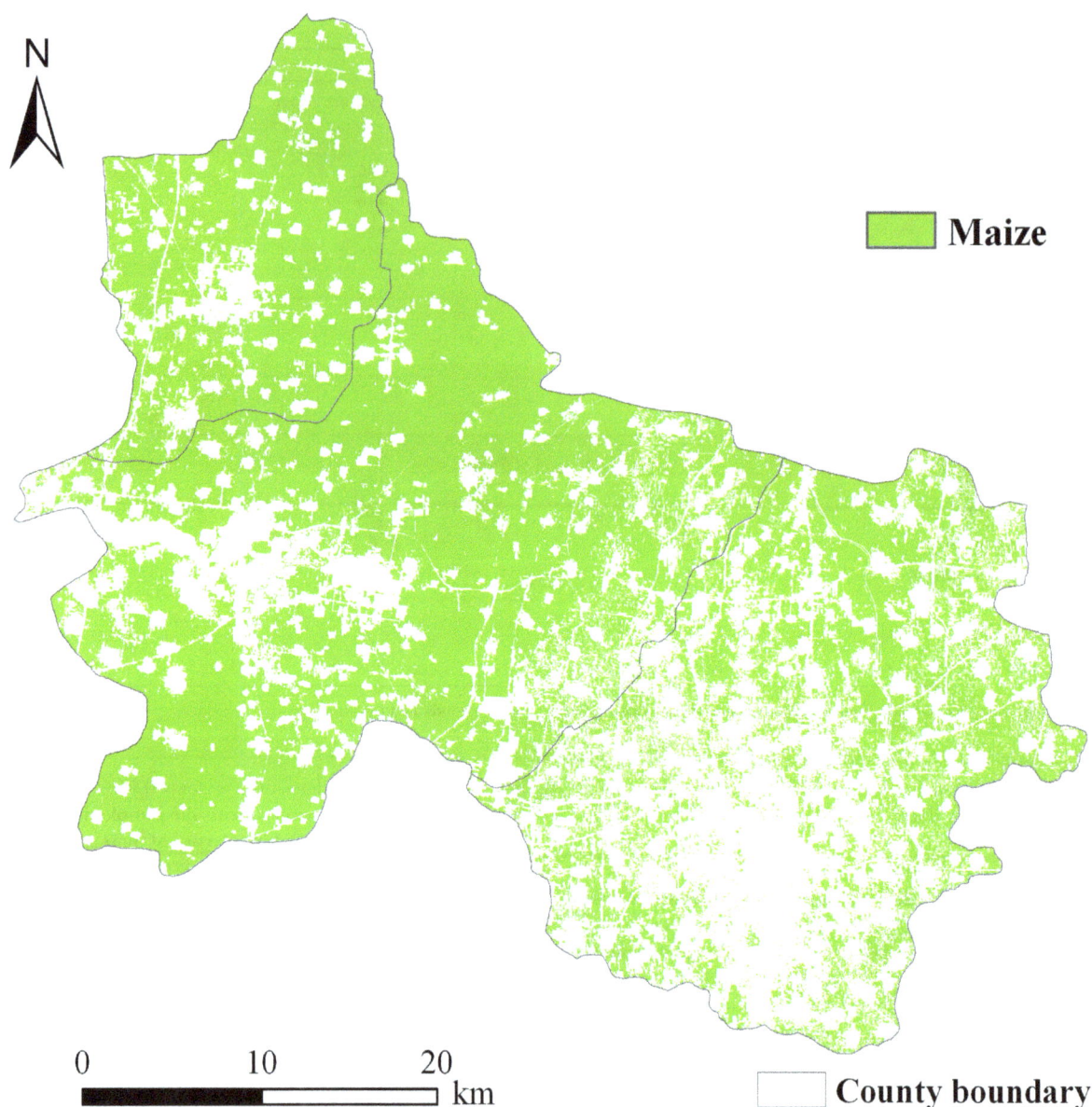

Figure 3. Maize classification map in three counties in North China.

Factors driving spatial variability of soil respiration

Based on Pearson's correlation analysis, five variables with significant correlation with R_s, namely, Chl_{canopy}, LAI, AGB, SOC content, and STN content, were selected (Table 2). However, the five selected variables were intercorrelated (Table 2), and their relationships with R_s combined both direct and indirect correlations. Thus, an SEM model was further used to evaluate the causal relationships among these interacting variables. The final SEM explained 79% of the variation in R_s (Fig. 2). The direct, indirect, and total effects of the variables are shown in Table 4. Among the five selected variables, LAI and SOC content directly affected R_s and can be used to predict R_s with relatively high accuracy ($R^2 = 0.79$). The other three variables (i.e., Chl_{canopy}, ABG, and STN content), despite having a significant correlation with R_s, only affected R_s indirectly through their direct relationship with SOC content and LAI. Thus, the two direct effect factors were used to estimate R_s, and the spatially distributed data proxies of

these two factors were used to quantify the spatial patterns of R_s in maize fields during the peak growing season.

Spatial data used for soil respiration estimation

Maize classification. The maize classification map of the study area is shown in Figure 3. The classification accuracy for maize at the study site could not be quantitatively assessed because of the limitation of the sample data. However, 53 sample plots were all located in the maize classification map, and the county-level maize patterns classified in the map were consistent with the general maize patterns across the three counties. In addition, the classified maize area was close to the maize area reported by the China County Statistical Yearbook [71]. Thus, the classification accuracy of maize was believed to be reasonable, and the maize classification map was then used to predict the spatial pattern of R_s during the peak growing season of maize.

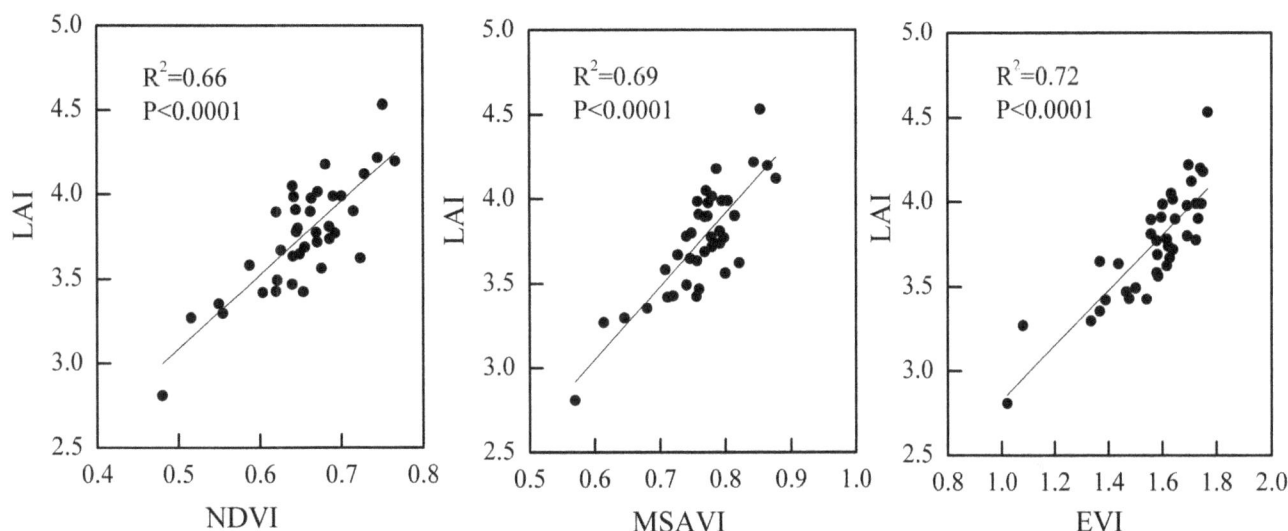

Figure 4. Linear relationships between three vegetation indices (VIs) and leaf area index (LAI) during the peak growing season of maize in three counties in North China (n = 38). The VIs are: normalized difference vegetation index (NDVI), enhanced vegetation index (EVI), and modified soil adjusted vegetation index (MSAVI).

LAI estimation from spectral vegetation index. Among the three greenness indices calculated from the optical image of HJ-1A satellite, EVI showed the best linear relationship with LAI, with a determination coefficient (R^2) of 0.72, followed by MSAVI and NDVI (Fig. 4). The explanation of LAI variance increased from 66% to 72% when EVI was used instead of NDVI for LAI estimation, and this increase was statistically significant (p<0.05). However, EVI and MSAVI did not exhibit a significant difference in explaining the variation in LAI, despite EVI having a slightly better relationship with LAI than MSAVI. Thus, EVI was used as a proxy for LAI to estimate R_s during the peak growing season of maize for simplicity. The spatially distributed EVI during the peak growing season of maize exhibited relatively small variability (Fig. 5). Overall, the EVI in the north and southwest parts of the study site (i.e., Baixiang and Longyao Counties) showed a high value. Relatively low EVI values mainly occurred in the southeast parts of the study site (i.e., Julu County), especially the northwest Julu County (Fig. 5).

Spatial distribution of SOC content. Kriging interpolation was performed by using ArcGIS 9.3 software to produce the spatial distribution map of the SOC content in maize fields of the study area. A cell size of 30 m×30 m was selected for the spatial interpolation to match the spatial resolution of images from OLI and HJ-1A/B. The final result of this spatial interpolation process is shown in Figure 6. Based on the spatial distribution map of the SOC content in maize fields, SOC content values were higher in the northwest and southwest parts of the study area than in the southeastern part.

Spatial distribution of soil respiration

The EVI and SOC content were used to estimate the spatial pattern of R_s during the peak growing season of maize on the basis of a simple exponential model. The geo-location information (latitude and longitude) of the 38 sample plots was used in the extraction of pixels. Pixels that contained these plots from the spatial distribution maps of EVI and SOC content data (Figs. 5 and 6) were extracted. These data were used to determine the model parameters by least-squares fitting. The resulting model was as follows:

$$R_s = 1.57 \times \exp(0.44 \times EVI + 0.05 \times SOC \ content) \qquad (2)$$

$$(n = 38, \ R^2 = 0.73)$$

where R_s refers to the daily mean soil respiration rate in μmol CO_2 m^{-2} s^{-1}; EVI refers to enhanced vegetation index, as a proxy for LAI; and SOC content is the soil organic carbon content (g kg^{-1}) in maize fields of the study area. Eq. (2) was employed to predict the spatial pattern of R_s from spatially distributed EVI and SOC content data during the peak growing season of maize (Figs. 5 and 6). The spatial variation in R_s showed a pattern similar to that in SOC content (Figs. 6 and 7).

Figure 8 shows the accuracy assessment result of the R_s prediction model. The field measured R_s was comparable with the spatial data predicted R_s. Based on the independent test dataset, EVI and SOC content accounted for 69% of the spatial variation in ground-measured R_s, and the RMSE was 0.51 μmol CO_2 m^{-2} s^{-1}. The result of the accuracy assessment suggests that the prediction model, which used EVI and SOC content as the dependent variables, was effective in estimating R_s in maize fields during the peak growing season.

Discussion

Relationships between LAI and three VIs

In this study, in situ measured data were obtained during the peak growing period of maize (corresponding to the tassel stage of maize). The effect of soil background on the spectral reflectance of remote sensing images was negligible during this period because the maize cover was higher with LAI ranging from 2.81 to 4.53. The difference in the capability of spectral vegetation index (VI) responding to LAI variation mainly depended on the sensitivity of VI to the canopy structural variation of maize. Thus, the VI modified the effect of soil reflectance (i.e. MSAVI) did not exhibit a significantly greater advantage than NDVI, which is strongly affected by soil reflectance in sparsely vegetated areas [50]. EVI, which is more sensitive to variation in dense vegetation than

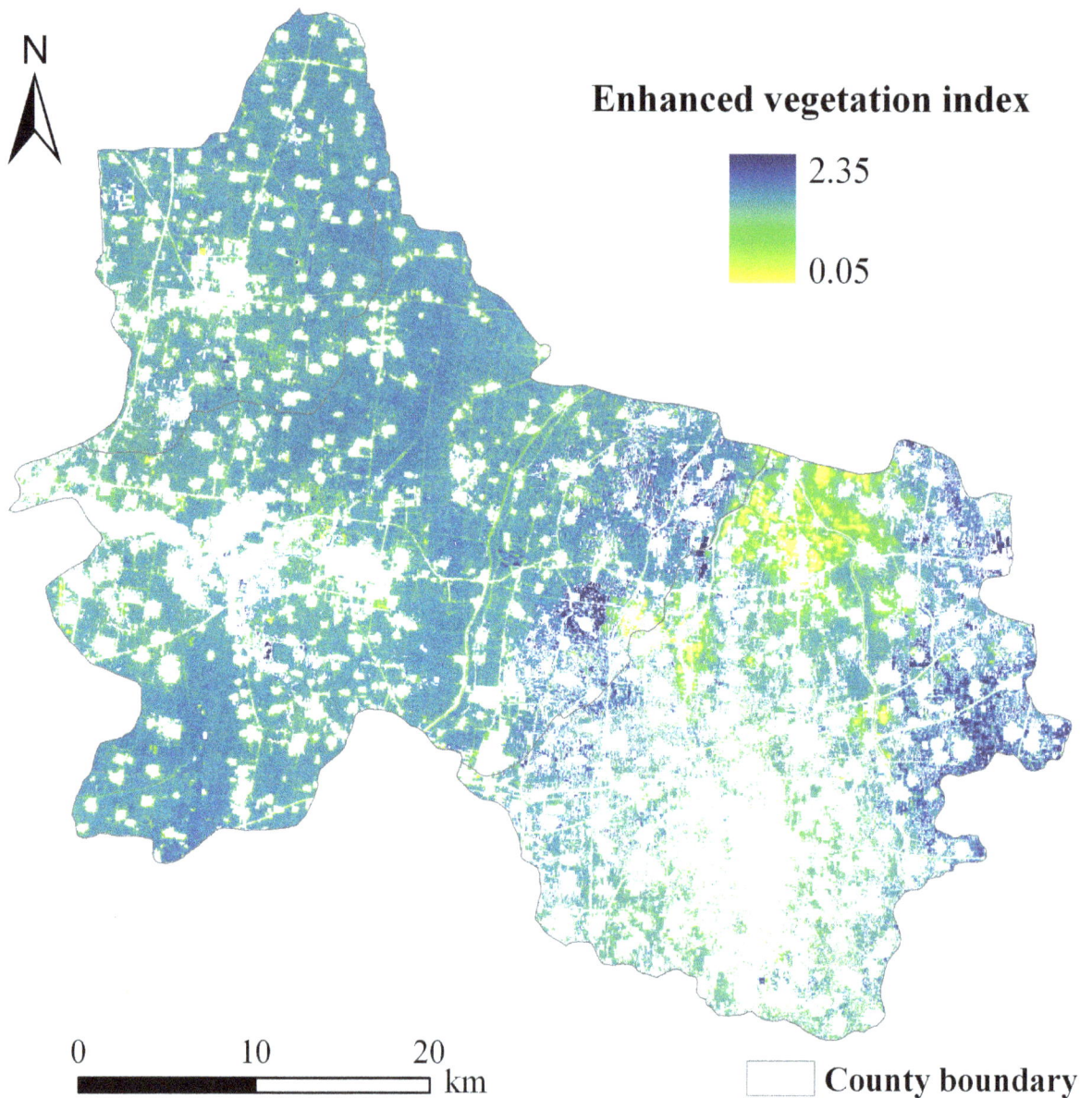

Figure 5. Spatial distribution map of enhanced vegetation index in maize fields in three counties in North China.

NDVI [50], showed the best relationship with the LAI of maize. This result was consistent with our previous study [58] that was conducted in irrigated and rainfed maize fields located at the University of Nebraska, Agricultural and Research Development Center, Mead, Eastern Nebraska, USA.

Measurement accuracy of SOC content

Field measurement data revealed that the SOC content at 0 cm to 20 cm depth in the maize fields ranged from 6.4 g kg^{-1} to 17.3 g kg^{-1}, and the mean value was 12.01 g kg^{-1}. For the mean dry land SOC content in North China, the value appeared to be higher than the previous estimate (0.83 from the average of 268 sample points) [72]. This difference was partly attributed to the fact that only the SOC content in maize fields, not in all dry land types, was considered. Most maize fields in the study site were on a winter wheat/maize rotation, and wheat straw was returned to the soil. The high productivity of maize crops contributed to the

development of a thick A horizon and high SOC content [73,74]. Additionally, only the SOC content in maize fields at 0 cm to 20 cm depth was analyzed, whereas previous studies estimated the SOC content on the basis of organic carbon content to a depth of 1 m [72,75,76]. In agricultural land, soil depth at 0 cm to 20 cm is located in the cultivation layer and has a higher SOC content than the SOC content at the deeper soil layers [34]. This condition contributed to the higher SOC content from the measured soil property data than the previous estimate.

Factors affecting spatial pattern of soil respiration

The spatial differences in R_s at the study site can be mainly attributed to the differences in vegetation productivity and soil property factors among the sample plots, whereas soil temperature and soil moisture served a minor function in regulating the spatial pattern of R_s. A previous study also demonstrated that site variables that reflect site productivity (e.g., LAI or aboveground

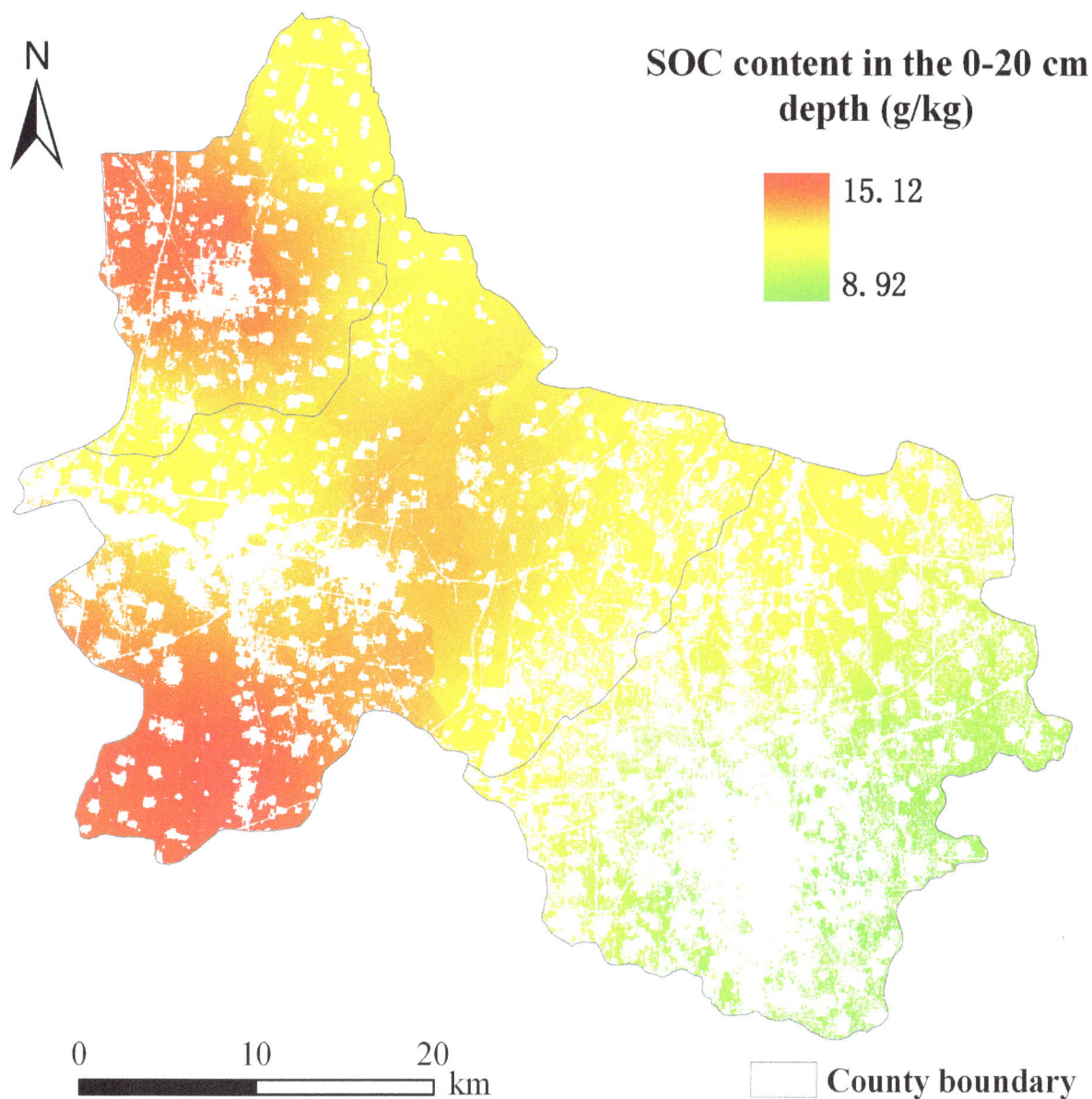

Figure 6. Spatial distribution map of soil organic carbon (SOC) content in the 0–20 cm depth in maize fields in three counties in North China.

net primary productivity) will provide a useful approach for large-scale estimates of regional R_s in terrestrial ecosystems [8]. Soil temperature evidently serves a predominant function in the spatial variations of R_s across sites of climatically contrasting environments [4]. However, at a local scale or under similar climatic conditions, other biological and biophysical factors, such as vegetation productivity and the size of organic carbon pools, may prevail as dominant drivers of R_s [4,77]. At a local scale, the spatial variation in T_{s10} in the study site was small (CV = 4.73%). Thus, soil temperature did not affect the spatial pattern of R_s. Although soil moisture in the maize fields showed a relatively large spatial variation (CV = 12.48%), this variation did not reach a degree that will affect the spatial dynamics of R_s. The soil C quantity and substrate quality factors (i.e., SOC and STN contents) were consistently and strongly correlated with one another and significantly affected the variation in R_s [5,12,13].

However, SEM results showed that the STN content only affected R_s indirectly through the direct effect on the SOC content at the study site.

During the peak growing season of maize, biophysical parameters, such as LAI, Chl_{canopy}, and AGB, were important variables that determined the size of the photosynthetic capacity [56,78]. However, these variables are not truly independent, and a correlation between one of them and R_s may lead to a correlation of the other with R_s. In this study, R_s was strongly correlated with LAI, Chl_{canopy} and AGB of maize fields, whereas LAI was the only variable directly related to R_s during the peak growing season of maize on the basis of SEM analysis.

The direct effect factors of R_s were used to estimate the spatial variability of R_s during the peak growing season of maize in three counties in North China. A simple exponential model, which included the corresponding spatial proxies from remote sensing

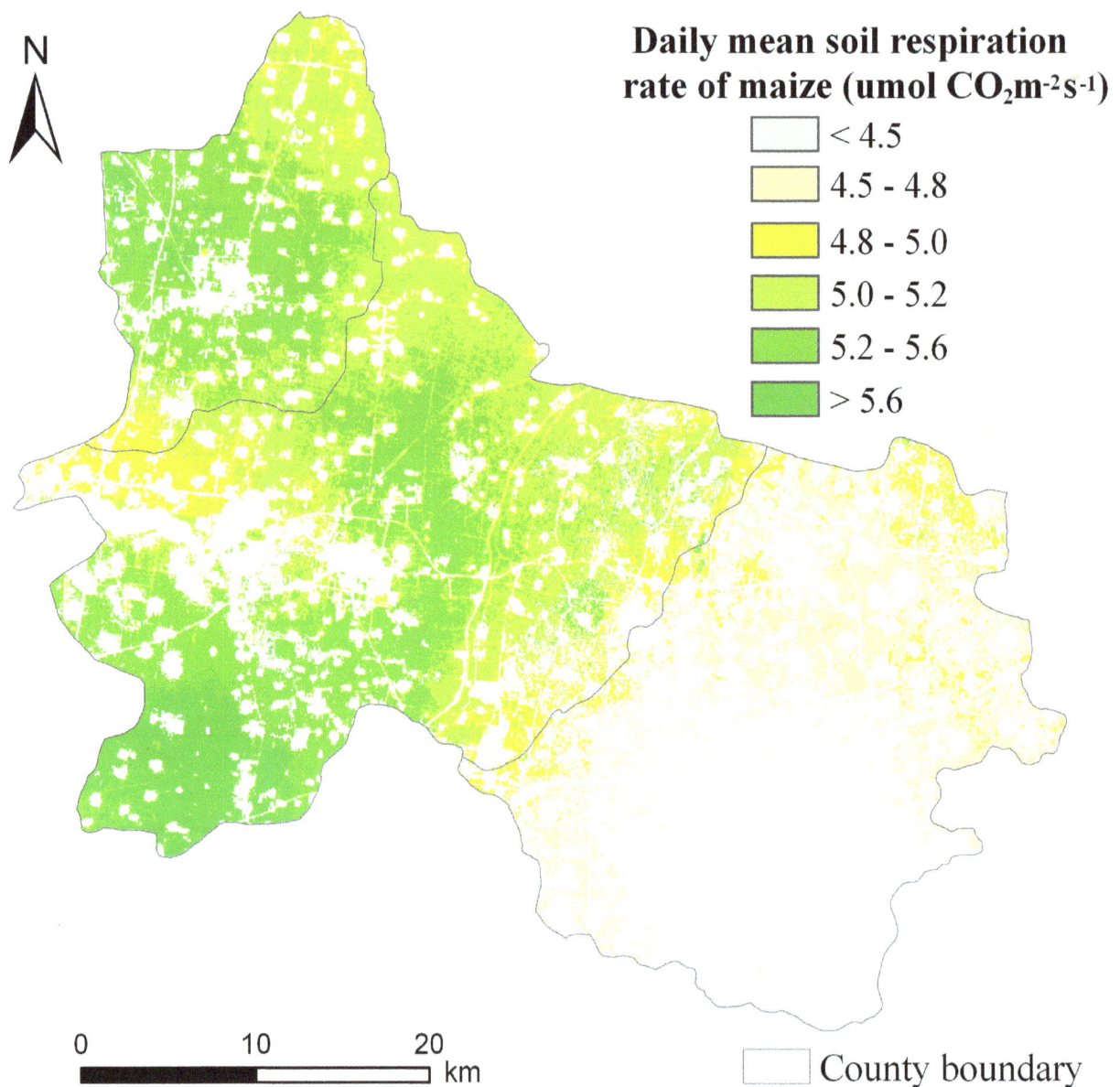

Figure 7. Spatial pattern of daily mean soil respiration rate during the peak growing season of maize in three counties in North China.

and GIS (i.e., EVI and spatially interpolated SOC content), was employed. A similar method was applied to a deciduous broadleaf forest site in the Midwest USA [79]. The independent test data also demonstrated the rationality of this method at the study site to a certain extent (Fig. 8). Regardless of the form of the R_s model, the relationship between LAI and EVI, as well as the kriging interpolation precision of the SOC content, affected the predictive accuracy of the R_s model. A moderate correlation between EVI and LAI (Fig. 4) affected the test accuracy of the exponential model with an R^2 value of 0.69 and an RMSE value of 0.51 μmol CO_2 m^{-2} s^{-1} (Fig. 8). The tendency of kriging to overestimate small values is supported by previous studies [80–82]. This tendency may help explain the bias toward overestimating R_s at low values (Fig. 8). Therefore, improving the accuracy of input parameters from remote sensing or GIS will increase the predictive capability of the R_s model.

Notably, the R_s model developed in this study was applicable to maize fields during the peak growth period in the three counties in North China. However, the model employed in this study does not consider temperature, a main driver of R_s that has high spatial variability. This model may be not used anywhere else or in other stages of the growing season. Furthermore, when spatially distributed data were used in the R_s model, a simple alternative method was employed to estimate the maize LAI by using the remotely sensed EVI, which may be problematic. Verstraeten et al. [83] highlighted that the assimilation of remotely sensed geophysical products into a carbon model is a complex process, and simply exchanging conventional input data for their remotely sensed counterparts is insufficient. Therefore, future research should focus on an integrating spatially distributed R_s datasets and geophysical products from remote sensing and GIS by using the data assimilation method, which has been extensively applied in

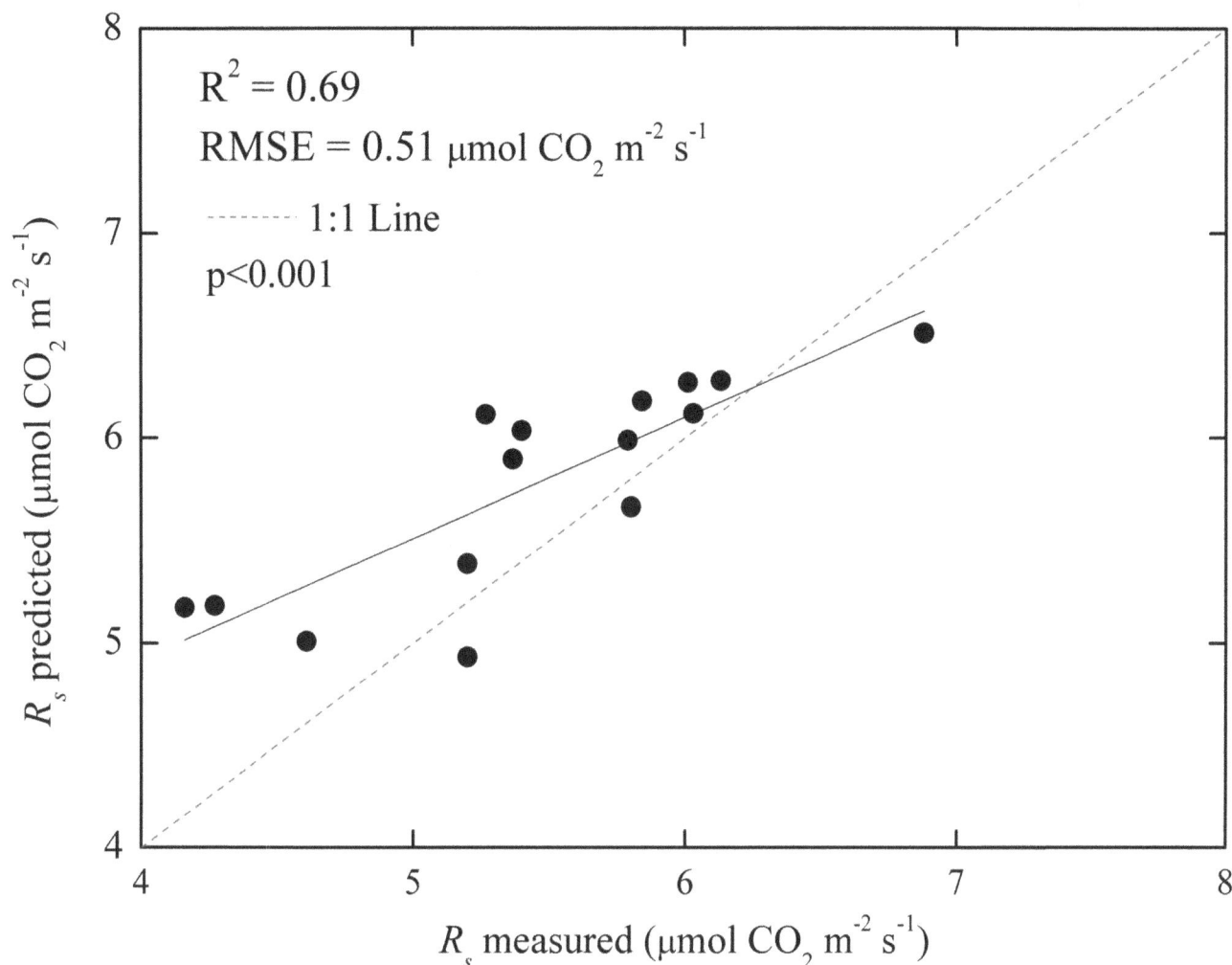

Figure 8. Spatial data predicted soil respiration (R_s) and corresponding ground-based measurements with R^2 and RMSE (μmol CO_2 m^{-2} s^{-1}) during the peak growing season of maize in three counties in North China ($n = 15$). The predicted soil respiration was attained with an exponential model that used EVI and SOC content as dependent variables.

terrestrial carbon cycle research [84–86]. However, this method lack the integration of R_s and spatially distributed data.

Conclusions

This study investigated the potential of spatial data from remote sensing and GIS for estimating the spatial patterns of R_s during the peak growing season of maize in three counties in North China. Based on in situ measurements, plant productivity (i.e., LAI) and soil property (i.e. SOC content) factors were identified as the most important determinants of spatial variability in R_s during the peak growing season of maize, and R_s was weakly related to soil temperature and soil moisture. Spectral VIs calculated from an HJ-1A CCD optical image were used to estimate LAI and EVI was found to be the best proxy for LAI. To derive the spatial pattern of R_s during the peak growing season of maize, a simple

exponential model, which included remotely sensed EVI and GIS spatially interpolated SOC content, was employed. This method was tested by using an independent sample dataset and was shown to be reasonable at the study site.

Acknowledgments

We sincerely thank the anonymous reviewers for their important and constructive revision advices on the manuscript.

Author Contributions

Conceived and designed the experiments: NH LW YQG. Performed the experiments: NH LW PYH. Analyzed the data: NH LW PYH. Contributed reagents/materials/analysis tools: NH LW YQG PYH ZN. Wrote the paper: NH.

References

1. Raich JW, Potter CS (1995) Global patterns of carbon dioxide emissions from soils. Global Biogeochemical Cycles 9: 23–36.
2. Davidson EA, Belk E, Boone RD (1998) Soil water content and temperature as independent or confound factors controlling soil respiration in a temperature mixed hardwood forest. Global Change Biology 4: 217–227.
3. Buchmann N (2000) Biotic and abiotic factors controlling soil respiration rates in Picea abies stands. Soil Biology and Biochemistry 32: 1625–1635.
4. Campbell JL, Sun OJ, Law BE (2004) Supply-side controls on soil respiration among Oregon forests. Global Change Biology 10: 1857–1869.

5. Webster KL, Creed IF, Skowronski MD, Kaheil YH (2009) Comparison of the performance of statistical models that predict soil respiration from forests. Soil Science Society of America Journal 73: 1157–1167.

6. Gaumont-Guay D, Black TA, Griffis TJ, Barr AG, Jassal RS, et al. (2006) Interpreting the dependence of soil respiration on soil temperature and water content in a boreal aspen stand. Agricultural and Forest Meteorology 140: 220–235.

7. Phillips SC, Varner RK, Frolking S, Munger JW, Bubier JL, et al. (2010). Interannual, seasonal, and diel variation in soil respiration relative to ecosystem respiration at a wetland to upland slope at Harvard Forest. Journal of Geophysical Research: Biogeosciences 115: G02019. doi: 10.1029/2008JG000858.

8. Reichstein M, Rey A, Freibauer A, Tenhunen J, Valentini R, et al. (2003) Modeling temporal and large-scale spatial variability of soil respiration from soil water availability, temperature and vegetation productivity indices. Global Biogeochemical Cycles 17: 1104. doi: 10.1029/2003GB002035.

9. Geng Y, Wang Y, Yang K, Wang S, Zeng H, et al. (2012) Soil respiration in Tibetan alpine grasslands: belowground biomass and soil moisture, but not soil temperature, best explain the large-scale patterns. PloS one 7: e34968. doi: 10.1371/journal.pone.0034968.

10. Huang N, Niu Z, Zhan YL, Tappertc MC, Wu CY, et al. (2012) Relationships between soil respiration and photosynthesis-related spectral vegetation indices in two cropland ecosystems. Agricultural and Forest Meteorology 160: 80–89.

11. Chen ST, Huang Y, Zou JW, Shen QR, Hu ZH, et al. (2010) Modeling interannual variability of global soil respiration from climate and soil properties. Agricultural and Forest Meteorology 150: 590–605.

12. Almagro M, Querejeta JI, Boix-Fayos C, Martínez-Mena M (2013) Links between vegetation patterns, soil C and N pools and respiration rate under three different land uses in a dry Mediterranean ecosystem. Journal of Soils and Sediments 13: 641–653.

13. Martin JG, Bolstad PV, Ryu SR, Chen J (2009) Modeling soil respiration based on carbon, nitrogen, and root mass across diverse Great Lake forests. Agricultural and Forest Meteorology 149: 1722–1729.

14. Longley P (Ed.) (2005) Geographic information systems and science. John Wiley & Sons.

15. Weng Q (2001) Modeling urban growth effects on surface runoff with the integration of remote sensing and GIS. Environmental Management 28: 737–748.

16. Lin ML, Chen CW (2011) Using GIS-based spatial geocomputation from remotely sensed data for drought risk-sensitive assessment. International Journal of Innovative Computing, Information and Control 7: 657–668.

17. Shalaby A, Tateishi R (2007) Remote sensing and GIS for mapping and monitoring land cover and land-use changes in the Northwestern coastal zone of Egypt. Applied Geography 27: 28–41.

18. Dewan AM, Yamaguchi Y (2009). Land use and land cover change in Greater Dhaka, Bangladesh: using remote sensing to promote sustainable urbanization. Applied Geography 29: 390–401.

19. Wang XD, Zhong XH, Liu SZ, Liu JG, Wang ZY, et al. (2008). Regional assessment of environmental vulnerability in the Tibetan Plateau: development and application of a new method. Journal of Arid environments 72: 1929–1939.

20. Ceccato P, Connor SJ, Jeanne I, Thomson MC (2005) Application of geographical information systems and remote sensing technologies for assessing and monitoring malaria risk. Parassitologia 47: 81–96.

21. Franklin J (1995) Predictive vegetation mapping: geographic modelling of biospatial patterns in relation to environmental gradients. Progress in Physical Geography 19: 474–499.

22. Raup B, Kääb A, Kargel JS, Bishop MP, Hamilton G, et al. (2007) Remote sensing and GIS technology in the Global Land Ice Measurements from Space (GLIMS) project. Computers & Geosciences 33: 104–125.

23. Keane RE, Burgan R, van Wagtendonk J (2001) Mapping wildland fuels for fire management across multiple scales: integrating remote sensing, GIS, and biophysical modeling. International Journal of Wildland Fire 10: 301–319.

24. He C, Wang S, Xu J, Zhou C (2002) Using remote sensing to estimate the change of carbon storage: a case study in the estuary of Yellow River delta. International Journal of Remote Sensing 23: 1565–1580.

25. Yuan W, Liu S, Yu G, Bonnefond JM, Chen J, et al. (2010) Global estimates of evapotranspiration and gross primary production based on MODIS and global meteorology data. Remote Sensing of Environment 114: 1416–1431.

26. Huang N, He JS, Niu Z (2013) Estimating the spatial pattern of soil respiration in Tibetan alpine grasslands using Landsat TM images and MODIS data. Ecological Indicators 26: 117–125.

27. Davidson EA, Richardson AD, Savage KE, Hollinger DY (2006) A distinct seasonal pattern of the ratio of soil respiration to total ecosystem respiration in a spruce-dominated forest. Global Change Biology 12, 230–239.

28. Vargas R, Baldocchi DD, Allen MF, Bahn M, Black TA, et al. (2010) Looking deeper into the soil: biophysical controls and seasonal lags of soil CO₂ production and efflux. Ecological Applications 20: 1569–1582.

29. Bondeau A, Smith PC, Zaehle S, Schaphoff S, Lucht W, et al. (2007) Modelling the role of agriculture for the 20th century global terrestrial carbon balance. Global Change Biology 13: 679–706.

30. Foley JA, DeFries RS, Asner GP, Barford C, Bonan G, et al. (2005) Global consequences of land use. Science 309: 570–574.

31. Krugh B, Bickham L, Miles D (1994) The solid-state chlorophyll meter: a novel instrument for rapidly and accurately determining the chlorophyll concentrations in seedling leaves. Maize Genetics Cooperation Newsletter 68: 25–27.

32. Wu CY, Wang L, Niu Z, Gao S, Wu MQ (2010) Nondestructive estimation of canopy chlorophyll content using Hyperion and Landsat/TM images. International Journal of Remote Sensing 31: 2159–2167.

33. Gao Y, Xie Y, Jiang H, Wu B, Niu J (2014) Soil water status and root distribution across the rooting zone in maize with plastic film mulching. Field Crops Research 156: 40–47.

34. Kou TJ, Zhu P, Huang S, Peng XX, Song ZW, et al. (2012) Effects of long-term cropping regimes on soil carbon sequestration and aggregate composition in rainfed farmland of Northeast China. Soil and Tillage Research 118: 132–138.

35. Nelson DW, Sommers LE (1982) Total carbon, organic carbon, and organic matter. In: Page AL, Miller RH, Keeney DR. (Eds.), Methods of Soil Analysis. American Society of Agronomy and Soil Science Society of American, Madison. 101–129.

36. Gallaher RN, Weldon CO, Boswell FC (1976) A semiautomated procedure for total nitrogen in plant and soil samples. Soil Science Society of America Journal 40: 887–889.

37. Wardlow BD, Egbert SL, Kastens JH (2007) Analysis of time-series MODIS 250 m vegetation index data for crop classification in the US Central Great Plains. Remote Sensing of Environment 108: 290–310.

38. Wilson EH, Sader SA (2002) Detection of forest harvest type using multiple dates of Landsat TM imagery. Remote Sensing of Environment 80: 385–396.

39. Zhong B, Ma P, Nie A, Yang A, Yao Y, et al. (2014) Land cover mapping using time series HJ-1/CCD data. Science China: Earth Sciences doi: 10.1007/s11430-014-4877-5.

40. Bian JH, Li AN, Jin HA, Lei GB, Huang CQ, et al. (2013) Auto-registration and orthorecification algorithm for the time series HJ-1A/B CCD images. Journal of Mountain Science 10: 754–767.

41. Liu Y, Li M, Mao L, Cheng L, Chen K (2013) Seasonal pattern of tidal-flat topography along the Jiangsu middle coast, China, using HJ-1 optical images. Wetlands 33: 871–886.

42. Wang SD, Miao LL, Peng GX (2012) An Improved Algorithm for Forest Fire Detection Using HJ Data. Procedia Environmental Sciences 13: 140–150.

43. Gamon JA, Field CB, Goulden ML, Griffin KL, Hartley AE, et al. (1995) Relationship between NDVI, canopy structure and photosynthesis in three Californian vegetation types. Ecological Applications 5: 28–41.

44. Hansen PM, Schjoerring JK (2003) Reflectance measurement of canopy biomass and nitrogen status in wheat crops using normalized difference vegetation indices and partial least squares regression. Remote Sensing of Environment 86: 542–553.

45. Yu X, Yan Q, Liu Z (2010) Atmospheric correction of HJ-1A multi-spectral and hyper-spectral data. Image and Signal Processing (CISP), 2010 3rd International Congress on. IEEE 5: 2125–2129.

46. Li P, Jiang L, Feng Z (2013) Cross-Comparison of Vegetation Indices Derived from Landsat-7 Enhanced Thematic Mapper Plus (ETM+) and Landsat-8 Operational Land Imager (OLI) Sensors. Remote Sensing 6: 310–329.

47. Rouse JW, Haas RH, Schell JA, Deering DW, Harlan JC (1974) Monitoring the vernal advancements and retrogradation of natural vegetation, In: NASA/GSFC, Final Report, Greenbelt, MD, USA, 1–137.

48. Gamon JA, Field CB, Goulden ML, Griffin KL, Hartley AE, et al. (1995) Relationship between NDVI, canopy structure and photosynthesis in three Californian vegetation types. Ecological Applications 5: 28–41.

49. Qi J, Chehbouni A, Huete AR, Kerr YH, Sorooshian S (1994) A modified soil adjusted vegetation index (MSAVI). Remote Sensing of Environment 48: 119–126.

50. Huete A, Didan K, Miura T, Rodriguez EP, Gao X, et al. (2002) Overview of the radiometric and biophysical performance of the MODIS vegetation indices. Remote Sensing of Environment 83: 195–213.

51. Broge NH, Leblanc E (2001) Comparing prediction power and stability of broadband and hyperspectral vegetation indices for estimation of green leaf area index and canopy chlorophyll density. Remote Sensing of Environment 76: 156–172.

52. Haboudane D, Miller JR, Tremblay N, Zarco-Tejada PJ, Dextraze L (2002) Integrated narrow-band vegetation indices for prediction of crop chlorophyll content for application to precision agriculture. Remote Sensing of Environment 81: 416–426.

53. Gitelson AA, Vina A, Ciganda V, Rundquist DC, Arkebauer TJ (2005) Remote estimation of canopy chlorophyll content in crops. Geophysical Research Letters 32: L08403. doi: 10.1029/2005GL022688.

54. Wu C, Niu Z, Tang Q, Huang W (2008) Estimating chlorophyll content from hyperspectral vegetation indices: Modeling and validation. Agricultural and Forest Meteorology 148: 1230–1241.

55. Hirose T, Ackerly DD, Traw MB, Ramseier D, Bazzaz FA (1997) CO₂ elevation, canopy photosynthesis, and optimal leaf area index. Ecology 78: 2339–2350.

56. Gitelson AA, Vina A, Verma SB, Rundquist DC, Arkebauer TJ, et al. (2006) Relationship between gross primary production and chlorophyll content in crops: Implications for the synoptic monitoring of vegetation productivity. Journal of Geophysical Research-Atmospheres 111: D08S11. doi: 10.1029/2005JD006017.

57. Glenn EP, Huete AR, Nagler PL, Nelson SG (2008) Relationship between remotely-sensed vegetation indices, canopy attributes and plant physiological

processes: what vegetation indices can and cannot tell us about the landscape. Sensors 8: 2136–2160.

58. Huang N, Niu Z (2013) Estimating soil respiration using spectral vegetation indices and abiotic factors in irrigated and rainfed agroecosystems. Plant and Soil 367: 535–550.

59. Chevallier T, Voltz M, Blanchart E, Chotte JL, Eschenbrenner V, et al. (2000) Spatial and temporal changes of soil C after establishment of a pasture on a long-term cultivated vertisol (Martinique). Geoderma 94: 43–58.

60. McGrath D, Zhang C (2003) Spatial distribution of soil organic carbon concentrations in grassland of Ireland. Applied Geochemistry 18: 1629–1639.

61. Liu D, Wang Z, Zhang B, Song K, Li X, et al. (2006) Spatial distribution of soil organic carbon and analysis of related factors in croplands of the black soil region, Northeast China. Agriculture, Ecosystems & Environment 113: 73–81.

62. Matheron G (1963) Principles of geostatistics. Economic geology 58: 1246–1266.

63. Webster R, Oliver MA (2007) Geostatistics for environmental scientists. John Wiley & Sons.

64. Raich JW, Tufekciogul A (2000) Vegetation and soil respiration: correlations and controls. Biogeochemistry 48: 71–90.

65. Schaefer DA, Feng W, Zou X (2009) Plant carbon inputs and environmental factors strongly affect soil respiration in a subtropical forest of southwestern China. Soil Biology and Biochemistry 41: 1000–1007.

66. Curiel Yuste J, Baldocchi DD, Gershenson A, Goldstein A, Misson L, et al. (2007) Microbial soil respiration and its dependency on carbon inputs, soil temperature and moisture. Global Change Biology 13: 2018–2035.

67. Pugesek BH, Tomer A, Von Eye A (Eds.) (2003) Structural equation modeling: applications in ecological and evolutionary biology. Cambridge University Press.

68. Iriondo JM, Albert MJ, Escudero A (2003) Structural equation modelling: an alternative for assessing causal relationships in threatened plant populations. Biological Conservation 113: 367–377.

69. Jonsson M, Wardle DA (2010) Structural equation modelling reveals plant-community drivers of carbon storage in boreal forest ecosystems. Biology Letters 6: 116–119.

70. Kim GS (2010) AMOS 18.0: Structural Equation Modeling. Seoul: Hannarae Publishing Co.

71. National Bureau of statistics of China (2006) China social-economic statistical yearbooks for China's counties and cities. China Statistics Press, Beijing.

72. Wang S, Tian H, Liu J, Pan S (2003) Pattern and change of soil organic carbon storage in China: 1960s–1980s. Tellus B 55: 416–427.

73. West TO, Post WM (2002) Soil organic carbon sequestration rates by tillage and crop rotation. Soil Science Society of America Journal 66: 1930–1946.

74. Wilhelm WW, Johnson JM, Karlen DL, Lightle DT (2007) Corn stover to sustain soil organic carbon further constrains biomass supply. Agronomy journal 99: 1665–1667.

75. Foley JA (1995) An equilibrium model of the terrestrial carbon budget. Tellus B 47: 310–319.

76. Lal R (1999) Soil management and restoration for C sequestration to mitigate the accelerated greenhouse effect. Progress in Environmental Science 1: 307–326.

77. Epron D, Bosc A, Bonal D, Freycon V (2006) Spatial variation of soil respiration across a topographic gradient in a tropical rain forest in French Guiana. Journal of Tropical Ecology 22: 565–574.

78. Suyker AE, Verma SB, Burba GG, Arkebauer TJ (2005) Gross primary production and ecosystem respiration of irrigated maize and irrigated soybean during a growing season. Agricultural and Forest Meteorology 131: 180–190.

79. Huang N, Gu L, Niu Z (2014) Estimating soil respiration using spatial data products: A case study in a deciduous broadleaf forest in the Midwest USA. Journal of Geophysical Research: Atmospheres 119. doi:10.1002/2013JD020515.

80. Hudak AT, Lefsky MA, Cohen WB, Berterretche M (2002) Integration of lidar and Landsat ETM+ data for estimating and mapping forest canopy height. Remote Sensing of Environment 82: 397–416.

81. Meng Q, Cieszewski C, Madden M (2009) Large area forest inventory using Landsat ETM+: a geostatistical approach. ISPRS Journal of Photogrammetry and Remote Sensing 64: 27–36.

82. Tsui OW, Coops NC, Wulder MA, Marshall PL (2013) Integrating airborne LiDAR and space-borne radar via multivariate kriging to estimate above-ground biomass. Remote Sensing of Environment 139: 340–352.

83. Verstraeten WW, Veroustraete F, Wagner W, Van Roey T, Heyns W, et al. (2010) Remotely sensed soil moisture integration in an ecosystem carbon flux model-The spatial implication. Climatic Change 103:117–136.

84. Rayner PJ, Scholze M, Knorr W, Kaminski T, Giering R, et al. (2005) Two decades of terrestrial carbon fluxes from a carbon cycle data assimilation system (CCDAS). Global Biogeochemical Cycles 19.

85. Chevallier F, Bréon FM, Rayner PJ (2007) Contribution of the Orbiting Carbon Observatory to the estimation of CO_2 sources and sinks: Theoretical study in a variational data assimilation framework. Journal of Geophysical Research: Atmospheres 112: D09307. doi: 10.1029/2006JD007375.

86. Knorr W, Kaminski T, Scholze M, Gobron N, Pinty B, et al. (2010) Carbon cycle data assimilation with a generic phenology model. Journal of Geophysical Research: Biogeosciences 115: G04017. doi: 10.1029/2009JG001119.

8

Interactions between Seagrass Complexity, Hydrodynamic Flow and Biomixing Alter Food Availability for Associated Filter-Feeding Organisms

Vanessa González-Ortiz[1], Luis G. Egea[1], Rocio Jiménez-Ramos[1], Francisco Moreno-Marín[1], José L. Pérez-Lloréns[1], Tjeed J. Bouma[2], Fernando G. Brun[1]*

1 Department of Biology, Faculty of Marine and Environmental Sciences of University of Cadiz, Puerto Real, Cadiz, Spain, **2** Department of Spatial Ecology, Netherlands Institute for Sea Research, Yerseke, The Netherlands

Abstract

Seagrass shoots interact with hydrodynamic forces and thereby a positively or negatively influence the survival of associated species. The modification of these forces indirectly alters the physical transport and flux of edible particles within seagrass meadows, which will influence the growth and survivorship of associated filter-feeding organisms. The present work contributes to gaining insight into the mechanisms controlling the availability of resources for filter feeders inhabiting seagrass canopies, both from physical (influenced by seagrass density and patchiness) and biological (regulated by filter feeder density) perspectives. A factorial experiment was conducted in a large racetrack flume, which combined changes in hydrodynamic conditions, chlorophyll *a* concentration in the water and food intake rate (FIR) in a model active filter-feeding organism (the cockle). Results showed that seagrass density and patchiness modified both hydrodynamic forces and availability of resources for filter feeders. Chlorophyll *a* water content decreased to 50% of the initial value when densities of both seagrass shoots and cockles were high. Also, filter feeder density controlled resource availability within seagrass patches, depending on its spatial position within the racetrack flume. Under high density of filter-feeding organisms, chlorophyll *a* levels were lower between patches. This suggests that the pumping activity of cockles (i.e. biomixing) is an emergent key factor affecting both resource availability and FIR for filter feeders in dense canopies. Applying our results to natural conditions, we suggest the existence of a direct correlation between habitat complexity (i.e. shoot density and degree of patchiness) and filter feeders density. Fragmented and low-density patches seem to offer both greater protection from hydrodynamic forces and higher resource availability. In denser patches, however, resources are allocated mostly within the canopy, which would benefit filter feeders if they occurred at low densities, but would be limiting when filter feeder were at high densities.

Editor: John F. Valentine, Dauphin Island Sea Lab, United States of America

Funding: This work was supported by the Junta de Andalucía Excellence Project PAMBIO (P08-RNM-03783) and by the Spanish Project Sea-Live (CTM2011-24482) from the Spanish Ministry of Science and Innovation. Gonzalez-Ortiz V. holds a regional fellowship from the Junta de Andalucía. The funders had no role in study design, data collection and analysis, decision to publish, or preparation of the manuscript.

Competing Interests: The authors have declared that no competing interests exist.

* Email: fernando.brun@uca.es

Introduction

Seagrasses are important ecosystem engineers, which can change the physical environment through their physical structures [1]. Such habitat modification can result in positive feedbacks, stabilizing seagrass meadows [2] as well as having either a positive (e.g. facilitation) or negative effects on the survival of associated species [3,4]. Numerous studies over the last decades have explored how physical and biological habitat modification promoted by seagrass meadows affects the occurrence of filter-feeding infauna (e.g. distribution, survival, growth, etc.), which constitutes an ecologically and economically important group of marine species [5–13].

In soft-bottom coastal areas, cockles have been shown to filter particles from the water column by raising their siphons up from the sediment and pumping water and food particles into their digestive system, resulting in important implications at the ecosystem level [15–19]. Some studies have highlighted the importance of the effects promoted by seagrass beds on the food supply, growth and survival of filter feeders [6,10,20]. Filter feeder activity is highly dependent on the physical transport of edible particles in the water and flow speed [21,22]. Classical theories suggest that filter and suspension feeders should benefit from higher water refreshment rates because of simultaneously higher mixing rates and particle fluxes (i.e. food availability) [23] within the benthic boundary layer [21,24,25]. In contrast, it has been suggested that calmer conditions would facilitate trapping of food particles [6,12] or increase their consumption by filter feeders, which are more stressed at high flow regimes; [21,23,26—29]. However, other studies have pointed out in turn that the reduction of particle fluxes associated with attenuated conditions within the canopy could fully deplete resources within the bottom benthic boundary layer and thus negatively affect the growth rates of filter feeders [8,11,14,30,31].

The spatial distribution of food particle concentration (particles · ml^{-1}) within a seagrass canopy will be the outcome of the balance between food input and consumption, with input depending strongly upon the volumetric flow rate within the canopy and the concentration of particles in the bulk water [32]. The degree to which seagrasses modify water flow depends largely on their own morphological and architectural complexity. From a physical perspective, dense canopies enhance both particle collision and sedimentation [33–37] because of the reduction in flow velocity and lost of momentum in the particles. Thus, in dense meadows with low particle flux and high competition for these resources (e.g. high filter feeder density) this could result in particle depletion (i.e. low concentration) within the canopy [38] and low survival rates [12]. In contrast, thinner canopies may allow for higher passage of water flow and constant renewing of particles across the bed, which could have potentially positive effects on the associated fauna [12,39]. In addition to vegetation density, landscape fragmentation also strongly affects water flow because of the enhancement of the "edge effect" in patchy meadows [40]. Edges are transitional zones, where substantial changes in hydrodynamic variables (e.g. vertical turbulence) and biological processes [14,32,41] co-occur, affecting distribution patterns of edible particles throughout the seagrass bed [42,43]. Although consumption rate mostly depends on the feeding activity [44], the effect of seagrass presence on filter feeder activity remains unclear, although it has been shown to reduce predation on filter feeders

[6]. This uncertainty may be due to the complexity of the overall effect, with possibly contrasting effects depending on the interaction of i) the flow rate within the meadow in relation to its structural characteristics (density and patchiness) and ii) the number of filter feeders in combination with their levels of activity in filter-feeding. Thus, the link between physical interactions (plant structure-water flow influenced by the complexity of the seagrass meadow) and biological ones (consumption and bio-mixing) will play a crucial role in determining the distribution and composition of the filter feeder community.

To our knowledge, none of the published studies on the effects of the presence of seagrasses on filter feeders used a full factorial design to explore the relationship between the physical effect (promoted by seagrasses) and the biological one (fostered by the faunal bio-mixing and consumption of food resources) on the food intake rate (FIR, μg Chla·l^{-1}·h^{-1}) of an active filter feeder. The present work contributes to gaining insights into the biological and physical mechanisms controlling food supply to filter feeders inhabiting seagrass canopies. Therefore, the specific objectives of our study were: (1) to establish the relationship between hydrodynamics (flow and turbulence) and seagrass complexity (shoot density and number of patches as a proxy for landscape patchiness), (2) to measure how variations in flow characteristics alter the availability of resources for filter feeders (within and outside the meadows), (3) to estimate if flow and food concentration were modified by the presence of high densities of an active filter feeder (i.e. the cockle, *Cerastoderma edule* Linnaeus, 1758), and (4) to check whether the interaction of both seagrass

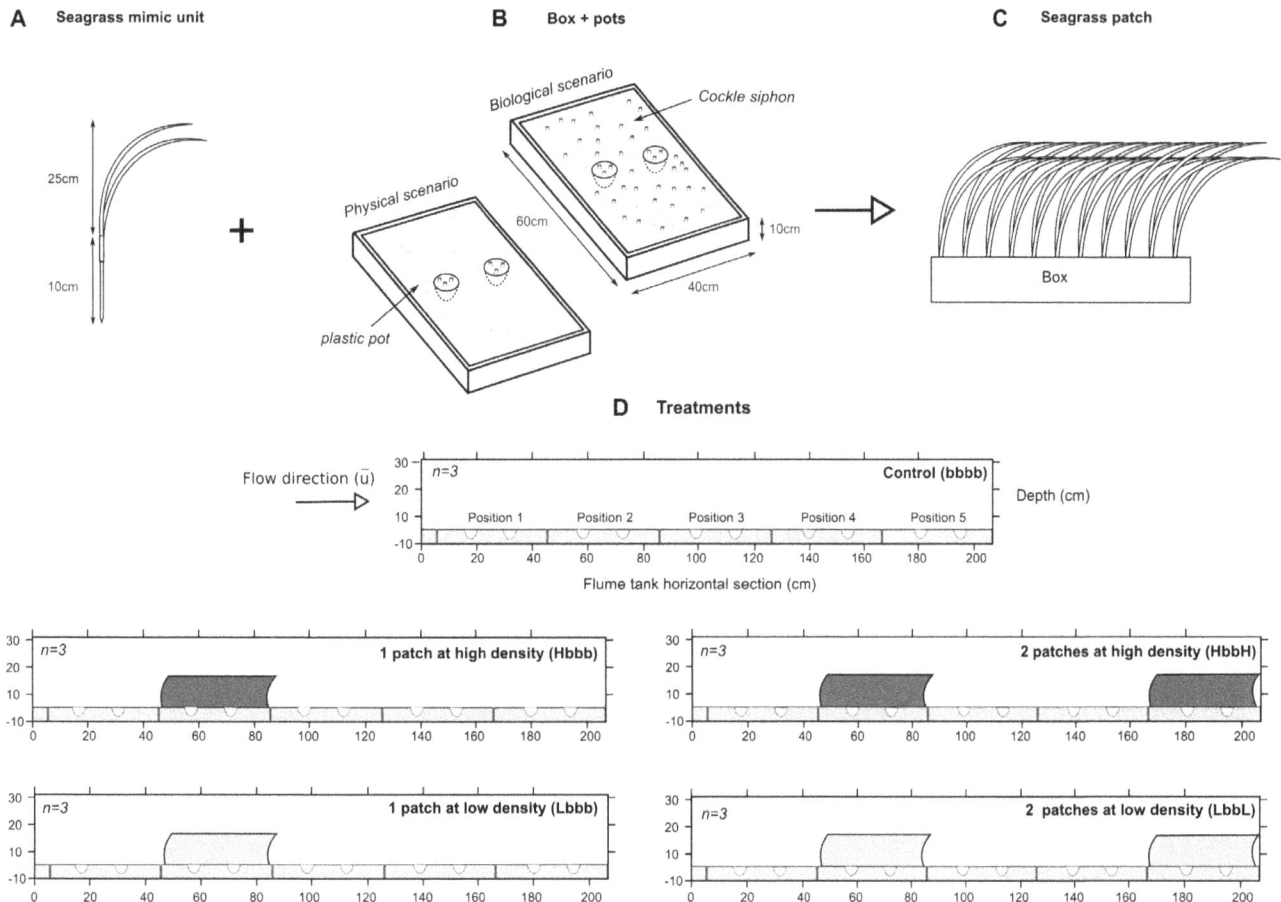

Figure 1. Drawing of the experimental set-up with views shown across the racetrack flume channel.

Table 1. Glossary: Summary table with the description of the most important terms used in this work.

Term	Definition	Equivalence
Pot	*Small plastic container hosting 3 experimental cockles.*	*1 Pot = 3 cockles*
Box	*Square wooden container (60×40×10 cm), filled with sediment, where two pots (6 cockles in total) were placed. Five boxes in total were used per experimental trial. In two of them, seagrass mimics were planted at low shoot density (L), while in the other two, they were planted at high shoot density (H). The rest were filled only with sediment (b)*	*1 box = 2 pots = 6 cockles*
Position	*Location of the boxes along the racetrack flume in a set order of 1–5, where box 1 was at the foreground flow direction (left) and box 5 at the downstream rearmost position (right). Details in Fig. 1.*	*Position = numbered box (box 1 to box 5) = 6 cockles.*
Shoot density	*Number of mimics (shoots) per square meter. Some boxes simulated low-density patches (L, 500 shoots m^{-2}), others high-density patches (H, 1,500 shoots m^{-2}) and also some boxes simulated bare sediment (b, 0 shoots m^{-2}).*	*L = box with 500 shoots m^{-2}. H = box with 1,500 shoots m^{-2}.*
Number of patches	*Number of boxes with mimics. Two levels were used in the treatments: one box (one patch; e.g. Lbbb) or 2 boxes (two patches; e.g. LbbL).*	*One patch = 1 box with mimics in position 2. Two patches = 2 boxes with mimics in positions 2 and 5.*
Physical scenario	*Experimental approach where physical structures of mimics were the main factor responsible for the changes in food availability within the canopy (e.g. changes in volumetric flow rate or turbulence).*	*Physical scenario = very low density of filter feeders.*
Biological scenario	*Experimental approach where physical structures of mimics plus the presence of high densities of filter feeders were the main factors responsible for the changes in food availability within the canopy (e.g. changes in volumetric flow rate, turbulence, biomixing and consumption).*	*Biological scenario = high density of filter feeders.*
Food availability	*Food particles available over a time interval (particles· cm^{-1}s^{-1}) controlled by the velocity of the flow and the complexity (i.e. number of structures) of the seagrass bed.*	
Food concentration	*Food particles in the water (particles·ml^{-1}) accumulated within the canopy, dependent on the balance between inputs (deposition) and consumption by filter feeders.*	

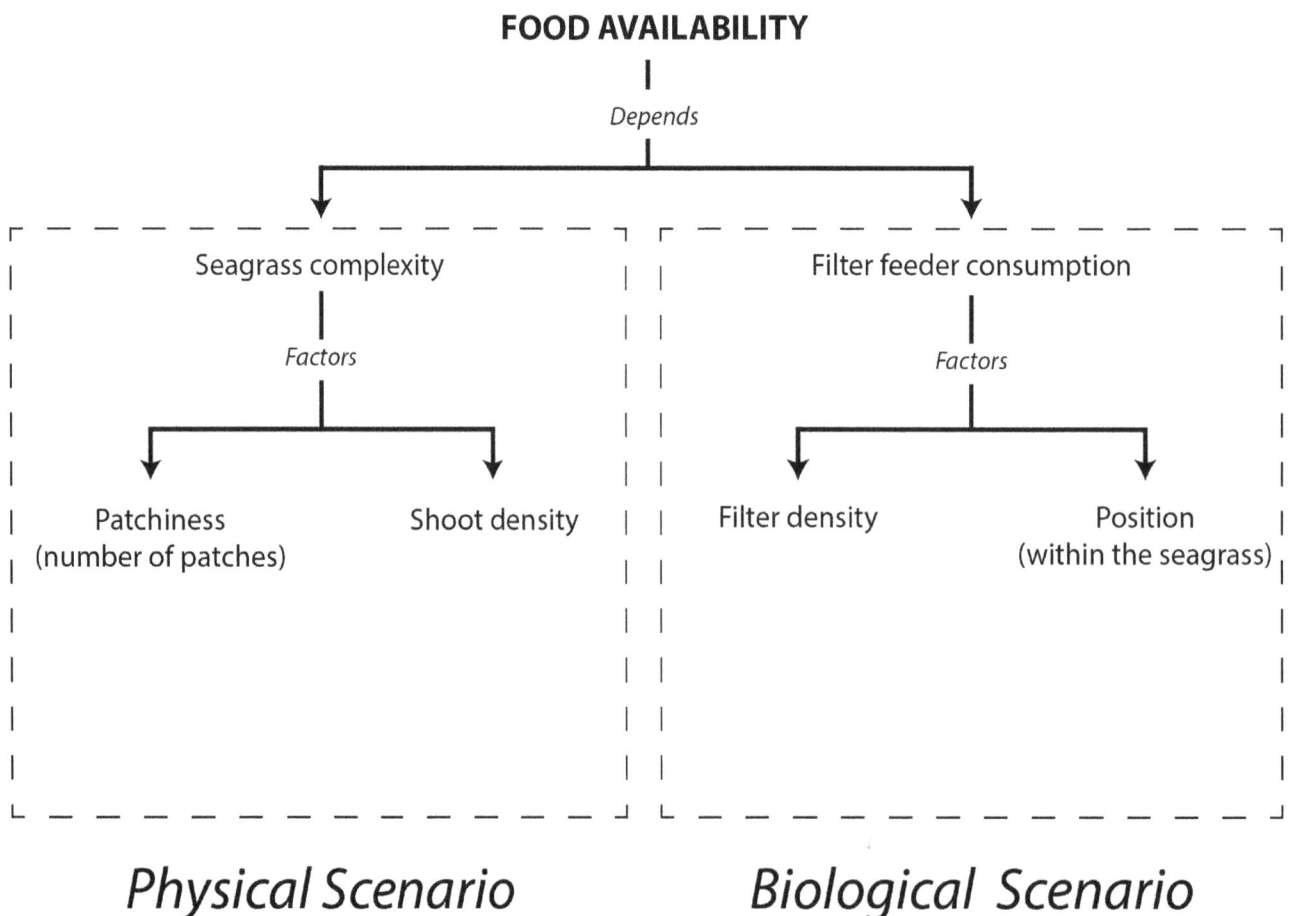

Figure 2. Tree diagram showing the relationships among all the measured factors.

complexity (i.e. a proxy for physical effect) and number of filter feeding organisms (i.e. a proxy for biological effect) could alter the FIR. To achieve these goals, a racetrack flume experiment combining different seagrass shoot densities (low and high) and degree of patchiness (1 patch and 2 patches) with the absence and presence (at high densities) of a filter feeder, the cockle, was carried out at the Netherlands Institute of Sea Research.

Materials and Methods

Artificial seagrass (mimics) design

Seagrass mimics were designed to simulate the main physical properties of the submerged vegetation of coastal areas. Manipulating the different treatments involving changes in architectural characteristics of the bed (e.g. leaf length, shoot density, patch size, etc.) was largely facilitated by the use of mimics [45,46]. Mimic structure was as follows: above-ground shoots were simulated by using a group of leaf-like plastic straps, which were attached to a wooden stick, simulating the rhizome-root system [47], through a 4×0.4 cm plastic straw filled with adhesive silicon (imitating the leaf sheath) (Figure 1A). Morphometric characteristics of the mimicked leaves (length, width and thickness; see Figure 1B) resembled those of the main species thriving in European Atlantic coasts, *Zostera noltei*, *Z. marina* and *Cymodocea nodosa*. The wooden stick kept the plastic straps anchored into the sediment,

somewhat mimicking the belowground function of real vegetation. This design ensured both buoyancy and sediment permeability through the belowground structures for the filter feeders avoiding any kind of biological interaction plant-animal (i.e. grazing, herbivory or chemical interactions).

Racetrack flume set-up

Experiments were run in a large unidirectional racetrack flume tank ([45], see Figure 1) with a test section of 200×60 cm and a total length of 1700 cm. Nine wooden boxes ($60 \times 40 \times 10$ cm^3) were constructed to create a kind of "seagrass puzzle system", where boxes (i.e. the puzzle pieces) with and without mimics could be placed interchangeably with each other at several positions (5 positions in total) within the tank to facilitate the run of the different treatments (Figure 1B). Two small plastic pots were fixed at the center of each box, then boxes were then filled with muddy sediment and mimics were planted inside (Figure 1C) using the following scheme: (1) two boxes representing low-density patches (L, 500 shoots m^{-2}); (2) two boxes representing high-density patches (H, 1500 shoots m^{-2}); and (3) five boxes without mimics to represent the bare spaces between patches (b). The flume tank was filled with natural seawater (water column height of 0.4 m) and a bare box (b) was always placed 40 cm in front of the test section (position 1). Thereafter, four of the aforementioned boxes were placed into the test section according to the following treatment

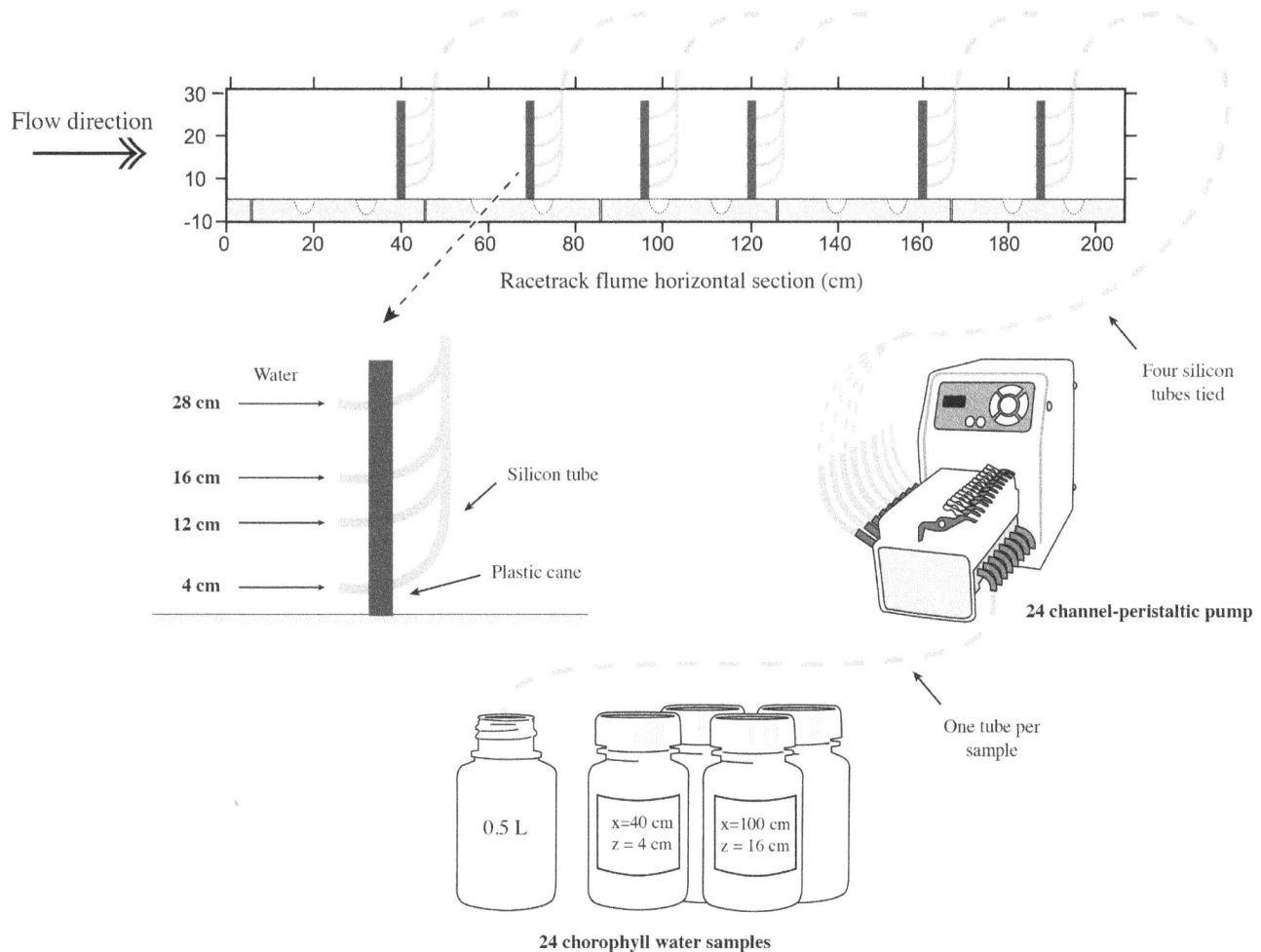

Figure 3. Detailed drawing of the chlorophyll sampling set-up.

combinations (treatments): (1) one L box (position 2) followed by three b boxes (positions 3 to 5) (Lbbb-treatment); (2) two L boxes (positions 2 and 5) separated by two b boxes (positions 3 and 4) (LbbL-treatment); (3) one H box (position 2) followed by three b boxes (positions 3 to 5) (Hbbb-treatment); (4) two H boxes (positions 2 and 5) separated by two b boxes (positions 3 and 4) (HbbH-treatment); and finally a control treatment with four b boxes (positions 2 to 5) (bbbb-control treatment) (full details in Figure **1** and **Table 1**). All the treatments were done in triplicate (n = 3).

To discriminate between physical (reduction of volumetric flow rate) and biological (filtering activity of organisms) factors controlling the resource availability and concentration for the filter feeders, treatments were carried out at two contrasting densities of filter feeders (Figure **2**). The very low density treatment of filter feeders was referred as the "physical scenario", since density was too low to explain shortages in food supply, whereas the high density treatment was called the "biological scenario" (see below).

Cockles sampling

The filter feeder cockle (*Cerastoderma edule*) was chosen due to its abundance and reliable behavior as model organism [12,48,49]. Adults were collected by hand from a muddy shore at KrabbenKreek (51° 37′ 17″N, 4° 6′ 59″E, Oosterschelde Estuary, the Netherlands) during low tide and transported within one hour

to the laboratory. As Oosterschelde Estuary is a Natura 2000 reserve, permission for scientific activities was obtained from the province of Zeeland and the "stichting Het Zeeuwse Landschap". Cockles had on average a shell length of 27.21±0.54 mm and a fresh weight of 1.68±0.09 g FW (only soft tissues, without the shell). Individuals were acclimatized to experimental conditions for a period of 7 days in a large reservoir with flowing water and constant temperature (21°C). A culture of *Isochrysis galbana* Parke, 1949 was used as food source during the period. Before any experimental run, some cockles (thereafter called "experimental cockles") were randomly selected from this large reservoir and starved for 24 hours in a 300 L oxygenated tank with circulating seawater. In both scenarios (physical and biological) 3 "experimental cockles" each placed in both small plastic pots at the center of each box ($3 \times 2 = 6$ cockles per position) and gently buried. This was done in all boxes from those in position 1 (leading edge of the flume) up to those in position 5 (downstream in the flume). In the biological scenario some starved cockles (30 per box) in addition to the "experimental cockles" were haphazardly distributed over all boxes to reach densities of 133 cockles·m^{-2} (recorded natural density in the area [50], see Figure **1A** for more details). In addition, 10 starved cockles were frozen before each run to estimate the initial chlorophyll stomach content.

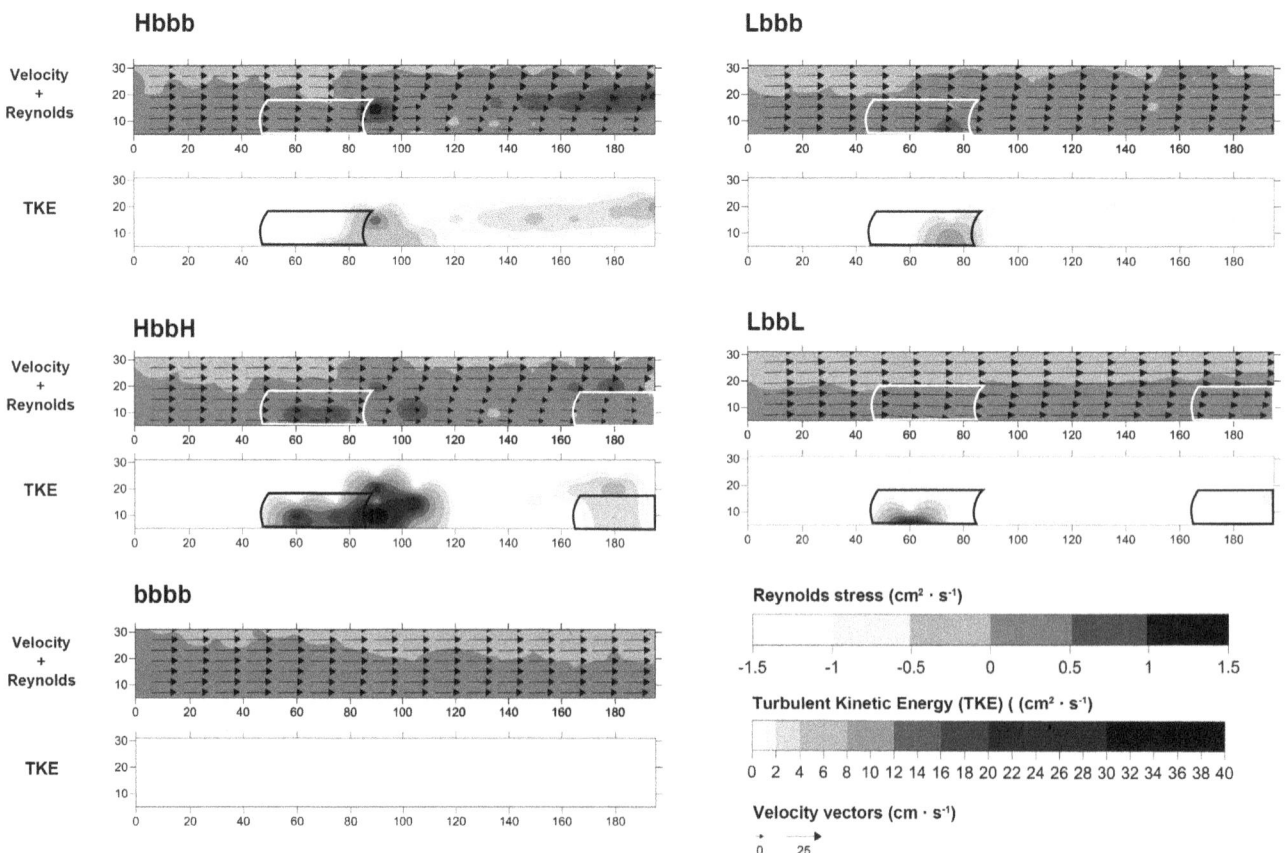

Physical scenario

Figure 4. Vector plots along the horizontal axis ($\bar{u}$) measured for different treatments in the physical scenario and Reynolds stress (τ_R) and TKE values. The graduated grey shading outlines the extent of the patch canopies.

Experimental set-up

A unidirectional flow velocity of 15 cm·s^{-1} was chosen according to values measured at the cockle sampling site (from 5 cm·s^{-1} to 25 cm·s^{-1}) [45]. Before starting each of the runs, the flow velocity of the racetrack flume was left to stabilize for 15 min, to the flow conditions, and an aliquot (50 mL) of concentrated *Isochrysis galbana* culture was added to the flume close to the drive belt to achieve an initial chlorophyll a concentration in the water of approximately 3.5 ± 1.5 µg Chl a L^{-1}. Regardless of the experimental treatment, the racetrack flume was run for 1 hour in each trial. Once the experimental period ended, "experimental cockles" were removed from pots and immediately frozen at $-20°C$ for further analysis. Subsequently, hydrodynamic characterization was done in all the trials (see below). In the biological scenario, the additional cockles added into the boxes were also removed and returned to the acclimation reservoir.

Water chlorophyll measurements in the racetrack flume

Three water samples (1 L per sample) were taken: i) one before adding the algal culture into the racetrack flume (Chla$_{flum}$), ii) one 10 min following the algal culture addition to estimate the initial chlorophyll a concentration at the beginning of each run (Chla$_{init}$), and iii) one additional water sample once the experiment ended to check the final chlorophyll a concentration in the flume (Chla$_{fin}$). Furthermore, to detect likely chlorophyll a gradients in the water (both x and z axis) resulting from cockle feeding activity, a set of

silicon tubes simulating cockle siphons were attached at different heights ($z = 4$, 12, 16 and 28 cm from the bottom) of a plastic cane [51]. Then, several sets were placed along the test section of the racetrack flume ($x = 40$, 70, 100, 120, 160 and 185 cm) in the centre of the channel ($y = 30$ cm) and connected to a 24 channel-peristaltic pump (Watson Marlow 205s) (Figure **3**). Water sampling (24 samples per set, approximately 0.5 L per sample) started 40 minutes after the experiment began and lasted for 20 min. Water samples were immediately filtered at low vacuum (Whatman GF/F, 0.7 µm filters), stored in labeled aluminum envelopes and frozen at $-80°C$ until chlorophyll a analysis. Chlorophyll a was extracted by soaking the samples in acetone (90%) in darkness (24 h at $4°C$), and the supernatant was measured using spectrophotometry [52]. Mean chlorophyll a values (n = 3) were interpolated along the tank sections (x and z axes) as chlorophyll percentage (%) with respect to the initial chlorophyll concentration (Chla$_{init}$) according to eq. (1). Thus, a value of 100% indicated that measured Chla concentration at the end of the experimental period exactly equaled Chla$_{init}$ and 0% denoted the total disappearance of initial Chla.

$$\%Chla(x,z) = \frac{\mu gChla(x,z)}{\mu gChla_{init}} * 100 \qquad (1)$$

Biological scenario

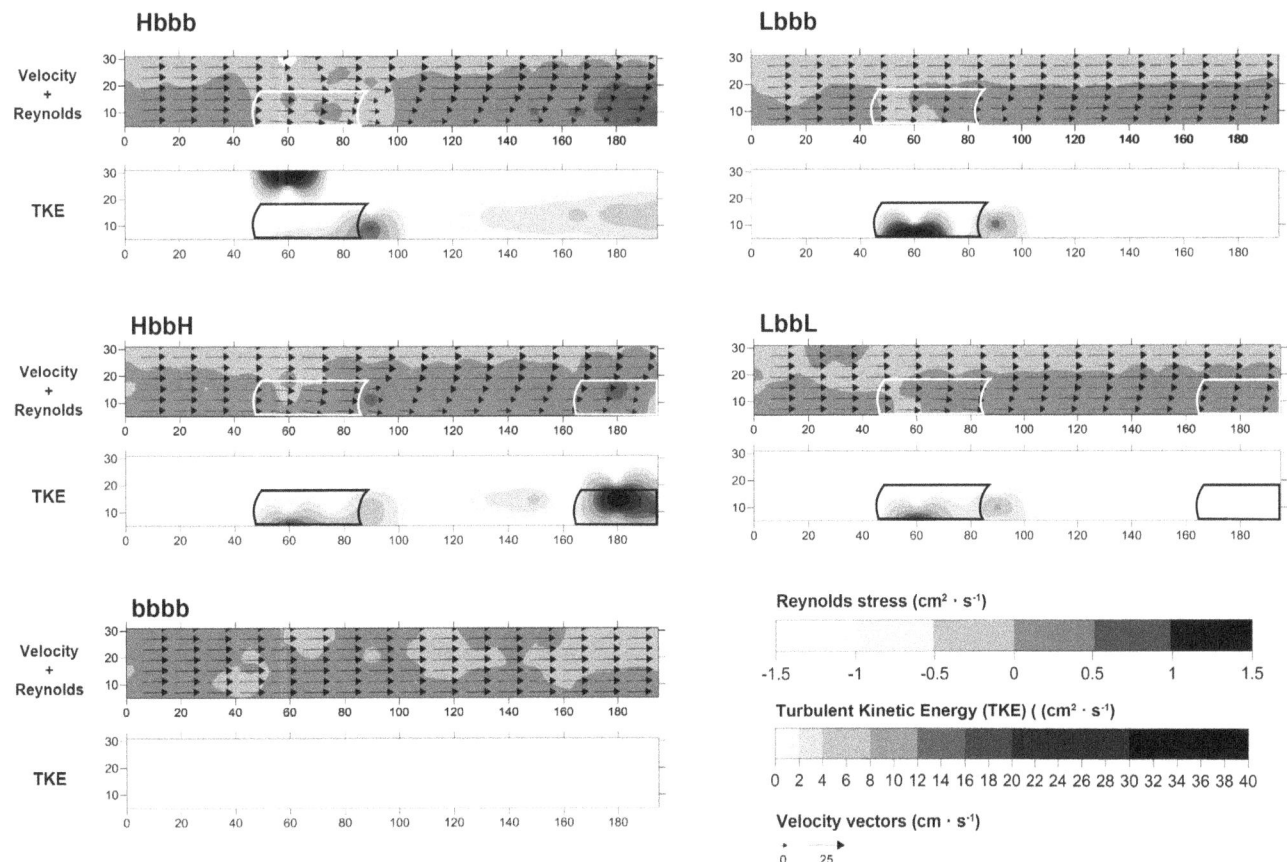

Figure 5. **Vector plots along the horizontal axis ($\bar{u}$), measured for different treatments in the biological scenario and Reynolds stress (τ_R) and TKE values.** The graduated grey shading outlines the extent of the patch canopies.

Table 2. Results of three-way ANOVA.

Factor	df	Physical scenario				Biological scenario			
		SS	MS	F	P	SS	MS	F	P
a) U velocity ($cm \cdot s^{-1}$)									
Shoot density	1	361.21	361.22	70.19	**>0.001**	258.87	258.87	27.19	**>0.001**
Patch number	1	34.92	34.93	6.78	**0.001**	16.374	16.37	1.72	0.191
Position	4	323.45	80.86	15.71	**>0.001**	314.83	78.71	8.26	**>0.001**
Shoot density×Patch number	1	0.79	0.79	0.15	0.695	12.579	12.58	1.32	0.251
Shoot density×Position	4	274.15	68.53	13.32	**>0.001**	215.53	53.88	5.66	**>0.001**
Patch number×Position	4	16.89	4.22	0.82	0.513	70.27	17.57	1.84	0.121
Shoot×Patch number×Position	4	5.23	1.30	0.25	0.906	25.08	6.27	0.65	0.621
Error		1029.1	5.14			1903.70	9.51		
b) TKE ($cm^{-2} \cdot s^{-2}$)									
Shoot density	1	233.37	233.37	9.10	**0.003**	51.97	51.97	0.74	0.390
Patch number	1	122.47	122.47	4.77	**0.029**	0.007	0.007	0.0001	0.991
Position	4	409.31	102.33	3.99	**0.004**	384.69	96.17	1.37	0.244
Shoot density×Patch number	1	105.96	105.96	4.13	**0.04**	122.22	122.27	1.74	0.187
Shoot density×Position	4	538.63	134.65	5.25	**0.0004**	723.48	180.87	2.58	**0.038**
Patch number×Position	4	298.57	74.64	2.91	**0.022**	248.17	62.04	0.88	0.473
Shoot×Patch number×Position	4	202.17	50.54	1.97	0.10	262.55	65.63	0.93	0.443
Error		5125.68				14010.12	70.05		
c) Reynolds stress (Pa)									
Shoot density	1	0.215	0.215	4.671	**0.03**	0.118	0.118	4.470	**0.035**
Patch number	1	0.005	0.005	0.114	0.735	0.015	0.015	0.582	0.446
Position	4	0.238	0.059	1.289	0.275	1.235	0.308	11.656	**>0.001**
Shoot density×Patch number	1	0.004	0.004	0.104	0.746	0.024	0.024	0.910	0.341
Shoot density×Position	4	0.121	0.030	0.657	0.622	0.213	0.053	2.012	0.09
Patch number×Position	4	0.052	0.013	0.285	0.887	0.421	0.105	3.977	**0.004**
Shoot×Patch number×Position	4	0.591	0.147	3.200	**0.01**	0.355	0.088	3.358	**0.01**
Error		9.242	0.046			5.295	0.026		

The hydrodynamic variables were tested with the factors "shoot density," "number of patches" and "position" in both the physical and biological scenarios. Significant differences are shown in bold.

Table 3. Flow velocity in the different scenarios (Phy = physical; Bio = biological), treatments and positions along the racetrack flume.

Scenario	Treatment	Velocity (cm·s−1)				
		Position 1	Position 2	Position 3	Position 4	Position 5
Phy	bbbb	15.82±0.37	16.36±0.21	16.38±0.30	16.46±0.22	16.06±0.56
	Lbbb	16.10±0.38	14.65±0.81	14.35±0.81	15.67±0.70	13.99±0.61
	LbbL	15.27±0.14	14.15±0.62	14.15±0.50	13.25±0.59	12.89±0.69
	Hbbb	15.69±0.17	14.64±0.71	11.87±1.97	9.33±0.86	8.41±0.26
	HbbH	14.57±0.20	14.29±0.82	12.11+1.65	8.38±0.81	6.85±0.93
Bio	bbbb	16.26±0.30	16.37±0.20	15.90±0.18	16.55±0.11	16.61±0.20
	Lbbb	14.43±0.27	13.59±0.68	12.59±1.19	14.04±0.75	13.93±0.79
	LbbL	15.46±0.13	14.13±0.71	13.69±0.63	12.98±0.46	12.71±0.52
	Hbbb	15.28±0.14	14.62±0.97	8.52±2.03	10.16±1.55	10.65±1.32
	HbbH	14.85±0.23	13.65±1.05	8.73±1.74	10.26±1.80	6.10±1.95

Data show the average for the canopy height (18 cm) ± SD. L, low shoot density; H, high shoot density and b, bare sediment.

Hydrodynamic measurements in the racetrack flume

Once an experimental period ended and "experimental cockles" were carefully collected, the flume was left running (at the same flow velocity) to characterize the hydrodynamic environment. The three components of velocity (u [horizontally parallel to the flume], v [vertical] and w [horizontally perpendicular to the flume]) were measured at 10 Hz with an acoustic Doppler velocimeter (Nortek ADV). The hydrodynamic variables estimated were: (1) the velocity profile ($\bar{u}$, cm·s^{-1}) including the vector of direction; (2) the turbulent kinetic energy (TKE, cm^{-2}·s^{-2}) as a proxy for the turbulent energy per mass of fluid; and (3) the Reynolds stress (τ_R, Pa) as a proxy for the vertical transfer of turbulence. In all the treatments, a 3D grid consisting of 98 points regularly distributed in 14 steps of 0.15 m along the x axis (from $x = 0$ to $x = 195$ cm), 7 steps of 0.05 m along the z axis (from $z = 1$ in the bottom to $z = 31$ cm) and 3 points along the y axis ($y = 19$, $y = 24$ and $y = 29$ cm) was used. At all points the ADV was positioned for 50 s, rendering 500 measurements per point (i.e. 10 measurements per second—10 Hz—during 50 s). The hydrodynamic parameters were corrected by removing those data with correlations below 70% (low correlation indicates unreliable data) as done by Morris et al. [32]. The velocity vector and the average velocity along the x axis ($\bar{u}$), turbulent kinetic energy (TKE = $0.5 \cdot (u'^2 + v'^2 + w'^2)$) and Reynolds stress ($\tau_R = \overline{\rho u' w'}$) were calculated according to published equations [46,53]. To get an overview of the hydrodynamic effects promoted by patchiness, hydrodynamic data were pooled within each of the 5 boxes along the x-axis (from position 1 to 5; Figure 1D).

Statistics analysis

Significant differences in hydrodynamic parameters (i.e. ($\bar{u}$), TKE and τ_R) were checked using a 3-way fixed-factor ANOVA (number of patches, shoot density and position) separately in both scenarios (physical and biological). This method was used in order to give a simple and comprehensible framework for the interrelation of multiple factors and their effect on the FIRs of the filter feeders. Data normality and homoscedasticity were checked before the ANOVA. To test for significant differences in the cockle FIR among different treatments and positions, a non-parametric Kruskal-Wallis test was applied, since data were not normally distributed even after applying several different trans-

formations. Data are presented as mean ± 1 standard error and significance levels were set at $p = 0.05$.

Results

Interaction between mimics and flow

Flow velocity ($\bar{u}$, cm·s^{-1}) showed a significant spatial reduction along the x-z plane of the racetrack flume with both increasing shoot density and with the number of patches (Hbbb and HbbH) regardless of the scenario (physical or biological, Table 2, Figures 4 and 5). In contrast, no differences were found in unvegetated controls (bbbb) under either biological or physical scenarios (K-W p-value<0.05; Table 3). Overall, a well-formed TKE wake behind the first patch (position 3) was observed, especially in the high shoot density treatments (Hbbb and HbbH). In the physical scenario, TKE values were modified by shoot density, number of patches and position (Table 2, Figure 4), whereas only the interaction between shoot density and position had a significant effect in the biological scenario (Table 2, Figure 4). In the physical scenario, TKE values increased with shoot density and number of patches: from 0.29±0.06 cm^{-2}·s^{-2} (control, bbbb) to 8.30±3.50 cm^{-2}·s^{-2} (HbbH) in the first patch (position 2), and from 0.38±0.17 cm^{-2}·s^{-2} (bbbb) to 1.47±0.51 cm^{-2}·s^{-2} (HbbH) in the second patch (position 5). A similar increase in TKE values were also recorded in the biological scenario (from 0.12±0.06 cm^{-2}·s^{-2} (bbbb) to 12.41±6.61 cm^{-2}·s^{-2} (HbbH)) but only in the second patch (position 5) (Table 4). Large fluctuations among treatments were recorded for τ_R, showing differences with the combination of shoot density, number of patches and box position in both scenarios (Table 2). In the biological scenario, τ_R values in the first patch (position 2) were lower than those recorded at adjacent positions (Figure 5).

Concentration and availability of resources

In all the treatments applied in the physical scenario, the final water chlorophyll a concentration (Chla$_{fin}$) was between 80–95% of the initial values, showing no noticeable differences with control treatment (bbbb). This trend varied slightly in the treatment LbbL where values decreased to 75% within the first patch (position 2). In the biological scenario, chlorophyll a remained between 80–95% in the control as well as in the low shoot density treatments

Table 4. TKE (cm^{-2}·s^{-2}) and Reynolds stress (τ_R, Pa) values in the different scenarios (Phy = physical; Bio = biological), treatments and positions along the racetrack flume (experimental details in Figure 1).

Scenario	Treatment	Position 1		Position 2		Position 3		Position 4		Position 5	
		TKE	τ_R	TKE	τ_R	TKE	τ_R	TKE	τ_R	TKE	τ_R
Phy	bbbb	0.25±0.076	0.15±0.036	0.29±0.06	0.15±0.02	0.28±0.10	0.14±0.039	0.28±0.12	0.13±0.04	0.38±0.17	0.14±0.05
	Lbbb	0.14±0.07	0.09±0.04	2.98±1.65	0.24±0.09	0.54+0.07	0.11±0.02	0.92±0.10	0.09±0.02	1.05±0.05	0.09±0.01
	LbbL	0.03±0.01	0.04±0.01	4.55±4.11	0.09±0.04	0.55±0.21	0.14±0.05	0.45±0.11	0.17±0.02	0.61±0.03	0.17±0.02
	Hbbb	0.15±0.07	0.09±0.03	4.67±1.78	0.12±0.04	0.88±0.7	0.29±0.16	2.11±0.83	0.25±0.12	1.54±0.66	0.27±0.13
	HbbH	0.10±0.04	0.08±0.02	8.30±3.50	0.34±0.14	14.77±5.82	0.30±0.09	0.63±0.13	0.13±0.06	1.47 ± 0.51	0.07±0.03
Bio	bbbb	0.03±0.01	0.02±0.01	0.16±0.07	0.03±0.02	0.05±0.27	0.03±0.02	0.04±0.02	0.02±0.01	0.12±0.06	0.05±0.03
	Lbbb	0.04±0.01	0.03±0.02	11.47±11.03	0.00±0.09	2.64±1.88	0.11±0.03	0.49±0.11	0.19±0.03	0.44±0.07	0.20±0.03
	LbbL	0.047±0.01	0.03±0.01	3.84±2.98	0.07±0.05	1.39±0.80	0.10±0.02	0.46±0.06	0.17±0.01	0.62±0.12	0.19±0.02
	Hbbb	0.04±0.01	0.04±0.01	1.55±0.85	−0.42±0.07	3.93±2.33	0.03±0.90	2.75±0.56	0.33±0.07	3.95±0.65	0.55±0.098
	HbbH	0.17±0.11	0.04±0.01	3.96±2.65	0.07±0.04	2.98±1.65	0.18±0.08	1.74±0.71	0.22±0.05	12.41±6.61	0.19±0.10

Data are the average for the canopy height (18 cm) ± SD. L, low shoot density; H, high shoot density and b, bare sediment.

(Lbbb and LbbL). However, a remarkable decrease (up to 50% of the initial value) was recorded in the high shoot density treatments (Hbbb and HbbH) (Figure **6**).

Interaction between mimics and cockles

Overall, "experimental cockles" ingested 25% more chlorophyll when occurring at high (biological scenario) than at low cockle density (physical scenario). Spatially, no significant differences in FIR were found among treatments (i.e. shoot density and number of patches) in the physical scenario, although cockles ingested more chlorophyll in the vegetated treatments than in the control ones in both scenarios (Figure **7**). Contrastingly, the chlorophyll stomach content of the individuals located either at position 1 (ahead of the forefront patch) or 3 (behind the forefront patch) of the high shoot density treatments (Hbbb and HbbH) of the biological scenario was significantly higher than that found in the adjacent positions. No significant differences among positions were found either in the control (bbbb) or in the low-density patches (Lbbb and LbbL; Figure **7**), but cockles ingested significantly more chlorophyll in treatment Lbbb of the biological scenario.

Discussion

This work provides the first quantitative evidence that food availability to filter feeders is modulated by seagrass patch complexity (i.e. shoot density and spatial patch arrangement). Furthermore, it also shows that filter feeder density is a key factor controlling food concentration for such organisms inhabiting seagrass meadows. Our experimental design created a matrix balancing both the hydrodynamic food supply rate (controlled by the seagrass characteristics) and the food consumption rate (controlled by cockle density), thereby generating different food supply gradients. These gradients affected the food intake rate (FIR) of individuals ("experimental cockles"), depending on their specific spatial position within the flume. This was especially noticeable in the biological scenario, where cockles at position 2 (ahead of the forefront patch) exhibited the highest FIRs with high shoot density treatments (Hbbb and HbbH). This could result from variations in TKE and τ_R associated with the pumping activity of cockles (i.e. biomixing) observed in this work.

Interaction between mimics and flow

In agreement with previous studies [33,34,45,46,54], our results showed the development of a turbulent patch–wake behind the forefront patch, an increase of TKE at the edges of the patch and a gradual reduction of the ($\bar{u}$) velocity within the seagrass canopy. The strongest reduction of ($\bar{u}$) velocity was recorded within the high-density shoot patches, with highest TKE-values at the edges. In contrast, in the low-density shoot patches, the edge effect was attenuated due to the greater permeability of the seagrass canopy and the higher values of volumetric flow rate [32,54]. A reduction in the TKE wake was observable between the Hbbb and HbbH treatments. Such effects have been previously described by Folkard [54], where the water velocity was quickly decreased between two high density patches, indicating that the wake was weakened by interaction with the second patch. When the gaps between the patches are small, there is not enough space for wake formation between them, and the hydrodynamic characteristics of a theoretically homogeneous meadow prevail. The reduction of ($\bar{u}$) velocity associated with the presence of a second patch (in position 5) at some distance downstream from the forefront one (in position 2) indicated that both density and number of patches modulated the flow speed. The differences in TKE and τ_R within the forefront patch observed between physical scenario and

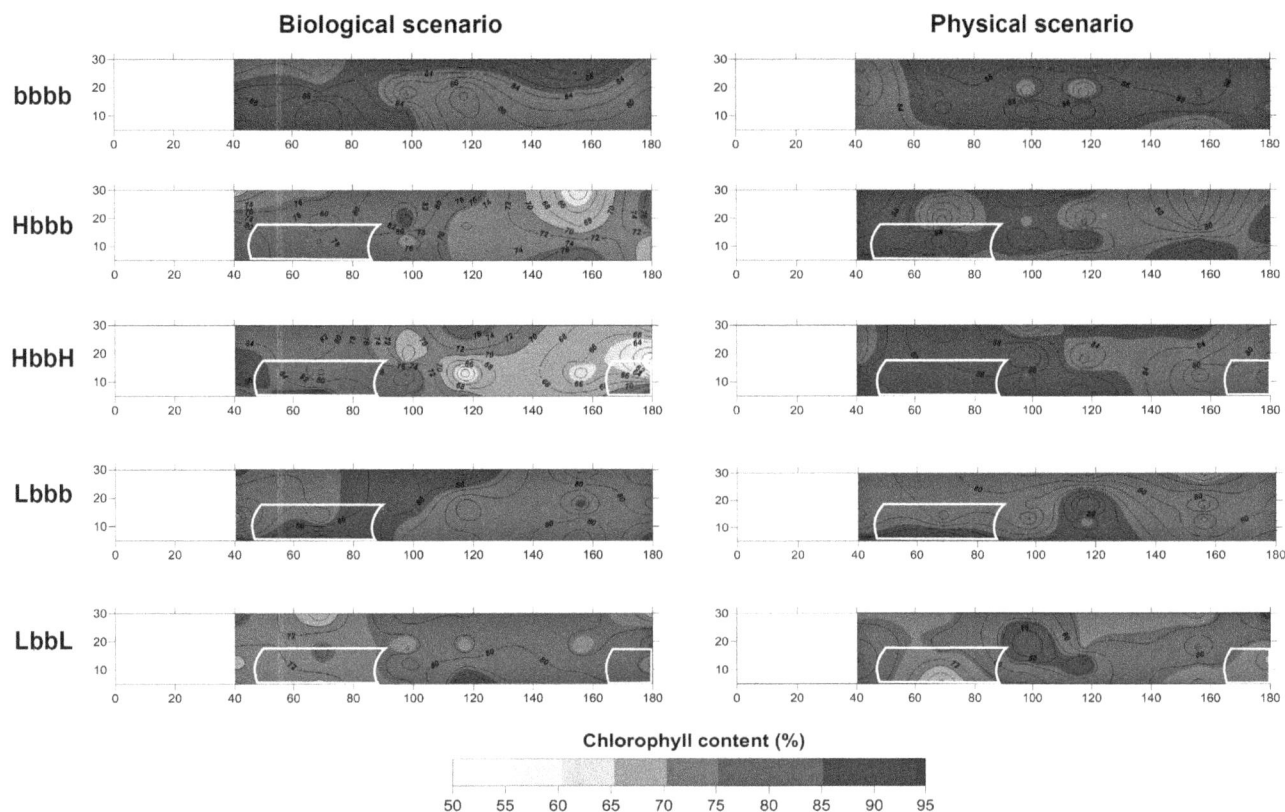

Figure 6. Water chlorophyll a content. Mean values (n = 3) were interpolated along the test section (x/z plane) as a percentage (%), where 100% is the initial concentration of chlorophyll *a* measured following the addition of the algae culture and 0% is the total absence of chlorophyll *a*.

biological scenario demonstrated how the presence of high densities of cockles influenced flow characteristics via the exhalant jets of their siphons (i.e. bio-mixing *sensu* [17,19,55]).

Resource concentration balance

Even though turbulence increased at the patch edges of the high shoot density treatment (i.e. HbbH) of the physical scenario, volumetric flow rate diminished. This concentrated a high percentage of chlorophyll *a* behind the forefront patch (position 2), while this percentage lessened when the mass of water moved downstream away from this patch. The edge effect seemed to enhance trapping and sedimentation of particles in accordance with previous studies [36]. Therefore, these areas contain plenty of edible particles (i.e. phytoplankton) available for filter feeders. By contrast, in the low shoot density canopies of the physical scenario, the attenuated edge effect and the high volumetric flow rate prevented any food depletion within the patch. Surprisingly, no differences in the FIRs were found among treatments or positions within the physical scenario (i.e. spatial differences) suggesting a large availability of resources for cockles in all treatments due to (1) high resource renewal exceeding the rate of consumption and (2) the absence of intraspecific competition. Such an idea was also supported by the absence of chlorophyll *a* gradients in the water, which indicated that food availability was not limiting in the physical scenario, although an increase in the filtering efficiency of the organisms due to reduction of the flow velocity had been expected [26,28,38].

Even though no strong variation in water chlorophyll *a* was observed in the low shoot density treatments (Lbbb and LbbL) of the biological scenario, depletion was clearly associated with high-

density patches (Hbbb and HbbH). In the low shoot density treatments—despite the higher resource consumption by the dense cockle bed—the unidirectional flow passed freely through the seagrass patch as in the control (unvegetated, bbbb) treatments, which refreshed the water across the canopy and avoided the formation of chlorophyll *a* gradients above cockles. This may explain why cockles in the low-density shoot treatments (Lbbb and LbbL) had similar FIRs, regardless of spatial location along the racetrack flume. In opposition, large variations in water chlorophyll *a* were detected in the high-density shoot treatments (Hbbb and HbbH), leading to a reduction of up to 50% of the initial chlorophyll *a* concentration behind the forefront patch. This pattern was correlated with the noticeable spatial differences found in the cockle FIR: two peaks of ingestion were observed with one ahead of (position 1) and another behind (position 3) the first patch. This agrees well with the increment in turbulence and the sharp water speed reduction observed within the canopy, where the organisms were able to capture the particles before total depletion of chlorophyll.

Filter feeder food intake rate

Cockle FIR was higher in all seagrass treatments than in control ones (i.e. unvegetated). Present findings are in agreement with previous studies reporting positive effects on growth and survival of clams inhabiting seagrass meadows [6,7,10]. For the biological scenario we expected lower FIRs due the strong intraspecific competition according to the reduction of chlorophyll *a* detected in Hbbb and HbbH treatments. However, cockles surprisingly ingested 25% more chlorophyll than under the physical scenario (i.e. very low cockle density). Such an increase in FIR could be

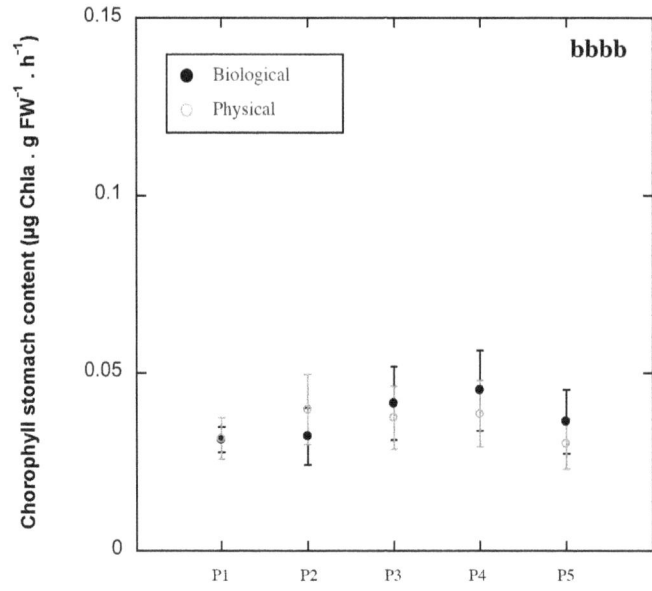

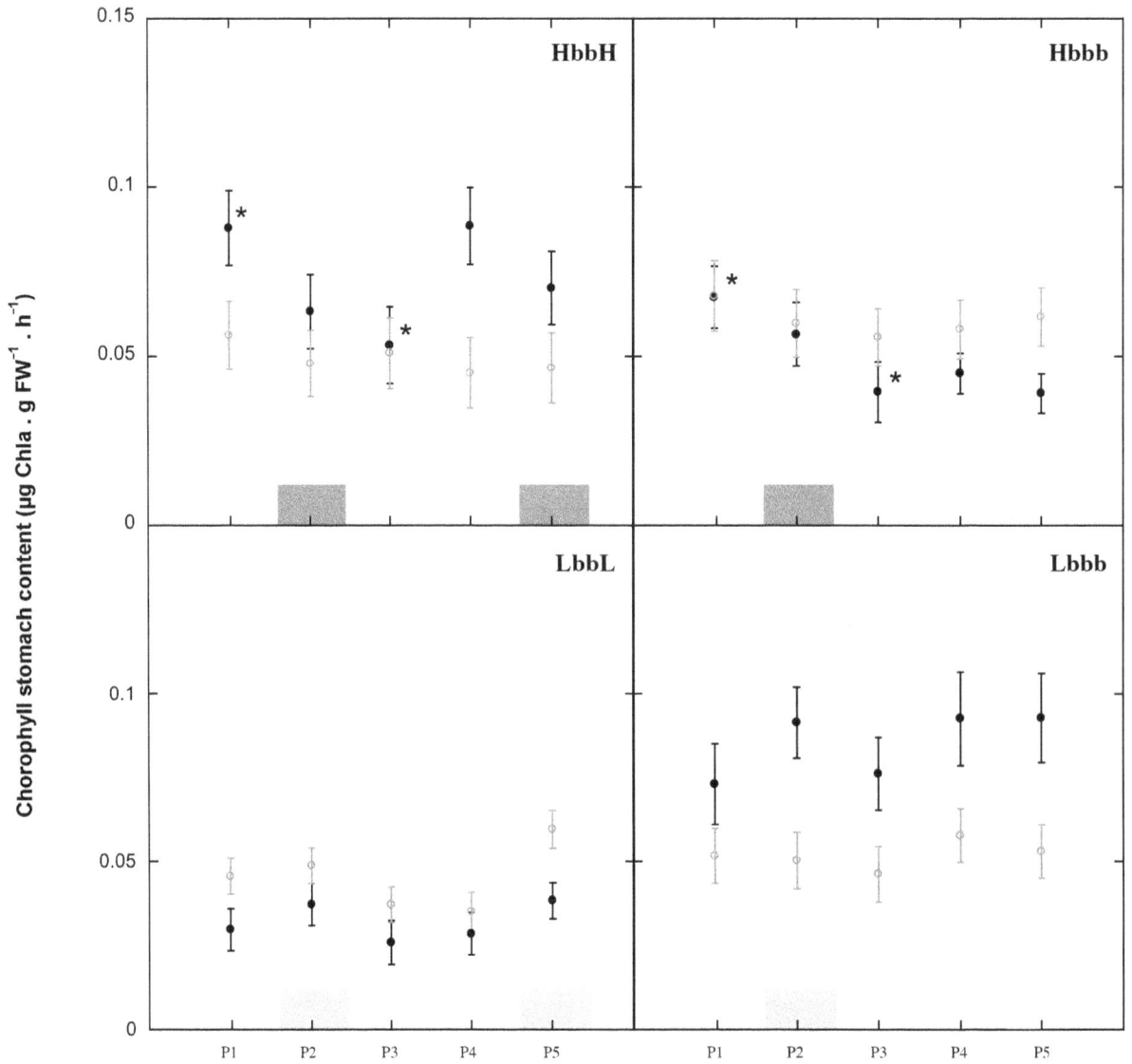

Figure 7. Mean chlorophyll stomach content of the cockles along the racetrack flume for different treatments and scenarios. Grey squares indicate the position of the patch (dark grey indicate high shoot density and light grey indicate low shoot density). Asterisks denote significant differences tested by the Kruskal-Wallis test (*p*-value<0.05). P (*1–5*) refers to the positions hosting the cockles along the racetrack flume.

initially explained by the alteration of their foraging behavior in response to resource depletion [44,56,57]. However, our results suggest that biomixing could also play an important role, allowing cockles to access additional resources. Filtering (FIR) has been shown to be affected—especially at low velocity regimes—by the biomixing generated by the pumping of other bivalves [19,55,58]. Siphon inhalation continuously withdraws particles from the bulk water, while exhalation feathers favor vertical water mixing, thus altering the structure of the bottom boundary layer and enhancing the refreshment of resources in depleted environments [19,44,59,60]. Considering that bivalves change their feeding behavior depending on particle concentration in the water column [38,44,56,57], or on the population density [61], food concentration and vertical mixing promoted by the exhalant feathers might

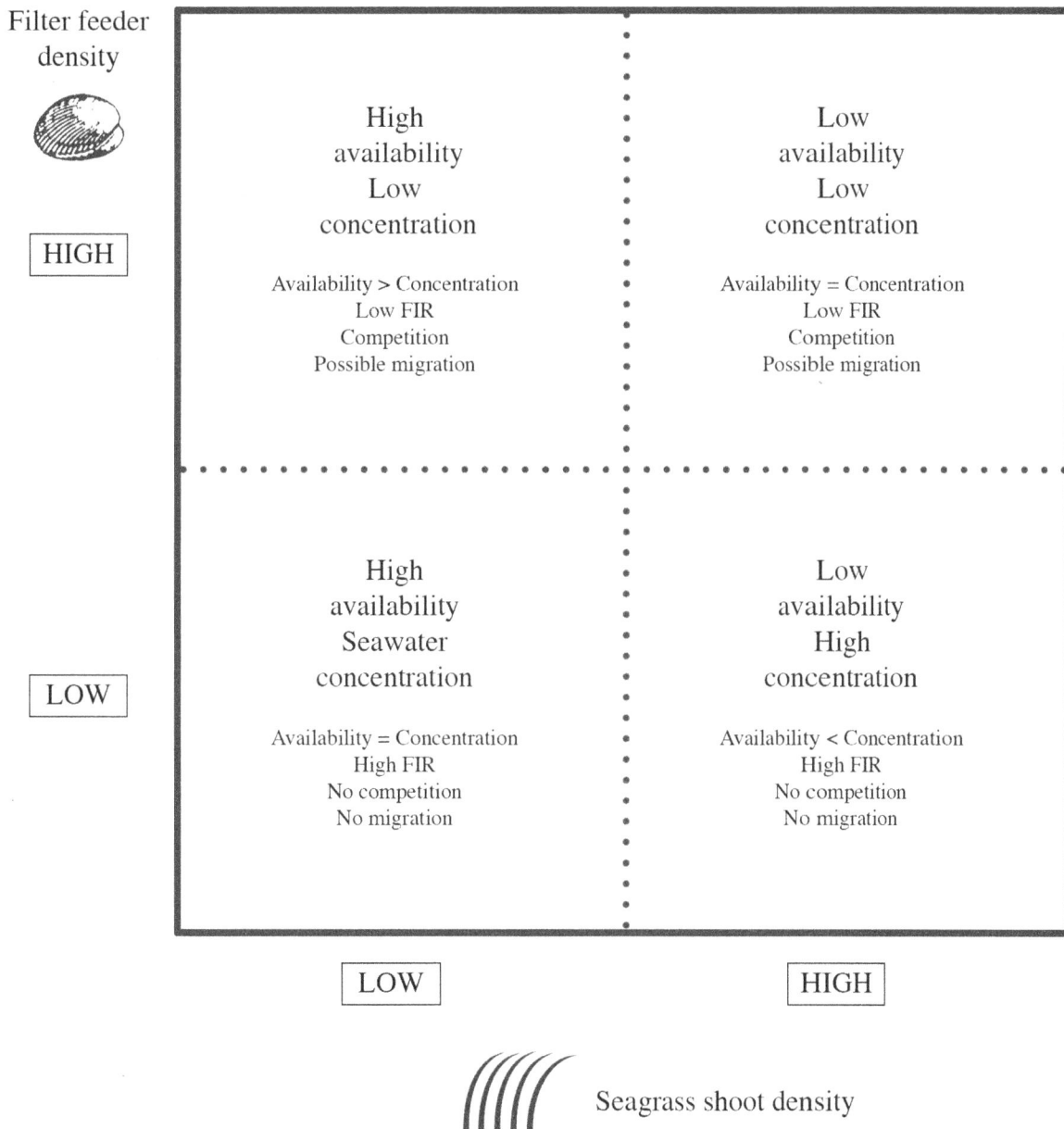

Filter feeder density

HIGH

LOW

High availability Low concentration

Availability > Concentration
Low FIR
Competition
Possible migration

Low availability Low concentration

Availability = Concentration
Low FIR
Competition
Possible migration

High availability Seawater concentration

Availability = Concentration
High FIR
No competition
No migration

Low availability High concentration

Availability < Concentration
High FIR
No competition
No migration

LOW   HIGH

Seagrass shoot density

Figure 8. Conceptual model showing the effects of filter feeders and shoot density on resource availability and concentration. Higher shoot densities reduce resource availability (e.g. lower volumetric flow rate) but may increase resource concentration (e.g. deposition or settling). Higher density of filter feeders will reduce resource concentration (e.g. active filtration by organisms) but also may increase biomixing. Thus, the balance between availability and concentration of resources may promote changes at the community levels (e.g. migration of species depending on resources availability).

be reciprocally modulated [19,44,58]. Accordingly, Sobral and Widdows [38] reported that the bivalve *Ruditapes decussatus* caused a significant depletion of the phytoplankton concentration at low water velocities but maintained high filtration rates by ejecting water at different heights in the water column (i.e. biomixing), avoiding the recirculation of algal cell-depleted water in the surroundings of the intake siphons. Also, Riisgård and Larsen [58] pointed out that biomixing enhanced the flow-induced down-mixing of phytoplankton, which will benefit the turnover of the low-chlorophyll concentration water layer and could be identified as peaks in profiles of turbulent shear stress and turbulent kinetic energy.

Ecological relevance

This work showed that food availability in seagrass meadows is the outcome of a complex interaction between hydrodynamic forces and habitat complexity together with intra-specific competition (consumption and bio-mixing). Our results suggest the existence of a direct correlation between plants (mimics) and bivalve density (Figure **8**): fragmented and sparse seagrass meadows could offer protection and increase food availability

(higher particles flux) for filter feeders without retaining the particles within the canopy. In opposition, continuous and dense meadows enhance the concentration of resources (e.g deposition or settling) favoring food intake but will limit resources at high filter feeder densities due to the lower turnover of the water and to the depletion of resources promoted actively by organisms. Under such scenario, the activity of the filter feeders demonstrated to play also an important role, contributing to reduce resource starvation.

Acknowledgments

We thank the Netherlands Institute of Marine Research (NIOZ) for its support. Also, we thank Jos van Soelen, Bas Koutstaal and Lowie Haazen for invaluable technical assistance. This is CEIMAR journal publication no. 68.

Author Contributions

Conceived and designed the experiments: VG FGB TB JLP. Performed the experiments: VG LGE RJ FM. Analyzed the data: VG LGE RJ FM FGB. Contributed reagents/materials/analysis tools: FGB TB JLP. Contributed to the writing of the manuscript: VG FGB TB JLP.

References

1. Jones CG, Gutierrez JL, Byers JE, Crooks JA, Lambrinos JG, et al. (2010) A framework for understanding physical ecosystem engineering by organisms. Oikos 119: 1862–1869.
2. Fonseca MS, Fisher JS (1986) A comparison of canopy friction and sediment movement between four species of seagrass with reference to their ecology and restoration. Mar Ecol Prog Ser 29: 15–22.
3. Bertness MD, Callaway R (1994) Positive interactions in communities. Trends Ecol Evol 9(5): 191–193.
4. Bruno JF, Stachowicz J, Bertness M (2003) Inclusion of facilitation into ecological theory. Trends Ecol Evol 18: 119–125.
5. Wilson DM (1991) The effect of submerged vegetation on the growth and the incidence of siphon nipping and incidence of siphon nipping of the northern quahog, *Mercenaria mercenaria* (Bivalvia), The University of Alabama at Tuscaloosa. 112 p.
6. Irlandi EA, Peterson CH (1991) Modification of animal habitat by large plants: mechanisms by which seagrasses influence clam growth. Oecologia 87(3): 307–318.
7. Irlandi EA (1996) The effects of seagrass patch size and energy regime on growth of a suspension-feeding bivalve. J Mar Res 54(1): 161–185.
8. Reusch TB, Williams SL (1999) Macrophyte canopy structure and the success of an invasive marine bivalve. Oikos 84: 398–416.
9. Irlandi EA, Orlando BA, Ambrose Jr WG (1999) Influence of seagrass habitat patch size on growth and survival of juvenile bay scallops, *Argopecten irradians concentricus* (Say). J Exp Mar Biol Ecol 235(1): 21–43.
10. Peterson BJ, Heck Jr KL (2001) Positive interactions between suspension-feeding bivalves and seagrass—a facultative mutualism. Mar Ecol Prog Ser 213: 143–155.
11. Allen BJ, Williams SL (2003) Native eelgrass *Zostera marina* controls growth and reproduction of an invasive mussel through food limitation. Mar Ecol Prog Ser 254: 57–67.
12. Brun FG, Van Zetten E, Cacabelos E, Bouma TJ (2009) Role of two contrasting ecosystem engineers (*Zostera noltii* and *Cymodocea nodosa*) on the filter feeding rate of *Cerastoderma edule*. Helgoland Mar Res 63: 19–25.
13. Dang C, De Montaudouin X, Gam M, Paroissin C, Bru N, Caill-Milly N (2010) The Manila clam population in Arcachon Bay (SW France): Can it be kept sustainable? J Sea Res 63(2): 108–118.
14. Carroll JM, Peterson BJ (2013) Ecological trade-offs in seascape ecology: bay scallop survival and growth across a seagrass seascape. Landscape Ecol 28: 1401–1413.
15. Jørgensen CB (1990) Bivalve filter feeding: hydrodynamics, bioenergetics, physiology and ecology. Olsen & Olsen. 140 p.
16. Jørgensen CB (1996) Bivalve filter feeding revisited. Mar Ecol Prog Ser 142(1): 287–302.
17. Riisgård HU (1998) Filter feeding and plankton dynamics in a Danish fjord: a review of the importance of flow, mixing and density-driven circulation. J Environ Manage 53(2): 195–207.
18. Widdows J, Lucas JS, Brinsley MD, Salkeld PN, Staff FJ (2002) Investigation of the effects of current velocity on mussel feeding and mussel bed stability using an annular flume. Helgoland Mar Res 56(1): 3–12.
19. Lassen J, Kortegård M, Riisgård HU, Friedrichs M, Graf G, et al. (2006) Downmixing of phytoplankton above filter-feeding mussels—interplay between water flow and biomixing. Mar Ecol Prog Ser 314: 77–88.
20. Danovaro R, Fabiano M (1997) Seasonal changes in quality and quantity of food available for benthic suspension-feeders in the Golfo Marconi (North-western Mediterranean. Est Coast Shelf Sci 44(6): 723–736.
21. Wildish D, Kristmanson D (2005) Benthic suspension feeders and flow. Cambridge University Press. 424 p.
22. Arkema KK (2009) Flow-mediated feeding in the field: consequences for the performance and abundance of a sessile marine invertebrate. Mar Ecol Prog Ser 388: 207–220.
23. Cahalan JA, Siddall SE, Luckenbach MW (1989) Effects of flow velocity, food concentration and particle flux on growth rates of juvenile bay scallops *Argopecten irradians*. J Exp Mar Biol Ecol 129(1): 45–60.
24. Fréchette M, Bourget E (1985) Food-limited growth of *Mytilus edulis* L in relation to the benthic boundary layer. Can J Fish Aquat Sci 42(6): 1166–1170.
25. Muschenheim DK (1987) The dynamics of near-bed seston flux and suspension-feeding benthos. J Mar Res 45(2): 473–496.
26. Wildish DJ, Miyares MP (1990). Filtration rate of blue mussels as a function of flow velocity: preliminary experiments. J Exp Mar Biol Ecol 142(3): 213–219.
27. Grizzle RE, Langan R, Huntting Howell W (1992) Growth responses of suspension-feeding bivalve molluscs to changes in water flow: differences between siphonate and nonsiphonate taxa. J Exp Mar Biol Ecol 162(2): 213–228.
28. Wildish DJ, Saulnier AM (1993) Hydrodynamic control of filtration in *Placopecten magellanicus*. J Exp Mar Biol Ecol 174(1): 65–82.
29. Kater BJ, Geurts van Kessel AJM, Baars JJMD (2006) Distribution of cockles *Cerastoderma edule* in the Eastern Scheldt: habitat mapping with abiotic variables. Mar Ecol Prog Ser 318: 221–227.
30. Butman CA, Frechette M, Geyer WR, Starczak VR (1994) Flume experiments on food supply to the blue mussel *Mytilus edulis* as a function of boundary-layer flow. Limnol Oceanogr 39(7): 1755–1768.
31. Tsai C, Yang S, Trimble AC, Ruesink JL (2010) Interactions between two introduced species: *Zostera japonica* (dwarf eelgrass) facilitates itself and reduces condition of *Ruditapes philippinarum* (Manila clam) on intertidal flats. Mar Bio 157(9): 1929–1936.
32. Morris EP, Peralta G, Brun FG, Van Duren LA, Bouma TJ, et al. (2008) Interaction between hydrodynamics and seagrass canopy structure: Spatially explicit effects on ammonium uptake rates. Limnol Oceanogr 53(4): 1531–1539.
33. Fonseca MS, Cahalan JA (1992) A preliminary evaluation of wave attenuation by four species of seagrass. Est Coast Shelf Sci 35(6): 565–576.
34. Gambi MC, Nowell ARM, Jumars PA (1990) Flume observations on flow dynamics in *Zostera marina* (eelgrass) beds. Mar Ecol Progr Ser 6: 159–169.
35. Koch EW, Ackerman JD, Verduin J, van Keulen M (2006) Fluid dynamics in seagrass ecology—from molecules to ecosystems. In: Larkum AWS, Orth RJ, Duarte CM, editors. Seagrasses: Biology, ecology and conservation. Springer Netherlands: pp. 193–225.
36. Hendriks IE, Sintes T, Bouma TJ, Duarte CM (2008) Experimental assessment and modeling evaluation of the effects of seagrass *Posidonia oceanica* on flow and particle trapping. Mar Ecol Prog Ser 356: 163–173.
37. Christianen MJ, van Belzen J, Herman PM, van Katwijk MM, Lamers LP, et al. (2013) Low-canopy seagrass beds still provide important coastal protection services. PloS one 8(5): e62413doi:101371/journalpone0062413
38. Sobral P, Widdows J (2000) Effects of increasing current velocity, turbidity and particle-size selection on the feeding activity and scope for growth of *Ruditapes*

decussatus from Ria Formosa, southern Portugal. J Exp Mar Biol Ecol 245(1): 111–125.

39. Widdows J, Brinsley M (2002) Impact of biotic and abiotic processes on sediment dynamics and the consequences to the structure and functioning of the intertidal zone. J Sea Res 48(2): 143–156

40. Fagan WF, Cantrell RS, Cosner C (1999) How habitat edges change species interactions. Am Nat 153(2): 165–182.

41. Ries L, Sisk TD (2004) A predictive model of edge effects. Ecology 85(11): 2917–2926.

42. Macreadie PI, Connolly RM, Jenkins GP, Hindell JS, Keough MJ (2010) Edge patterns in aquatic invertebrates explained by predictive models. Mar Freshw Res 61(2): 214–218.

43. Macreadie PI, Hindell JS, Keough MJ, Jenkins GP, Connolly RM (2010) Resource distribution influences positive edge effects in a seagrass fish. Ecology 91(7): 2013–2021.

44. Riisgård HU, Kittner C, Seerup DF (2003) Regulation of opening state and filtration rate in filter-feeding bivalves (*Cardium edule*, *Mytilus edulis* and *Mya arenaria*) in response to low algal concentration. J Exp Mar Biol Ecol 284(1): 105–127.

45. Bouma TJ, De Vries MB, Low E, Peralta G, Tanczos IC, Van de Koppel J, et al. (2005) Trade-offs related to ecosystem engineering: a case study on stiffness of emerging macrophytes. Ecology 86(8): 2187–2199.

46. Peralta G, Duren LA, Morris EP, Bouma TJ (2008) Consequences of shoot density and stiffness for ecosystem engineering by benthic macrophytes in flow dominated areas: a hydrodynamic flume study. Mar Ecol Prog Ser 368: 103–115.

47. González-Ortiz V, Alcazar P, Vergara JJ, Pérez-Lloréns JL, Brun FG (2014) Effects of two antagonistic ecosystem engineers on infaunal diversity. Est Coast Shelf Sci 139: 20–26.

48. André C, Jonsson PR, Lindegarth M (1993) Predation on settling bivalve larvae by benthic suspension feeders—the role of hydrodynamics and larval behavior. Mar Ecol Prog Ser 97(2): 183–192.

49. Jonsson PR, Petersen JK, Karlsson O, Loo L, Nilsson S (2005) Particle depletion above experimental bivalve beds: In situ measurements and numerical modeling of bivalve filtration in the boundary layer. Limnol Oceanogr 50(6): 1989–1998.

50. Bouma TJ, De Vries MB, Low E, Kusters L, Herman PMJ, Tanczos IC, et al. (2005) Flow hydrodynamics on a mudflat and in salt marsh vegetation:

identifying general relationships for habitat characterizations. Hydrobiologia 540 (1–3): 259–274.

51. Judge ML, Coen LD, Heck Jr KL (1993) Does *Mercenaria mercenaria* encounter elevated food levels in seagrass beds? Results from technique to collect suspended food resources. Mar Ecol Prog Ser 92: 141–150.

52. Ritchie RJ (2008) Universal chlorophyll equations for estimating chlorophylls a, b, c, and d and total chlorophylls in natural assemblages of photosynthetic organisms using acetone, methanol, or ethanol solvents. Photosynthetica 46(1): 115–126.

53. Jonsson PR, van Duren LA, Amielh M, Asmus R, et al. (2006) Making water flow: a comparison of the hydrodynamic characteristics of 12 different benthic biological flumes. Aquat Ecol 40: 409–428.

54. Folkard AM (2005) Hydrodynamics of model *Posidonia oceanica* patches in shallow water. Limnol Oceanogr 50: 1592–1600.

55. van Duren LA, Herman PM, Sandee AJ, Heip CH (2006) Effects of mussel filtering activity on boundary layer structure. J Sea Res 55(1): 3–14.

56. Foster-Smith RL (1975) The effect of concentration of suspension on the filtration rates and pseudofaecal production for *Mytilus edulis* L, *Cerastoderma edule* (L) and *Venerupis pullastra* (Montagu). J Exp Mar Biol Ecol 17(1): 1–22.

57. Iglesias JIP, Navarro E, Alvarez Jorna P, Armentina I (1992) Feeding, particle selection and absorption in cockles *Cerastoderma edule* (L) exposed to variable conditions of food concentration and quality. J Exp Mar Biol Ecol 162(2): 177–198.

58. Riisgård HU, Larsen PS (2007) Viscosity of seawater controls beat frequency of water-pumping cilia and filtration rate of mussels *Mytilus edulis*. Mar Ecol Prog Ser 343: 141–150.

59. Fréchette M, Butman CA, Geyer WR (1989) The importance of boundary-layer flows in supplying phytoplankton to the benthic suspension feeder, *Mytilus edulis*. Limnol Oceanogr 34: 19–36.

60. Plew DR, Enright MP, Nokes RI, Dumas JK (2009) Effect of mussel bio-pumping on the drag on and flow around a mussel crop rope. Aquacult Eng 40(2): 55–61.

61. Ólafsson EB (1989) Contrasting influences of suspension-feeding and deposit-feeding populations of *Macoma balthica* on infaunal recruitment. Mar Ecol Prog Ser 55(2): 171–179.

9

Pectinmethylesterases (PME) and Pectinmethylesterase Inhibitors (PMEI) Enriched during Phloem Fiber Development in Flax (*Linum usitatissimum*)

David Pinzon-Latorre*, Michael K. Deyholos

Department of Biological Sciences, University of Alberta, Edmonton, Alberta, Canada

Abstract

Flax phloem fibers achieve their length by intrusive-diffusive growth, which requires them to penetrate the extracellular matrix of adjacent cells. Fiber elongation therefore involves extensive remodelling of cell walls and middle lamellae, including modifying the degree and pattern of methylesterification of galacturonic acid (GalA) residues of pectin. Pectin methylesterases (PME) are important enzymes for fiber elongation as they mediate the demethylesterification of GalA *in muro*, in either a block-wise fashion or in a random fashion. Our objective was to identify PMEs and PMEIs that mediate phloem fiber elongation in flax. For this purpose, we measured transcript abundance of candidate genes at nine different stages of stem and fiber development and found sets of genes enriched during fiber elongation and maturation as well as during xylem development. We expressed one of the flax PMEIs in *E. coli* and demonstrated that it was able to inhibit most of the native PME activity in the upper portion of the flax stem. These results identify key genetic components of the intrusive growth process and define targets for fiber engineering and crop improvement.

Editor: David D. Fang, USDA-ARS-SRRC, United States of America

Funding: This work was funded by the Genome Canada project: Total Utilization of Flax Genomics (TUFGEN) www.genomecanada.ca. The funder had no role in study design, data collection and analysis, decision to publish, or preparation of the manuscript.

Competing Interests: The authors have declared that no competing interests exist.

* Email: david.pinzon@ualberta.ca

Introduction

Flax phloem fibers achieve their remarkable length through an extended period of intrusive growth. Intrusive elongation requires that fibers extend themselves through the middle lamellae of hundreds of cells, even destroying plasmodesmata in the process [1]. Once intrusive growth ceases, fibers begin to thicken their walls. The transition from elongation to thickening occurs around the snap point, a mechanically defined region of the stem described by Gorshkova and collaborators [2].

The demethylesterification of the cell wall plays a major role in the elongation and development of the phloem fibers of flax. Within the flax genome, 105 putative flax pectin methylesterases (LuPMEs) and 95 putative pectin methylesterase inhibitors (LuPMEIs) have been identified. The majority of these genes (77 LuPMEs and 83 LuPMEIs) have been demonstrated to be transcribed in at least one of the following tissues and developmental stages: floral buds, flowers, green capsules, early cortical peels, early fibers, late fibers, shoot apices, xylem, roots, leaf, senescent leaves [3]. Having thus defined the LuPME and LuPMEI families, we now have the opportunity to more precisely characterize these genes in the context of flax bast fiber development.

Heterologous expression is one tool that can be used to characterize gene function. PMEIs from different species have been successfully expressed in various systems. The mature proteins (i.e. without the signal peptide) of the Arabidopsis PMEIs

AtPMEI-1 and AtPMEI-2 were both expressed in *Escherichia coli* strain Rosetta-gami (DE3) [4] and in *Pichia pastoris* strain X-33 [5] producing in both cases functional inhibitors. Also, the complete and mature protein of BoPMEI1, a PMEI from *Brassica oleracea*, was effectively expressed in *E. coli* strain ER2566. On the other hand, the heterologous expression of PMEs has produced less consistent results. The complete proteins of the type-2 PMEs QUARTET1 [6] and AtPME31 [7] were successfully expressed in *E. coli* strains BL21(DE3) and JM101, respectively. However, the mature portion (removing signal peptide and pro-region) of a type-1 PME (At1g11580) was expressed in *E. coli* strain M15 but was not functional compared to the native protein from Arabidopsis [8]. One explanation for these results is that post-translational modifications, such as glycosylation, may be necessary for the correct activity of some proteins as has been demonstrated for PMEs and PMEIs from kiwi fruit (*Actinidia chinensis*) [9,10] and PMEs from mandarin orange (*Citrus sp.*) [11].

Our previous transcript profiling report of LuPMEs and LuPMEIs [3] did not provide sufficient spatial and temporal resolution to support efforts to identify PMEs and PMEIs that are involved in specific stages of fiber development. Consequently, to determine the expression of genes of interest at key developmental stages along the stem, we measured the activity of the PMEs, and assessed the transcript expression of 21 LuPMEs and 9 LuPMEIs in nine stages of fiber development. Our analysis allowed definition of a set of candidate PMEs and PMEIs with roles in fiber development during elongation, and during secondary cell wall deposition and maturation, and also a set of genes that could have important roles in xylem development.

Materials and Methods

Plant material

Plants were grown in a growth chamber at 22°C, with 16 hours day length, and were fertilized with 3 g/L of a 20–20–20 water soluble fertilizer (Plant-Prod) every two weeks. The soil was left to almost dry before watering the plants again.

Tissue was collected when plants reached between 46 and 48 cm, which occurred approximately five weeks after germination. At the time of harvest, the snap point was at an average distance of 7.1 cm from the shoot apex. In all cases, the leaves were removed. Sections 1-cm long were collected from positions along the stem as either whole stem, or as stem peels. Sections were collected at nine positions along the stem, based on the stage of development of the fiber [2], as follows: 0 to 1 cm (SA), 1 to 2 cm (1–2), 2 to 3 cm (2–3), 3 to 4 cm (3–4), 4 to 5 cm (A), 11.5 to 12.5 cm (B), 18 to 19 cm (C), 32 to 33 cm (D), and 44 to 45 cm (E). For microscopy, 100 μm cross sections for points A, B, C, D and E were obtained using a vibratome and 10 μm cross sections were obtained for position SA by wax-embedding in a Leica TP1020 tissue processor.

RNA extraction

Fifteen 1 cm-fragments per tissue were used for the RNA extraction. The RNA was extracted using Trizol extraction coupled with the RNeasy Plant Mini Kit (QIAGEN). Tissue was ground, and 2 ml of Trizol (Sigma) were added, followed by incubation at 60°C for 5 min with vortexing. The supernatant was transferred to a new tube by centrifugation at 12000 rcf at 4°C for 15 min. 0.2 volumes of chloroform were added, mixed, and centrifuged at 12000 rcf at 4°C for 20 min. The supernatant was obtained, and from here the extraction was coupled with the RNeasy Plant Mini Kit. 0.25 volumes of solution RLT plus 0.5 volumes of cold ethanol were added. These were applied to RNeasy columns, and the kit manufacturer's instructions were followed.

The RNA was tested for DNA contamination using a set of primers that flank an intron, producing differential product sizes for gDNA amplicons and for cDNA amplicons (Fw: 5′-TGCATATGCTCAGACCGACT-3′, Rv: 5′-TGGTGTA-GATTTTCGGAAGAGAC-3). The RNA quality and concentration were assessed using the Agilent 2100 BioAnalyzer (Agilent Technologies, Inc.).

cDNA synthesis and quantitative real time PCR

1 μg of RNA was used to synthesize cDNA using RevertAid H Minus Reverse Transcriptase (Thermo Scientific) and oligo(dT)18 primers following the manufacturer's protocol.

The Applied Biosystems 7500 Fast Real-Time PCR System was used to conduct quantitative real time PCR (qRT-PCR) on the stem peel tissues, in 96 well-plates. For the whole stem tissues, we used the Applied Biosystems 7900 HT Fast Real-Time PCR System, in 384 well-plates. Three biological replicates and three technical replicates were used per sample. The cDNA was diluted 1:40. The 10 μL sample mix consisted of 2.5 μL of diluted cDNA, 0.4 μM of each primer, and 1X MBSU buffer Tris (pH 8.3), containing KCl, MgCl$_2$, Glycerol, Tween 20, DMSO, dNTPs, ROX as a normalizing dye, SYBR Green (Molecular Probes) as the detection dye, and an antibody-inhibited Taq polymerase.

Primers used are the same used in Pinzon-Latorre and Deyholos [3].

Gene clustering based on expression

The STEM (Short Time-series Expression Miner) software package [12], was used to cluster the genes according to their transcript expression patterns. The negative of the dCT values were used as input in STEM, which was run using the "normalize data" option, so the values of the first tissue were transformed to 0. We also used a minimum correlation of 0.8, with a maximum of nine model profiles for whole stem samples and five model profiles for stem peel samples, and also the minimum absolute expression change was adjusted to 2, so those genes in which there was less than a 4 fold difference between the highest and the lowest expression value were not used to generate the clustering.

Heterologous expression

The coding region of the mature protein (i.e. excluding the signal peptide) of LuPMEI45 was used for heterologous expression. It was synthesized (Bio Basic Inc.) with codon optimization for *E. coli* (File S1) and was transformed into pET22b(+) (Novagen, Madison, WI, USA) via the restriction sites XhoI and NcoI. This plasmid was then transformed into *E. coli* Rosetta-Gami B(DE3)pLysS (Novagen, Madison, WI, USA). The empty pET22b(+) vector without inserts was used as a negative control in the various assays.

A single colony was grown overnight at 37°C in 2XYT medium plus chloramphenicol (34 μg/ml), tetracycline (12.5 μg/ml), kanamycin (15 μg/ml) and ampicillin (50 μg/ml). From this, 1 mL was transferred into 1 L of medium, and grown at 37°C until OD$_{600 \text{ nm}}$ 0.5–0.6, which was cooled on ice. IPTG at a final concentration of 1 mM was added, followed by growth for 18 hours at 20°C. Cells were pelleted at 4°C at 8000 rpm for 20 min. All subsequent manipulations were performed at 4°C unless otherwise indicated. The pellet was then mixed with 5% v/v of the original volume of 300 mM NaCl Tris HCl (the pH was at least one unit away from the predicted pI of the protein). This solution was left for at least 4 hours at −20°C, and was then sonicated at 55% for 30 seconds five times, with the intermediate tip of a Sonic Dismembrator model 300 (Fisher), with at least 1 min on ice between pulses. It was then centrifuged at 15000 rpm for 30 min at 4°C. The supernatant was incubated with 2% v/v of Ni-NTA agarose (QIAGEN) and rocked overnight prior to purification.

The His-tagged protein was purified using a Poly-Prep chromatography column (0.8×4 cm) which was prepared by adding 2 volumes of 50 mM Tris-HCl and 300 mM NaCl at the selected pH. The protein extract was then added, and it was washed with two volumes of 50 mM Tris-HCl, 1.5 M NaCl, then with 50 mM Tris-HCl, 300 mM NaCl, 20 mM imidazole, and then with 50 mM Tris-HCl, 300 mM NaCl, 40 mM imidazole. The protein was eluted with 5 ml of 50 mM Tris HCl, 1 M NaCl and 250 mM imidazole, containing one cOmplete ULTRA protease inhibitor (Roche) tablet per 10 ml. Five 1 ml-fractions were obtained, which were dialyzed against 50 mM Tris-HCl, 300 mM NaCl using a Amicon Ultra 3K centrifugal filter unit (Millipore).

LC MS/MS

The protein observed with the expected size in the Coomassie-stained polyacrylamide gel was confirmed by in-gel tryptic digestion and identification by LC MS/MS analysis in the Institute for Biomolecular Design (University of Alberta).

Protein extraction for PME activity

Proteins were extracted from three biological replicates according to the protocol of Hongo and collaborators [13]. Seven fragments (1 cm length each) obtained from equivalent positions along stems of different individuals were pooled for each extraction. Tissues were ground in liquid nitrogen, and 1 mL of extraction buffer, containing 12.5 mM citric acid, 50 mM phosphate buffer pH 7.0, with 1 M NaCl plus one tablet per 10 mL of cOmplete ULTRA protease inhibitor (Roche), was added. The sample was incubated at 4°C on a rocker and was then centrifuged at 15,000 rcf for 15 min, and the supernatant was collected. The protein concentration was determined using Qubit Fluorometric Quantitation (Life Technologies).

Radial activity assay

The radial assay was done with three biological replicates and three technical replicates as described by Downie and collaborators [14], with modifications [13]. 2% (w/v) agarose was dissolved in McIlvaine buffer with pH adjusted to 6.0 and 7.0 and was autoclaved, after which 0.1% (w/v) of highly methylesterified pectin (Sigma-Aldrich, P9561) was added and dissolved. From this mixture, 13 mL was poured into 90 mm petri dishes. After cooling, wells with a diameter of 4 mm were punched in the agarose using a micropipette tip. 10 μL of freshly extracted protein (396 μg/mL) plus 10 μL of 50 mM Tris HCl 300 mM NaCl buffer were dispensed into each well. This was incubated for 18 h at 28°C, and the gel was stained with an aqueous solution of 0.05% (w/v) ruthenium red for 1 h and washed with distilled water. The plates were photographed immediately and the area of the halo was measured using ImageJ [15].

PMEI inhibitory activity

The ability of recombinant PMEI to inhibit native PME activity in proteins extracted from flax stems was tested as in Raiola and collaborators [5]. For this purpose, PME activity was assayed as described above. For inhibition assays, 10 μL of flax cell wall proteins (396 μg/mL) were mixed with 10 μL of heterologous LuPMEI45 dialyzed solution (146 μg/mL) and incubated for 30 min at room temperature, and then the mixture was added to a well in the assay plate (20 μl per well).

Results

Tissues corresponding to the different stages of development

Gorshkova and collaborators [2] defined different stages of development of flax fibers relative to a mechanically defined "snap point" on the stem. In general, fiber specification and elongation occur apically to the snap point, and fiber cell wall thickening occurs basally. With this frame of reference, we examined the stem anatomy of linseed flax (variety CDC Bethune), in plants 46 to 48 cm long, ~5 weeks after germination, just before flowering. Based on our observations, and with reference to the precedent established by Gorshkova, we identified nine positions along the stem that represented progressive stages of fiber development (Figure 1). Five of these positions (points SA to A) were apical to the snap point, and four positions (B through E) were basal to the snap point. A 1 cm segment of whole stem was harvested at each of the nine positions. Additional 1 cm segments were obtained from positions A through E, and these were peeled to obtain only the outer tissues of the stem (epidermis, cortex, phloem, and some cambial zone cells), while excluding xylem. The four, apical-most segments (SA to 3–4) were too delicate to effectively peel.

Selection of candidate genes

To identify genes that affect the development and extractability of flax fibers, we selected 21 LuPMEs and 9 LuPMEIs for detailed characterization (Table S1). The selection of these genes was based on two previous studies: (i) previously published Fluidigm qRT-PCR expression data that showed the selected genes to be enriched in fiber-bearing tissues [3], and (ii) oligonucleotide microarray data that showed transcripts of the selected genes to be enriched in at least one of the points of the stem [16].

We tested three genes for their suitability as endogenous controls in the qRT-PCR assays. These three genes (GAPDH, ETIF1, ETIF5A) were selected for evaluation based on the results from Huis and collaborators [17]. We used BestKeeper software [18] to evaluate the expression stability of these genes in the tissues used in this study. ETIF1 had the least overall variation with a standard deviation of 0.67, followed by ETIF5A (0.74) and GAPDH (0.75). The best correlation between BestKeeper index and candidate reference gene was for ETIF5A (0.995), followed by GAPDH (0.992), and then ETIF1 (0.984). All three genes were therefore considered suitable as endogenous references for the qRT-PCR experiments described here, and the geometric mean of their Ct value was used to calculate the delta-C_T.

We measured relative transcript abundance of 21 LuPMEs and 9 LuPMEIs in fourteen stem segments and stem peels (Figure 2, and Figure S1). Transcripts of four genes (LuPME3, LuPME96, LuPMEI27, and LuPMEI60) could not be reliably detected in stem peel and these data were therefore not included in the results presented here. Because we were interested in identifying genes that were dynamically expressed during stem and fiber development, we calculated the maximum fold-change in transcript abundance between any two tissues (i.e. difference between the highest and the lowest mean dCT values (Table 1). We found 10 genes in the whole stem and 6 genes in the stem peel that differed at least 20-fold in transcript abundance between any two positions along the stem. Among them, the three highest fold-changes in the whole stem tissues were observed in LuPME85 (419 fold higher at A compared to SA), LuPME61 (307 fold higher at A respect to SA), and LuPME1 (191 fold higher at A compared to segment 2–3). Among outer stem peels, the three genes with the greatest difference in transcript abundance were LuPME79 (1085 fold higher at A respect to C), LuPME67 (153 fold higher at A respect to C), and LuPMEI66 (37 times higher at E respect to B).

We also compared the expression of the genes in the same position in the whole stem and the stem peel (Tables 2, 3, and S2). Five genes (LuPME67, LuPME79, LuPME92, LuPMEI45, and LuPMEI66) showed an expression at least 20 times higher in at least one of the whole stem tissues in comparison to the corresponding position in the stem peel; all of these observations of differential expression were made in tissues below the snap point (B to E). Meanwhile, transcripts of four genes (LuPME1, LuPME45, LuPME85, and LuPMEI65) were at least 20 times more abundant in stem peel tissue than in a whole stem tissue. In all the cases the 20-fold change in expression was observed in tissues below the snap point.

Clustering of transcript expression data

To identify shared patterns of transcript expression among the genes surveyed, we clustered the qRT-PCR results using STEM (Short Term Expression Miner) software [12]. STEM was designed specifically for time-series expression data and is therefore well-suited to clustering the developmental series represented by the stem and peel segments we analyzed.

Three genes for the whole stem tissues were filtered out and not used for the clustering because the difference in expression

SA 0-1 cm
    1-2 cm
    2-3 cm
    3-4 cm
A   4-5 cm

    7.1 cm
    Snap Point

B   11.5-12.5 cm

C   18-19 cm

D   32-33 cm

E   44-45 cm

Figure 1. Location in the stem of the tissues used for the experiments used in this study. Plants were 5 weeks old, in vegetative stage, and their height from the hypocotyls was between 46 to 48 cm. Tissues were harvested and rapidly frozen with liquid nitrogen for RNA and protein extraction. 100 μm (points A to E) and 10 μm (point SA) cross sections. Bar: 50 μm. e: epidermis, f: fibers, p: phloem.

between the lowest value and the highest value was less than four-fold. Using STEM we identified five broad patterns among the transcript expression data from segments of the whole stem (Figure 3). In Group 1, which contained nine LuPMEs and three LuPMEIs, expression was highest in positions undergoing intrusive growth (SA though A), and decreased as the fibers matured (positions B through E). In Group 2, which contained seven LuPMEs and one LuPMEI, we observed an expression peak just above the snap point, at point A. In Group 3, which contained two LuPMEs and three LuPMEIs, expression was highest in positions below the snap point (B through E), which represent secondary cell wall deposition. In Group 4, which includes only one LuPMEI,

Table 1. Tissue enrichment of selected PME and PMEIs.

Gene	Highest fold change in whole stem					Highest fold change in stem peel				
	Fold change	CI	Max at	Min at	p-value	Fold change	CI	Max at	Min at	p-value
LuPME1	191.4	97.8–374.6	A	2to3	****	20.3	6.2–66.3	B	A	****
LuPME3	13.4	4.8–36.9	E	SA	***					
LuPME5	3.4	2–5.6	1to2	C	***	4.9	3.5–6.6	A	B	**
LuPME7	6.5	5.1–8.3	SA	E	***	2.1	2–2.3	E	B	ns
LuPME10	15.7	9.4–26.1	SA	E	****	2.7	1.4–5.1	D	B	*
LuPME11	3	1.8–5	1to2	E	**	2.3	1.7–3.3	A	D	*
LuPME28	7.8	3.9–15.5	SA	D	***	2.3	1.6–3.4	E	B	*
LuPME30	45.1	33.4–61	A	E	****	9.2	6.3–13.4	A	D	****
LuPME31	40.2	3.1–527.3	A	SA	****	4.7	2.9–7.7	A	B	**
LuPME45	185.9	88–392.6	A	SA	****	19.5	12.9–29.4	B	E	****
LuPME46	4.6	2.9–7.3	A	D	*	5.2	3.6–7.4	A	B	**
LuPME61	307	171.5–549.4	A	SA	**	24.3	13.5–43.9	B	A	****
LuPME67	19.9	17.7–22.4	3to4	E	****	152.6	94.3–246.8	A	C	****
LuPME71	6	1.2–30.4	SA	E	**	3	1.2–7.8	C	A	**
LuPME73	7.9	5.1–12.3	B	E	***	11.8	10.9–12.7	A	C	****
LuPME79	26	7.3–92.3	1to2	E	****	1084.9	707.9–1662.6	A	C	****
LuPME85	419.4	110.4–1593.1	A	SA	****	31.5	22.5–44.2	B	E	****
LuPME92	3.4	1.7–6.6	D	E	***	2.6	2–3.4	A	D	*
LuPME96	15.8	7.8–32.3	1to2	2to3	****					
LuPME102	8.8	5.5–14	SA	E	****	2.2	0.8–6.1	A	D	ns
LuPME105	49.3	10.2–237.8	1to2	D	****	3.6	3–4.3	D	C	***
LuPMEI27	6.2	3.3–11.7	B	2to3	*					
LuPMEI44	11.8	6.8–20.5	A	SA	***	1.4	1–2.1	B	C	ns
LuPMEI45	17	7.1–40.8	SA	C	****	4.8	2.4–9.8	A	D	****
LuPMEI59	6.5	1.6–26.8	C	D	***	6.6	3.3–13.3	E	B	****
LuPMEI60	13.6	1.3–137.8	B	2to3	ns					
LuPMEI65	91.8	39–215.7	1to2	E	****	18.2	7.4–44.9	A	B	****
LuPMEI66	6.5	3.9–11	SA	A	****	37.4	21.1–66.4	E	B	****
LuPMEI67	4.4	3.5–5.4	D	2to3	ns	2	1.7–2.4	E	C	ns
LuPMEI73	74.7	43.9–127	A	SA	****	2.9	2.2–3.9	A	D	**

The fold-enrichment between the tissue sample with the highest transcript abundance and the lowest transcript abundance was calculated for each gene. This calculation was done separately for whole stem (WS) and stem peel (SP) samples. Fold enrichment is shown in a linear scale and is the mean of 3 measurements from 3 biologically independent samples. The p-value of the difference between the two points denoted was obtained by an ANOVA test that was followed by a Tukey's multiple comparisons test using GraphPad Prism version 6.00 for Windows The asterisks denote the p-value as follows. *0.01–0.05; **0.001–0.01, ***0.0001–0.001; ****<0.0001. ns: non-significant difference (p>0.05). The values not shown are genes that were not detected in those tissues. The confidence interval (CI) was calculated by using one standard deviation of the difference of the dCT between the two tissues compared.

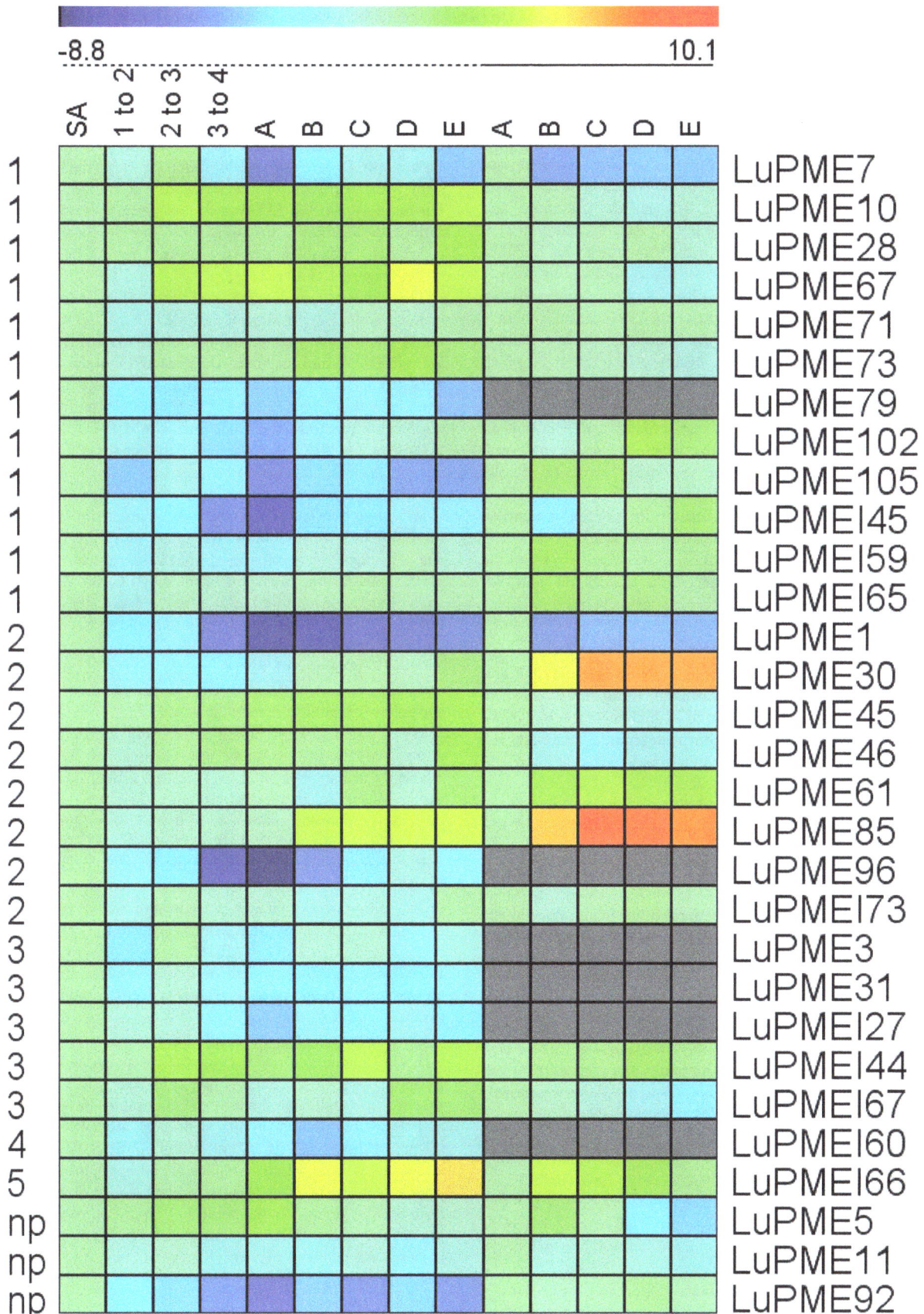

Figure 2. Transcript expression of genes from whole stem and stem peel tissues (ddCT). dCT was obtained by subtracting the geometric mean of the three endogenous controls used to the Ct value of the genes studied for every biological replicate. Here we show the average of the

three biological replicates. The negative of the dCT value was used to calculate the ddCT, so a higher value represents higher transcript abundance. The ddCT was obtained by substracting the dCT value from the SA for whole stem tissues, and from point A, for stem peel tisues. The tissues below the dotted line are whole stem tissues, and below the solid line are stem peel tissues.

peak expression occurred at point B. Finally in Group 5, one LuPMEI showed its lowest transcript expression at point A.

We also applied the same clustering method to transcript expression data from the outer stem peels. We eliminated 11 genes from clustering as the difference between the minimum and maximum dCT value was less than 2. Four different patterns were established (Figure 4). In Group 1, which contained four LuPMEs and one LuPMEI, a peak in expression was observed at point A (representing intrusive growth). In Group 2, which contained two LuPMEs and two LuPMEIs, transcript abundance generally increased as the fiber matured. In Group 3, which contained two LuPMEs, peak in expression occurred in position B, which was associated with the onset of secondary cell wall thickening, and expression decreased rapidly below this point. Group 4 contained three LuPMEs and one PMEI, and showed an expression minimum at point B, when secondary cell wall deposition started, and then the expression increased basally (points C, D, and E) as the fibers matured.

To assess the statistical significance of the differences between tissues in a given gene, we performed an ANOVA statistical analysis for the expression of the genes in the whole stem and the stem peel tissues, which is depicted in Tables S3 and S4, respectively.

The PME activity is lower at the top of the plant

PME activity in the nine different segments of whole stem and five segments of stem peels was assessed, using three biological replicates, which were each measured in three technical replicates. In this assay, proteins extracted from stem segments were allowed to radially diffuse from a well into an agarose gel containing pectin and ruthenium red, and PME activity was detected by the development of a dark halo around the well. Measurement of the area of the halo allowed for a semi-quantitative estimate of PME activity.

We used standard curves with different concentrations of proteins extracted from a flax stem and pectinesterase from orange peel (Sigma) to determine if the area of the halo is directly proportional with the concentration of protein. We found that the area of the halo was positively correlated with PME concentration ($R^2 = 0.96$ for proteins extracted from flax and $R^2 = 0.94$ for the commercial PME; Figure S2), which supports the use of the radial assay to quantify the activity of the PMEs.

We assayed PME activity at both pH 7.0 and pH 6.0. These pH values were chosen based on the results of a pilot study of flax stem PME activity at pHs 5.0, 6.0, 7.0 and 8.0, which showed maximum activity at pH was 7.0 (data not shown). We also conducted the full assay at pH 6.0, since this was representative of the pH of the natural cell wall. In whole stem tissues at either pH 6.0 or 7.0, PME activity was significantly lower ($p < 0.05$) at position SA relative to almost all other tissues (Figure 5, Table S5). This was also true in the stem peel, where PME activity was lower in SA than in any other tissues tested. Furthermore, the activity of SA (whole stem) was significantly lower ($p < 0.05$) compared to the stages A to E of the stem peel tissues (Figure 5 panels D to F). Thus, PME activity (as a proportion of the total proteins extracted) appears to be highest in tissues below the apical-most 1 cm of the stem, with a peak around position A, and is higher in stem peels than in whole stem.

Heterologous expression

LuPMEI45, whose transcript abundance peaked in expression during intrusive growth and diminished towards the bottom of the stem was selected for heterologous expression in *E. coli*. Furthermore, LuPMEI45 was chosen because it was one of the LuPMEIs, together with LuPMEI65, that showed a significant enrichment in expression during intrusive growth in the stem peel tissues (Figure 4, Table S4).

Three different strains of *E. coli* were tested for their suitability in heterologous expression of LuPMEI45: BL21(DE3), Rosetta(-DE3)pLysS, and Rosetta-Gami B(DE3)pLysS (Novagen, Madison, WI, USA). We also evaluated different IPTG inducer concentrations (0.5 and 1 mM), induction times (2 hours, 4 hours, 18 hours), and induction temperatures (20°C, 30°C, and 37°C), and purification methods. We found that 1 mM IPTG, and 18 hours of induction at 20°C, were the best parameters for induction (data not shown).

LuPMEI45 expression was successfully detected in all of the strains, but the concentration was highest in Rosetta-Gami B(DE3)pLysS, so this strain was used in further experiments. The His-tagged heterologous LuPMEI45 protein was partially purified, and its identity was confirmed by LC MS/MS (Table S6) analysis and assayed in a radial diffusion assay. The recombinant LuPMEI45 protein was not purified to homogeneity and therefore the extract still contained some residual *E. coli* protein (Figure S3). Therefore an empty pET22b(+) vector expressed under the same conditions in Rosetta-Gami B(DE3)pLysS was used as a negative control in subsequent functional assays.

We found that recombinant LuPMEI45 successfully inhibited native PME activity of flax stem protein extracts, while no inhibitory activity was observed from the proteins extracted from the vector control or the dialysis buffer (Figure S4). The purified protein at a concentration of 7310 μg/mL was diluted at 1:12.5, 1:25, 1:50, 1:75, and 1:100. We tested volumes of 10 μL of the different dilutions against proteins extracted from the top of the stem (first 5 cm), middle (11 to 16 cm from apex), and bottom (40 to 45 cm from apex), all at a concentration of 396 μg/mL (10 μL added). We determined that at both pH 6.0 and pH 7.0, a 1:50 dilution, 146 μg/mL of LuPMEI45, was sufficient to reduce native LuPMEs activity by approximately 50%, while a 1:12.5 dilution, 585 μg/mL, was sufficient to achieve a 100% inhibition in all the tissues (Figure 6).

Once we knew the necessary concentration of heterologous LuPMEI45 to inhibit ~50% of the PME activity, we expanded the assessment of the inhibition capacity of LuPMEI45 to cell wall proteins extracted from the nine different points in the whole stem and 5 different points in the stem peel used along this study (Figure 1). It showed significant inhibition ($p < 0.05$) at pH 6.0 at all the points in the whole stem, and all, except point E, in the stem peel, and at pH 7.0 it inhibited at points SA, 1–2, 2–3, A, B, and E from the whole stem, and at points C, D and E form the stem peel, the activity of the PMEI on the whole stem SA in the stem peel tissues is shown as a reference (Figure 7).

Discussion

To identify PMEs and PMEIs that were expressed dynamically during fiber development, we calculated the maximum fold difference in transcript abundance for any two tissues set, for

Table 2. Tissue enrichment in whole stem compared to stem peels of genes with higher expression in whole stem.

Gene/Point	Fold change					CI of genes with higher expression in whole stem				
	A	B	C	D	E	A	B	C	D	E
LuPME1	3.8					1.6–9.3				
LuPME3										
LuPME5	1	3.6	1.8	2.6	1.6	0.7–1.5	2.6–4.9	1.2–2.6	1.6–4.3	0.9–2.6
LuPME7	1.3	1.6	1.2			1.1–1.5	0.7–3.7	0.9–1.5		
LuPME10	3.6	8.5	3.7	1.9	1.8	3–4.2	6–12.2	2.1–6.6	0.7–5	1.1–2.8
LuPME11	1.2	2.8	1.4	1.8		0.8–1.8	2.3–3.4	1.1–1.8	1.7–1.9	
LuPME28	1.4	1.3	1.2			0.9–2.2	1–1.5	0.9–1.6		
LuPME30	2.3					1.5–3.4				
LuPME31	1					0.1–9.1				
LuPME45	3.1					1.6–5.9				
LuPME46										
LuPME61	7.6					3.6–16.1				
LuPME67		5.7	17.8	22.9	9.5		5–6.4	13.4–23.7	14.5–36	6.5–14
LuPME71										
LuPME73		2.1		1.2			1.4–3.3		0.6–2.5	
LuPME79		2.1	21.8	8.6	4		1.5–2.9	13.9–34.3	4.2–17.4	1.0–16
LuPME85	5.8					1.8–18.8				
LuPME92	5.6	12.2	16.3	26.5	5.5	3.1–10.2	10.4–14.3	13.2–20.1	14–50	4.2–7.1
LuPME96										
LuPME102				1					0.7–1.5	
LuPME105	1.6	4.2	7.1		1	1.2–2.1	2.8–6.2	4.9–10.3		0.7–1.4
LuPMEI27										
LuPMEI44	4.2		1.9		1.5	2.8–6.1		1.5–2.4		1–2.4
LuPMEI45	9.1	21.3	6.9	56.4	8.3	5.1–16.1	9.3–49	3.6–13.2	26.5–119.9	3.1–22
LuPMEI59	13.7	12.7	15	1.2		9.6–19.6	5.4–29.8	6.9–32.3	0.3–3.9	
LuPMEI60										
LuPMEI65										
LuPMEI66	3.5	36.9	23.9	4.3	2.1	2.4–5.3	21.3–63.8	19.3–29.4	2.5–7.4	1.5–3
LuPMEI67	14.3	12.9	9.2	17.2	5.1	9.4–21.6	9.8–17	6.8–12.5	16.6–17.8	4.3–6.2
LuPMEI73	3		2.4	1	2	2.2–4.1		1.8–3.1	0.7–1.5	1.2–3.3

The fold-enrichment between the tissue sample with the highest transcript abundance and the lowest transcript abundance was calculated for each gene. This calculation was done separately for whole stem (WS) and stem peel (SP) samples. Fold enrichment is shown in a linear scale and is the mean of 3 measurements from 3 biologically independent samples. The values not shown are genes that were not detected in those tissues. The confidence interval (CI) was calculated by using one standard deviation of the difference of the dCT between the two tissues compared. The significance of the difference between the points in the stem in stem peel and whole stem is shown in Table S2.

Table 3. Tissue enrichment in whole stem compared to stem peels of genes with higher expression in stem peel.

Gene/Point	Fold change					Confidence Interval				
	A	B	C	D	E	A	B	C	D	E
LuPME1		53.2	121.2	86.3	15.2		20.6–137.2	99.1–148.1	22.1–337.2	9.5–24.4
LuPME3										
LuPME5										
LuPME7				1.4	1.7				1.1–1.7	1.5–2
LuPME10					1					0.6–1.7
LuPME11				1.8	1.5				1–3.4	0.5–4.8
LuPME28										
LuPME30		2.8	3.4	1.2	4.9		1.6–4.7	2.4–4.7	0.4–3.6	3.7–6.6
LuPME31		1	1.4	1.4	1.4		0.4–2.5	1.0–2.0	0.6–3.4	1–1.9
LuPME45		40.5	6.7	9.6	4.7		26.3–62.3	4.9–9.1	5.9–15.8	3–7.5
LuPME46	2.9	1.5	2.5	6.5	3.1	2–4.3	1.2–1.9	1.7–3.4	3.2–13.3	1.9–5
LuPME61		4	7	9.8	18.2		2.9–5.5	4.4–10.9	4.7–20.5	12.2–27.1
LuPME67	1.9					1.3–2.9				
LuPME71	4.9	8.3	15.4	3.3	17.1	1.7–14.1	2–33.9	3.6–66.2	1–11.6	3.5–85
LuPME73	10.8		2.1	3.4	3.4	9.3–12.6		1.6–2.7		2.8–4.1
LuPME79	3.2					1.6–6.4				
LuPME85		12.5	12.8	34.9	3.1		2.9–53.6	6.8–24.2	15.5–78.5	1.3–7.2
LuPME92										
LuPME96										
LuPME102	2.4	2.5	2		6.9	0.9–6.7	0.9–7.2	0.9–4.5		4.4–10.8
LuPME105				2.7					1.3–5.7	
LuPMEi27										
LuPMEi44		1.3		1.4			0.8–2		0.9–2.2	
LuPMEi45										
LuPMEi59					1.7					1.4–2.1
LuPMEi60										
LuPMEi65	16.6	5.9	4.5	15.7	56.1	11.2–24.4	2.3–14.7	1.4–14.7	6.7–37.1	25.4–123.7
LuPMEi66										
LuPMEi67										
LuPMEi73		1.4					1.2–1.6			

The fold-enrichment between the tissue sample with the highest transcript abundance and the lowest transcript abundance was calculated for each gene. This calculation was done separately for whole stem (WS) and stem peel (SP) samples. Fold enrichment is shown in a linear scale and is the mean of 3 measurements from 3 biologically independent samples. The values not shown are genes that were not detected in those tissues. The confidence interval (CI) was calculated by using one standard deviation of the difference of the dCT between the two tissues compared. The significance of the difference between the points in the stem in stem peel and whole stem is shown in Table S2.

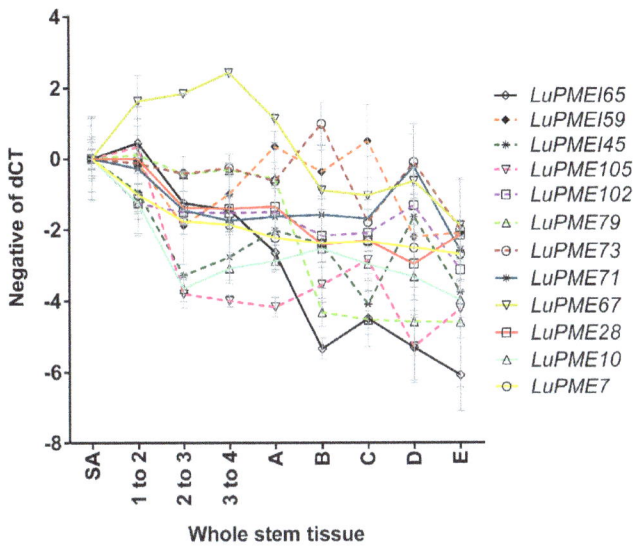

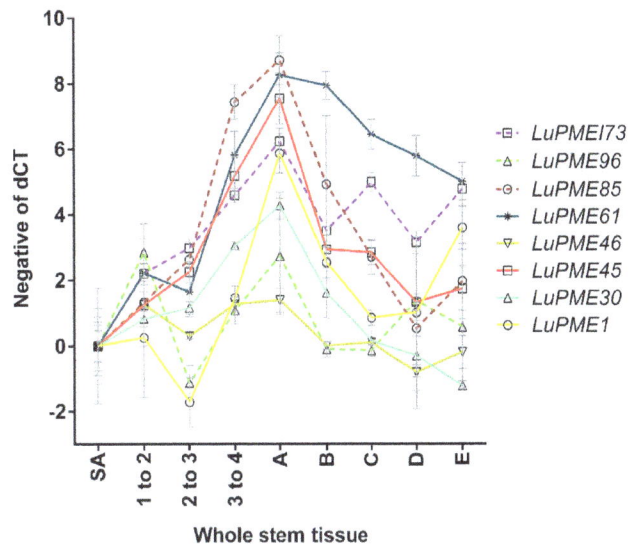

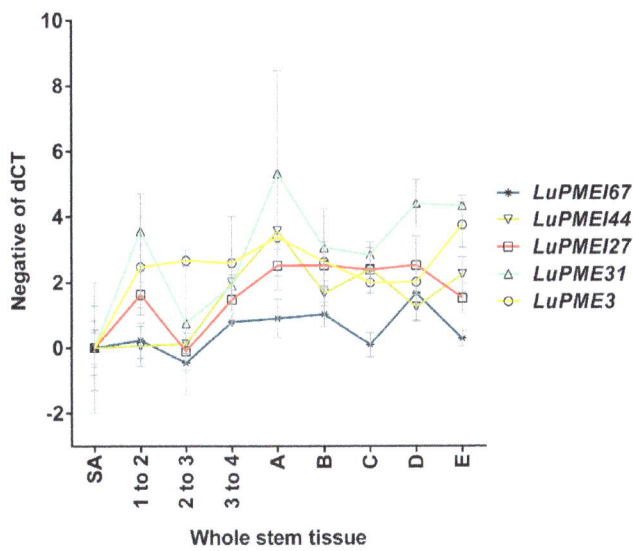

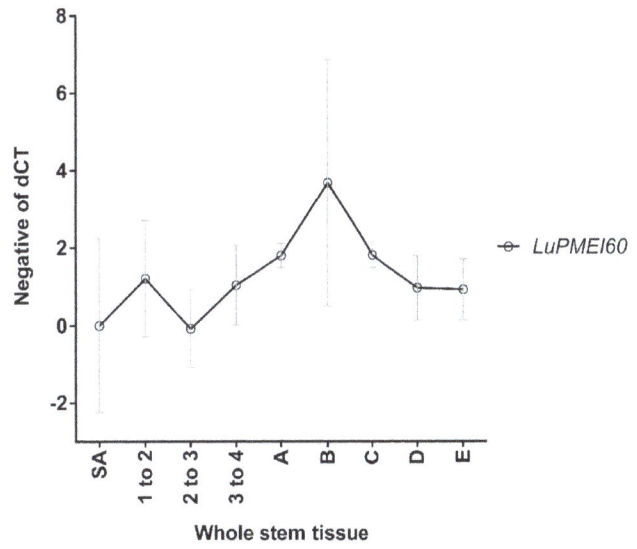

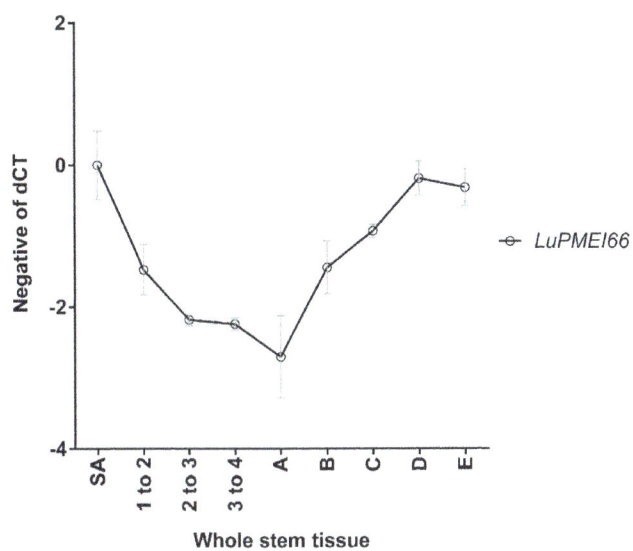

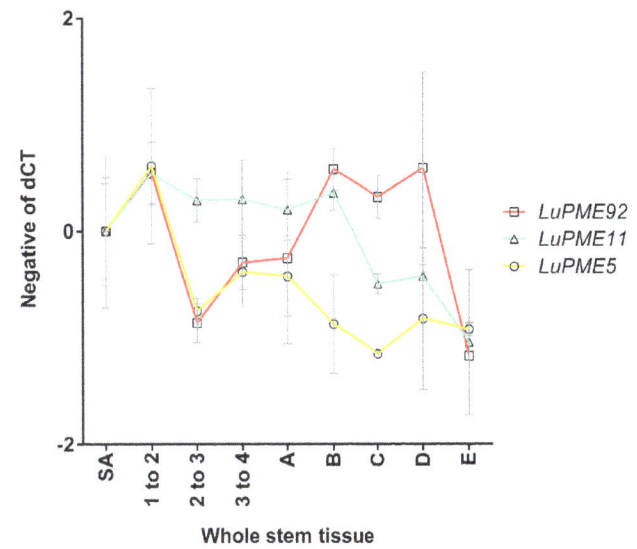

Figure 3. Clusters of transcript expression patterns in segments of whole stem tissues. Stems positions (SA, 1–2, 2–3, 3–4, A through E) are as defined in Figure 1. Transcript expression (y-axis) is the normalized negative dCT with point SA transformed to 0. Clusters are as defined by STEM software, using genes a minimum fold change of 4 between any two tissues.

each gene assayed (Table 1). This was done separately for both whole stems and stem peels. We also calculated the fold difference in transcript expression at equivalent positions in whole stems and stem peels (Table 2). If a gene was important for fiber development, the magnitude of enrichment was expected to be at least as high in stem peels as it was in whole stems. On the other hand, if a pattern was only observed in the whole stem, or if the magnitude of the change was significantly higher in the whole stem than in the stem peel, then that gene could be rather implicated in xylem development.

Genes enriched in fiber containing tissues during fiber elongation

All the genes that had a similar pattern of transcript expression in the whole stem and the stem peel showed enrichment during intrusive growth. As the fold change was of at least the same magnitude in the stem peel compared to the whole stem, this meant that the expression of these genes may be specific to fibers and surrounding tissues. Those genes were *LuPME46*, *LuPME67*, *LuPME73*, *LuPME79*, *LuPMEI45*, and *LuP-MEI65* (Figures 3 and 4). Here we will discuss the genes that are most likely to have important roles in fiber development, based on the magnitude of their transcript enrichment. We note that transcript abundance is not necessarily correlated with protein abundance or ultimate expression and activity, and that this important limitation must be considered when interpreting the data we present here.

LuPME67 and *LuPME79* showed the largest change between the lowest and highest dCT values in the stem peel (152 and 1082 fold respectively, Figure 3) and a comparatively low change in the whole stem (20 and 26 fold respectively), which is evidence of fiber-specific enrichment of these genes. *LuPME67* expression drastically diminished (p<0.05) below the snap point in stem peel tissues, in relation to point A (Figure 4, Table S3). Based on the whole stem results, it can be concluded that the expression was constant above the snap point (p>0.05), only presenting a difference between points SA to 3–4, where 3–4 was significantly larger (p<0.05) (Table S3), which might indicate that as fibers increased in number in a given section [2], the gene expression also increased. *LuPME67* is a type 2 PME, and one of the few LuPMEs with a predicted acidic isoelectric point (pI 5.63), which implies that its mode of action might be random, leading to cell wall loosening as the pectin becomes a substrate for polygalacturonases and pectate lyases [19,20]. Consequently, this is a gene that can be inferred to be active in the dissolution of the middle lamella between cells that the fibre penetrates during intrusive growth.

LuPME79 showed a drastic decrease in expression below the snap point, and its expression was constant above the snap point, there was no difference (p>0.05) detected between stages of development SA to A in whole stem tissue (Table S3). *LuPME79* is a type 1 PME, which interestingly does not have a predicted cleavage for separation of the PMEI-like domain from the PME domain [3]. It has a basic isoelectric point (pI predicted 9.04), which indicates *LuPME79* may demethylesterify the homogalacturonan (HG) in a blockwise fashion [19], leading to calcium cross linking between HG domains [21], and ultimately to cell wall stiffening. It has been shown that type 1 PMEs are retained in the Golgi until the pro-region is cleaved out by subtilisin proteases [22]

which are co-expressed with the PMEs [23]. However, it was shown that *LuPME3* can be secreted to the cell wall without processing the pro-region [24], so *LuPME79* may likewise be secreted without processing. The persistence of the pro-region (PMEI-like domain) may affect the PME activity, of this protein, so it will be informative to achieve its heterologous expression in the future.

LuPMEI45 and *LuPMEI65* were the LuPMEIs found to have similar patterns in both the stem peel and in whole stem tissues; they both had significantly higher expression in stem peel tissues (p<0.05) in point A as compared to the tissues below the snap point, meaning that they are genes involved in the regulation of LuPMEs expression in the stem peel above the snap point. *LuPMEI45* was chosen for heterologous expression.

The expression of *LuPME5*, a type-1 PME with a predicted pI of 9.53, did not display major changes among the tissues in the whole stem (less than four-fold change), however, in the stem peel we did observe a higher expression in A, than in B, C, and D (p<0.05), although the largest fold change was only 5-fold. This is consistent with observations of Al-Qsous and collaborators [25], who determined that its highest expression occurs in the elongating parts of the hypocotyl, the apex and the root tip.

Genes enriched in fiber-containing tissues below the snap point

From the stem peel expression data, we identified two genes that showed increased expression below the snap point (points B to E) respect to A (p<0.05) and their expression was stable between points B to E (p>0.05). *LuPME1* had a 20-fold change and *LuPME61* had a 24-fold change, between the minimum point (A) and the maximum point (B). The expression pattern of these genes in the stem peel was not similar to their expression in the whole stem, in which the expression, oppositely, diminished from A to B in *LuPME1*, and did not change from A to B in *LuPME61*. This means that the expression observed in the stem peel for these genes is a specific to this tissue, and indeed expression of *LuPME1* was 53, 121, and 86 times higher in points B, C, and D, respectively, in stem peels as compared to whole stem tissues (Table 1). As described above, the mode of action of these genes might be blockwise demethylesterification, so they would aid in the strengthening of the cell wall once the cells stop elongating (below the snap point) [2]. The role suggested for these genes is based on analysis of an orthologous gene from Arabidopsis, AtPME35 [3], which was found to strengthen the inflorescence stem by a blockwise demethylesterification action [13].

Genes enriched in the xylem

We found seven genes that showed a peak in expression in point A of the stem. These included the four genes that showed the highest fold change (between any two stem points) among any of the genes analyzed in the whole stem tissue: *LuPME85* (419-fold), *LuPME61* (306-fold), *LuPME1* (191-fold), and *LuPME45* (186-fold). The other three genes were *LuPME30* (45-fold), *LuPME31* (40-fold), and *LuPME96* (16-fold) (Table 1). As this expression pattern was unique for the whole stem and was not observed in the stem peel tissues, we concluded that these genes may play a role in xylem or pith development. The predicted isoelectric point of all of these proteins is basic, so blockwise demethylesterification is expected to occur [19] leading to cell wall rigidification.

Group 1

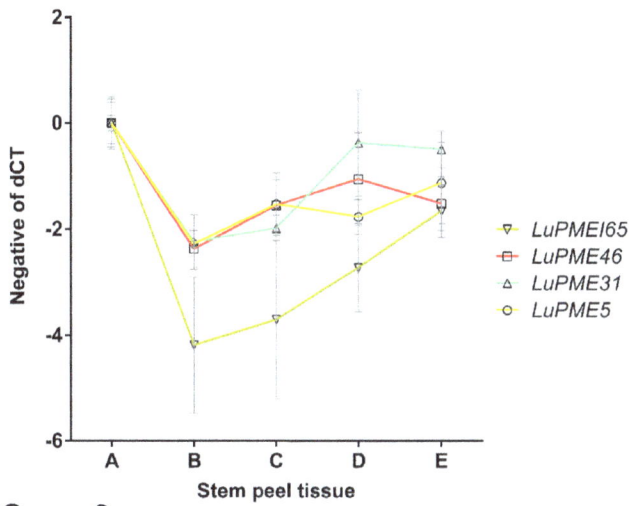

Group 2

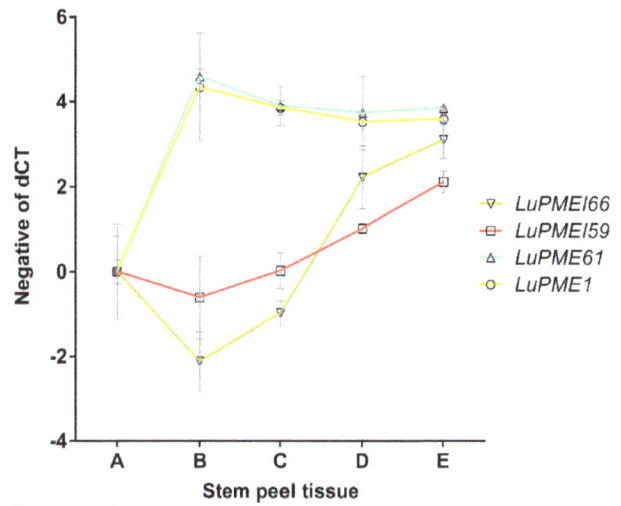

Group 3

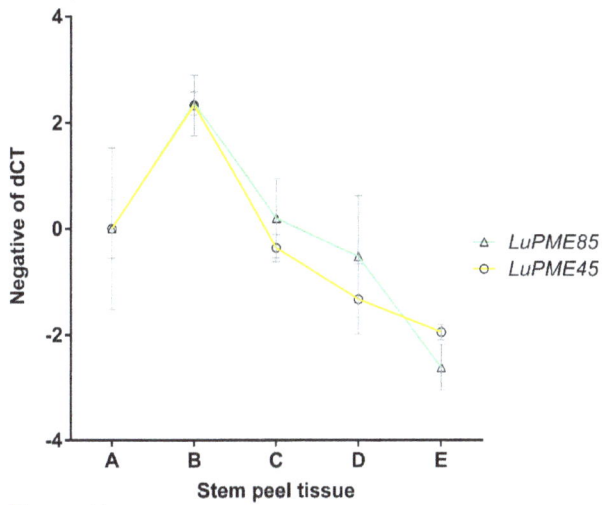

Group 4

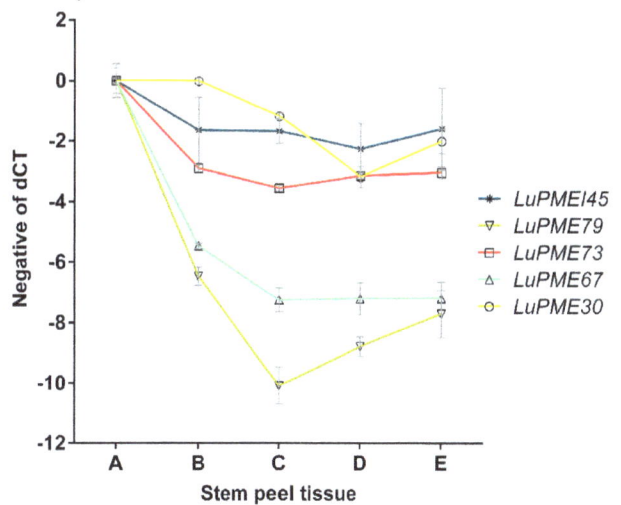

No pattern

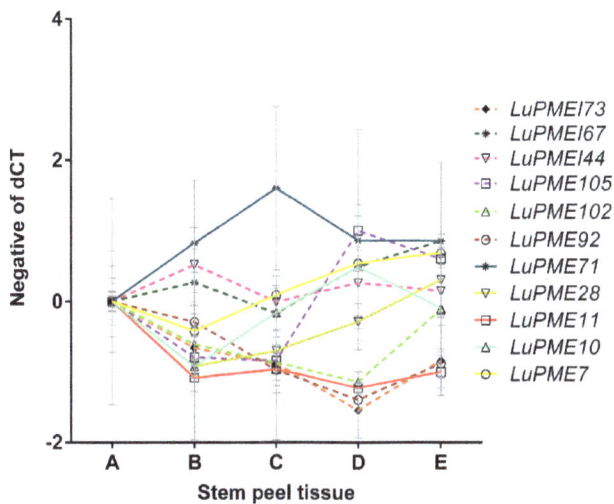

Figure 4. Clusters of transcript expression patterns in segments of stem peels. Stems positions (A through E) are as defined in Figure 1. Transcript expression (y-axis) is the normalized negative dCT with point SA transformed to 0. Clusters are as defined by STEM software, using genes a minimum fold change of 4 between any two tissues.

A

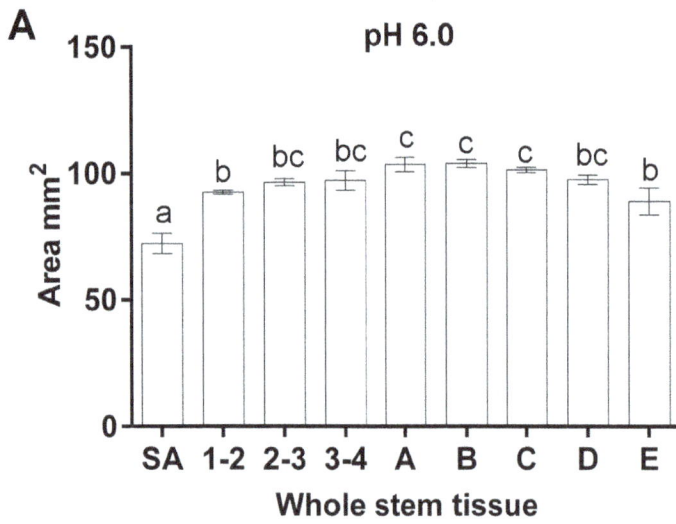

D

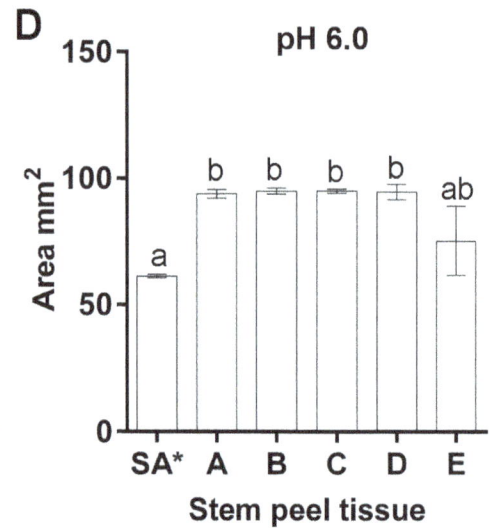

B

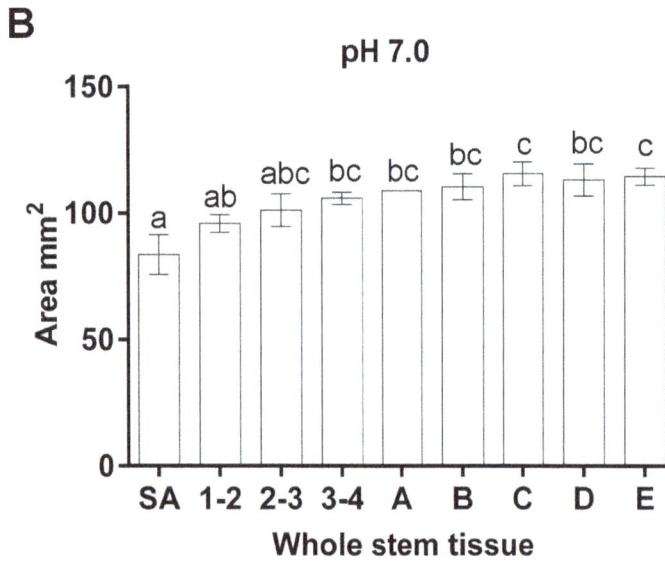

E

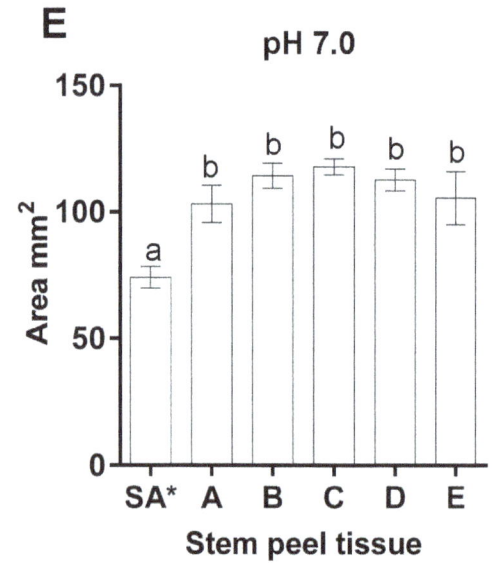

C

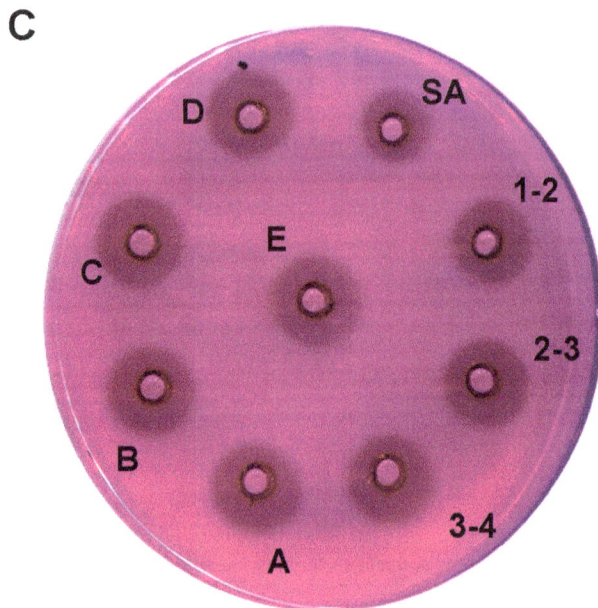

F

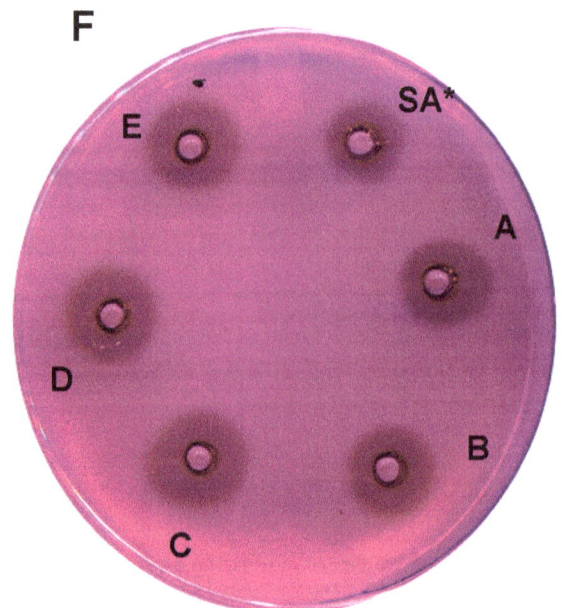

Figure 5. PME activity of native flax proteins. Proteins extracted from whole stems or stem peels at different developmental stages (as defined in Figure 1) were assayed for PME activity using a radial diffusion assay. In this semi-quantitative assay, the area of the halo formed was proportional to PME activity. The bar graphs show results of an ANOVA followed by Tukey's multiple comparisons test from three technical replicates of each of three biological replicates. The plates show results of a representative diffusion assay at pH 7.0. Panels A to C correspond to the activity of proteins extracted from the whole stem. Panels D to F correspond to proteins extracted from the stem peel, plus stage SA of whole stem (SA*), for comparison purposes.

Furthermore, one PMEI, *LuPMEI73*, was observed with this pattern in the whole stem but not in the stem peel; its high expression in point A, respect to the SA (74-fold), leads us to speculate that it has an important role in regulating PME activities at this point in the inner tissues, presumably within the xylem.

LuPME3 (type-1 PME with a predicted pI of 9.8) expression was not detected in the stem peel, while it was detected in low amounts in the whole stem tissues where its expression was significantly lower in SA (p<0.05) with respect to the rest of the tissues (1–2 to E), where the expression was not significantly different (p>0.05). The xylem undergoes differentiation, expansion, and maturation. In the vicinity of the shoot apex, very little vascular tissue maturation is expected to occur and it is only at

node 3–5 that thickening starts [26], so if *LuPME3* is involved in the cell wall stiffening of the xylem, it is then expected that its expression is lower in point SA, which we found, and then as more xylem is produced along the stem the maturation of the xylem is a constant process which is observed in the expression of this gene. *LuPME3* was previously found to have detectable expression in the vascular tissue of stems and leaves, and in the root meristem [27], and was found to have similar expression in the whole extension of the hypocotyl and the root in a 10 days old seedling [25]; they did not find lower expression at the top of the seedling, however, their detection method (RT–PCR Southern blot) is not as sensitive as qRT-PCR. Based on the phylogenetic analysis done by Pinzon-Latorre and Deyholos [3], it was established that

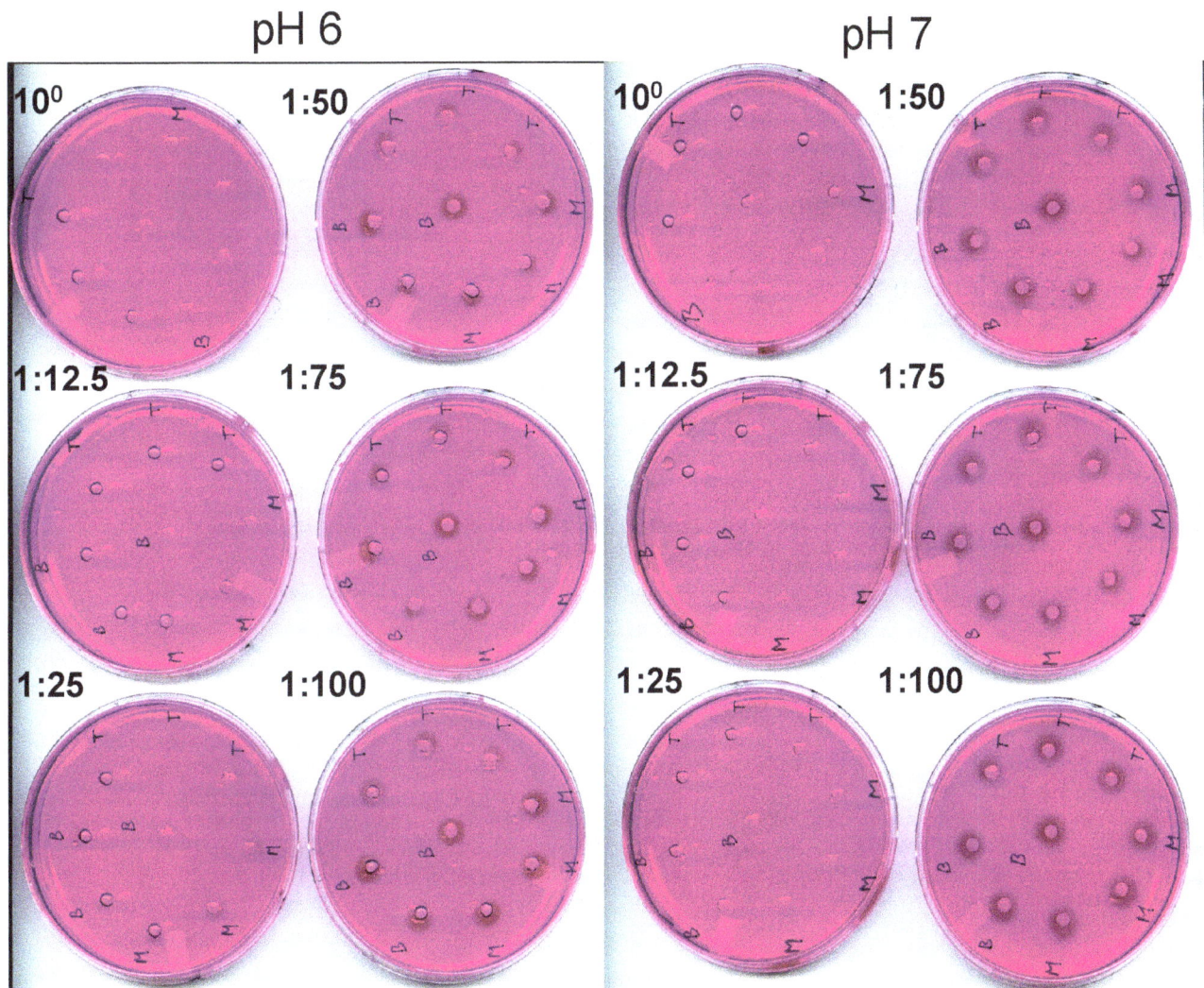

Figure 6. Radial assay of inhibitory capacity of LuPMEI45. Different dilutions of the purified proteins (7310 µg/mL) were assessed to establish the concentration at which ~50% of the PME activity (396 µg/mL of flax proteins) was inhibited. The letters in the plates denote the position of the stem were the proteins were extracted: Bottom (B), Medium (M), and Top (T).

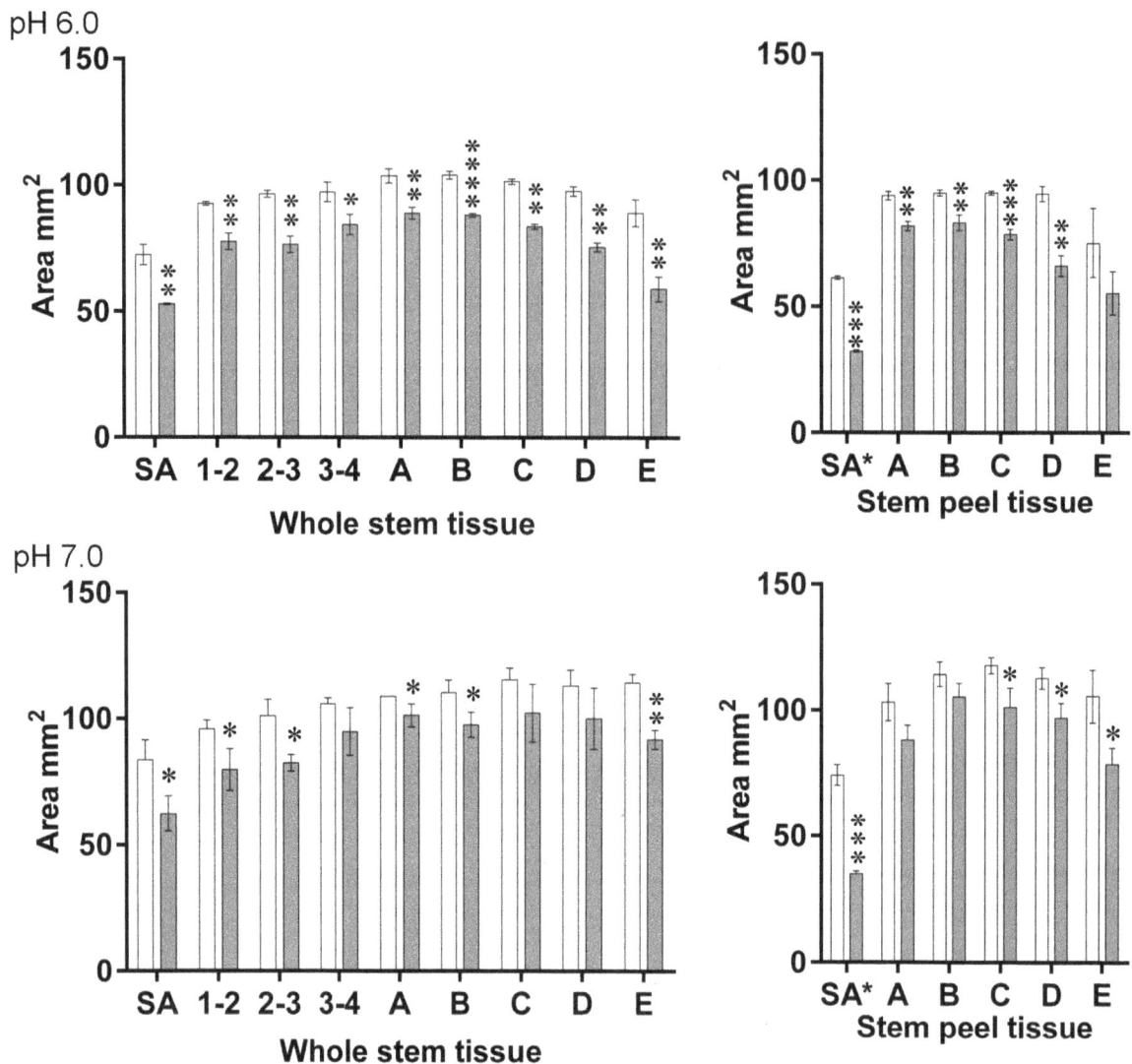

Figure 7. LuPMEI45 inhibitory activity on flax proteins extracted from whole stem and cortical peel. 10 µL of proteins extracted from the different tissues (396 µg/mL) were mixed with 10 µL of LuPMEI45 (146 µg/mL) (grey filled bars) or with 10 µL of the buffer (empty bars) in which LuPMEI45 was dialyzed. A t-test was done to determine if the activity of LuPMEI45 significantly reduce the PME activity at the different tissues. The asterisk denotes the p-value as follows. *0.01–0.05; **0.001–0.01, ***0.0001–0.001; ****<0.0001.

LuPME3 is one of the most similar genes to *PttPME1* [3], a PME in hybrid aspen that Siedlecka and collaborators [28] determined that when it was downregulated the xylem fibers elongation increased, so it was suggested that PttPME1 strengthens cellular adhesion, hindering intrusive growth. As the expression of *LuPME3* in flax occurs in the xylem, it is possible that the same situation is occuring in flax.

LuPME7 and *LuPME92* are the other LuPMEs closely related to *PttPME1*. The expression of *LuPME7* was significantly higher (p<0.05) in the whole stem in point SA respect to A, B, C, D, and E (Table S3), while in the stem peel there was no significant difference between the tissues (Table S4). *LuPME92* expression did not show a difference higher than 4-fold between any of the tissues in the whole stem and the stem peel, however the expression was higher in the whole stem tissues (A to E). With respect to stem peel, point D was 26 times higher in the whole stem than in stem peel (Table 2), suggesting a role in xylem maturation, as the one observed for PttPME1 [28].

PMEI inhibitory activity

LuPMEI45 was found to effectively inhibit the action of flax LuPMEs along the stem (Figure S4). As the expression of LuPMEI45 is higher during intrusive growth (p<0.05) (Tables S3 and S4), it could be expected that its inhibitory capacity is higher at the tissues undergoing intrusive growth, however the inhibitory capacity was not significantly different along the stem (data not shown). The preferred target(s) of LuPMEI45 will be important to determine, so its activity can be correlated with the mode of action of a PME.

Conclusion

We were able to characterize in detail the transcript expression of selected LuPMEs and LuPMEIs along the stem, in relation to stages of development of flax fibers. Candidate genes with expression patterns involving them in specific processes of phloem fibers and xylem development were presented, and a functional heterologous expression of one of them was achieved. The detailed study of these genes by the subcellular localization of the proteins, mutagenesis, silencing and/or overexpression techniques will allow the finding of genes relevant for the improvement of the crop, either by producing longer and easier to extract fibers or by obtaining plants with shorter fibers avoiding the obstruction of the machinery.

Supporting Information

Figure S1 Transcript expression of genes from whole stem and stem peel tissues (dCT). dCT was obtained by subtracting the geometric mean of the three endogenous controls used to the Ct value of the genes studied for every biological replicate. Here we show the average of the three biological replicates. The tissues below the dotted line are whole stem tissues, and below the solid line are stem peel tissues.

Figure S2 Standard curve of PME activity by radial assay. Proteins extracted from the whole stem (a) or pectinesterase from orange peel (b) were used at different concentration in a radial assay to assess the correlation with the area of the halo they produced.

Figure S3 Purification of LuPMEI45 expressed in E. coli. *Excised band of LuPMEI45 (~ 20.7 KDa) successfully identified by LC MS/MS analysis. Left: Protein ladder. FT: Flow through; W: Wash; E: Elution. W1:50 mM Tris-HCl 1.5 M NaCl. W2:50 mM Tris-HCl, 300 mM NaCl, 20 mM Imidazole. W3:50 mM Tris-HCl, 300 mM NaCl, 40 mM Imidazole. E: 50 mM Tris HCl, 1 M NaCl and 250 mM Imidazole

Figure S4 Inhibitory capacity of LuPMEI45 expressed in E. coli. The activity of LuPMEI45 was assessed in a radial assay measured as its capacity of blocking the activity of flax cell wall proteins extracted from the top 5 cm of a ~5 weeks old plant. Two different controls were used: The buffer used for the dialysis of the protein after purification, and the purified proteins from the empty vector, pET22b(+), expressed in the same system under the same conditions.

Table S1 Selected LuPMEs and LuPMEIs for expression profiling at different fiber developmental stages.

Table S2 Significance of the difference on the expression of the genes in the same point of the stem in stem peels and whole stem tissues. Statistical significance was determined using the Holm-Sidak method, correcting for multiple comparisons. The asterisks denote the significance of the test based on the p-value as follows. *$0.01–0.05$; **$0.001–0.01$, ***$0.0001–0.001$; ****<0.0001. ns: non-significant difference (p>0.05). np: no pattern.

Table S3 Tukey's multiple comparisons test of dCT expression values between whole stem tissues. An ANOVA test was followed by a Tukey's multiple comparisons test using GraphPad Prism version 6.00 for Windows The asterisks denote the p-value as follows. *$0.01–0.05$; **$0.001–0.01$, ***$0.0001–0.001$; ****<0.0001. ns: non-significant difference (p>0.05). np: no pattern.

Table S4 Tukey's multiple comparisons test of dCT expression values between stem peel tissues. An ANOVA test was followed by a Tukey's multiple comparisons test using GraphPad Prism version 6.00 for Windows. The asterisks denote the p-value as follows. *$0.01–0.05$; **$0.001–0.01$, ***$0.0001–0.001$; ****<0.0001. ns: non-significant difference (p>0.05). np: no pattern.

Table S5 Tukey's multiple comparisons test of Pectin Methylesterase activity along the flax stem. An ANOVA test was followed by a Tukey's multiple comparisons test using GraphPad Prism version 6.00 for Windows. SA*: SA tissue from whole stem compared to stempeel tissue. The asterisks denote the p-value as follows. *$0.01–0.05$; **$0.001–0.01$, ***$0.0001–0.001$; ****<0.0001. ns: non-significant difference (p>0.05).

Table S6 LC MS/MS analysis results of excised band at the expected size for LuPMEI45. A coverage of 57.06 was obtained. 305 peptide spectrum matches (PSMs) corresponding to 9 unique peptides matching LuPMEI45 were identified.

File S1 Codon optimized sequence of LuPMEI45 expressed in E. coli.

Acknowledgments

We thank Walid El Kayal for training on the Applied Biosystems 7900 HT Fast Real-Time PCR System.

Author Contributions

Conceived and designed the experiments: DPL MKD. Performed the experiments: DPL. Analyzed the data: DPL. Contributed reagents/materials/analysis tools: DPL. Contributed to the writing of the manuscript: DPL MKD.

References

1. Ageeva MV, Petrovska B, Kieft H, Sal'nikov VV, Snegireva AV, et al. (2005) Intrusive growth of flax phloem fibers is of intercalary type. Planta 222: 565–574.

2. Gorshkova TA, Sal'nikova VV, Chemikosova SB, Ageeva MV, Pavlencheva NV, et al. (2003) The snap point: a transition point in *Linum usitatissimum* bast fiber development. Industrial Crops and Products 18: 213–221.

3. Pinzon-Latorre D, Deyholos MK (2013) Characterization and transcript profiling of the pectin methylesterase (PME) and pectin methylesterase inhibitor (PMEI) gene families in flax (*Linum usitatissimum*). BMC Genomics 14: 742.

4. Wolf S, Grsic-Rausch S, Rausch T, Greiner S (2003) Identification of pollen-expressed pectin methylesterase inhibitors in Arabidopsis. FEBS Lett 555: 551–555.

5. Raiola A, Camardella L, Giovane A, Mattei B, De Lorenzo G, et al. (2004) Two *Arabidopsis thaliana* genes encode functional pectin methylesterase inhibitors. FEBS Lett 557: 199–203.

6. Francis KE, Lam SY, Copenhaver GP (2006) Separation of Arabidopsis pollen tetrads is regulated by QUARTET1, a pectin methylesterase gene. Plant Physiol 142: 1004–1013.

7. Dedeurwaerder S, Menu-Bouaouiche L, Mareck A, Lerouge P, Guerineau F (2009) Activity of an atypical *Arabidopsis thaliana* pectin methylesterase. Planta 229: 311–321.

8. De-la-Peña C, Badri DV, Vivanco JM (2008) Novel role for pectin methylesterase in Arabidopsis: a new function showing ribosome-inactivating protein (RIP) activity. Biochim Biophys Acta 1780: 773–783.

9. Giovane A, Quagliuolo L, Castaldo D, Servillo L, Balestrieri C (1990) Pectin methyl esterase from *Actinidia-chinensis* fruits. Phytochemistry 29: 2821–2823.

10. Balestrieri C, Castaldo D, Giovane A, Quagliuolo L, Servillo L (1990) A glycoprotein inhibitor of pectin methylesterase in kiwi fruit (*Actinidia chinensis*). Eur J Biochem 193: 183–187.

11. Rillo L, Castaldo D, Giovane A, Servillo L, Balestrieri C, et al. (1992) Purification and properties of pectin methylesterase from mandarin orange fruit. Journal of Agricultural and Food Chemistry 40: 591–593.

12. Ernst J, Bar-Joseph Z (2006) STEM: a tool for the analysis of short time series gene expression data. BMC Bioinformatics 7: 191.

13. Hongo S, Sato K, Yokoyama R, Nishitani K (2012) Demethylesterification of the primary wall by Pectin Methylesterase35 provides mechanical support to the Arabidopsis stem. Plant Cell 24: 2624–2634.

14. Downie B, Dirk LM, Hadfield KA, Wilkins TA, Bennett AB, et al. (1998) A gel diffusion assay for quantification of pectin methylesterase activity. Anal Biochem 264: 149–157.

15. Schneider CA, Rasband WS, Eliceiri KW (2012) NIH Image to ImageJ: 25 years of image analysis. Nat Methods 9: 671–675.

16. To L (2013) Genetics of Seed Coat and Stem Development in Flax (*Linum usitatissimum* L.). Edmonton, AB, Canada: University of Alberta. 109 p.

17. Huis R, Hawkins S, Neutelings G (2010) Selection of reference genes for quantitative gene expression normalization in flax (*Linum usitatissimum* L.). BMC Plant Biol 10: 71.

18. Pfaffl MW, Tichopad A, Prgomet C, Neuvians TP (2004) Determination of stable housekeeping genes, differentially regulated target genes and sample integrity: BestKeeper - Excel-based tool using pair-wise correlations. Biotechnology Letters 26: 509–515.

19. Micheli F (2001) Pectin methylesterases: cell wall enzymes with important roles in plant physiology. Trends Plant Sci 6: 414–419.

20. Koch JL, Nevins DJ (1989) Tomato fruit cell-wall 1. Use of purified tomato polygalacturonase and pectinmethylesterase to identify developmental changes in pectins. Plant Physiology 91: 816–822.

21. Liners F, Letesson JJ, Didembourg C, Van Cutsem P (1989) Monoclonal antibodies against pectin: recognition of a conformation induced by calcium. Plant Physiol 91: 1419–1424.

22. Wolf S, Rausch T, Greiner S (2009) The N-terminal pro region mediates retention of unprocessed type-I PME in the Golgi apparatus. Plant J 58: 361–375.

23. Sénéchal F, Graff L, Surcouf O, Marcelo P, Rayon C, et al. (2014) Arabidopsis Pectin Methylesterase17 is co-expressed with and processed by SBT3.5, a subtilisin-like serine protease. Annals of Botany.

24. Mareck A, Lamour R, Schaumann A, Chan P, Driouich A, et al. (2012) Analysis of LuPME3, a pectin methylesterase from *Linum usitatissimum*, revealed a variability in PME proteolytic maturation. Plant Signal Behav 7: 59–61.

25. Al-Qsous S, Carpentier E, Klein-Eude D, Burel C, Mareck A, et al. (2004) Identification and isolation of a pectin methylesterase isoform that could be involved in flax cell wall stiffening. Planta 219: 369–378.

26. Esau K (1977) Anatomy of Seed Plants. New York: Wiley. xx, 550 p. p.

27. Roger D, Lacoux J, Lamblin F, Gaillet D, Dauchel H, et al. (2001) Isolation of a flax pectin methylesterase promoter and its expression in transgenic tobacco. Plant Sci 160: 713–721.

28. Siedlecka A, Wiklund S, Peronne MA, Micheli F, Lesniewska J, et al. (2008) Pectin methyl esterase inhibits intrusive and symplastic cell growth in developing wood cells of *Populus*. Plant Physiol 146: 554–565.

10

Bioinformatic Indications That COPI- and Clathrin-Based Transport Systems Are Not Present in Chloroplasts: An Arabidopsis Model

Emelie Lindquist[9], Mohamed Alezzawi[9], Henrik Aronsson*

Department of Biological and Environmental Sciences, University of Gothenburg, Gothenburg, Sweden

Abstract

Coated vesicle transport occurs in the cytosol of yeast, mammals and plants. It consists of three different transport systems, the COPI, COPII and clathrin coated vesicles (CCV), all of which participate in the transfer of proteins and lipids between different cytosolic compartments. There are also indications that chloroplasts have a vesicle transport system. Several putative chloroplast-localized proteins, including CPSAR1 and CPRabA5e with similarities to cytosolic COPII transport-related proteins, were detected in previous experimental and bioinformatics studies. These indications raised the hypothesis that a COPI- and/or CCV-related system may be present in chloroplasts, in addition to a COPII-related system. To test this hypothesis we bioinformatically searched for chloroplast proteins that may have similar functions to known cytosolic COPI and CCV components in the model plants *Arabidopsis thaliana* and *Oryza sativa* (subsp. *japonica*) (rice). We found 29 such proteins, based on domain similarity, in Arabidopsis, and 14 in rice. However, many components could not be identified and among the identified most have assigned roles that are not related to either COPI or CCV transport. We conclude that COPII is probably the only active vesicle system in chloroplasts, at least in the model plants. The evolutionary implications of the findings are discussed.

Editor: Steven M. Theg, University of California - Davis, United States of America

Funding: This work was supported by Olle Engkvist Byggmästare Foundation (to H.A.), and a PhD student fellowship from the Libyan Higher Education (to M.A.). The funders had no role in study design, data collection and analysis, decision to publish, or preparation of the manuscript.

Competing Interests: The authors have declared that no competing interests exist.

* Email: henrik.aronsson@bioenv.gu.se

9 These authors contributed equally to this work.

Introduction

Chloroplasts, the most fully characterised plastids, contain photosynthetically active thylakoids located in an aqueous stroma, surrounded by a double membrane. In addition to the stroma they have two other aqueous compartments: the intermembrane space between the double membrane's outer and inner envelopes, and the lumen enclosed by the thylakoids. Some chloroplast-localized proteins are encoded by the chloroplast genome. However, most (ca. 95%) are encoded by the nuclear genome, processed in the cytoplasm then transferred to chloroplasts [1]. These proteins are translocated across the outer and inner envelope membranes to the stroma via two translocons, designated TOC and TIC, respectively, mostly aided by cleavable transit peptides [2]. However, some non-canonical proteins may enter the chloroplast without a transit peptide. After entering the chloroplast, proteins are further targeted to specific sub-compartments. Thylakoid targeted proteins are transferred from the stroma via one of four pathways: the Secretory (Sec) pathway, the Signal Recognition Particle (SRP) pathway, the Twin Arginine Translocation (Tat) pathway, or the spontaneous pathway. Proteins transported across the thylakoid membrane into the lumen are using the Sec or the Tat pathway, whereas integral thylakoid membrane proteins are using the SRP or the spontaneous pathway [3,4]. All of these pathways are energy-dependent and mediated by specific combi-nations of proteins except the spontaneous pathway, which requires no energy inputs or specific proteins for protein transport [5].

Although thylakoid membranes contain proteins their main components are lipids, transferred to the thylakoids after synthesis in the envelope [6,7]. Several studies indicate that the lipids could be transported by vesicles [8–10], but as yet there is no clear evidence of protein transport via vesicles in chloroplasts. In contrast, three coated vesicle transport systems have been characterized in the plant cytosol: the COPII (coat protein complex II), COPI (coat protein complex I) and CCV (clathrin coated vesicle) systems, all similar to corresponding systems in yeast and mammals [11–13]. Cytosolic vesicles are known to deliver both soluble and membrane-bound proteins to target membranes, leading to the hypothesis that the vesicle system in chloroplasts may deliver not only lipids, but also proteins [14]. If so, it would represent an uncharacterized fifth pathway for thylakoid-targeted proteins, in addition to the four already identified.

Vesicle transport in chloroplasts has been observed mainly at low temperatures in *Pisum sativum* (pea), *Glycine max* (soybean), *Spinacia oleracea* (spinach) and *Nicotiana tabacum* (tobacco) [15,16]. Proteins required for vesicle transport in the chloroplast are so far suggested to be similar to those of the well-characterized COPII vesicle transport system in the cytosol [14].

COPII, COPI vesicles and CCV in the cytosol have similar functions, but distinct protein and lipid compositions, and recognize different sets of cargo, which make each transport specific [17,18]. COPII-coated vesicles appears to be involved exclusively in transport from ER to Golgi [19,20]. The COPII coat comprises five subunits: Sec23/24, Sec13/31 and Sar1 [11,21]. Formation of a vesicle starts with activation and recruitment of the small GTPase Sar1 to the donor membrane with the help of Sec12p acting as a guanine nucleotide exchange factor (GEF) at ribosome-free ER membranes in the cytosol [22,23]. Subsequently coat proteins are gathered and the vesicle is formed. Most cytosol localized coat subunits of COPII have predicted homologs in chloroplasts [14,24].

Homologues of two important proteins for vesicle transport in the cytosol, RabA5e and Sar1, respectively named CPRabA5e and CPSAR1 (CP = chloroplast localized), have been identified in the chloroplast [9,14,25]. CPSAR1 (which has been detected in the envelopes, stroma and stromal vesicles) is required for thylakoid biogenesis, and is more abundant in the envelopes than the stroma at low temperature (4°C), supporting the hypothesis that it participates in a chloroplast vesicle transport system similar to the cytosolic COPII system [9]. CPRabA5e was subsequently identified in chloroplasts showing an attenuation of vesicles and alteration of thylakoid morphology, under oxidative stress [25].

COPI vesicles primarily mediate transport within the Golgi and between the Golgi and ER [11,26]. The COPI coat (sometimes called coatomer) consists of two main subcomplexes: a cargo-selective F-COPI subcomplex (with β, ∂, γ and ζ subunits), and B-COPI subcomplex (with α, β' and ε subunits) [11,26]. The active form of the GTPase ADP-ribosylation factor 1 (Arf1) is needed to initiate coatomer recruitment to Golgi membranes, similarly to the Sar1 requirement for initiation of COPII coat recruitment. Thus, Arf1 and Sar1 act as triggers for COPI- and COPII-coated vesicle maturation, respectively [27].

CCVs play a key role in membrane and protein transport between the trans-Golgi network, plasma membrane and endosomes [26,28] through the endocytic and late secretory pathways [29]. Their coats consist of clathrin triskelions, structures composed of three "legs" consisting of three heavy chains (each ~190 kDa) and three light chains (each ~25 kDa). They form a basket-like lattice of pentagons and hexagons [30,31] assembled in coordination with other proteins and Arf1. In contrast to COPII and COPI vesicles, adaptor proteins (APs) — including five AP complexes (designated AP1–5), various monomeric adaptors (GGAs) and cargo-specific adaptors — rather than the coat *per se*, are the cargo selectors in CCV vesicles [11,26,29]. They bind to membranes and collect cargo to be transported with the vesicles, sometimes forming networks enabling different kinds of cargo to be transported simultaneously [32,33]. Here we focus on the AP complexes. AP1 and AP2 are dependent on clathrin for vesicle formation, whereas AP3 and AP4 appear to be clathrin-independent [32]. The fifth adaptor protein complex, AP5, was recently discovered in human (HeLa) cells, where it localizes to the endosome and is believed to act independently of clathrin. In this paper vesicles containing these components are collectively referred to as clathrin coated vesicles (CCVs), regardless of their clathrin dependence/independence.

Land plants are known to possess AP1–4, and recent homology analysis suggests they also have AP5 [33]. AP complexes generally consist of four subunits: two large, one medium, and one small [11]. One of the large subunits is called γ, α, δ, ε or ζ, depending on the associated AP complex. The second large subunit is called β and numbered 1–5 depending on the AP complex. Similarly, the medium and small subunits are named μ1–5 and σ1–5,

respectively [33]. However, in plants a single subunit called β1/2 probably functions in both AP1 and AP2 complexes, whereas there are distinct β1 and β2 proteins in mammals [34,35] and σ5 is predicted to be missing from AP5 in Arabidopsis [33]. Like COPI, all AP complexes need Arf proteins for recruitment to membranes [32,36].

There are numerous similarities in the three vesicle systems, e.g. the requirement for activation of small GTPases (Sar1 in the COPII system, Arf1 in the COPI and CCV systems) for recruitment of the coat and additional proteins [11]. There are also similarities in structural architecture of the coats and the domains they possess. For example, α and β' subunits of the B-COPI subcomplex form a triskelion similar to clathrin, generating a curved structure. In terms of domain configuration, the β' subunit of COPI has high similarities to Sec13/31 of COPII, indicating that COPI has similarities to the coats of both CCVs and COPII vesicles [37]. Further similarities include the presence of N-terminal β propellers (enabling binding to AP complexes) and α solenoid legs in the heavy chains of clathrin triskelions of CCVs [37–39], Sec13/31 of the COPII coat [40,41] and the B-COPI subcomplex [26,37]. In addition, γ and β subunits of the F-COPI subcomplex [26] have similarities to "appendages" of the AP complexes of CCVs, and thus are considered to be cargo-binding [11,37].

The similarities in, and differences between, the vesicle systems pose intriguing questions about their origin and evolution. COPII is hypothetically the most ancestral system, since it is an essential biosynthetic pathway in all investigated organisms [26], while COPI (which has strong similarities with both COPII and CCV in domain organization and coat structure, respectively) is putatively an intermediate system [37]. If so, the clathrin system evolved most recently.

Current knowledge of chloroplast vesicles indicates that they are most strongly related to the putatively ancestral COPII system [9,14]. However, the possibility that homologues to cytosolic COPI and CCV systems may be present in chloroplasts has not been systematically explored previously. Thus, we addressed this possibility using the model plant *Arabidopsis thaliana* and yet another model plant *Oryza sativa* (subsp. *japonica*) (rice) to support our findings in Arabidopsis.

Methods

Multiple *in silico* approaches were used to search for proteins in *Arabidopsis thaliana* chloroplasts that could have homologous functions to COPI and CCV proteins in the cytosol of various organisms. The workflow is presented in Figure 1 and described below.

Identifying domains, patterns and motifs

Protein sequences matching cytosolic COPI and CCV subunits in *Arabidopsis thaliana*, *Saccharomyces cerevisiae* (Baker's yeast), *Homo sapiens* (human) and *Mus musculus* (mouse) were retrieved from literature and Uniprot (http://www.uniprot.org) (Figure 1, Tables S1, S2, S3, S4, S5, S6, S7, S8, S9). COPII-related proteins were omitted since they were recently investigated [14].

The collected proteins were compiled and searched for characteristic domains, patterns or motifs, using Prosite release 20.95 (http://prosite.expasy.org) [42] and the Pfam database 26.0 (http://pfam.sanger.ac.uk) [43] (Tables S1, S2, S3, S4, S5, S6, S7, S8, S9). Each identified domain, pattern or motif is denoted by either a PS (Prosite) or PF (Pfam) entry.

After converting the dataset of Arabidopsis chloroplast proteins (GO:0009507) (retrieved from TAIR version 10, www.arabidopsis.

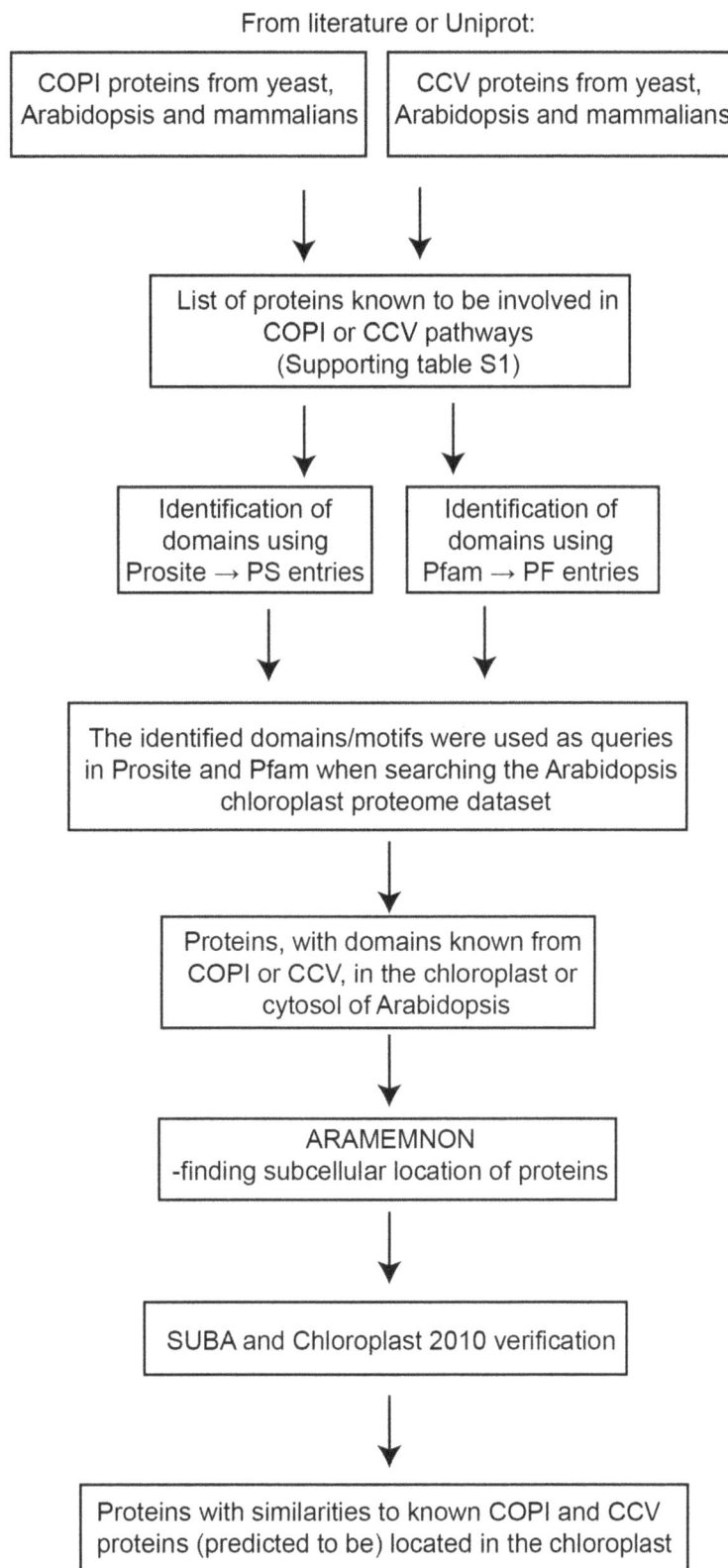

Figure 1. Identification of putative chloroplast COPI and CCV transport components in Arabidopsis. Schematic work flow of the bioinformatics methods used to find putative COPI- and CCV-related transport proteins in chloroplasts. Cytosolic COPI and CCV proteins were retrieved from the literature or Uniprot, and their characteristic domains were identified using Prosite or Pfam. Identified domains were used to search a database of chloroplast-localized proteins, and the localization of proteins found in the chloroplast with relevant domains was further checked using ARAMEMNON, SUBA and Chloroplast 2010. Finally, a list of chloroplast proteins with similarities to known COPI and CCV proteins was compiled.

org) [44] into fasta file format the dataset was searched for the identified Prosite entries using ScanProsite (http://prosite.expasy. org/scanprosite). The Pfam database does not offer corresponding tools so proteins containing the requested Pfam entries were sought manually. The searches generated a list of proteins in Arabidopsis chloroplasts with identical combinations of domains to proteins known to participate in cytosolic COPI or CCV pathways.

As mentioned above, the CCV AP5 complex was first identified in human (HeLa) cells, and subsequently predicted to be present in Arabidopsis, although the degree of conservation is low [33]. Thus, for AP5 we used both human proteins and the predicted proteins in Arabidopsis to identify characteristic domains, which were later used to search the chloroplast dataset.

Subcellular localization of identified proteins

To further check that identified proteins are localized in chloroplasts we used the ARAMEMNON plant membrane protein database [45] to retrieve their names and predict their subcellular localizations, applying 17 tools provided by the host website (http://aramemnon.uni-koeln.de): BaCelLo [46], Chlor-oP_v1.1 [47], iPSort [48], Mitopred [49], Mitoprot_v2 [50], MultiLoc [51], PA-SUB_v2.5 [52], PCLR_v0.9 [53], PProw-ler_v1.1 [54], PrediSi [55], Predotar_v1 [56], PredSL [57], SignalP_HMM_v3 [58,59], SignalP_NN_v3 [60], SLP-Local [61], TargetP_v1 [60,62], and WoLF PSort [63]. Using these tools, a Bayesian consensus (SigConsens) score was obtained from ARAMEMNON for each protein with patterns of interest. A score ≥10 was considered reliable, providing a strong prediction of subcellular location [64]. To corroborate the findings we used the SUBA database version 2.21 (http://suba.plantenergy.uwa.edu. au) [65], which in addition to bioinformatics predictions contains information from proteomic and GFP experiments on subcellular localizations of Arabidopsis proteins. Finally, the Chloroplast 2010 database (http://bioinfo.bch.msu.edu/2010_LIMS) [66] was used to confirm the validity of predictions and/or experiments that identified proteins are present in chloroplasts (Figure 1).

Complementary studies using rice

Complementary investigations of rice (*Oryza sativa* subsp. *japonica*) proteins were conducted to assess the validity and generality of the findings from the Arabidopsis analysis. Domains of relevant Arabidopsis, yeast, mouse and human proteins were retrieved from the Prosite and Pfam websites (Tables S1, S2, S3, S4, S5, S6, S7, S8, S9). Proteins with corresponding combinations of domains were identified in the rice subsp. *japonica* dataset of the National Center for Biotechnology Information (NCBI, TaxID39947) using ScanProsite release 20.102, and their subcellular localizations were predicted using Target P 1.1.

Entries were also searched using Pfam, and proteins with domains of interest were identified manually using a rice dataset downloaded from Phytozome v9.1 (http://www.phytozome.net). To ensure that each hit was from rice subsp. *japonica* they were checked using the Rice genome annotation project (http://rice. plantbiology.msu.edu) or RiceChip Annotation Site (http://www. ricechip.org). Subcellular locations of the hits were then identified using Target P 1.1. Names of identified proteins were retrieved from Uniprot, or the Rice genome annotation project if not identified in Uniprot.

Results

To find putative components of a hypothetical COPI or CCV system in Arabidopsis chloroplasts, known COPI or CCV proteins

from the cytosol of various organisms were retrieved and analysed to identify characteristic domains (Figure 1, Tables S1, S2, S3, S4, S5, S6, S7, S8, S9). The domains were used to search a dataset of protein sequences of chloroplast-localized proteins to identify chloroplast proteins with COPI or CCV domains. Diverse tools, based on differing principles, for predicting the likelihood of proteins having transit peptides were used to strengthen the localization. Several proteins identified in the chloroplast dataset were identical to proteins known to act in the cytosol, raising doubts about their true locations. Occasionally, an identified domain or a domain combination was found in several of the cytosolic proteins, and subsequently in several different chloroplast proteins, hence generating chloroplast proteins which could function as several of the cytosolic subunits. These chloroplast proteins are described below as having commonly occurring domains.

Putative Clathrin triskelion related chloroplast proteins in Arabidopsis

A protein named Putative heavy chain of clathrin complex (AtCHC2)/At3g08530 was found in the TAIR chloroplast dataset (Table 1) with a Clathrin propeller repeat (PF01394), a Clathrin, heavy-chain linker (PF09268), a Clathrin-H-link (PF13838), a Region in Clathrin and VPS domain (PF00637), and a Clathrin heavy-chain (CHCR) repeat profile (PS50236) as identified in yeast (Table S1). The chloroplast localization of AtCHC2 was also supported by SUBA and Chloroplast 2010 (Table 1).

Similarly, Putative light chain of clathrin complex (AtCLC1/ At2g40060; Table 1) was found to have a Clathrin light chain domain (PF01086), identified with known vesicle proteins from both yeast and Arabidopsis (Table S1). Chloroplast localization for this protein was supported by SUBA and Chloroplast 2010 (Table 1).

Putative Clathrin AP1–5 related chloroplast proteins in Arabidopsis

In clathrin-coated vesicles five AP complexes are known, designated AP1–5. Five proteins similar to the AP1 complex γ subunit were found in the chloroplast dataset: Putative ascorbate peroxidase/At1g07890, Putative thylakoid-bound ascorbate per-oxidase (AttAPX)/At1g77490, RNase E/G-type endoribonuclease (AtRNEE/G)/At2g04270, Stromal ascorbate peroxidase (At-sAPX)/At4g08390, and Putative peroxisomal ascorbate peroxi-dase (AtAPX3)/At4g35000 (Table 2). These five proteins all have the same Peroxidases proximal heme-ligand signature domain (PS00435) as the γ subunit of AP1 in yeast (Table S2). Chloroplast localization was supported for the Putative ascorbate peroxidase and AtAPX3 by Chloroplast 2010, and for AtAPX3 also by SUBA. The other three identified proteins (AttAPX, AtRNEE/G, and AtsAPX) had ARAMEMNON consensus scores >10, indicating a chloroplast location, supported by SUBA and Chloroplast 2010 (Table 2).

Considering AP2 homologues, five proteins similar to the β2 subunit in yeast were identified: Putative large subunit of carbamoyl phosphate synthetase VEN3 (AtCarB)/At1g29900, Putative H-protein of glycine decarboxylase/At1g32470, Acetyl-CoA carboxylase (AtACC2)/At1g36180, Biotin carboxylase subunit of plastidic acetyl-coenzyme A carboxylase complex (At-CAC2)/At5g35360, and Putative RimM-like protein involved in 16S rRNA processing/At5g46420 (Table 3). These five proteins all had a Carbamoyl-phosphate synthase subdomain signature 2 (PS00867) identified using Prosite (Table S3).

Table 1. Putative chloroplast localized CCV triskelion components identified using characteristic domains in searches of the TAIR chloroplast dataset.

Name (ARAMEMNON), Accession No	Role of chloroplast protein (TAIR)	SigConsens (ARAMEMNON)			SUBA	Chloroplast 2010
		CP	MT	SEC		
Putative clathrin heavy chain						
AtCHC2, At3g08530	Protein binding, vesicle transport, endocytosis	0.0	0.0	2.7	Yes (MS/MS)	Yes
Putative clathrin light chain						
AtCLC1, At2g40060	Vesicle transport	0.0	0.0	2.0	Yes (MS/MS)	Yes

CP, chloroplast; MT, mitochondria; SEC, secretory pathway.

Of the five proteins predicted to be chloroplast localized AP2 β2 subunits, three (AtCarB, AtCAC2 and the Putative RimM-like protein involved in 16S rRNA processing) had ARAMEMNON consensus scores >10 and support for this localization from both SUBA and Chloroplast 2010. ARAMEMNON also strongly predicted chloroplast localization for AtACC2, but a mitochondrial location for the Putative H-protein of glycine decarboxylase, although chloroplast localization for the latter was supported by SUBA and Chloroplast 2010 (Table 3).

For AP3, AP4 and AP5 only subunits with commonly occurring domains were identified (Table 4, Tables S4, S5, S6). Further details regarding these proteins are presented below in a separate paragraph.

Putative B-COPI subcomplex related chloroplast proteins in Arabidopsis

COPI vesicle coats consist of a B-COPI subcomplex and an F-COPI subcomplex, both composed of several subunits (Tables S7, S8). Our searches detected eight proteins similar to the β′ subunit of the B-COPI subcomplex, with a Trp-Asp (WD) repeats circular profile (PS50294) and a Trp-Asp (WD) repeats profile (PS50082), which identifiey the β′ subunit in both Arabidopsis and human cytosol: Receptor for activated C kinase (AtRACK1A)/At1g18080, Putative U-box-type E3 ubiquitin ligase (AtPUB60)/At2g33340, Putative Cdc20-like mitotic specificity factor for anaphase-promoting complex (AtFZR2/AtCCS52A1)/At4g22910, Putative Cdc20-like mitotic specificity factor for anaphase-promoting complex (AtFZR3/AtCCS52B)/At5g13840, WD40 repeat protein, functions in chromatin assembly (AtMSI1)/

Table 2. Putative chloroplast localized CCV AP1 complex components identified using characteristic domains in searches of the TAIR chloroplast dataset.

Name (ARAMEMNON), Accession No., subunit	Role of chloroplast protein (TAIR)	SigConsens (ARAMEMNON)			SUBA	Chloroplast 2010
		CP	MT	SEC		
Putative clathrin AP1 complex protein						
Putative gamma subunit of coatomer adaptor complex At4g34450*, β1	Cytoskeleton organization, protein transport, catabolic processes, vesicle transport	0.4	0.0	4.0	No	No
Unknown protein At1g51350*, β1	Unknown	20.4	0.0	3.9	No	Yes
Putative ascorbate peroxidase At1g07890, γ	Golgi organization, glycolysis, hyperosmotic response, photorespiration, protein folding (is a ascorbate peroxidase)	0.8	9.0	0.0	No	Yes
AttAPX At1g77490, γ	Chloroplast-nucleus signalling, thylakoid membrane organization (is a ascorbate peroxidase)	22.8	5.0	4.9	Yes (MS/MS)	Yes
AtRNEE/G At2g04270, γ	Chloroplast mRNA processing, chloroplast organisation, thylakoid membrane organization (is a ribonuclease)	11.2	4.3	0.4	Yes (MS/MS)	Yes
AtsAPX At4g08390, γ	Oxidation-reduction processes (is a ascorbate peroxidase)	17.0	5.3	2.2	Yes (MS/MS and GFP)	Yes
AtAPX3 At4g35000, γ	Oxidation-reduction processes (is a ascorbate peroxidase)	0.0	5.8	0.0	Yes (MS/MS)	Yes

*contains common occurring domain(s); CP, chloroplast; MT, mitochondria; SEC, secretory pathway.

Table 3. Putative chloroplast localized CCV AP2 complex components identified using characteristic domains in searches of the TAIR chloroplast dataset.

Name (ARAMEMNON), Accession No., subunit	Role of chloroplast protein (TAIR)	SigConsens (ARAMEMNON)			SUBA	Chloroplast 2010
		CP	MT	SEC		
Putative clathrin AP2 complex protein						
Putative gamma subunit of coatomer adaptor complex At4g34450*, β2	Cytoskeleton organization, protein transport, catabolic processes, vesicle transport	0.4	0.0	4.0	No	No
Unknown protein At1g51350*, β2	Unknown	20.4	0.0	3.9	No	Yes
Unknown protein At5g57460*, μ2	Unknown	6.6	3.0	4.4	Yes (MS/MS)	Yes
AtCarB, At1g29900, β2	Response to phosphate starvation, chromatin silencing, gluconeogenesis, metabolic processes	20.1	3.9	0.0	Yes (MS/MS)	Yes
Putative H-protein of glycine decarboxylase, At1g32470, β2	Glycine processes, PSII assembly, rRNA processing, biosynthesis of cysteine	8.4	16.2	2.3	Yes (MS/MS)	Yes
AtACC2, At1g36180, β2	Fatty acid and metabolic processes (is a acetyl CoA carboxylase)	17.7	6.9	0.2	No	No
AtCAC2, At5g35360, β2	Fatty acid and metabolic processes, brassinosteroid and polysaccharide biosynthesis (is a acetyl CoA carboxylase)	20.8	0.0	0.0	Yes (MS/MS)	Yes
Putative RimM-like protein involved in 16S rRNA processing, At5g46420, β2	Virus defence, metabolic processes, gene silencing, ribosome biogenesis	14.4	2.7	3.1	Yes (MS/MS)	Yes

*contains common occurring domain(s); CP, chloroplast; MT, mitochondria; SEC, secretory pathway.

At5g58230; and three Unknown proteins/At1g24130/At4g02660/At1g15850 (Table 5). Out of these eight proteins only AtFZR2/AtCCS52A1, AtFZR3/AtCCS52B and one of the Unknown proteins (At1g24130) had scores above 10 using ARAMEMNON, and support by Chloroplast 2010. The other five proteins; AtRACK1A, AtPUB60, AtMSI1, and the other two Unknown proteins (At4g02660 and At1g15850), had scores below

Table 4. Putative chloroplast-localized CCV AP3, AP4 and AP5 complex components identified using characteristic domains in searches of the TAIR chloroplast dataset.

Name (ARAMEMNON), Accession No.	Role of protein (TAIR)	SigConsens (ARAMEMNON)			SUBA	Chloroplast 2010
		CP	MT	SEC		
Putative clathrin AP3 complex protein						
δ and β3 subunit						
Putative gamma subunit of coatomer adaptor complex, At4g34450*	Cytoskeleton organization, protein transport, catabolic processes, vesicle transport	0.4	0.0	4.0	No	No
Unknown protein, At1g51350	Unknown	20.4	0.0	3.9	No	Yes
Putative clathrin AP4 complex protein						
ε subunit						
Unknown protein, At5g57460*	Unknown	6.6	3.0	4.4	Yes (MS/MS)	Yes
Putative gamma subunit of coatomer adaptor complex, At4g34450*	Cytoskeleton organization, protein transport, catabolic processes, vesicle transport	0.4	0.0	4.0	No	No
μ4 and σ4 subunit						
Unknown protein, At5g57460	Unknown	6.6	3.0	4.4	Yes (MS/MS)	Yes
Putative clathrin AP5 complex protein						
μ5 subunit						
Unknown protein, At5g57460	Unknown	6.6	3.0	4.4	Yes (MS/MS)	Yes

*contains common occurring domain(s); CP, chloroplast; MT, mitochondria; SEC, secretory pathway.

Table 5. Putative chloroplast localized B-COPI components identified using characteristic domains in searches of the TAIR chloroplast dataset.

Name (ARAMEMNON), Accession No.	Role of protein (TAIR)	SigConsens (ARAMEMNON)			SUBA	Chloroplast 2010
		CP	MT	SEC		
Putative B-COPI subcomplex protein (β′ subunits)						
AtRACK1A, At1g18080	Response to ABA, GA signalling, glycolysis, translation, salt stress, ribosome biogenesis, seed germination	0.0	0.0	0.0	Yes (MS/MS)	No
AtPUB60, At2g33340	Nucleotide binding	0.0	0.0	0.0	Yes (MS/MS)	Yes
AtFZR2 (AtCCS52A1), At4g22910	Protein binding, cell growth, proteasome assembly, regulation of cell division	11.6	0.0	0.0	No	Yes
AtFZR3 (AtCCS52B), At5g13840	Protein binding, DNA methylation, gamete generation, microtubule organization, proteasome assembly, cell division	18.1	0.0	0.0	No	Yes
Unknown protein, At1g24130	Nucleotide binding	10.8	0.0	0.5	No	Yes
Unknown protein, At4g02660	Signal transduction	0.0	0.0	0.0	Yes (MS/MS)	No
AtMSI1, At5g58230	Protein binding, cell proliferation, chromatin modification, seed development, DNA replication	0.4	0.0	0.0	Yes (MS/MS)	Yes
Unknown protein, At1g15850	Nucleotide binding	9.2	0.0	2.9	No	Yes

CP, chloroplast; MT, mitochondria; SEC, secretory pathway.

10 in ARAMEMNON but were supported as chloroplastic by SUBA and/or Chloroplast 2010 (Table 5).

Putative F-COPI subcomplex related chloroplast proteins in Arabidopsis

For the F-COPI subcomplex ζ subunit two proteins were found in the chloroplast: Component of magnesium-protoporphyrin IX chelatase complex (AtCHLD)/At1g08520, and Unknown protein/At1g67120 (Table 6), both having a VWFA domain profile (PS50234) (Table S8). ARAMEMNON strongly predicted chloroplast localization for AtCHLD, but not for the Unknown protein At1g67120, although it was supported for both of these proteins by SUBA and Chloroplast 2010 (Table S8).

Proteins with commonly occurring domains in Arabidopsis chloroplasts

Seven proteins in the chloroplast dataset (four of which were potential Coat GTPases) were found to be possible homologues of two or more components of the COPI and CCV system, since some vesicle proteins from the cytosol share the same domain(s) (Tables S2, S3, S4, S5, S6, S8, S9), which thus identify the same proteins in the chloroplast dataset (Figure 2, Tables 2–4, 6–7).

The first is the Putative gamma subunit of coatomer adaptor complex/At4g34450, which has domain homology with the following subunits: AP1 β1, AP2 β2, AP3 δ, AP3 β3, AP4 ε, and F-COPI γ. All these subunits have the same identifying domain, the Adaptin N terminal region (PF01602) (Tables S2, S3, S4, S5, S6, S8), except the F-COPI γ subunit, which also contains the Coatomer gamma subunit appendage platform subdomain (PF08752) (Table S8). Chloroplast localization was very weakly predicted for this protein by ARAMEMNON (consensus score 0.4), and not supported by either SUBA or Chloroplast 2010 (Tables 2–4, 6).

The second protein, an Unknown protein/At1g51350 also contains the PF01602 domain and could function homologously to the AP1 β1, AP2 β2, AP3 δ, AP3 β3 and AP4 ε subunits (Tables 2–4, S2, S3, S4, S5). It has strongly predicted chloroplast localization according to ARAMEMNON, supported by Chloroplast 2010 (Tables 2–4).

The third protein found in the chloroplast dataset that could have several functions was another Unknown protein/At5g57460 with a Mu homology domain (MHD) profile (PS51072), an identifier of AP2 μ2, AP4 μ4, AP4 σ4, AP5 μ5 and F-COPI coat δ subunits (Tables S3, S5, S6, S8). The Unknown protein (At5g57460) is located in chloroplasts according to SUBA and Chloroplast 2010 (Tables 3–4, 6).

Putative Coat GTPase related chloroplast proteins in Arabidopsis

The last four proteins with commonly occurring domains were identified as putative homologues to Arf proteins. The Arf proteins used as queries in this search were from yeast and the Arabidopsis cytosol, where the latter are divided into four groups (A, B, D and B2) (Bassham et al, 2008). Regardless of their origin, all identified Arf proteins have a small GTPase Arf family profile (PS51417), and an ADP-ribosylation factor family (PF00025) domain (Table S9). Since there was no distinction in the domains identifying the known proteins, the chloroplast search recognized the same four proteins (At1g09180, At1g05810, At4g35860 and At5g57960), regardless of which Arf protein used as a query (Table 7, Table S9). Three of these proteins are already known to be involved in vesicle systems of the Arabidopsis cytosol. At1g09180 is described as a Secretion-associated RAS 1 protein (AtSARA1A) GTPase functioning in COPII transport [11], whereas At1g05810 and At4g35860 are listed as Rab proteins, namely the putative RAB-A-

Table 6. Putative chloroplast localized F-COPI components identified using characteristic domains in searches of the TAIR chloroplast dataset.

Name (ARAMEMNON), Accession No.	Role of protein (TAIR)	SigConsens (ARAMEMNON)			SUBA	Chloroplast 2010
		CP	MT	SEC		
Putative F-COPI subcomplex protein						
ζ subunit						
AtCHLD, At1g08520*	Chlorophyll biosynthesis, cytokinin metabolic process, photosynthesis	22.8	0.4	0.0	Yes (MS/MS)	Yes
Unknown protein, At1g67120	Cytoskeleton organization, embryo sac development, gluconeogenesis	0.0	1.1	3.2	Yes (MS/MS)	Yes
γ subunit						
Putative gamma subunit of coatomer adaptor complex, At4g34450	Cytoskeleton organization, protein transport, catabolic processes, vesicle transport	0.4	0.0	4.0	No	No
δ subunit						
Unknown protein, At5g57460	Unknown	6.6	3.0	4.4	Yes (MS/MS)	Yes

*contains common occurring domain(s); CP, chloroplast; MT, mitochondria; SEC, secretory pathway.

class small GTPase (AtRabA5e) and the putative RAB-B-class small GTPase (AtRabB1c), respectively (Table 7).

Chloroplast localization was supported for AtSARA1A and AtRabB1c by both SUBA and Chloroplast 2010 (Table 7). For AtRabA5e, a transit peptide directing the protein to the chloroplast has been previously suggested [14], its chloroplast location — supported by ARAMEMNON and Chloroplast 2010 (Table 7) — was recently confirmed and it was renamed CPRabA5e to better reflect its location [25]. The only one of these four proteins not already recorded as part of the secretory system [11] is the putative GTPase of unknown function/At5g57960, assigned a chloroplast location by ARAMEMNON, supported by SUBA and Chloroplast 2010, rendering it a candidate Arf in chloroplasts (Table 7).

Putative CCV, COPI and Coat GTPase related chloroplast proteins in rice

In total, 15 proteins in *O. sativa* (subsp. *japonica*) chloroplasts were found to have domains, or combinations of domains, characteristic of CCV, COPI including Coat GTPases (Table 8). Nine of these proteins correspond only to a single subunit. Two, Clathrin heavy chain 1/LOC_Os11g01380 and Clathrin heavy chain 2/LOC_Os12g01390, were identified as possible clathrin heavy chain proteins with a predicted chloroplast location (Table 8). Both have Clathrin propeller repeat (PF01394), Clathrin, heavy-chain linker (PF09268), Clathrin-H-link (PF13838), Region in Clathrin and VPS (PF00637), Clathrin heavy-chain (CHCR) repeat profile (PS50236) domains and an Orn/DAP/Arg decarboxylases family 2 pyridoxal-P attachment site (PS00878) (Table S10).

Two other proteins were identified as putative AP1 γ subunits: Probable L-ascorbate peroxidase 7 (APX7)/LOC_Os04g35520 and Probable L-ascorbate peroxidase 8 (APX8)/LOC_Os02g34810 (Table 8), both of which have a Peroxidases proximal heme-ligand signature (PS00435) (Table S10). Two putative AP2 β2 subunits in rice chloroplasts were also identified: Acetyl-CoA carboxylase 2 (ACC2)/LOC_Os05g22940 and Carbamoyl-phosphate synthase large chain (CARB)/LO-

C_Os01g38970 (Table 8), both containing a Carbamoyl-phosphate synthase subdomain signature 2 domain (PS00867) (Table S10).

Further, a Regulatory-associated protein of TOR 1 (RAPTOR1)/LOC_Os12g01922 was found to have the required domains — a Trp-Asp (WD) repeats profile (PS50082) and a Trp-Asp (WD) repeats circular profile (PS50294) —for a functional B-COPI β′ subunit, whereas PPR repeat-containing protein/LOC_Os07g14530 has the Coatomer epsilon subunit domain (PF04733) required for B-COPI ε subunits (Table 8, Table S10). In addition, Magnesium-chelatase subunit ChlD (CHLD)/LOC_Os03g59640 was found as a putative F-COPI ζ subunit, with a VWFA domain profile (PS50234), in rice chloroplasts (Table 8, Table S10).

In contrast, the remaining six proteins have commonly occurring domains identifying them as possible homologues for several subunits (Table S10). The Adaptin N terminal region (PF01602) domain was found in Armadillo/beta-catenin-like repeat family protein/LOC_Os11g41990, Adaptin, putative/LOC_Os01g43630 and the protein AP3 complex subunit delta/LOC_Os01g32880, identifying them as candidate AP1 β1, AP2 β2, AP3 δ, AP3 β3 and AP4 ε (Table 8, Table S10). In addition, Adaptin, putative/LOC_Os01g43630 has a Beta2-adaptin appendage C-terminal sub-domain (PF09066), providing the domains needed to be a putative AP4 β4 subunit (Table 8, Table S10). Further, Adaptor complexes medium subunit family protein/LOC_Os12g34370 was identified as a putative AP2 μ2, AP3 μ3, AP5 μ5 and F-COPI δ subunit, having Adaptor complexes medium subunit family domain (PF00928), and the Guanine nucleotide-binding protein subunit beta/LOC_Os03g46650 was identified as a possible B-COPI α or B-COPI β′ subunit, with a Trp-Asp (WD) repeats profile PS50082, Trp-Asp (WD) repeats circular profile (PS50294) and Trp-Asp (WD) repeats signature (PS00678). Finally, a candidate for all ARF groups was identified: the Mitochondrial Rho GTPase/LOC_Os03g59590 (Table 8), having an ADP-ribosylation factor family domain (PF00025) (Table 8, Table S10).

a.) Clathrin triskelion

✓ Clathrin light chain
⌐ Clathrin heavy chain

b.) Clathrin coat components

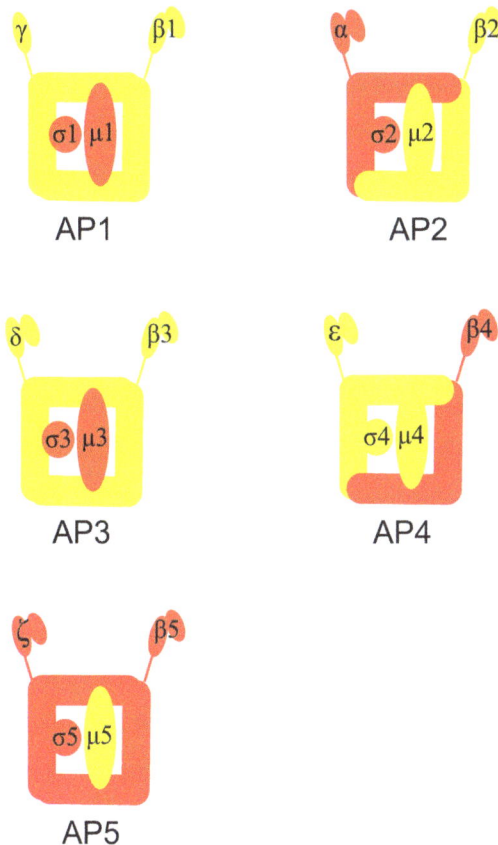

AP1

AP2

AP3

AP4

AP5

c.) COPI coat components and Arf proteins

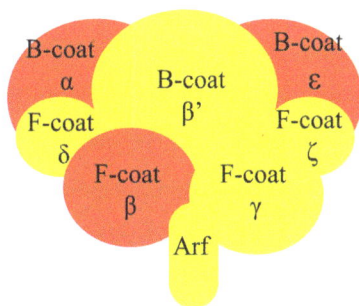

Figure 2. A model of putative CCV and COPI components in Arabidopsis chloroplasts. The figure is based on homologies to components of cytosolic systems (http://www.endocytosis.org/Adaptors/index.html) [130,131]. Red, no proteins with domains of interest detected in chloroplasts. Yellow, proteins with domains of interest identified in chloroplasts, but known to have other roles than vesicle transport and/or the proteins identified had commonly occurring domains, and thus predicted as different subunits in the chloroplasts, and unknown proteins. Green, proteins with domains of interest found in chloroplasts and previously known to have a vesicle transport role.

Discussion

We identified 22 proteins in Arabidopsis that may function as parts of a functional COPI or CCV transport system in chloroplasts: a putative clathrin heavy chain component, clathrin light chain, five AP1 γ subunits, five AP2 β2 subunits, eight B-COPI β′ subunits, and two F-COPI ζ subunit proteins (all having characteristic domains, patterns or motifs of cytosolic counterparts; Table 9).

In addition, seven chloroplast proteins with commonly occurring domains were identified as several putative subunits, possibly with multiple functions (Table 9). Four were identified as similar to Arf proteins, while two others could function as an AP1 β1, AP2 β2, AP3 δ, AP3 β3 and/or AP4 ε subunit. One of these two could also function as an F-COPI γ subunit. The seventh protein was found to have a potential function as an AP2 μ2, AP4 μ4, AP4 σ4, AP5 μ5 and/or F-COPI δ subunit (Table 9).

Thus, various possible components of a COPI or CCV system have been detected in chloroplasts, but several of these have other assigned roles, whereas some required components could not be identified at all. Hence, a key question is whether sufficient components are present to form a functional COPI- or CCV-like transport system.

Evidence for clathrin-coated vesicle system components in Arabidopsis chloroplasts

Triskelion proteins. Concerning triskelion proteins AtCHC2 and AtCLC1 were identified as putative clathrin heavy and light chain respectively inside chloroplast, but have previously been assigned same roles in the cytosol [11,67–69] but both SUBA and Chloroplast 2010 indicate a chloroplast location (Table 1). A possible explanation for this apparent discrepancy, supported by mass spectrometry experiments [70–73], is that they have dual locations. However, further tests of this hypothesis are required.

AP1. Known Arabidopsis AP1 complex γ subunit proteins have specific domains, or combinations of domains, detected in none of the chloroplast localized proteins (Table S2). However, the AP1 complex γ subunit in yeast contains a domain called Peroxidases proximal heme-ligand signature (PS00435), which was also found in five proteins in the Arabidopsis chloroplast dataset: Putative ascorbate peroxidase, AttAPX, AtsAPX, and AtAPX3 (all ascorbate peroxidases), and the RNAse AtRNEE/G (Table 2). AttAPX, AtRNEE/G, and AtsAPX were predicted to be chloroplast-targeted, but Putative ascorbate peroxidase and AtAPX3 do not have unambiguously chloroplast locations (Table 2). Regardless of the localization it could be argued that even if having the same domain as the yeast AP1 complex γ subunit the proteins are not likely to act as components in the AP complex since they act as peroxidases rather than as clathrin-related components.

AP1 and AP2 complexes in the Arabidopsis cytosol are believed to share a β1/β2 subunit. Since the constitution of a hypothetical COPI or CCV system in the chloroplast is inevitably unknown, we included the separate β1 subunit of AP1 and β2 subunit of AP2 from yeast, in addition to the β1/β2 subunit from Arabidopsis as queries in our searches. As for AP1, this made no difference since

Table 7. Putative chloroplast localized CCV and COPI Coat GTPase components identified using characteristic domains in searches of the TAIR chloroplast dataset.

Name (ARAMEMNON), Accession No.	Role of chloroplast protein (TAIR)	SigConsens (ARAMEMNON)			SUBA	Chloroplast 2010
		CP	MT	SEC		
Putative Coat GTPases						
AtSARA1A, At1g09180*	Intracellullar transport	0.4	0.0	19.9	Yes (MS/MS)	Yes
AtRabA5e, At1g05810*	Protein transport, GTP mediated signalling	19.2	1.2	2.8	No	Yes
AtRabB1c, At4g35860*	Protein transport, vesicle transport, protein targeting to the vacuole	0.0	0.0	5.9	Yes (MS/MS)	Yes
Putative GTPase of unknown function, At5g57960*	GTP binding	17.3	1.4	0.0	Yes (MS/MS)	Yes

*contains common occurring domain(s); CP, chloroplast; MT, mitochondria; SEC, secretory pathway.

we found no chloroplast proteins similar to either β1/β2 or β1 (Table S2).

AP2. Regarding the AP2 complex, we found a characteristic domain of the β2 subunit (PS00867) in five chloroplast proteins: Putative H-protein of glycine decarboxylase, AtCarB, AtACC2, AtCAC2, and Putative RimM-like protein involved in 16S rRNA processing (Table S3). This indicates that these proteins are more similar to the yeast subunits than those in the Arabidopsis cytosol.

Table 8. Overview of putative chloroplast localized proteins in rice (*subsp. Japonica*) with characteristic domains of CCV and COPI subunit counterparts.

Subunits	Putative chloroplast localized proteins holding domains characteristics for each subunit respectively
CCV	
Clathrin heavy chain	Clathrin heavy chain 1 (LOC_Os11g01380), Clathrin heavy chain 2 (LOC_Os12g01390)
AP1 γ subunit	Probable L-ascorbate peroxidase 7 (APX7) (LOC_Os04g35520), Probable L-ascorbate peroxidase 8 (APX8) (LOC_Os02g34810)
AP1 β1 subunit	Armadillo/beta-catenin-like repeat family protein (LOC_Os11g41990)*, Adaptin, putative (LOC_Os01g43630)*, AP-3 complex subunit delta (LOC_Os01g32880)*
AP2 β2 subunit	Armadillo/beta-catenin-like repeat family protein (LOC_Os11g41990)*, Carbamoyl-phosphate synthase large chain (CARB) (LOC_Os01g38970), Acetyl-CoA carboxylase 2 (ACC2) (LOC_Os05g22940), Adaptin, putative (LOC_Os01g43630)*, AP-3 complex subunit delta (LOC_Os01g32880)*
AP2 μ2 subunit	Adaptor complexes medium subunit family protein (LOC_Os12g34370)*
AP3 δ subunit	Armadillo/beta-catenin-like repeat family protein (LOC_Os11g41990)*, Adaptin, putative (LOC_Os01g43630)*, AP-3 complex subunit delta (LOC_Os01g32880)*
AP3 β3 subunit	Armadillo/beta-catenin-like repeat family protein (LOC_Os11g41990)*, Adaptin, putative (LOC_Os01g43630)*, AP-3 complex subunit delta (LOC_Os01g32880)*
AP3 μ3 subunit	Adaptor complexes medium subunit family protein (LOC_Os12g34370)*
AP4 β4 subunit	Adaptin, putative (LOC_Os01g43630)*
AP4 ε subunit	Armadillo/beta-catenin-like repeat family protein (LOC_Os11g41990)*, Adaptin, putative (LOC_Os01g43630)*, AP-3 complex subunit delta (LOC_Os01g32880)*
AP5 μ5 subunit	Adaptor complexes medium subunit family protein (LOC_Os12g34370)*
COPI	
B-COPI α-subunit	Guanine nucleotide-binding protein subunit beta (LOC_Os03g46650)*
B-COPI β'-subunit	Guanine nucleotide-binding protein subunit beta (LOC_Os03g46650)*, Regulatory-associated protein of TOR 1 (RAPTOR1) (LOC_Os12g01922)
B-COPI ε-subunit	PPR repeat containing protein (LOC_Os07g14530)
F-COPI δ subunit	Adaptor complexes medium subunit family protein (LOC_Os12g34370)*
F-COPI ζ-subunit	Magnesium-chelatase subunit ChID (LOC_Os03g59640)
Coat GTPases	
ArfA group	Mitochondrial Rho GTPase (LOC_Os03g59590)*
ArfB group	Mitochondrial Rho GTPase (LOC_Os03g59590)*
ArfB2 group	Mitochondrial Rho GTPase (LOC_Os03g59590)*
ArfD group	Mitochondrial Rho GTPase (LOC_Os03g59590)*

* = proteins with several assigned roles, having common occurring domains.

Table 9. Overview of putative chloroplast-localized proteins with characteristic domains of CCV and COPI subunit counterparts.

Subunits	Putative chloroplast localized proteins holding domains characteristics for each subunit respectively
CCV	
Clathrin heavy chain	AtCHC2
Clathrin light chain	AtCLC1
AP1 γ subunit	Putative ascorbate peroxidase, AttAPX, AtRNEE/G, AtsAPX, AtAPX3
AP1 β1 subunit	Putative gamma subunit of coatomer adaptor complex*, Unknown protein* (At1g51350)
AP2 β2 subunit	AtCarB, Putative H-protein of glycine decarboxylase, AtACC2, AtCAC2, Putative RimM-like protein involved in 16S rRNA processing, Putative gamma subunit of coatomer adaptor complex*, Unknown protein* (At1g51350)
AP2 μ2 subunit	Unknown protein* (At5g57460)
AP3 δ subunit	Putative gamma subunit of coatomer adaptor complex*, Unknown protein* (At1g51350)
AP3 β3 subunit	Putative gamma subunit of coatomer adaptor complex*, Unknown protein* (At1g51350)
AP4 ε subunit	Putative gamma subunit of coatomer adaptor complex*, Unknown protein* (At1g51350)
AP4 μ4 subunit	Unknown protein* (At5g57460)
AP4 σ4 subunit	Unknown protein* (At5g57460)
AP5 μ5 subunit	Unknown protein* (At5g57460)
COPI	
B-COPI β' subunit	AtRACK1A, AtPUB60, AtFZR2/AtCCS52A1, AtFZR3/AtCCS52B, AtMSI1, Unknown proteins (At1g24130, At4g02660, At1g15850)
F-COPI γ subunit	Putative gamma subunit of coatomer adaptor complex*
F-COPI δ subunit	Unknown protein* (At5g57460)
F-COPI ζ subunit	AtCHLD, Unknown protein (At1g67120)
Coat GTPases	
ArfA group	AtSARA1A*, AtRabA5e*, AtRabB1c*, Putative GTPase of unknown function*
ArfB group	AtSARA1A*, AtRabA5e*, AtRabB1c*, Putative GTPase of unknown function*
ArfD group	AtSARA1A*, AtRabA5e*, AtRabB1c*, Putative GTPase of unknown function*
ArfB2 group	AtSARA1A*, AtRabA5e*, AtRabB1c*, Putative GTPase of unknown function*

* = proteins with several assigned roles, having common occurring domains.

Furthermore, although they have a predicted chloroplast location they have already been assigned roles that are not related to vesicle transport (Figure 2). The Putative H-protein of glycine decarboxylase is a component of the glycine decarboxylase complex (GDC) that decarboxylates and deaminates glycine, a step in photorespiration occurring in mitochondria [74], despite experimental indications of a chloroplast location [72] (Table 3). AtCarB is part of a Carbamoyl phosphate synthase involved in arginine synthesis, likely in the chloroplast [75]. AtACC2 and AtCAC2 are both acetyl-CoA carboxylases (AACs) [76,77]. The role of AACs is to convert acetyl-CoA to malonyl-CoA during fatty acid synthesis, and plants generally have two types: heterotrimeric and homomeric ACCs. The heterotrimeric ACC in the chloroplast has four subunits: a biotin carboxyl carrier protein, a biotin carboxylase and two carboxyl transferases (α and β) [78]. The homomeric ACC is encoded by two genes, *ACC1* and *ACC2*, and has been considered to be cytosolic, but it was recently shown that the *ACC2* protein product is located in plastids of Arabidopsis [78]. The last putative protein to be discussed as an AP2 β2 subunit is the Putative RimM-like protein involved in 16S rRNA processing, which has been found in the stroma and is involved in RNA processing [70]. Thus, none of the proteins identified as putative AP2 β2 subunits are likely to act in this manner in the chloroplast based on their proposed functions, which are not related to vesicle transport (Figure 2).

In contrast to AP1 and AP2 subunit candidates, putative AP3, AP4 and AP5 subunits identified only had commonly occurring domains, making them weaker candidates as true vesicle transport system components (Figure 2).

Evidence for COPI vesicle system components in Arabidopsis chloroplasts

B-COPI subcomplex. Eight candidate proteins for the B-COPI coat β' subunit of COPI vesicles with a predicted chloroplast location were detected. The first is AtRACK1 (Table 5), which plays various roles in plants. Plant mutants defective in this protein have reduced sensitivity to various hormones and impairments in developmental processes, including leaf production [79]. AtRACK1A is also a negative regulator of abscisic acid (ABA) responses [80], and recently a number of proteins have been suggested to interact with AtRACK1A, including proteins involved in photosynthesis and stress responses [81]. Its location is ambiguous; chloroplast localization lacks support from ARAMEMNON and Chloroplast 2010, but it has been found experimentally in chloroplasts according to SUBA (Table 5).

The next three identified proteins (AtPUB60, AtFZR2/CCS52SA1 and AtFZR3/CCS52B) have demonstrated roles in protein degradation (Table 5). AtPUB60 is a U-box protein similar to E3 ubiquitin ligases in yeast and humans, involved in plant innate immunity and plant pathogen resistance [82], in addition to its role in the ubiquitin degradation pathway [83]. In Arabidopsis, the cell cycle process is regulated by a number of cyclins, grouped into A, B, D and H cyclins. Some group B cyclins are degraded during mitosis by a specific ubiquitin E3 ligase, known as anaphase promoting complex (APC), following activation by subunits, which include AtFZR2/CCS52A1 and AtFZR3/CCS52B [84,85]. Further, both of these proteins are involved in endoreduplication, which increases ploidy by inhibiting mitosis [86]. Considering the roles of AtFZR2/CCS52A1 and AtFZR3/CCS52B one might assume a cytosolic location, but ARAMEMNON assigns a chloroplast location, but this has not been experimentally proven according to SUBA.

Further, AtMSI1 was identified as a putative β′ subunit (Table 5). Together with FAS1 and FAS2 this is a member of the Chromatin assembly factor-1 (CAF-1) complex in Arabidopsis, which functions as a histone chaperone in chromatin assembly [87,88], and is also important for additional processes such as seed development [89].

The last three proteins identified in this category are unknown and largely uncharacterized (Table 5). Two, Unknown proteins At1g15850 and At1g24130 are only known to be nucleotide binding, having WD40 domains. The third (At4g02660), is a putative transport protein with a BEACH domain, found in trichome cells [90] and chloroplasts [71]. The function of the BEACH domain is unknown, but appears to be crucial for a number of proteins involved in e.g. vesicle transport [90,91]. However this domain could not be identified using Prosite or Pfam. Thus, only the three Unknown proteins can be considered as likely candidates for hypothesized β′ subunits, as the only ones lacking other assigned roles.

F-COPI subcomplex. Two proteins were identified as putative ζ subunits of the F-COPI subcomplex: AtCHLD and an Unknown protein (At1g67120) (Table 6). AtCHLD has already been identified as involved in the secretory system [11]. Closer examination revealed that AtCHLD significantly differs from known ζ subunits in Arabidopsis and yeast secretory systems (Table S8). It has a VWFA domain profile (PS50234), similar to von Willebrand factor type A domain (PF13519), and a Magnesium chelatase subunit ChlI domain (PF01078), which are not present in any other known ζ subunits. Other ζ subunits have a Clathrin adaptor complex small chain domain (PF01217) lacked by AtCHLD. Thus, AtCHLD appears to be the Magnesium-chelatase subunit ChlD (Uniprot) of Magnesium chelatase, a complex with three subunits [92,93]. This complex is involved in chlorophyll biosynthesis, mediating insertion of magnesium ions into protoporphyrin IX, thereby generating Mg-protoporphyrin IX, and is located in chloroplasts [94,95].

The other putative ζ subunit identified in the chloroplast dataset was the Unknown protein (At1g67120) (Table 6), likely to be chloroplastic according to Chloroplast 2010 and SUBA [73].

Given the distinct differences between AtCHLD and other known ζ subunits, previous reports that AtCHLD functions as a magnesium chelatase in the chloroplast [92,93], and the finding that the Unknown protein has the same domains as AtCHLD, there are probably no homologues of the ζ subunit in chloroplasts (Figure 2).

Commonly occurring domains in Arabidopsis chloroplast proteins

Some proteins are reported to perform several roles, such as the AP4 μ4 subunit in Arabidopsis cytosol which has also been noted as the σ4 subunit in the same complex [11] and the newly identified AP5 ζ subunit which has been previously designated a DNA helicase [96]. Thus, the possibility that some of the putative subunits identified here could play several roles and/or other roles than previously reported should not be excluded. We found three proteins that all correspond to several known subunits: Putative

gamma subunit of coatomer adaptor complex, and Unknown proteins At1g51350 and At5g57460 (Table 9). The Putative gamma subunit of the coatomer adaptor complex has been ambiguously called both Sec21 and a COPI γ subunit [24,97–99], but is considered to be a γ subunit located in Golgi and ER membranes in Arabidopsis [99]. It has even been used experimentally as a Golgi marker [98,100] and shown to be involved in cytosolic vesicle transport in Arabidopsis [11], raising doubts about a true chloroplast location, and thus the likelihood of its involvement in vesicle transport in chloroplasts (Figure 2).

It has been suggested that the Unknown protein At1g51350 is a homologue of the human ARMC8α [101], and involved in endosomal sorting and trafficking [102]. However, ARAMEM-NON strongly indicates that it is chloroplast localized. The other Unknown protein, At5g57460, has no clear assigned function yet. Thus, the two Unknown proteins could be involved in some of the suggested functions, but further confirmation is needed (Figure 2).

Coat GTPases in Arabidopsis chloroplasts

Four other proteins with commonly occurring domains were found in the Arabidopsis chloroplast, sharing domains with the previously described cytosolic Arf proteins: AtSARA1A, AtRabA5e, AtRabB1c and the Putative GTPase of unknown function (Table 7). All but one of these four proteins has been ascribed other functions, showing that searches for proteins with this domain will not detect only Arf proteins. The Putative GTPase of unknown function, strongly predicted to be chloroplastic by all the databases and experimental data [70,71], is downregulated in a cold-resistant bri1 (brassinosteroid-insensitive 1) Arabidopsis mutant [103]. Thus, its possible involvement in vesicle transport is not clear (Figure 2).

AtSARA1A has both a small GTPase Arf family profile domain (PS51417) and an ADP-ribosylation factor family domain (PF00025) and has already been identified in the secretory system of Arabidopsis as a Sar1 protein [11] (Table S9). It acts as a GTPase, regulating COPII coat assembly in the cytosol [104,105]. However, SARA1A has also been detected in chloroplasts [106], thus it has an ambiguous or possibly dual localization. Interestingly, another Sar1 protein, CPSAR1, identified in the chloroplast has been shown to affect vesicle transport [9].

Two Rab proteins were identified, CPRabA5e and RabB1c (Table 7). CPRabA5E has previously been predicted as an Arf protein [24], but was recently shown to be a Rab protein with a chloroplast location involved in thylakoid biogenesis. It was affected by oxidative stress, accumulating vesicles at the envelope in chloroplasts when incubated at low temperature under oxidative stress [25].

RabB1c is assumed to participate in vesicle transport according to Uniprot, and is a member of the AtRabB family, which is related to human Rab2 GTPases that are involved in COPI transport in mammalian cells [107,108], and may play a similar role in the Arabidopsis secretory system [109]. RabB1c lacks a transit peptide [14], but has been detected in chloroplasts experimentally [71]. Thus, the only plausible candidate Arf in the chloroplast is the Putative GTPase of unknown function, but GTP binding is apparently not sufficient for a functional Arf, thus further confirmation that it acts as one is required (Figure 2).

Evidence for clathrin-coated vesicle system components in rice

Six proteins were identified in rice with CCV relevant domains and chloroplast localization according to Target P. Clathrin heavy chain 1 and Clathrin heavy chain 2 (Table 8) are referred to as clathrin heavy chains [110,111], based on their similarity to other clathrin components (Uniprot) but have not been characterized. If the predicted chloroplast localization is correct further investigation is warranted since their homologies and designations clearly imply a role in vesicle transport.

Two ascorbate peroxidases were identified, APX7 and APX 8 (Table 8). In plants, ascorbate peroxidases use ascorbate as an electron donor to convert H_2O_2 to H_2O. In rice there are eight known *APX* genes, and four of which are believed to be chloroplast localized (*APX5-APX8*) [112,113]. However, they are unlikely to act as AP1 γ subunits in chloroplasts due to their role as peroxidases.

Two AP2 β2 subunit candidates, ACC2 and CARB (Table 8), were identified. However, as discussed above in the Arabidopsis analysis, ACC2 and CARB are involved in fatty acid and arginine synthesis; hence they are unlikely to be subunits of chloroplast vesicles.

Evidence for COPI system components in rice

With COPI relevant domains and chloroplast location according to Target P, three proteins were identified. RAPTOR1 was identified as a putative B-COPI β′ subunit (Table 8). However, in Arabidopsis RAPTOR1 is known as to regulate TOR1 (TARGET OF RAPAMYCIN), a kinase involved in growth signalling pathways, and interacts with a putative substrate of TOR, S6K1, in vivo [114]. The role of RAPTOR1 in the TOR pathway in Arabidopsis makes it an unlikely candidate as possible B-COPI β′ subunit also in rice.

As a putative B-COPI ε subunit, the PPR repeat containing protein was identified (Table 8). Pentatricopeptide repeat proteins (PPR proteins) are RNA-binding proteins involved in various post-transcriptional processes in both mitochondria and chloroplasts [115]. The PPR family is defined by a tandem 35 amino acid motif. The proteins are predicted to have multiple α helices, placing them in the α-solenoid superfamily together with e.g. HEAT domain proteins [115,116]. One of the PPR proteins in rice, OsPPR1, is located in chloroplasts, essential for chloroplast biogenesis, and its suppression results in chlorophyll deficiency [117].

As also found in the Arabidopsis analysis, the only protein corresponding to F-COPI ζ identified in rice chloroplasts was a magnesium chelatase, CHLD [118] (Table 8). This again raises doubts about its function as a COPI component, which was previously indicated [11].

Commonly occurring domains of proteins including Coat GTPases in rice chloroplasts

Six proteins of rice chloroplasts were identified with commonly occurring domains found in multiple subunits, but due to the low specificity of the identifying domains they are less robust candidates. One domain, the Adaptin N terminal region (PF01602), was detected in AP-3 complex subunit delta, Armadillo/beta-catenin-like repeat family protein and Adaptin, putative (Table S10). Little is known about these proteins; AP-3 complex subunit delta has a name implying a role in vesicle transport, but has not yet been characterized. The Armadillo/beta-catenin-like repeat family protein has Armadillo repeats, placing it in the ARM repeat superfamily together with AP-3 complex subunit delta, according to Uniprot and Superfamily 1.75 [119]. Armadillo repeats are found in proteins with various roles, for instance β-catenin [120]. They are about 40 amino acids long and usually tandemly repeated, forming an armadillo domain. Adaptin, putative has an Adaptin N terminal region (PF01602), but also a Beta2-adaptin appendage C-terminal sub-domain (PF09066) (Table S10). The protein has not yet been characterized in rice.

Since two of these proteins have names related to vesicle transport, and all three share domain PF01602, they could potentially all be true subunits.

Adaptor complexes medium subunit family protein/LOC_Os12g34370 has, similarly to AP-3 complex subunit delta and Adaptin putative not either been characterized but a name implying a role in vesicle transport. In Arabidopsis, At1g56590 was annotated as Clathrin adaptor complexes medium subunit family protein [121] and is considered as the AP3 μ3 subunit [11]. Hence, a role in vesicle transport in chloroplasts cannot be excluded.

Guanine nucleotide-binding protein subunit beta was identified as a putative B-COPI α and B-COPI β′ subunit. Its name implies a role as a subunit of a heterotrimeric G-protein, but it has not yet been characterized. The domains identified in this protein (PS50294, PS50082 and PS00678) all refer to WD repeats (Table S10). WD repeat proteins have four of more repetitive subunits, each consisting of about 40–60 amino acids and usually ending with tryptophan (W) and aspartic acid (D) [122]. It has been assumed that all WD repeat proteins form β propellers, and the best characterized is the β subunit of the heterotrimeric G protein [39,122,123]. WD repeat proteins have known importance in various processes, including vesicle transport [122,123], but as shown here simply detecting WD repeats in a protein is not sufficient to elucidate a protein's functions completely.

The Ras superfamily of small GTPases is divided into five families: Rab, Arf/Sar, Ran, Ras and Rho. In plants, no representatives of the Ras family have been found [124]. One protein was identified in a search for proteins with an ADP-ribosylation factor family domain (PF00025): Mitochondrial Rho GTPase. Its Uniprot name indicates a mitochondrial location, but its Rice Genome Annotation project designation is less specific (ATP/GTP/Ca++ binding protein, putative, expressed), and it is located in the chloroplast according to Target P (Table S10). Thus, future experiments are needed to resolve its location.

Conclusion

The acquired data indicate that no transport system resembling cytosolic CCV or COPI systems is present in Arabidopsis chloroplasts. Several putative subunits identified in the chloroplast dataset were shown to be located elsewhere according to previous studies or various tools, having a possible dual location and/or roles unrelated to vesicle transport. Out of 29 proteins identified in Arabidopsis, the majority had either commonly occurring domains, vesicle unrelated or unknown function (Figure 2). Only two proteins among the suggested, Putative heavy chain of clathrin complex (AtCHC2/At3g08530) and Putative light chain of clathrin complex (AtCLC1/At2g40060), could be considered likely subunits in the chloroplast, having known roles related to vesicle transport. Several subunits could not be identified at all in the chloroplast, when searching for relevant domains (Figure 2). The findings indicate that if a CCV- or COPI-like vesicle system is present in chloroplasts it probably differs substantially from the cytosolic counterpart. However, the possible presence of a different and/or simplified CCV or COPI system cannot be excluded. The occurrence of a putative AP2 β2 subunit supports the possible presence of a unique system, since this homologue is present in yeast but not Arabidopsis cytosol, and many of the putative subunits identified have greater resemblance to yeast counterparts than Arabidopsis counterparts (Tables S1, S2, S3, S4, S5, S6, S7, S8, S9).

Considering rice, most of the subunits that could be identified are uncharacterized and named by their similarity to other proteins. As in Arabidopsis chloroplasts, many proteins were also found to have commonly occurring domains. Only two proteins (still uncharacterized in rice) have names indicating a role in vesicle transport, a predicted chloroplast location and domains that are not commonly occurring: Clathrin heavy chain 1 and Clathrin heavy chain 2. It is interesting to note that the results in rice support the findings in Arabidopsis i.e. not many proteins can be clearly said to be chloroplast localized and involved in vesicle transport.

No prokaryote vesicle transport system has been reported [16,125,126], but a few examples of prokaryotic structures analogous to vesicles have been observed [35]. The vesicle system in eukaryotes has been hypothesized as a trait that developed soon after the divergence from prokaryotes and thereafter further specialized as adaptations to new environments [35]. Chloroplasts, believed to have resulted via endosymbiosis of early eukaryotes with cyanobacteria, have vesicles with properties resembling other eukaryotic vesicles, including probable regulation of their formation by GTPases, and inhibition of fusion by microcystin LR and low temperature [16]. However, two proteins of prokaryotic origin have suggested involvement in vesicle formation in chloroplasts: CPSAR1 and Vipp1 [9,127]. Vesicles have also been found in representatives of embryophytes, including bryophytes, pteridophytes, spermatophytes (gymnosperms and angiosperms), but not in other groups including cyanobacteria, glaucocystophytes, rhodophytes, chlorophytes and charophytes. Hence, it has been proposed that the vesicles in chloroplasts evolved after the division of embryophytes from charophytes as an adaptation to land colonization [125].

Plastids occur in several forms, in diverse organisms, and their broad variation in thylakoid organization is assumed to have arisen via evolution in different hosts after the ancestral endosymbiosis [126]. Regarding the three known vesicle transport systems, the COPII system is likely to ancestral, since it is used in essential biosynthetic pathways in all eukaryotes, while the COPI and CCV systems could be later specializations involved in recycling resources to the ER and endocytosis [26,128,129].

Taken together, the available evidence indicates that a vesicle system arose in early eukaryotes, COPII is the ancestral machinery, chloroplast vesicles show clear eukaryotic traits and first evolved during land colonization. In addition, we conclude that no COPI- or CCV-like vesicle system is likely to be found in chloroplasts, in contrast to a COPII-like system, for which a chloroplast location has bioinformatic support [14]. Speculatively, early eukaryotes gained a COPII-like vesicle system, engulfed cyanobacteria and developed plastids, to which the system was transferred. If so, since some photosynthetic eukaryotes do not have vesicles in their plastids, a major speciation event was presumably involved, separating those that form COPII-like chloroplast vesicles from others, before all lines continued to develop the COPI and CCV systems in the cytosol. Alternatively, all three vesicle systems may have already developed in the cytosol of the ancestral eukaryotes when cyanobacteria were engulfed, but only the COPII system was transferred to the chloroplast, or the other two were lost during subsequent evolution. Thus, future experimental evidence is needed to solve the intriguing questions how, when and why a suggested COPII system emerged as the sole vesicle system in chloroplasts.

Supporting Information

Table S1 CCV triskelion proteins from Arabidopsis (*A. thaliana*) cytosol (retrieved from Bassham et al, 2008) and yeast (*S. cerevisiae*), mouse (*M. musculus*) and human (*H. sapiens*) cytosol

(retrieved from Uniprot). Domains of these proteins were extracted using Prosite and Pfam, then run against the chloroplast protein dataset to identify proteins putatively involved in vesicle transport in chloroplasts.

Table S2 CCV AP1 complex proteins from Arabidopsis (*A. thaliana*) cytosol (retrieved from Bassham et al, 2008) and yeast (*S. cerevisiae*), mouse (*M. musculus*) and human (*H. sapiens*) cytosol (retrieved from Uniprot). Domains of these proteins were extracted using Prosite and Pfam, then run against the chloroplast protein dataset to identify proteins putatively involved in vesicle transport in chloroplasts.

Table S3 CCV AP2 complex proteins from Arabidopsis (*A. thaliana*) cytosol (retrieved from Bassham et al, 2008) and yeast (*S. cerevisiae*), mouse (*M. musculus*) and human (*H. sapiens*) cytosol (retrieved from Uniprot). Domains of these proteins were extracted using Prosite and Pfam, then run against the chloroplast protein dataset to identify proteins putatively involved in vesicle transport inside chloroplasts.

Table S4 CCV AP3 complex proteins from Arabidopsis (*A. thaliana*) cytosol (retrieved from Bassham et al, 2008) and yeast (*S. cerevisiae*), mouse (*M. musculus*) and human (*H. sapiens*) cytosol (retrieved from Uniprot). Domains of these proteins were extracted using Prosite and Pfam, then run against the chloroplast protein dataset to identify proteins putatively involved in vesicle transport inside chloroplasts.

Table S5 CCV AP4 complex proteins from Arabidopsis (*A. thaliana*) cytosol (retrieved from Bassham et al, 2008) and yeast (*S. cerevisiae*), mouse (*M. musculus*) and human (*H. sapiens*) cytosol (retrieved from Uniprot). Domains of these proteins were extracted using Prosite and Pfam, then run against the chloroplast protein dataset to identify proteins putatively involved in vesicle transport in chloroplasts.

Table S6 CCV AP5 complex proteins from Arabidopsis (*A. thaliana*) cytosol (retrieved from Hirst et al, 2011), and human (*H. sapiens*) cytosol (retrieved from Uniprot). Domains of these proteins were extracted using Prosite and Pfam, then run against the chloroplast protein dataset to identify proteins putatively involved in vesicle transport inside chloroplasts.

Table S7 B-COPI subcomplex proteins from Arabidopsis (*A. thaliana*) cytosol (retrieved from Bassham et al, 2008) and yeast (*S. cerevisiae*), mouse (*M. musculus*) and human (*H. sapiens*) cytosol (retrieved from Uniprot). Domains of these proteins were extracted using Prosite and Pfam, then run against the chloroplast protein dataset to identify proteins putatively involved in vesicle transport inside chloroplasts.

Table S8 F-COPI subcomplex proteins from Arabidopsis (*A. thaliana*) cytosol (retrieved from Bassham et al, 2008) and yeast (*S. cerevisiae*), mouse (*M. musculus*) and human (*H. sapiens*) cytosol (retrieved from Uniprot). Domains of these proteins were extracted using Prosite and Pfam, then run against the chloroplast protein dataset to identify proteins putatively involved in vesicle transport inside chloroplasts.

Table S9   Coat GTPase proteins from Arabidopsis (*A. thaliana*) cytosol (retrieved from Bassham et al, 2008) and yeast (*S. cerevisiae*), mouse (*M. musculus*) and human (*H. sapiens*) cytosol (retrieved from Uniprot). Domains of these proteins were extracted using Prosite and Pfam, then run against the chloroplast protein dataset to identify proteins putatively involved in vesicle transport inside chloroplasts.

Table S10   CCV and COPI proteins from Arabidopsis (*A. thaliana*) cytosol (retrieved from Bassham et al, 2008) and yeast (*S. cerevisiae*), mouse (*M. musculus*) and human (*H. sapiens*) cytosol (retrieved from Uniprot). Domains of these proteins were extracted using Prosite and Pfam, run against the rice (subsp. *japonica*) protein dataset to identify proteins with the same domains, then

those putatively involved in vesicle transport in chloroplasts were identified using Target P, and listed.

Acknowledgments

The authors thank Jenny Carlsson and Sazzad Karim for critical reading of the manuscript, Kemal Sanli for valuable help with organising Pfam data, Nadir Zaman Khan, Selvakumar Sukumuran, and Mageshwaran Rajasekar for initial assistance.

Author Contributions

Conceived and designed the experiments: EL MA HA. Performed the experiments: EL MA HA. Analyzed the data: EL MA HA. Contributed reagents/materials/analysis tools: EL MA HA. Wrote the paper: EL MA HA.

References

1. Abdallah F, Salamini F, Leister D (2000) A prediction of the size and evolutionary origin of the proteome of chloroplasts of Arabidopsis. Trends in plant science 5: 141–142.
2. Aronsson H, Jarvis P (2008) The chloroplast protein import apparatus, its components, and their roles. In: Sandelius AS, Aronsson H (eds) Plant Cell Monograph "Chloroplast - interactions with the environment" Springer Verlag, Berlin, Germany.
3. Jarvis P, Robinson C (2004) Mechanisms of protein import and routing in chloroplasts. Current Biology 14: R1064–R1077.
4. Robinson C, Thompson SJ, Woolhead C (2001) Multiple pathways used for the targeting of thylakoid proteins in chloroplasts. Traffic 2: 245–251.
5. Spetea C, Aronsson H (2012) Mechanisms of Transport Across Membranes in Plant Chloroplasts. Current Chemical Biology 6: 230–243.
6. Kelly AA, Dörmann P (2004) Green light for galactolipid trafficking. Current opinion in plant biology 7: 262–269.
7. Shimojima M, Ohta H, Iwamatsu A, Masuda T, Shioi Y, et al. (1997) Cloning of the gene for monogalactosyldiacylglycerol synthase and its evolutionary origin. Proceedings of the National Academy of Sciences 94: 333–337.
8. Andersson MX, Kjellberg JM, Sandelius AS (2001) Chloroplast biogenesis. Regulation of lipid transport to the thylakoid in chloroplasts isolated from expanding and fully expanded leaves of pea. Plant physiology 127: 184–193.
9. Garcia C, Khan NZ, Nannmark U, Aronsson H (2010) The chloroplast protein CPSAR1, dually localized in the stroma and the inner envelope membrane, is involved in thylakoid biogenesis. Plant Journal 63: 73–85.
10. Räntfors M, Evertsson I, Kjellberg JM, Stina Sandelius A (2000) Intraplastidial lipid trafficking: Regulation of galactolipid release from isolated chloroplast envelope. Physiologia Plantarum 110: 262–270.
11. Bassham DC, Brandizzi F, Otegui MS, Sanderfoot AA (2008) The secretory system of Arabidopsis. The Arabidopsis Book/American Society of Plant Biologists 6.
12. Donaldson JG, Cassel D, Kahn RA, Klausner RD (1992) ADP-ribosylation factor, a small GTP-binding protein, is required for binding of the coatomer protein beta-COP to Golgi membranes. Proceedings of the National Academy of Sciences 89: 6408–6412.
13. Kirchhausen T (2000) Three ways to make a vesicle. Nature Reviews Molecular Cell Biology 1: 187–198.
14. Khan NZ, Lindquist E, Aronsson H (2013) New putative chloroplast vesicle transport components and cargo proteins revealed using a bioinformatics approach: an Arabidopsis model. PLoS One 8: e59898.
15. Morré DJ, Selldén G, Sundqvist C, Sandelius AS (1991) Stromal low temperature compartment derived from the inner membrane of the chloroplast envelope. Plant Physiology 97: 1558–1564.
16. Westphal S, Soll J, Vothknecht UC (2001) A vesicle transport system inside chloroplasts. FEBS Letters 506: 257–261.
17. Rothman JE (1994) Mechanisms of intracellular protein transport. Nature 372: 55–63.
18. Schekman R, Orci L (1996) Coat proteins and vesicle budding. Science 271: 1526–1533.
19. Lee MCS, Miller EA, Goldberg J, Orci L, Schekman R (2004) Bi-directional protein transport between the ER and Golgi. Annual Review of Cell and Developmental Biology 20: 87–123.
20. Bethune J, Wieland F, Moelleken J (2006) COPI-mediated transport. Journal of Membrane Biology 211: 65–79.
21. Bednarek SY, Ravazzola M, Hosobuchi M, Amherdt M, Perrelet A, et al. (1995) COPI-and COPII-coated vesicles bud directly from the endoplasmic reticulum in yeast. Cell 83: 1183–1196.
22. Barlowe C, Schekman R (1993) SEC12 encodes a guanine-nucleotide-exchange factor essential for transport vesicle budding from the ER. Nature 365: 347–349.
23. Yoshihisa T, Barlowe C, Schekman R (1993) Requirement for a GTPase-activating protein in vesicle budding from the endoplasmic reticulum. Science 259: 1466–1468.
24. Andersson MX, Sandelius AS (2004) A chloroplast-localized vesicular transport system: a bio-informatics approach. BMC Genomics 5: 40.
25. Karim S, Alezzawi M, Garcia-Petit C, Solymosi K, Khan NZ, et al. (2014) A novel chloroplast localized Rab GTPase protein CPRabA5e is involved in stress, development, thylakoid biogenesis and vesicle transport in Arabidopsis. Plant Molecular Biology 84: 675–692.
26. McMahon HT, Mills IG (2004) COP and clathrin-coated vesicle budding: different pathways, common approaches. Current Opinion in Cell Biology 16: 379–391.
27. Lee M, Orci L, Hamamoto S, Futai E, Ravazzola M, et al. (2005) Sar1p N-terminal helix initiates membrane curvature and completes the fission of a COPII vesicle. Cell 122: 605–617.
28. Bonifacino JS, Glick BS (2004) The mechanisms of vesicle budding and fusion. Cell 116: 153–166.
29. Dell'Angelica EC (2001) Clathrin-binding proteins: Got a motif? Join the network! Trends in Cell Biology 11: 315–318.
30. Kirchhausen T (2000) Clathrin. Annual Review of Biochemistry 69: 699–727.
31. Wilbur JD, Hwang PK, Brodsky FM (2005) New faces of the familiar clathrin lattice. Traffic 6: 346–350.
32. Robinson MS (2004) Adaptable adaptors for coated vesicles. Trends in Cell Biology 14: 167–174.
33. Hirst J, Barlow LD, Francisco GC, Sahlender DA, Seaman MN, et al. (2011) The fifth adaptor protein complex. PLoS Biology 9: e1001170.
34. Boehm M, Bonifacino JS (2001) Adaptins The Final Recount. Molecular Biology of the Cell 12: 2907–2920.
35. Dacks JB, Field MC (2007) Evolution of the eukaryotic membrane-trafficking system: origin, tempo and mode. Journal of Cell Science 120: 2977–2985.
36. Paleotti O, Macia E, Luton F, Klein S, Partisani M, et al. (2005) The small G-protein Arf6GTP recruits the AP-2 adaptor complex to membranes. Journal of Biological Chemistry 280: 21661–21666.
37. Lee C, Goldberg J (2010) Structure of coatomer cage proteins and the relationship among COPI, COPII, and clathrin vesicle coats. Cell 142: 123–132.
38. Fotin A, Cheng Y, Sliz P, Grigorieff N, Harrison SC, et al. (2004) Molecular model for a complete clathrin lattice from electron cryomicroscopy. Nature 432: 573–579.
39. Ter Haar E, Harrison SC, Kirchhausen T (2000) Peptide-in-groove interactions link target proteins to the β-propeller of clathrin. Proceedings of the National Academy of Sciences 97: 1096–1100.
40. Fath S, Mancias JD, Bi X, Goldberg J (2007) Structure and organization of coat proteins in the COPII cage. Cell 129: 1325–1336.
41. Stagg SM, LaPointe P, Razvi A, Gürkan C, Potter CS, et al. (2008) Structural basis for cargo regulation of COPII coat assembly. Cell 134: 474–484.
42. Gattiker A, Gasteiger E, Bairoch A (2002) ScanProsite: a reference implementation of a PROSITE scanning tool. Applied Bioinformatics 1: 107–108.
43. Ortiz-Zapater E, Soriano-Ortega E, Marcote MJ, Ortiz-Masiá D, Aniento F (2006) Trafficking of the human transferrin receptor in plant cells: effects of tyrphostin A23 and brefeldin A. Plant Journal 48: 757–770.
44. Lamesch P, Berardini TZ, Li D, Swarbreck D, Wilks C, et al. (2012) The Arabidopsis Information Resource (TAIR): improved gene annotation and new tools. Nucleic Acids Research 40: D1202–D1210.
45. Schwacke R, Schneider A, van der Graaff E, Fischer K, Catoni E, et al. (2003) ARAMEMNON, a novel database for Arabidopsis integral membrane proteins. Plant Physiology 131: 16–26.
46. Pierleoni A, Martelli PL, Fariselli P, Casadio R (2006) BaCelLo: a balanced subcellular localization predictor. Bioinformatics 22: e408–e416.

47. Emanuelsson O, Nielsen H, Heijne GV (1999) ChloroP, a neural network-based method for predicting chloroplast transit peptides and their cleavage sites. Protein Science 8: 978–984.

48. Bannai H, Tamada Y, Maruyama O, Nakai K, Miyano S (2002) Extensive feature detection of N-terminal protein sorting signals. Bioinformatics 18: 298–305.

49. Guda C, Fahy E, Subramaniam S (2004) MITOPRED: a genome-scale method for prediction of nucleus-encoded mitochondrial proteins. Bioinformatics 20: 1785–1794.

50. Claros MG, Vincens P (1996) Computational method to predict mitochondrially imported proteins and their targeting sequences. European Journal of Biochemistry 241: 779–786.

51. Höglund A, Dönnes P, Blum T, Adolph H-W, Kohlbacher O (2006) MultiLoc: prediction of protein subcellular localization using N-terminal targeting sequences, sequence motifs and amino acid composition. Bioinformatics 22: 1158–1165.

52. Lu Z, Szafron D, Greiner R, Lu P, Wishart DS, et al. (2004) Predicting subcellular localization of proteins using machine-learned classifiers. Bioinformatics 20: 547–556.

53. Schein AI, Kissinger JC, Ungar LH (2001) Chloroplast transit peptide prediction: a peek inside the black box. Nucleic Acids Research 29: e82–e82.

54. Bodén M, Hawkins J (2005) Prediction of subcellular localization using sequence-biased recurrent networks. Bioinformatics 21: 2279–2286.

55. Hiller K, Grote A, Scheer M, Münch R, Jahn D (2004) PrediSi: prediction of signal peptides and their cleavage positions. Nucleic Acids Research 32: W375–W379.

56. Small I, Peeters N, Legeai F, Lurin C (2004) Predotar: A tool for rapidly screening proteomes for N-terminal targeting sequences. Proteomics 4: 1581–1590.

57. Petsalaki EI, Bagos PG, Litou ZI, Hamodrakas SJ (2006) PredSL: a tool for the N-terminal sequence-based prediction of protein subcellular localization. Genomics, Proteomics and Bioinformatics 4: 48–55.

58. Nielsen H, Brunak S, von Heijne G (1999) Machine learning approaches for the prediction of signal peptides and other protein sorting signals. Protein Engineering 12: 3–9.

59. Nielsen H, Krogh A (1998) Prediction of signal peptides and signal anchors by a hidden Markov model. ISMB-98 Proceedings 6: 122–130.

60. Nielsen H, Engelbrecht J, Brunak S, Heijne GV (1997) A neural network method for identification of prokaryotic and eukaryotic signal peptides and prediction of their cleavage sites. International Journal of Neural Systems 8: 581–599.

61. Matsuda S, Vert JP, Saigo H, Ueda N, Toh H, et al. (2005) A novel representation of protein sequences for prediction of subcellular location using support vector machines. Protein Science 14: 2804–2813.

62. Emanuelsson O, Nielsen H, Brunak S, von Heijne G (2000) Predicting subcellular localization of proteins based on their N-terminal amino acid sequence. Journal of Molecular Biology 300: 1005–1016.

63. Horton P, Park K-J, Obayashi T, Nakai K (2006) Protein Subcellular Localisation Prediction with WoLF PSORT. APBC: 39–48.

64. Schwacke R, Fischer K, Ketelsen B, Krupinska K, Krause K (2007) Comparative survey of plastid and mitochondrial targeting properties of transcription factors in Arabidopsis and rice. Molecular Genetics and Genomics 277: 631–646.

65. Heazlewood JL, Verboom RE, Tonti-Filippini J, Small I, Millar AH (2007) SUBA: the Arabidopsis subcellular database. Nucleic Acids Research 35: D213–D218.

66. Lu Y, Savage LJ, Larson MD, Wilkerson CG, Last RL (2011) Chloroplast 2010: a database for large-scale phenotypic screening of Arabidopsis mutants. Plant Physiology 155: 1589–1600.

67. Holstein SE (2002) Clathrin and plant endocytosis. Traffic 3: 614–620.

68. Chen X, Irani NG, Friml J (2011) Clathrin-mediated endocytosis: the gateway into plant cells. Current Opinion in Plant Biology 14: 674–682.

69. Ito E, Fujimoto M, Ebine K, Uemura T, Ueda T, et al. (2012) Dynamic behavior of clathrin in Arabidopsis thaliana unveiled by live imaging. Plant Journal 69: 204–216.

70. Olinares PDB, Ponnala L, van Wijk KJ (2010) Megadalton complexes in the chloroplast stroma of Arabidopsis thaliana characterized by size exclusion chromatography, mass spectrometry, and hierarchical clustering. Molecular and Cellular Proteomics 9: 1594–1615.

71. Zybailov B, Rutschow H, Friso G, Rudella A, Emanuelsson O, et al. (2008) Sorting signals, N-terminal modifications and abundance of the chloroplast proteome. PLoS One 3: e1994.

72. Kleffmann T, Russenberger D, von Zychlinski A, Christopher W, Sjolander K, et al. (2004) The Arabidopsis thaliana chloroplast proteome reveals pathway abundance and novel protein functions. Current Biology 14: 354–362.

73. Froehlich JE, Wilkerson CG, Ray WK, McAndrew RS, Osteryoung KW, et al. (2003) Proteomic Study of the Arabidopsis t haliana Chloroplastic Envelope Membrane Utilizing Alternatives to Traditional Two-Dimensional Electrophoresis. Journal of Proteome Research 2: 413–425.

74. Maurino VG, Peterhansel C (2010) Photorespiration: current status and approaches for metabolic engineering. Current Opinion in Plant Biology 13: 248–255.

75. Slocum RD (2005) Genes, enzymes and regulation of arginine biosynthesis in plants. Plant Physiology and Biochemistry 43: 729–745.

76. Yanai Y, Kawasaki T, Shimada H, Wurtele ES, Nikolau BJ, et al. (1995) Genomic organization of 251 kDa acetyl-CoA carboxylase genes in Arabidopsis: tandem gene duplication has made two differentially expressed isozymes. Plant and Cell Physiology 36: 779–787.

77. Han X, Yin L, Xue H (2012) Co-expression Analysis Identifies CRC and AP1 the Regulator of Arabidopsis Fatty Acid Biosynthesis. Journal of Integrative Plant Biology 54: 486–499.

78. Sasaki Y, Nagano Y (2004) Plant acetyl-CoA carboxylase: structure, biosynthesis, regulation, and gene manipulation for plant breeding. Bioscience, Biotechnology, and Biochemistry 68: 1175–1184.

79. Wen W, Chen L, Wu H, Sun X, Zhang M, et al. (2006) Identification of the yeast R-SNARE Nyv1p as a novel longin domain-containing protein. Molecular Biology of the Cell 17: 4282–4299.

80. Guo J, Wang J, Xi L, Huang W-D, Liang J, et al. (2009) RACK1 is a negative regulator of ABA responses in Arabidopsis. Journal of Experimental Botany 60: 3819–3833.

81. Kundu N, Dozier U, Deslandes L, Somssich IE, Ullah H (2013) Arabidopsis scaffold protein RACK1A interacts with diverse environmental stress and photosynthesis related proteins. Plant Signaling and Behavior 8: e24012.

82. Monaghan J, Xu F, Gao M, Zhao Q, Palma K, et al. (2009) Two Prp19-like U-box proteins in the MOS4-associated complex play redundant roles in plant innate immunity. PLoS Pathogens 5: e1000526.

83. Wiborg J, O'Shea C, Skriver K (2008) Biochemical function of typical and variant Arabidopsis thaliana U-box E3 ubiquitin-protein ligases. Biochemical Journal 413: 447–457.

84. Fülöp K, Tarayre S, Kelemen Z, Horváth G, Kevei Z, et al. (2005) Arabidopsis anaphase-promoting complexes: multiple activators and wide range of substrates might keep APC perpetually busy. Cell Cycle 4: 4084–4092.

85. Gutierrez C (2009) The Arabidopsis cell division cycle. The Arabidopsis book/ American Society of Plant Biologists 7.

86. Larson-Rabin Z, Li Z, Masson PH, Day CD (2009) FZR2/CCS52A1 expression is a determinant of endoreduplication and cell expansion in Arabidopsis. Plant Physiology 149: 874–884.

87. Kaya H, Shibahara K-i, Taoka K-i, Iwabuchi M, Stillman B, et al. (2001) FASCIATA genes for chromatin assembly factor-1 in Arabidopsis maintain the cellular organization of apical meristems. Cell 104: 131–142.

88. Zhu Y, Dong A, Shen W-H (2012) Histone variants and chromatin assembly in plant abiotic stress responses. Biochimica et Biophysica Acta (BBA)-Gene Regulatory Mechanisms 1819: 343–348.

89. Köhler C, Hennig L, Bouveret R, Gheyselinck J, Grossniklaus U, et al. (2003) Arabidopsis MSI1 is a component of the MEA/FIE Polycomb group complex and required for seed development. EMBO Journal 22: 4804–4814.

90. Wienkoop S, Zoeller D, Ebert B, Simon-Rosin U, Fisahn J, et al. (2004) Cell-specific protein profiling in Arabidopsis thaliana trichomes: identification of trichome-located proteins involved in sulfur metabolism and detoxification. Phytochemistry 65: 1641–1649.

91. Jogl G, Shen Y, Gebauer D, Li J, Wiegmann K, et al. (2002) Crystal structure of the BEACH domain reveals an unusual fold and extensive association with a novel PH domain. EMBO Journal 21: 4785–4795.

92. Eckhardt U, Grimm B, Hörtensteiner S (2004) Recent advances in chlorophyll biosynthesis and breakdown in higher plants. Plant Molecular Biology 56: 1–14.

93. Papenbrock J, Gräfe S, Kruse E, Hänel F, Grimm B (1997) Mg-chelatase of tobacco: identification of a Chl D cDNA sequence encoding a third subunit, analysis of the interaction of the three subunits with the yeast two-hybrid system, and reconstitution of the enzyme activity by co-expression of recombinant CHL D, CHL H and CHL I. Plant Journal 12: 981–990.

94. Masuda T (2008) Recent overview of the Mg branch of the tetrapyrrole biosynthesis leading to chlorophylls. Photosynthesis Research 96: 121–143.

95. Solymosi K, Aronsson H (2013) Etioplasts and Their Significance in Chloroplast Biogenesis. Plastid Development in Leaves during Growth and Senescence: Springer. pp. 39–71.

96. Słabicki M, Theis M, Krastev DB, Samsonov S, Mundwiller E, et al. (2010) A genome-scale DNA repair RNAi screen identifies SPG48 as a novel gene associated with hereditary spastic paraplegia. PLoS Biology 8: e1000408.

97. Gao C, Christine K, Qu S, San MWY, Li KY, et al. (2012) The Golgi-localized Arabidopsis endomembrane protein12 contains both endoplasmic reticulum export and Golgi retention signals at its C terminus. Plant Cell 24: 2086–2104.

98. Kleine-Vehn J, Dhonukshe P, Swarup R, Bennett M, Friml J (2006) Subcellular trafficking of the Arabidopsis auxin influx carrier AUX1 uses a novel pathway distinct from PIN1. Plant Cell 18: 3171–3181.

99. Movafeghi A, Happel N, Pimpl P, Tai G-H, Robinson DG (1999) Arabidopsis Sec21p and Sec23p homologs. Probable coat proteins of plant COP-coated vesicles. Plant Physiology 119: 1437–1446.

100. Vanhee C, Zapotoczny G, Masquelier D, Ghislain M, Batoko H (2011) The Arabidopsis multistress regulator TSPO is a heme binding membrane protein and a potential scavenger of porphyrins via an autophagy-dependent degradation mechanism. Plant Cell 23: 785–805.

101. Kobayashi N, Yang J, Ueda A, Suzuki T, Tomaru K, et al. (2007) RanBPM, Muskelin, p48EMLP, p44CTLH, and the armadillo-repeat proteins ARMC8α and ARMC8β are components of the CTLH complex. Gene 396: 236–247.

102. Tomaru K, Ueda A, Suzuki T, Kobayashi N, Yang J, et al. (2010) Armadillo Repeat Containing 8α Binds to HRS and Promotes HRS Interaction with Ubiquitinated Proteins. The Open Biochemistry Journal 4: 1.

103. Kim SY, Kim BH, Nam KH (2010) Reduced expression of the genes encoding chloroplast–localized proteins in a cold-resistant bri1 (brassinosteroid-Insensitive 1) mutant. Plant Signaling and Behavior 5: 458–463.

104. Robinson DG, Herranz M-C, Bubeck J, Pepperkok R, Ritzenthaler C (2007) Membrane dynamics in the early secretory pathway. Critical Reviews in Plant Sciences 26: 199–225.

105. Cevher-Keskin B (2013) ARF1 and SAR1 GTPases in Endomembrane Trafficking in Plants. International Journal of Molecular Sciences 14: 18181–18199.

106. Joyard J, Ferro M, Masselon C, Seigneurin-Berny D, Salvi D, et al. (2010) Chloroplast proteomics highlights the subcellular compartmentation of lipid metabolism. Progress in Lipid Research 49: 128–158.

107. Tisdale EJ, Bourne JR, Khosravi-Far R, Der CJ, Balch W (1992) GTP-binding mutants of rab1 and rab2 are potent inhibitors of vesicular transport from the endoplasmic reticulum to the Golgi complex. Journal of Cell Biology 119: 749–761.

108. Rutherford S, Moore I (2002) The Arabidopsis Rab GTPase family: another enigma variation. Current Opinion in Plant Biology 5: 518–528.

109. Moore I, Diefenthal T, Zarsky V, Schell J, Palme K (1997) A homolog of the mammalian GTPase Rab2 is present in Arabidopsis and is expressed predominantly in pollen grains and seedlings. Proceedings of the National Academy of Sciences 94: 762–767.

110. Wei Z, Hu W, Lin Q, Cheng X, Tong M, et al. (2009) Understanding rice plant resistance to the brown planthopper (Nilaparvata lugens): a proteomic approach. Proteomics 9: 2798–2808.

111. Park C-J, Sharma R, Lefebvre B, Canlas PE, Ronald PC (2013) The endoplasmic reticulum-quality control component SDF2 is essential for XA21 mediated immunity in rice. Plant Science 210: 53–60.

112. Teixeira FK, Menezes-Benavente L, Margis R, Margis-Pinheiro M (2004) Analysis of the molecular evolutionary history of the ascorbate peroxidase gene family: inferences from the rice genome. Journal of Molecular Evolution 59: 761–770.

113. Teixeira FK, Menezes-Benavente L, Galvão VC, Margis R, Margis-Pinheiro M (2006) Rice ascorbate peroxidase gene family encodes functionally diverse isoforms localized in different subcellular compartments. Planta 224: 300–314.

114. Mahfouz MM, Kim S, Delauney AJ, Verma DPS (2006) Arabidopsis TARGET OF RAPAMYCIN interacts with RAPTOR, which regulates the activity of S6 kinase in response to osmotic stress signals. Plant Cell 18: 477–490.

115. Schmitz-Linneweber C, Small I (2008) Pentatricopeptide repeat proteins: a socket set for organelle gene expression. Trends in Plant Science 13: 663–670.

116. Small ID, Peeters N (2000) The PPR motif–a TPR-related motif prevalent in plant organellar proteins. Trends in Biochemical Sciences 25: 45–47.

117. Gothandam KM, Kim E-S, Cho H, Chung Y-Y (2005) OsPPR1, a pentatricopeptide repeat protein of rice is essential for the chloroplast biogenesis. Plant Molecular Biology 58: 421–433.

118. Zhang H, Li J, Yoo J-H, Yoo S-C, Cho S-H, et al. (2006) Rice Chlorina-1 and Chlorina-9 encode ChlD and ChlI subunits of Mg-chelatase, a key enzyme for chlorophyll synthesis and chloroplast development. Plant Molecular Biology 62: 325–337.

119. Gough J, Karplus K, Hughey R, Chothia C (2001) Assignment of homology to genome sequences using a library of hidden Markov models that represent all proteins of known structure. Journal of Molecular Biology 313: 903–919.

120. Tewari R, Bailes E, Bunting KA, Coates JC (2010) Armadillo-repeat protein functions: questions for little creatures. Trends in Cell Biology 20: 470–481.

121. Niihama M, Takemoto N, Hashiguchi Y, Tasaka M, Morita MT (2009) ZIP genes encode proteins involved in membrane trafficking of the TGN–PVC/vacuoles. Plant and Cell Physiology 50: 2057–2068.

122. Smith TF, Gaitatzes C, Saxena K, Neer EJ (1999) The WD repeat: a common architecture for diverse functions. Trends in Biochemical Sciences 24: 181–185.

123. Li D, Roberts R (2001) Human Genome and Diseases: WD-repeat proteins: structure characteristics, biological function, and their involvement in human diseases. Cellular and Molecular Life Sciences 58: 2085–2097.

124. Vernoud V, Horton AC, Yang Z, Nielsen E (2003) Analysis of the small GTPase gene superfamily of Arabidopsis. Plant Physiology 131: 1191–1208.

125. Westphal S, Soll J, Vothknecht UC (2003) Evolution of chloroplast vesicle transport. Plant and Cell Physiology 44: 217–222.

126. Vothknecht UC, Westhoff P (2001) Biogenesis and origin of thylakoid membranes. Biochimica et Biophysica Acta (BBA)-Molecular Cell Research 1541: 91–101.

127. Kroll D, Meierhoff K, Bechtold N, Kinoshita M, Westphal S, et al. (2001) VIPP1, a nuclear gene of Arabidopsis thaliana essential for thylakoid membrane formation. Proceedings of the National Academy of Sciences 98: 4238–4242.

128. Cavalier-Smith T (2009) Predation and eukaryote cell origins: a coevolutionary perspective. International Journal of Biochemistry and Cell Biology 41: 307–322.

129. Cavalier-Smith T (2000) Membrane heredity and early chloroplast evolution. Trends in Plant Science 5: 174–182.

130. Nickel W, Brugger B, Wieland FT (2002) Vesicular transport: the core machinery of COPI recruitment and budding. Journal of Cell Science 115: 3235–3240.

131. Bonifacino JS, Lippincott-Schwartz J (2003) Coat proteins: shaping membrane transport. Nature Review Molecular Cell Biology 4: 409–414.

11

Genomes and Transcriptomes of Partners in Plant-Fungal- Interactions between Canola (*Brassica napus*) and Two *Leptosphaeria* Species

Rohan G. T. Lowe[1], Andrew Cassin[2], Jonathan Grandaubert[3], Bethany L. Clark[1], Angela P. Van de Wouw[1], Thierry Rouxel[3], Barbara J. Howlett[1]*

1 School of Botany, The University of Melbourne, Parkville, Victoria, Australia, **2** ARC Centre of Excellence in Plant Cell Walls, School of Botany, The University of Melbourne, Parkville, Victoria, Australia, **3** INRA-Bioger, UR1290, Thiverval-Grignon, France

Abstract

Leptosphaeria maculans 'brassicae' is a damaging fungal pathogen of canola (*Brassica napus*), causing lesions on cotyledons and leaves, and cankers on the lower stem. A related species, *L. biglobosa* 'canadensis', colonises cotyledons but causes few stem cankers. We describe the complement of genes encoding carbohydrate-active enzymes (CAZys) and peptidases of these fungi, as well as of four related plant pathogens. We also report dual-organism RNA-seq transcriptomes of these two *Leptosphaeria* species and *B. napus* during disease. During the first seven days of infection *L. biglobosa* 'canadensis', a necrotroph, expressed more cell wall degrading genes than *L. maculans* 'brassicae', a hemi-biotroph. *L. maculans* 'brassicae' expressed many genes in the Carbohydrate Binding Module class of CAZy, particularly CBM50 genes, with potential roles in the evasion of basal innate immunity in the host plant. At this time, three avirulence genes were amongst the top 20 most highly upregulated *L. maculans* 'brassicae' genes *in planta*. The two fungi had a similar number of peptidase genes, and trypsin was transcribed at high levels by both fungi early in infection. *L. biglobosa* 'canadensis' infection activated the jasmonic acid and salicylic acid defence pathways in *B. napus*, consistent with defence against necrotrophs. *L. maculans* 'brassicae' triggered a high level of expression of isochorismate synthase 1, a reporter for salicylic acid signalling. *L. biglobosa* 'canadensis' infection triggered coordinated shutdown of photosynthesis genes, and a concomitant increase in transcription of cell wall remodelling genes of the host plant. Expression of particular classes of CAZy genes and the triggering of host defence and particular metabolic pathways are consistent with the necrotrophic lifestyle of *L. biglobosa* 'canadensis', and the hemibiotrophic life style of *L. maculans* 'brassicae'.

Editor: Richard A Wilson, University of Nebraska-Lincoln, United States of America

Funding: The following sources of funding supported this work: Grains Research and Development Corporation for funding RL AW BC BH, the Australian Research Council for funding AC, the Victoria Life Sciences Computation Initiative (VLSCI) for computational resources via grant RAS990 for RL, the University of Melbourne for an Early Career Researcher award to RL, and the French agency Agence Nationale de la Recherche, contract ANR-09-GENM-028 ('FungIsochores') for funding JG and TR. The funders had no role in study design, data collection and analysis, decision to publish, or preparation of the manuscript.

Competing Interests: The authors have declared that no competing interests exist.

* Email: bhowlett@unimelb.edu.au

Introduction

As more fungal genome sequences become available, it is apparent that their complement of genes and transcriptomes reflects fungal lifestyles. Lifestyles of plant pathogenic fungi are often classified into three broad categories: biotrophy, where the pathogen feeds from live host cells, necrotrophy, where the host cells are killed ahead of colonisation, and hemi-biotrophy, where the pathogen feeds from living cells before switching to a necrotrophic style of growth. These designations are imprecise and as the mechanisms of pathogenicity in a range of fungi are elucidated, lifestyle boundaries become more blurred [1].

The fungal genus *Leptosphaeria* belongs to the class Dothideomycetes, which includes a number of economically important plant pathogens that have a range of lifestyles on their hosts [2]. The *Leptosphaeria* species complex has two species *L. maculans* and *L. biglobosa* [3] and several sub-species or clades including *L. maculans* 'brassicae' and 'lepidii', and *L. biglobosa* 'canadensis', 'brassicae', 'australensis' and 'occiaustralensis' [4]. The nomenclature for these fungi is currently under review [5]. *L. maculans* 'brassicae', a hemibiotroph, causes blackleg, the most important disease of *Brassica napus* (canola) worldwide. Airborne sexual spores (ascospores) released from infested crop residues from the previous year's crop land on seedlings. Hyphae of germinated spores enter plant tissue via stomatal apertures and asymptomatically colonise the apoplastic spaces, between the plant cells. After eight to ten days, plant cells collapse and asexual sporulation begins within the necrotic leaf lesion. Hyphae then grow along the petiole and the stem, often resulting in a canker that girdles the stem causing lodging of the plant [6]. In contrast, *L. biglobosa* 'canadensis' is not well-characterised. Although it causes cotyledonary lesions, stem cankers are rarely produced [7].

Leptosphaeria maculans 'brassicae' has numerous 'gene for gene' interactions with *B. napus* whereby an avirulence allele in the fungus renders it unable to attack cultivars with the corresponding resistance gene. This 'gene for gene' resistance is expressed in seedlings; five avirulence genes have been cloned so far - *AvrLm1*, *AvrLm4-7*, *AvrLm6 AvrLm1* and *AvrLmJ1* [8–12]. Only one resistance gene, *LepR3*, has been cloned from *B. napus* [13].

During infection fungi derive nutrition from the host plant, often by enzymatic degradation of proteins and carbohydrates. These latter enzymes are classed as Carbohydrate-Active enZymes (CAZys) and they often have well characterised domains. As well as providing nutrition, CAZy activity releases cell wall products that can act as DAMPs (Damage Associated Molecular Patterns) that activate the host immune system [14]. The biotrophic plant pathogens *Blumeria graminis*, *Puccinia graminis*, *Melampsora laricus-populina*, *Hyaloperonospora arabidopsis* and *Ustilago maydis* have fewer CAZy genes than necrotrophic plant pathogens do, perhaps because biotrophs do not need to digest plant cell walls for nutrition and they must evade the host immune system [15–17].

Genomic sequences are now available for many Dothideomycetes and an extensive comparative analysis of 18 of them, including *L. maculans* 'brassicae' has been published [18]. These include the *Brassica*-infecting pathogen, *Alternaria brassicicola*, as well as three wheat-infecting pathogens, *Stagonospora nodorum*, *Pyrenophora tritici-repentis*, and *Zymoseptoria tritici*. The former three fungi, like many members of the order Pleosporales, have a necrotrophic lifestyle releasing toxins soon after invasion, whilst *Zymoseptoria tritici*, like most members of the order Capnodiales, is a hemibiotroph, with an extended period as a biotroph before causing necrosis [19–21].

The *L. maculans* 'brassicae' genome is compartmentalised into AT-rich -that are gene-poor comprising up to 35% of the genome, and gene- rich regions that have a high GC content [2], while the genome of *L. biglobosa* 'canadensis' is 30 Mb and lacks AT-rich regions (Grandaubert et al. manuscript submitted). A limited amount of oligo-array transcriptome data have been produced for *L. maculans* 'brassicae' during *in vitro* culture and infection of *B. napus* [11], but transcriptome data have not been reported for *L. biglobosa* or any closely related Brassica pathogen.

Patterns of global gene expression can be generated by RNA-seq, a technique that enables analysis of dual transcriptomes; for instance, during a plant pathogen interaction. Furthermore RNA-seq can be exploited to analyse non-model organisms for which genomic resources are not well developed [22]. Few dual transcriptomes for plants and pathogenic fungi have been reported. Two recent reports of dual RNA-seq analysis of fungal diseases are of rice blast [23] and target leaf spot of sorghum [24]. Here we describe the genomes and transcriptomes of *L. biglobosa* 'canadensis' and *L. maculans* 'brassicae' and canola (*Brassica napus*), during infection and *in vitro*. Our aim is to characterise genes that each pathogen uses to evade detection by the host, or to derive nutrition from the host, viz. the carbohydrate-active enzymes and peptidases. We also examine genes upregulated by the host during infection by each pathogen.

Methods

Fungal isolates and culture conditions

L. maculans 'brassicae' isolate IBCN18 and *L. biglobosa* 'canadensis' isolate 06J154, hereafter referred to as Lmb and Lbc, respectively, were subcultured on 10% Campbell's V8 juice agar at 22°C with a 12 h light/12 h dark light cycle. Conidia

(5×10^6) were added to 30 mL of liquid medium and incubated in still culture in a petri dish (15 cm diameter) at 22°C in the dark. Culture media were either 10% Campbell's V8 juice, or oilseed rape medium. The latter medium was prepared by homogenising leaves (200 g) of *B. napus* cv. Westar in a waring blender in a final volume of 1 L of water. The homogenate was centrifuged at 2000 g for 20 min and the resulting supernatant was filter sterilised (0.22 μm Millipore stericup filter).

Plant growth and infection conditions

Brassica napus cv. Westar was used for all infection assays; it has no known resistance genes. Seedlings were grown in a glasshouse maintained at 25°C under natural lighting. Wounded cotyledons were infected with conidia (10 uL of a 1×10^5 spores/mL suspension) or water (mock inoculum), at 10 days post sowing as described previously [25].

Extraction of RNA and gene expression analysis

For RNA-seq analysis, *B. napus* cotyledons were infected with Lmb, Lbc or water (control mock inocula), and at 7 and 14 days post inoculation (dpi) tissue around the inoculation site was harvested using a cork borer (0.5 cm diameter) and then placed into liquid nitrogen before freeze drying and subsequent grinding under liquid nitrogen. Tissue samples were prepared in biological triplicate. RNA was extracted using Trizol reagent from infected tissue and from mycelia of Lmb and Lbc from 7-day still cultures grown in oilseed rape medium. RNA was then DNAase-treated (Life Technologies) and cleaned up.

The two biological replicates of each sample with the highest RNA integrity number values (>6) were sequenced with Illumina TruSeq version 3 chemistry on an Illumina HiSeq2000 sequencer at the Australian Genome Research Facility. *In vitro* derived RNA was sequenced with 100 bp paired-end reads in order to aid gene annotation, and *in planta* derived RNA was sequenced with 100 bp single-end reads. A total of 15.5 Gbp sequence was generated from the *in vitro* libraries of the two fungi (7.75 Gbp per sample), and 72 Gbp sequence was generated from 12 *in planta* libraries (6 Gbp per sample) (Table S1 in file S2). Reads were trimmed to a minimum phred quality score of 20 using Nesoni sequence software [26], orphaned members of pairs were retained, adaptor sequences were removed, and reads shorter than 20 bp were rejected. Trimmed reads were aligned to a reference genome sequence with Tophat v1.4.1 splice-junction mapper [26]. Reference genomes were Lmb isolate v23.1.3 [2], Lbc isolate J154 (Grandaubert et al., submitted), and a *Brassica* exon array curated unigene set representing 135,201 gene models from *B. napus*, *B. rapa* and *B. oleracea* [27]. Aligned reads were quantified using Cufflinks v1.0.3 transcript assembly and quantification software and denoted as average expression levels (FPKM - fragments per kilobase of exon per million mapped reads) [28]. Cufflinks was run with bias correction to reduce variance due to sequence positional bias during Illumina sequencing. Gene expression FPKM values are listed in the relevant supplementary tables. All of the aligned RNA-seq reads have been deposited at the NCBI Sequence Read Archive (SRA), accessible under bioproject accession PRJNA230885 or sequence read archive SRP035525. SRA files may be read using the NCBI SRA toolkit (http://www.ncbi.nlm.nih.gov/sra).

Quantitative RT-PCR experiments were carried out to determine levels of expression of genes containing CBM50 (LysM) domains. Cotyledon tissue from 32 *B. napus* cv. Westar seedlings was harvested at 3, 7 and 14 days after inoculation with Lmb isolate IBCN18. RNA (2 μg) was treated with DNase 1 (Life Technologies) for 1 h at 25°C, and reverse-transcribed using oligo-

Table 1. *B. napus* defence genes analysed [50].

Gene	Full Name	Defence Signalling Pathway	*B. napus* RRES unigene ID	GenBank accession of CDS
NCED3	9-cis-epoxycarotenoid dioxygenase 3	Abscisic acid	rres040714.v1	EV137674
ACS2a	1-amino-cyclopropane-1-carboxylate synthase 2	Ethylene	rres079827.v1	HM450312
CHI	Chitinase	Ethylene/salicylic acid//jasmonic acid	rres036231.v1	X61488
HEL	Hevein-like protein	Ethylene/jasmonic acid	rres036743.v1	FG577475
ICS1	Isochorismate synthase 1	Salicylic acid	rres059693.v1	EV225528
PR-1	Pathogenesis related protein 1	Salicylic acid	rres112514.v1	BNU21849
WRKY70	WRKY transcription factor 70	Salicylic acid	rres038189.v1	EV113862
PDF1.2	Plant Defensin 1.2	Jasmonic acid	rres071321.v1	KC967203

Defence signalling pathways indicate regulatory pathway to which the gene belongs. RRES unigenes are listed from the Brassica exon array [27], to which RNA-seq reads were aligned. Corresponding GenBank accession numbers are given for each gene. The response of each gene (with the exception of PDF1.2) to its corresponding signalling pathway was confirmed by Sasek et al [50].

dT primer and Superscript III (Life Technologies) at 50°C for 1 h. Levels of gene expression were determined by qPCR using SensiMix (dT) SYBR Green PCR kit (Bioline) in a Corbett Rotor-Gene 3000 machine. Transcript levels of the gene of interest were normalized to that of Lmb actin as described previously (Gardiner et al., 2004). Primers are listed in Table S2 in file S2.

Annotation of genes and domains

Genes encoding CAZys were identified in six dothideomycetes (Lmb, Lbc, *A. brassicicola*, *S. nodorum*, *P. tritici-repentis* and *Z. tritici*) using www.cazy.org [29] and the dbCAN v3.0 HMM-based CAZy annotation server (http://csbl.bmb.uga.edu/dbCAN/) [30]. The major CAZy classes are Polysaccharide Lyases (PL), Glycosyl Transferases (GT), Glycosyl Hydrolases (GH), Carbohydrate Esterases (CE), Carbohydrate Binding Modules (CBM) and Auxiliary Activities (AA). Secretion signal peptides were predicted using the SignalP 4.0 algorithm [31]. Pfam domains (Pfam A and B) were identified using profile hidden Markov models with HMMER3.0 [32] and Pfam_scan.pl software; the e value cut off was set to 1e-5 [33]. LysM-containing genes (with CBM50 domain) were examined in more detail. Predicted gene models and aligned RNA-seq reads were viewed with the IGV browser [34], predicted intron splice sites were verified and translation start sites were checked for congruence with observed RNA-seq transcript boundaries. They were initially identified by comparison to the Pfam database [33], and then aligned to ECP6, the well

characterised LysM-containing protein from *Cladosporium fulvum* [35]. Other *Leptosphaeria* proteins with LysM motifs were identified and characterised as described in supporting information (Figure S1 in file S1).

Peptidase domains were identified by BlastP [36] comparison of predicted protein sequences to the MEROPS 'peptidase database pepunit.lib dataset' of database of peptidase and inhibitor units (http://merops.sanger.ac.uk) [37].

Analysis of expression of fungal and *B. napus* genes

The 100 most highly up regulated genes of Lmb and Lbc at 7 and 14 dpi, compared to *in vitro* growth, were identified from the RNA-seq data (Table S 3–6 in file S2). Averages of quantile-normalised log10-transformed FPKM values were calculated for each dataset for direct comparison of CAZy gene expression between Lmb and Lbc. Genes containing a CBM, GH, PL or AA domain and predicted to be secreted were analysed further. The sum of the expression values of each CAZy gene across the three treatments (*in vitro*, 7 and 14 dpi) was determined and the top 100 genes were identified. Quantile-normalisation was applied so that expression values of genes of both fungi could be directly compared and log10-transformed FPKM values were graphed. For each treatment, a heat map based on expression values was generated.

B. napus defence genes, 9-cis-epoxycarotenoid dioxygenase 3 (NCED), 1-amino-cyclopropane-1-carboxylate synthase 2(ACS2),

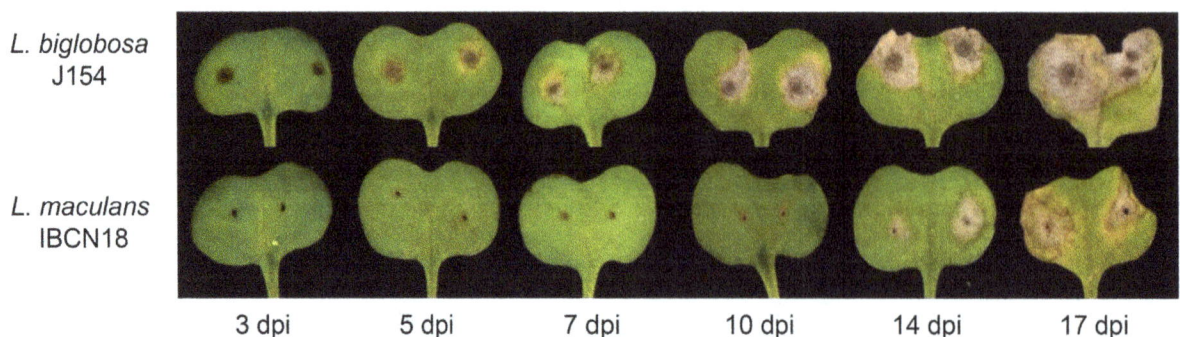

Figure 1. Symptoms on cotyledons of *B. napus* cv. Westar infected with *L. maculans* 'brassicae' or *L. biglobosa* 'canadensis'. *B. napus* cv. Westar cotyledons were wounded and inoculated with Lmb or Lbc spores and disease allowed to progress for 17 days post inoculation (dpi). Cotyledons were harvested and photographed at 3, 5, 7, 10, 14, and 17 dpi to track lesion development by the two pathogens.

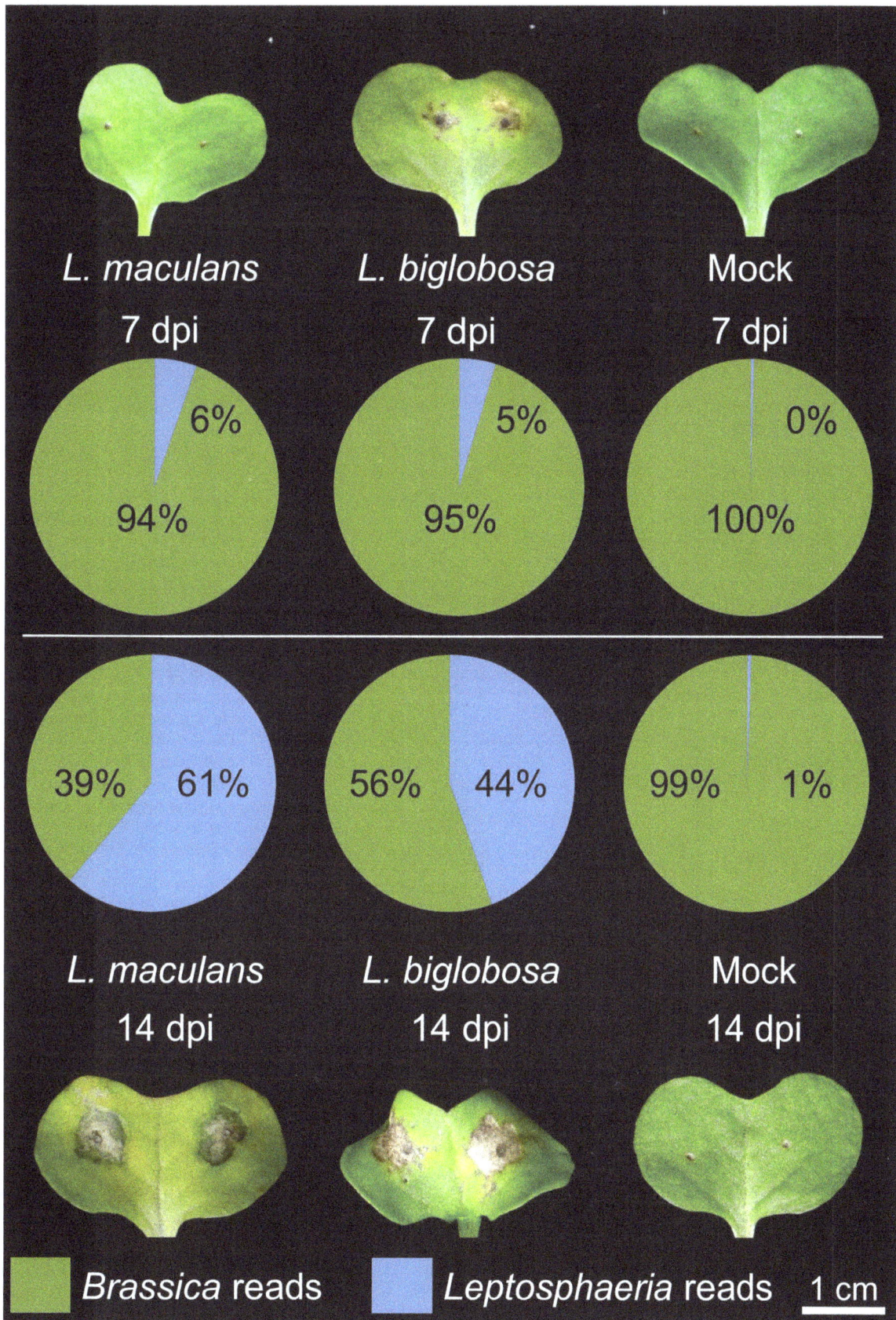

Figure 2. Percentages of plant and fungal transcripts in *B. napus* **cotyledons, uninoculated or at 7 and 14 days post-inoculation dpi with** *L. maculans* **'brassicae' (Lmb) or** *L. biglobosa* **'canadensis' (Lbc).** *B. napus* cotyledons were infected with Lmb, Lbc or water (control mock inoculum). Total RNA was extracted from lesion tissue at 7 and 14 dpi for Illumina RNA-seq sequencing. The percentage of reads aligned to the

Brassica exon array unigene set (green) or the reference genomes for Lmb or Lbc (blue) is presented, as well as a photo of each representative infection. All of the aligned sequence reads were deposited at the NCBI Sequence Read Archive (SRA), accessible under bioproject accession SRP035525.

chitinase (CHI), hevein-like protein (HEL), isochorismate synthase 1 (ICS1), Pathogenesis related protein 1(PR1), WRKY transcription factor 70, and plant defensin 1 (PDF1-2) [38] were identified (Table 1). Their RNA-seq expression was analysed at 7 and 14 dpi in inoculated and uninoculated cotyledons. Additionally, expression of *Brassica* genes involved in metabolic pathways was compared at 7 dpi by either Lmb or Lbc, and then analysed using MapMan, software that processes large gene expression datasets into metabolic pathways or other processes [39]. A Wilcoxon rank sum test with Bonferroni correction for multiple tests was used within MapMan to identify 20 functional categories of *B. napus* genes that were significantly regulated in response to infection by Lmb or Lbc.

Results and Discussion

Symptoms and lifestyles of *L. biglobosa* 'canadensis' and *L. maculans* 'brassicae' on cotyledons of *B. napus*

Leptosphaeria maculans 'brassicae' (Lmb) and *L. biglobosa* 'canadensis' (Lbc) exhibited different timing of symptom development on *B. napus* cotyledons (Figure 1). Lmb had a visually asymptomatic phase until after 7 days post inoculation (dpi), when lesions became visible. Lbc produced initial darkening of the cotyledon tissue at 3 dpi, and cell death and necrosis were apparent by 5 dpi and lesions increased in size until 17 dpi. At this time lesions caused by Lmb were of a similar size. Thus Lmb at 7 dpi appeared to be growing biotrophically, but at 14 dpi was growing necrotrophically, whilst Lbc was necrotrophic from 5 days onwards, although as previously described, its growth is usually arrested before it can colonise the stem [7]. The rapid *in planta* necrosis caused by Lbc was similar to that previously described for *L. biglobosa* 'brassicae' [40].

General features of fungal and plant transcriptomes

RNA-seq was used to define the transcriptomes of both pathogens and host during infection. A total of 87.5 Gbp of raw sequence was generated across all libraries (Table S1 in file S2). In spite of the differences in disease symptoms at 7 dpi, the percentage of Lmb and Lbc reads aligned to the Lmb and Lbc reference genomes was similar (5 and 6% of the total), while 61 and 44% of reads aligned to the reference genomes at 14 dpi (Figure 2). These data reflect that by 14 dpi the plant tissue is heavily colonised by both fungi. As expected, very few reads (less than 1%) from the mock-infected plant libraries aligned to the *Leptosphaeria* reference genomes. These libraries represent highly complex large datasets and Principal Component Analysis was carried out to compare the overall features of the transcriptomes, particularly gene identity and expression level. FPKM values were calculated for each gene in the three organisms. The duplicate sets of RNA-seq data were very similar, with the exception of the data for Lbc at 7 dpi. Gene expression of Lmb in the three conditions (7 and 14 dpi, and *in vitro* growth) was clearly distinguishable one from another (Figure 3), and expression profiles of both fungi *in vitro* were more distinctive than *in planta*. Most of the variance (around 80%) was captured in PC1 for Lmb (Figure 3A) and Lbc infections (Figure 3B), where the 7 and 14 dpi time points were distinguished from each other. The difference between *in vitro* and *in planta* samples was captured in PC2 in for both Lmb and Lbc plots. *B. napus* gene expression was markedly different at 7 dpi

after infection by Lmb compared to Lbc, but similar at 14 dpi. The response of *B. napus* to Lmb at 7 dpi was similar to that of mock inoculation, implying that at this time the plant had not mounted a strong response to infection (Figure 3C).

Since microarray expression (Nimblegen) data for another isolate of Lmb (v23.1.3) were available [2], the relative levels of expression of three avirulence genes (*AvrLm1*, *AvrLm4-7* and *AvrLm11*), present in both Lmb isolates were compared to check for broad agreement between the two technologies (microarrays and RNA-seq). These avirulence genes were ranked in the top 100 most highly expressed genes *in planta* for both microarray and RNA-seq, and both techniques reported a top 10 ranking for *AvrLm1* and *AvrLm4-7* (see below).

Only eight of the 20 most highly upregulated *in planta* genes of Lmb had Pfam domains (Table 2). These included one gene with five CBM50 (LysM) domains (see later); other genes included three cytochrome P450 monooxygenases, and a transferase present in a gene cluster containing a polyketide synthase. Several hypotheticals were amongst the top 20 genes, as well as three avirulence genes, *AvrLm1*, *AvrLm4-7* and *AvrLmJ1* [12]. In contrast, 19 of the 20 most highly upregulated *in planta* genes of Lbc had Pfam domains (Table 3), although five of these were conserved domains without functional annotation (Pfam-Bs, or Domain of Unknown Function (DUF)). Two cellulases, four other glycosyl hydrolases, as well as three peptidases were present.

At 14 dpi, 14 of the top 20 *in planta*-expressed genes of Lmb were hypothetical genes lacking any functional annotation (Table 4). Few were also in the top 20 at 7 dpi, with the exception of *AvrLm1* and LemaP114790.1, which has no Pfam domains and was the most highly expressed gene at both time points. Even though there was a high degree of necrosis on cotyledons at 14 dpi, only two hydrolytic enzymes (trypsin and a glycohydrolase 7) were included in the top 20 most highly expressed Lmb genes. Two of the top 20 genes of Lbc were CAZys; two peptidases and three dehydrogenases, genes that may have degrading roles were also present (Table 5).

Small secreted proteins (SSPs) were well represented in the most highly in planta upregulated genes of Lmb. These included hypothetical, avirulence and CAZy genes. Ohm et al (2012) compared the small secreted proteins of 18 Dothideomycete fungi including Lmb isolate 23.1.3, and found that 21.3% of all Lmb SSPs identified were singletons with no homologue in any of the other genomes [18]. A recent study found 30 of the 100 most highly expressed SSPs *in planta* at 7 dpi were unique to Lmb [52].

In summary, amongst the top 20 most highly expressed genes *in planta*, fewer encoding cell wall degrading enzymes were expressed by Lmb than by Lbc at 7 and 14 dpi. Cell wall degrading enzymes not only provide nutrition for the pathogen by hydrolysing carbohydrates, but facilitate fungal progression through the plant apoplast (intercellular spaces), during biotrophic stages of infection. The repertoire and expression of fungal CAZy genes presumably reflects the cognate carbohydrate present in the host plant and it is notable that enzymes such as those degrading cellulose are very highly expressed. The cell wall of leaves of *Arabidopsis thaliana*, which like *B. napus* is a crucifer, contain polysaccharides including rhamnogalacturonan I and II (as pectin), xyloglucan, glucuronoarabinoxylan, and cellulose (14%). Presumably the cotyledon has a similar polysaccharide profile. An additional 14% of the wall is composed of protein [41]. More than

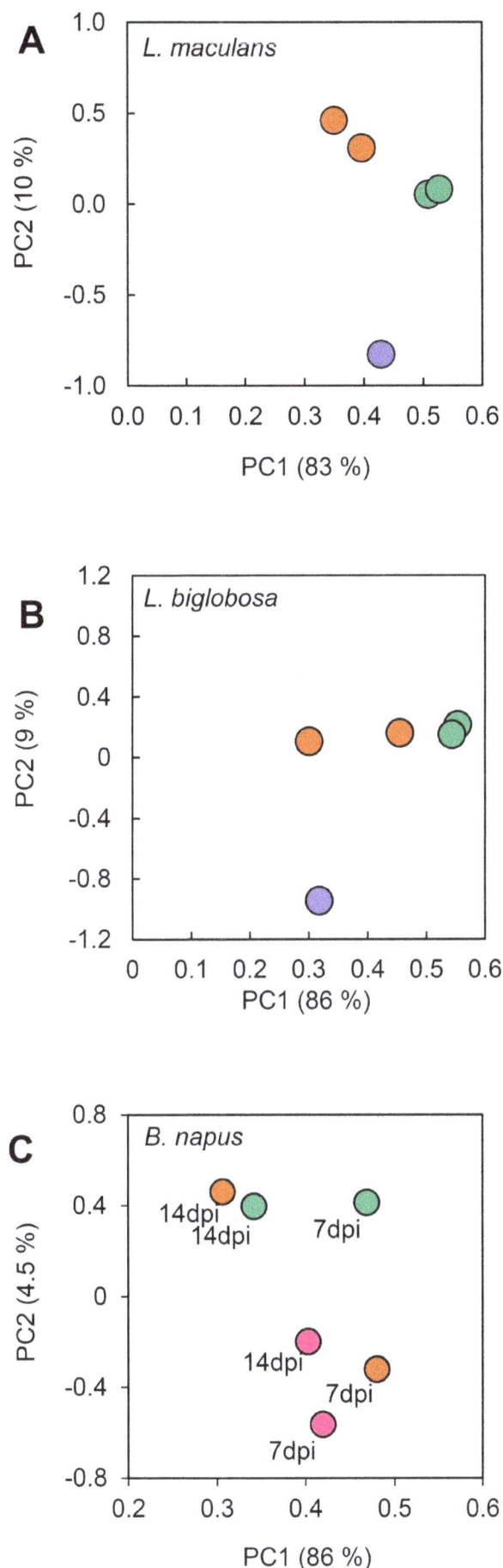

A *L. maculans*

B *L. biglobosa*

C *B. napus*

Figure 3. Principal Component Analyses of duplicate sets of RNA-seq data for *L. maculans* 'brassicae' (Lmb), *L. biglobosa* 'canadensis' (Lbc) and *B. napus*. Scores plots for principal component analysis of expression values (FPKM) of genes of Lmb IBCN18 (A), Lbc J154 (B), or *B. napus* (C). Gene expression values during infections of cotyledons are plotted in orange (7 dpi) and green (14 dpi). Gene expression values during growth *in vitro* are plotted in purple; those in mock-infected cotyledons are plotted in magenta. The percentage of the total variance explained is listed on each axis label.

a third of the carbohydrate is soluble in phosphate buffered saline, including half of the total pectin, suggesting it would be readily available to a pathogen growing in the apoplast [41].

Distribution of CAZy domains and their expression

In view of the dominance of CAZys in the 20 most highly expressed genes of Lbc after seven days *in planta*, CAZy domains were sought in genome sequences of Lbc, Lmb, and also in four other Dothideomycetes for comparison. The repertoire of domains derived from the dbCAN database and its HMM-based sequence similarity search was consistent with CAZy classifications carried out previously on Dothideomycetes (Table S7 in file S2) [18,42], and the www.cazy.org annotation for Lmb isolate v23.1.3. In general, the numbers of carbohydrate-binding module (CBM), glycosyl hydrolase (GH), glycosyl transferase (GT), and polysaccharide lyase (PL) domains were in agreement with previous studies (Table 6). CAZys are divided into sub classes, generally based on substrate specificity or ligand; some of these are described later. Two CAZy classes in Lmb were identified more frequently using our analyses; CE domains (109 vs 34) and AA (78 vs 27). The difference in numbers of CE domains was probably because 27 CE1 and 42 CE10 domains had been previously excluded from the CAZy expert curated dataset on the basis that they were likely to act upon non-carbohydrate substrates [43]. AA domains are a newly formed CAZy classification for ligninolytic or lytic polysaccharide monooxygenases. Our automated approach may have included closely related monooxygenases that do not degrade lignin or polysaccharides.

Most of the CAZy containing-genes had only one CAZy domain, but several had multiple CBM domains, particularly of the subclass CBM50 (see later and Figure S1 in file S1). Five of the six dothideomycetes had between 580 and 700 CAZy domains, but *Z. tritici* only had 489, which is in general agreement with previous findings [44]. Overall the two *Leptosphaeria* species had similar complements of CAZy genes. *Alternaria brassicicola* had the next most similar profile. *Pyrenophora tritici-repentis* had a profile more similar to *Stagonospora nodorum* than to *Leptosphaeria*, consistent with the host of the two former fungi being a monocotyledonous cereal, rather than a dicotyledonous oilseed plant. Dicot cell walls generally contain a higher proportion of pectin, compared to glucuronoarabinoxylan, which is predominant in a typical monocot cell wall and believed to partially substitute for low levels of pectin in monocot cell walls [15]. *Zymoseptoria tritici* had a distinct profile with many fewer CAZy genes. This is congruent with previously reported characteristics of the members of order Capnodiales, versus the Pleosporales, to which the other five fungi belong. *L maculans* 'brassicae' had fewer AA domains than Lbc, *S. nodorum* and *P. tritici-repentis* did. The number of PL domains ranged from 19 to 24 in Lmb, Lbc and *A. brassicicola*. In contrast, the three wheat pathogens, *S. nodorum*, *P. tritici repentis* and *Z. tritici* had many fewer – between 4 and 10.

Although the complement of CAZys in the two *Leptosphaeria* species was generally similar, there were some interesting

Table 2. Top 20 upregulated genes of *L. maculans* 'brassicae' isolate IBCN18 seven days after inoculation of *B. napus* cv. Westar.

Gene ID	Gene name	7dpi *in planta* (FPKM)	14dpi *in planta* (FPKM)	*in vitro* (FPKM)	Log2 (7dpi FPKM/in vitro FPKM)	Pfam description and accession	Pfam expect score
Lema_P114790.1		12645.1	1163.5	0.1	16.9		
Lema_P114800.1		4066.8	728.9	0.1	15.3		
Lema_P092260.1		1375.5	30.2	0.1	13.7		
Lema_P086290.1	AvrLm1	1831.5	38.6	0.1	13.7		
Lema_P049660.1	AvrLm4-7	2540.0	70.8	0.3	13.1		
Lema_uP037480.1		793.8	42.8	0.1	13.0		
Lema_P084480.1		2645.1	207.0	0.5	12.5	Pfam-B_14615 [PB014615]	4.2E-08
Lema_P054900.1		567.4	24.7	0.1	12.5		
Lema_uP070880.1	AvrLmJ1	1349.9	59.3	0.3	12.4		
Lema_uP082260.1		521.8	3.4	0.1	12.3		
Lema_P087720.1		498.9	3.9	0.1	12.3	Cytochrome P450 [PF00067.17]	9.0E-57
Lema_P070100.1	Lm5LysM	2672.9	105.4	0.8	11.7	LysM domain [PF01476.15]	3.9E-08
Lema_uP121480.1		295.4	19.0	0.1	11.5		
Lema_P006160.1		5011.4	1356.8	1.8	11.5	Pfam-B_2613 [PB002613]	2.1E-33
Lema_P087710.1		282.5	6.2	0.1	11.5	Transferase family [PF02458.10]	5.7E-23
Lema_P087700.1		279.5	2.1	0.1	11.4	Cytochrome P450 [PF00067.17]	6.4E-57
Lema_uP002340.1		1624.1	96.0	0.7	11.2		
Lema_P087750.1		223.7	2.1	0.1	11.1	Cytochrome P450 [PF00067.17]	4.1E-43
Lema_P037680.1		1289.1	6.9	0.6	11.0	Pfam-B_8517 [PB008517]	8.0E-140
Lema_uP123070.1		388.9	34.8	0.2	10.9		

The top 100 *in planta* upregulated genes are listed in Table S3 in file S2.

Table 3. Top 20 upregulated genes of *L. biglobosa* 'canadensis' isolate J154 seven days after inoculation of *B. napus* cv. Westar.

Gene name	Gene name	7dpi in planta (FPKM)	14dpi in planta (FPKM)	*in vitro* (FPKM)	Log2 (7dpi FPKM/in vitro FPKM)	Pfam description and accession	Pfam expect score
Lb_J154_P000524		886.3	18.0	0.01	17.0	Protein of unknown function (DUF3678) [PF12435.3]	1.20E-01
Lb_J154_P003557		515.7	63.8	0.01	16.2	Pfam-B_18451 [PB018451]	3.60E-69
Lb_J154_P001652		1083.8	151.2	0.1	13.5	Cellulase (glycosyl hydrolase family 5) [PF00150.13]	7.70E-20
Lb_J154_P009247		1012.1	146.1	0.1	13.2	Glycosyl hydrolases family 12 [PF01670.11]	1.20E-31
Lb_J154_P006347		341.8	76.1	0.04	13.2	Deuterolysin metalloprotease (M35) family [PF02102.10]	5.00E-86
Lb_J154_P002168		878.3	9.0	0.1	12.8	Endoribonuclease L-PSP [PF01042.16]	7.30E-11
Lb_J154_P001276		317.1	24.8	0.05	12.6	Pfam-B_19830 [PB019830]	2.60E-36
Lb_J154_P001225		605.1	114.3	0.5	10.4	Putative cyclase [PF04199.8]	2.00E-13
Lb_J154_P010735		5991.3	6460.7	4.7	10.3	Alcohol dehydrogenase GroES-like domain [PF08240.7]	1.60E-23
Lb_J154_P008673		1467.7	220.5	1.2	10.3	Pfam-B_19830 [PB019830]	1.10E-38
Lb_J154_P004138		7216.7	1370.4	6.0	10.2	Trypsin [PF00089.21]	6.60E-64
Lb_J154_P004093		3826.7	456.8	3.2	10.2	Glycosyl hydrolases family 39 [PF01229.12]	3.20E-05
Lb_J154_P006834		410.0	148.0	0.4	10.1		
Lb_J154_P005286		953.8	53.9	0.9	10.0	Pectate_lyase [PF03211.8]	1.50E-58
Lb_J154_P001719		1410.8	291.4	1.4	9.9	Trypsin [PF00089.21]	7.90E-66
Lb_J154_P000527		12641.8	8785.2	13.1	9.9	Pfam-B_14072 [PB014072]	4.30E-34
Lb_J154_P009204		694.9	99.8	0.7	9.9	Cellulase (glycosyl hydrolase family 5) [PF00150.13]	6.00E-13
Lb_J154_P006775		307.5	11.8	0.4	9.7	Putative amidotransferase (DUF4066) [PF13278.1]	1.10E-26
Lb_J154_P001246		1340.6	88.9	1.9	9.5	Glycosyl hydrolases family 43 [PF04616.9]	1.20E-59
Lb_J154_P001995		617.0	66.2	0.9	9.4	Sugar (and other) transporter [PF00083.19]	1.10E-95

The top 100 *in planta* up regulated genes are listed in Table S4 in file S2.

Table 4. Top 20 upregulated genes of *L. maculans* 'brassicae' isolate IBCN 18 fourteen days after inoculation of *B. napus* cv. Westar.

Gene ID	Gene name	7dpi in planta (FPKM)	14dpi in planta (FPKM)	*in vitro* (FPKM)	Log2 (14dpi FPKM/in vitro FPKM)	Pfam description and accession	Pfam expect score
Lema_P114790.1		12645.1	1163.5	0.1	13.51		
Lema_P114800.1		4066.8	728.9	0.1	12.83		
Lema_P082270.1		861.2	3339.4	0.9	11.86	Trypsin [PF00089.21]	3.30E-66
Lema_P043000.1		106.6	762.8	0.7	10.12	Pfam-B_11894 [PB011894]	3.90E-35
Lema_P006160.1		5011.4	1356.8	1.8	9.59	Pfam-B_2613 [PB002613]	2.10E-33
Lema_P110730.1		3.4	66.5	0.1	9.38	Flavin-containing amine oxidoreductase [PF01593.19]	6.90E-27
Lema_uP085940.1		22.6	56.5	0.1	9.14		
Lema_P085920.1		55.2	74.6	0.1	8.97		
Lema_P084480.1		2645.1	207.0	0.5	8.84	Pfam-B_14615 [PB014615]	4.20E-08
Lema_P117020.1		211.3	85.9	0.2	8.81	Pfam-B_7042 [PB007042]	6.50E-17
Lema_uP037480.1		793.8	42.8	0.1	8.74		
Lema_P077490.1		19.1	45.9	0.1	8.74	Cutinase [PF01083.17]	8.00E-49
Lema_P006340.1		14.3	249.7	0.6	8.71	Sugar (and other) transporter [PF00083.19]	2.70E-88
Lema_P060160.1		191.1	60.0	0.1	8.71		
Lema_P036670.1		113.9	213.6	0.6	8.58	Endoribonuclease L-PSP [PF01042.16]	1.10E-10
Lema_P013700.1		515.5	628.0	1.9	8.35	Glycoside hydrolase family 7 [PF00840.15]	3.30E-190
Lema_P085930.1		24.6	88.9	0.3	8.33		
Lema_P092260.1		1375.5	30.2	0.1	8.24		
Lema_P077180.1		24.1	30.2	0.1	8.24		
Lema_P086290.1	AvrLm1	1831.5	38.6	0.1	8.17		

The top 100 *in planta* up regulated genes are listed in Table S5 in file S2.

Table 5. Top 20 upregulated genes of *L. biglobosa* 'canadensis' isolate J154 fourteen days after inoculation of *B. napus* cv. Westar.

Gene ID	Gene name	7dpi in planta (FPKM)	14dpi in planta (FPKM)	*in vitro* (FPKM)	Log2 (14dpi FPKM/in vitro FPKM)	Pfam description and accession	Pfam expect score
Lb_j154_P003075		156.1	116.5	0.01	14.0		
Lb_j154_P003557		515.7	63.8	0.01	13.2	Pfam-B_18451 [PB018451]	3.6E-69
Lb_j154_P000524		886.3	18.0	0.01	11.3	Protein of unknown function (DUF3678) [PF12435.3]	1.2E-01
Lb_j154_P006347		341.8	76.1	0.04	11.0	Deuterolysin metalloprotease (M35) family [PF02102.10]	5.0E-86
Lb_j154_P001652		1083.8	151.2	0.1	10.6	Cellulase (glycosyl hydrolase family 5) [PF00150.13]	7.7E-20
Lb_j154_P009247		1012.1	146.1	0.1	10.4	Glycosyl hydrolase family 12 [PF01670.11]	1.2E-31
Lb_j154_P010735		5991.3	6460.7	4.7	10.4	Alcohol dehydrogenase GroES-like domain [PF08240.7]	1.6E-23
Lb_j154_P001707		63.1	9.2	0.01	10.4		
Lb_j154_P006327		113.6	8.5	0.01	10.2		
Lb_j154_P008556		1.4	4.9	0.01	9.4	Mediator complex subunit 27 [PF11571.3]	1.8E-01
Lb_j154_P000527		12641.8	8785.2	13.1	9.4	Pfam-B_14072 [PB014072]	4.3E-34
Lb_j154_P000350		32.9	4.3	0.01	9.3		
Lb_j154_P006232		3710.7	5194.8	8.5	9.2		
Lb_j154_P005193		1399.2	1588.4	2.8	9.2		
Lb_j154_P003830		27.8	38.5	0.1	9.1	Serine carboxypeptidase [PF00450.17]	4.2E-87
Lb_j154_P009122		5.3	3.5	0.01	9.0		
Lb_j154_P003076		158.0	118.9	0.2	9.0	Uncharacterized protein conserved in bacteria (DUF2321) [PF10083.4]	2.9E-01
Lb_j154_P001276		317.1	24.8	0.05	9.0	Pfam-B_19830 [PB019830]	2.6E-36
Lb_j154_P008991		3272.4	3034.9	6.1	9.0	Short-chain dehydrogenase [PF00106.20]	9.0E-27
Lb_j154_P003555		1776.8	1658.6	3.5	8.9	Alcohol dehydrogenase GroES-like domain [PF08240.7]	2.2E-22

The top 100 *in planta* up regulated genes are listed in Table S6 in file S2

Table 6. Distribution of CAZy domains in predicted proteins of six Dothideomycetes.

Number of domains

CAZy domain Fungus (lifestyle)	AA	CBM	CE	GH	GT	PL	Total
Leptosphaeria biglobosa (necrotroph)	98	85	117	234	106	23	663
Leptosphaeria maculans (hemibiotroph)	78	63	109	217	100	19	586
Alternaria brassicicola (necrotroph)	92	66	120	233	92	24	627
Stagonospora nodorum (necrotroph)	122	66	143	267	96	10	704
Pyrenophora tritici-repentis (necrotroph)	114	56	124	246	105	10	655
Zymoseptoria tritici (hemibiotroph)	55	26	97	199	108	4	489

The classes of CAZy domains are Auxiliary Activities (AA) Carbohydrate Binding Modules (CBM), Carbohydrate Esterases (CE), Glycosyl Hydrolases (GH), Glycosyl Tranferases (GT) Polysaccharide lyases (PL). Some genes contain multiple CAZy domains.

differences. The major difference was in the CBM class with > 30% more domains present in Lbc than in Lmb. *L maculans* 'brassicae' had fewer CBM18-containing-genes than Lbc, but a similar number of CBM18 domains. CBM18 domains in Lmb were in a more compact gene structure (28 domains in 11 genes); indeed this domain was always present in multiple copies within a gene, five genes contained homopolymers of only CBM18. *L. biglobosa* 'canadensis' had 32 domains in 21 genes, eight of which had a single CBM18 domain. CBM50 (LysM), CE4, and GH18 CAZy domains occurred frequently in genes that had CBM18 domains. (Table S8 in file S2). In eukaryotes this domain often has a chitin-binding function and modules are often adjacent to chitinase catalytic domains, but this domain is also present in non-catalytic proteins either singly or as multiple repeats. CBM18 domains may be involved in targeting the degradative domains to particular carbohydrates, or may enhance catalysis by ensuring retention of the substrate between catalytic cycles.

About half of the genes with CAZy domains in the two *Leptosphaeria* species did not have a signal peptide and thus were not predicted as being secreted. This may be due to incorrect annotation at the 5' end of genes, genes incorrectly merged during annotation, or a false negative result from SignalP. Pectate lyases were the most frequently secreted CAZy (100% for Lbc, 89% for Lmb), while GTs were the least frequently secreted (11% for Lbc, 12% for Lmb). This is consistent with the role of GTs in synthesising carbohydrates for the fungal cell wall, and their membrane or intracellular location.

We then examined the RNA-seq data to determine expression levels of CAZys of Lbc and Lmb during infection of *B. napus* cotyledons (Table 7, Table S7 in file S2). The expression profiles of the top 100 most highly expressed (summed FPKM values at 7 dpi, 14 dpi, and *in vitro*) secreted CAZy genes are presented in Figure 4. At 7 dpi, Lmb had a high level of expression of genes with CBM domains, particularly CBM50/LysM (for instance, genes A and C in Figure 4); this level decreased by more than 50% by 14 dpi and was even lower during growth *in vitro*.

Genes with CBM domains were also among the most highly expressed CAZy genes of Lbc *in planta*, but the overall level of expression was much lower than that of Lmb genes, at the same time point. To reveal the most *in planta* specific CAZy classes, expression ratios between *in planta* and *in vitro* growth were calculated for each CAZy class (Table S9 in file S2). For Lmb, the CBM50 (chitin-binding) domain was highly upregulated *in planta* at 7 dpi, whereas at the same time point Lbc upregulated CBM6 (cellulose binding) domains more highly (Table S9 in file S2). The *in planta* upregulated RNA-seq expression profile of the CBM50-containing genes of Lmb was validated by quantitative RT-PCR (Figure S2 in file S1). These experiments showed peak expression of each LysM gene at 7 dpi *in planta* and much lower expression levels *in vitro*. At 3 dpi, LysM gene expression was higher than *in vitro* levels, but lower than levels at 7 dpi. For the CBM18-containing genes, expression was varied, with >3 orders of magnitude between the highest and lowest expressed genes. *L. biglobosa* 'canadensis' had generally higher expression compared

Table 7. Expression of CAZy genes in *L. maculans* 'brassicae' and *L. biglobosa* 'canadensis.'

	L. maculans			*L. biglobosa*		
CAZy class	7dpi *in planta* (FPKM)	14dpi *in planta* (FPKM)	*in vitro* (FPKM)	7dpi *in planta* (FPKM)	14dpi *in planta* (FPKM)	*in vitro* (FPKM)
AA	86.5	109.9	117.3	104.5	89.1	159.7
CBM	508.7	203.6	158.5	163.7	174.3	341.4
CE	78.1	69.5	63.6	78.7	55.8	99.9
GH	110.2	122.3	88.1	122.7	81.3	125.5
GT	54.8	60.9	57.1	46.3	50.0	112.3
PL	56.3	23.9	6.9	121.3	37.7	29.4

RNA-seq derived gene expression values were mapped onto the identified CAZy domains for Lmb and Lbc. Average expression levels (FPKM) were then calculated for each major class of CAZy.

Key to category colours

| | High expression *in vitro* | | Medium expression *in planta* and *in vitro* | | High expression *in planta* at 7dpi |
| | High expression *in planta* | | High expression *in planta* at 14 dpi | | High expression *in planta* and *in vitro* |

Figure 4. Expression profiles of secreted CAZys of *L. maculans* **'brassicae' (Lmb),** *L. biglobosa* **'canadensis' (Lbc)** *in planta* **and** *in vitro.* The top 100 genes expressed across the three treatments (7, 14 dpi and *in vitro*) and predicted to be secreted and to contain a CAZy domain (CBM, GH, PL, or AA) were selected. Quantile-normalisation was applied and log10-transformed FPKM values were graphed. The intensity of blue shading is proportional to the expression level. The gene order is based on a dendrogram created from a Euclidean similarity matrix with average group distance. Letters with a triangle point to genes that were amongst the top 20 most highly expressed genes *in vitro* or *in planta* (Tables 2, 3, 4, or 5). A/ Lema_P102640.1 (Lm2LysM); B/Lema_P013700.1 (Glycoside hydrolase family 7); C/Lema_P070100.1 (Lm5LysM); D/Lb_j154_P005286 (pectate lyase); E/Lb_j154_P009204 (cellulase); F/Lb_j154_P001246 (Glycosyl hydrolase family 43); G/Lb_j154_P001652 (cellulase); H/Lb_j154_P009247 (Glycosyl hydrolases family 12); I/Lb_j154_P004093 Glycosyl hydrolases family 39). Categories of gene expression (High expression *in vitro*, High expression *in planta*, Medium expression *in planta* and *in vitro*, High expression *in planta* at 14 dpi, High expression *in planta* at 7dpi, High expression *in planta* and *in vitro*) were manually assigned and indicated by coloured polygons linking the genes in each category across Lmb and Lbc; the number of genes in each category is indicated on the vertical sides of the polygon.

to that of Lmb (Table S8 in file S2). For Lbc, the average expression value of all CBM genes was highest during *in vitro* growth, which was due to high expression of several genes that had both CBMs and hydrolytic CAZY domains (Table 7). Such genes included Lb_j154_P004089, which has three CBM18 and one CE4 domains (Table S8 in file S2).

Polysaccharide lyases (PL) were expressed highly at 7 dpi, and expression decreased by 14 dpi in both species. At 7 dpi, the expression levels of PLs in Lbc were twice those of Lmb. Expression levels then dropped by 75% at 14 dpi and were low during growth *in vitro*. *L. biglobosa* 'canadensis' had higher levels of expression of degrading CAZys such as AAs, GHs and PLs (genes D,E,F,G,H and I; Figure 4) at 7 dpi, than at 14 dpi, when lesion growth had slowed. The PL3 class (pectate lyase) was highly upregulated by Lbc at 7 dpi (Table S9 in file S2), and was also the most upregulated PL class in Lmb. This may be a reflection of the abundance of pectin in the cotyledon.

Glycosyl transferase (GT) -containing genes were expressed at similar levels in all conditions in both *Leptosphaeria* species, except for a two-fold increase in Lbc grown *in vitro* compared to *in planta* (Table 7). The class GT21, which encodes biosynthetic enzymes that glycosylate lipids, was the most *in planta* upregulated class of glycosyltransferases. Of the carbohydrate esterase enzymes, CE8 (pectin methylesterase) and CE12 (pectin acetylesterase) were the most *in planta* upregulated classes for Lbc and Lmb, respectively (Table S9 in file S2).

At 14 dpi, of the six CAZy classes, expression of only PLs was higher in Lbc than in Lmb. By 14 dpi, expression of CBM and PL-containing genes by Lmb had decreased, but expression of genes with AA and GH domains had increased. *In vitro* growth was characterised by low CBM expression and very low PL expression (Table 7). The CAZy expression profile for Lmb *in planta* may reflect early avoidance of chitin- triggered immunity via high expression of CBMs such as LysM genes, whilst the fungus feeds from soluble pectin via PL genes. At 14 dpi elevated expression of

the oxidative AA class and the hydrolytic GH class may facilitate degradation of lignin and cellulose in dead cells.

In many fungi, expression of cell wall degrading enzymes CAZys is regulated by carbon catabolite repression, a global mechanism that ensures readily assimilated carbon sources such as glucose are preferentially used. A Lmb homologue of the *Saccharomyces cerevisiae* sucrose non-fermenting protein kinase 1 *(SNF1)* carbon catabolite regulator has been recently shown to regulate expression of several CAZy genes encoding pectate lyases, beta-1,3-glucanase, and a glucosidase [45]. These CAZy genes were upregulated four days after inoculation of canola cotyledons with a wild type strain of Lmb; whereas an LmSNF1 knockout strain had significantly reduced expression of these CAZys during growth on pectin *in vitro*. We recorded similar upregulation of the CAZy genes encoding two pectate lyases and the carbohydrate esterase *in planta* at 7 dpi, but not significant upregulation of the chitin deacetylase. This latter difference could be due to differing time points at which tissue was analysed and the more robust normalisation used in an RNA-seq analysis compared to a qPCR method.

In both Lmb and Lbc, the top 5% most highly expressed CAZy genes accounted for approximately 50% of the sum total CAZy expression (FPKM) at each growth stage (7 and 14 dpi and *in vitro* growth) that was analysed (data not shown). This suggests that the vast majority of CAZys are expressed at low levels or only turned on at specific times in the fungal life cycle. This phenomenon has been reported previously for *Z. tritici*, which differentially expresses CAZy genes, such as cutinases (CE5 subclass) according to biotrophic, necrotrophic or saprotrophic stages of growth [44]. The hemi-biotroph *Colletotrichum higginsianum* upregulates expression of effectors and secondary metabolite biosynthetic enzymes both before penetration and during biotrophic growth on *Arabidopsis*. *C. higginsaneum* then upregulates hydrolases and transporters at a later stage, each wave delivered according to the stage of pathogenic transition [46]. This again matches our

Table 8. Distribution of peptidase domains in six Dothideomycete genomes.

Peptidase type	A	C	G	I	M	S	T	U	Total
Species (Lifestyle)									
Leptosphaeria biglobosa (necrotroph)	18	73	1	8	144	145	23	1	413
Leptosphaeria maculans (hemibiotroph)	16	78	1	10	140	140	25	1	411
Alternaria brassicicola (necrotroph)	17	77	1	6	146	163	23	1	434
Stagonospora nodorum (necrotroph)	23	80	0	7	151	192	23	1	477
Pyrenophora tritici-repentis (necrotroph)	22	78	1	9	155	163	22	1	451
Zymoseptoria tritici (hemibiotroph)	33	67	4	8	130	191	23	1	457

Peptidase domains are classified according to the catalytic type, or inhibitor activity. Peptidase types are Aspartic (A), Cysteine (C), Glutamic (G), Inhibitor, Mettallo-(M), Serine (S), Threonine (T), Unknown (U).

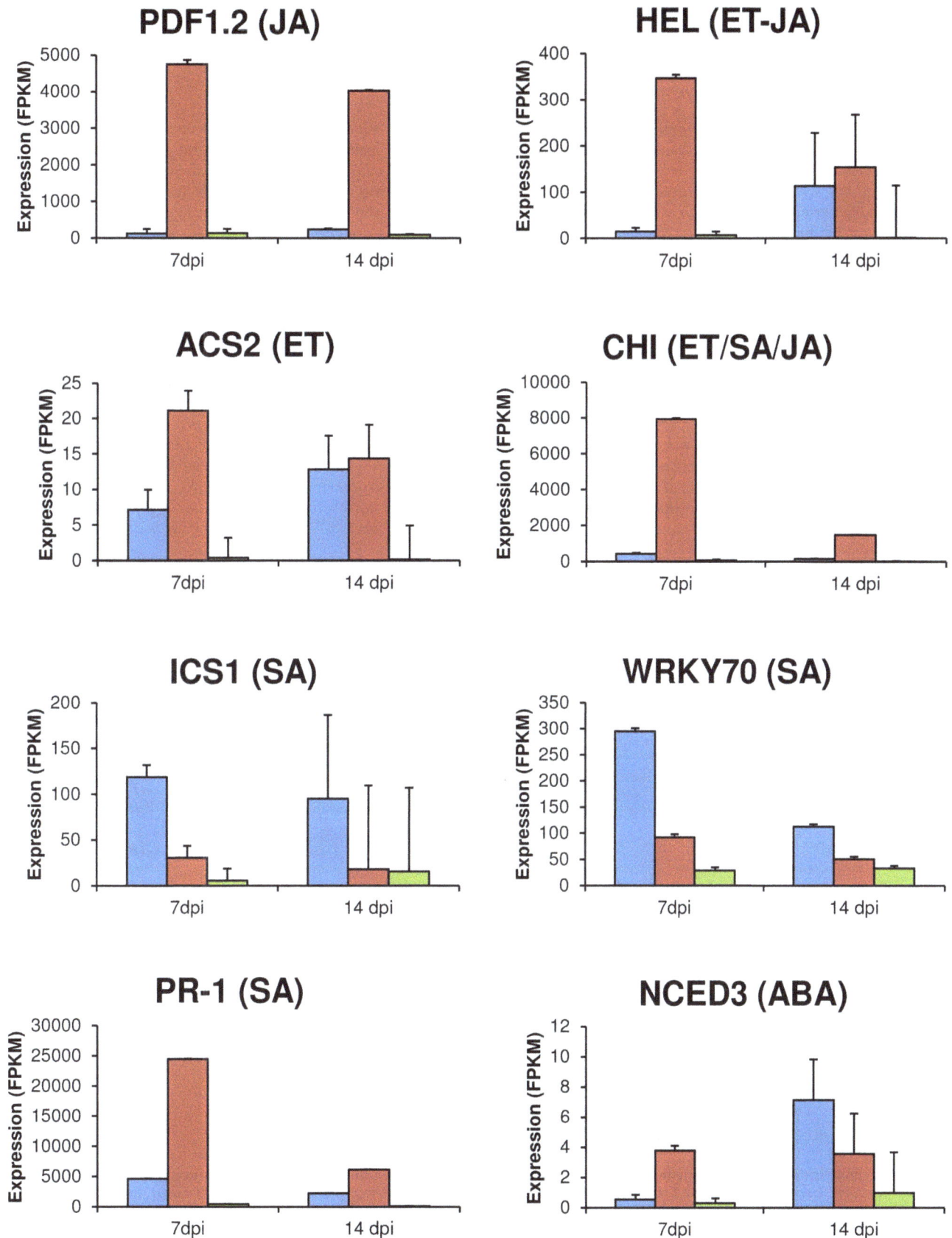

Figure 5. Expression of key defence genes of *B. napus* **at 7 or 14 dpi with either** *L. maculans* **'brassicae' (Lmb) or** *L. biglobosa* **'canadensis' (Lbc).** RNA-seq data for eight key Brassica defence genes were determined at 7 and 14 dpi. Average expression (FPKM) after infection by *L. maculans* (blue), *L. biglobosa* (red), and mock inoculum (green) is plotted. Error bars indicate the Cufflinks 95% confidence interval for each FPKM

value. The genes assayed were 9-cis-epoxycarotenoid dioxygenase 3 (NCED3), 1-amino-cyclopropane-1-carboxylate synthase 2 (ACS2), chitinase (CHI), hevein-like protein (HEL), isochorismate synthase 1 (ICS1), Pathogenesis related protein 1 (PR-1), WRKY transcription factor 70 (WRKY70), and plant defensin 1 (PDF1.2). Each gene was also classified according to hormone(s) abscisic acid (ABA), ethylene (ET), jasmonic acid (JA) and salicylic acid (SA) that induced higher expression.

observation that *L. biglobosa* lacks or only has a short biotrophic stage and expresses hydrolases earlier than Lmb does. CAZy genes of the two *Leptosphaeria* species that have low expression during development of cotyledonary lesions might be expressed more highly in another growth situation, such as earlier in penetration of the tissue, or during colonisation of petiole, stem, or during saprophytic growth on woody stubble of *B. napus*.

Peptidase distribution and expression

Since the two *Leptosphaeria* species grow between the plant cells in the apoplast, protein from the plant cell wall is a potential source of nitrogen for nutrition. A trypsin-like peptidase was highly expressed during necrotic stages of infection for both Lmb (14 dpi) and Lbc (7 dpi) (Table 2, Table 3), therefore we characterised the peptidase content of both *Leptosphaeria* genomes and compared them to four other Dothideomycete plant pathogens.

The MEROPS peptidase classifications categorises peptidases by their catalytic nucleophile [37]. For example, aspartic (A), cysteine (C), serine (S), threonine (T) peptidases are named for the corresponding catalytic residue. Exceptions are for metallo (M)

peptidases, which use a coordinated metal ion, and the classifications for unknown activities (U) and peptidase inhibitors (I). Peptidase composition and number was very similar for Lmb, Lbc and *A. brassicicola* (Table 8). Metallo- and serine-peptidase genes comprised almost 70% of the total peptidases. The next most abundant classes were the cysteine, threonine and aspartic peptidases, in that order. *L. biglobosa* 'canadensis' had seven C56 cysteine peptidases (7) whilst Lmb only had three. C56 peptidases act on peptides <20 amino acids, presumably requiring prior digestion of the substrate by another peptidase. *P. tritici-repentis* and *S. nodorum* had more peptidases in the metallo and serine classes than Lmb or Lbc. The hemi-biotroph *Z. tritici* had a larger complement of serine and aspartic peptidases than the two *Leptosphaeria* species did.

In general, the expression levels of peptidase at 7 and 14 dpi *in planta* were similar in Lmb and Lbc (Pearson correlation coefficient of 0.76 for Lmb IBCN18 and 0.85 for Lbc J154), with a few notable exceptions. A trypsin-like serine peptidase (MEROPS S01A) was the most highly *in planta* up-regulated peptidase (500-fold higher at 7 dpi *in planta* than *in vitro*) in both Lbc

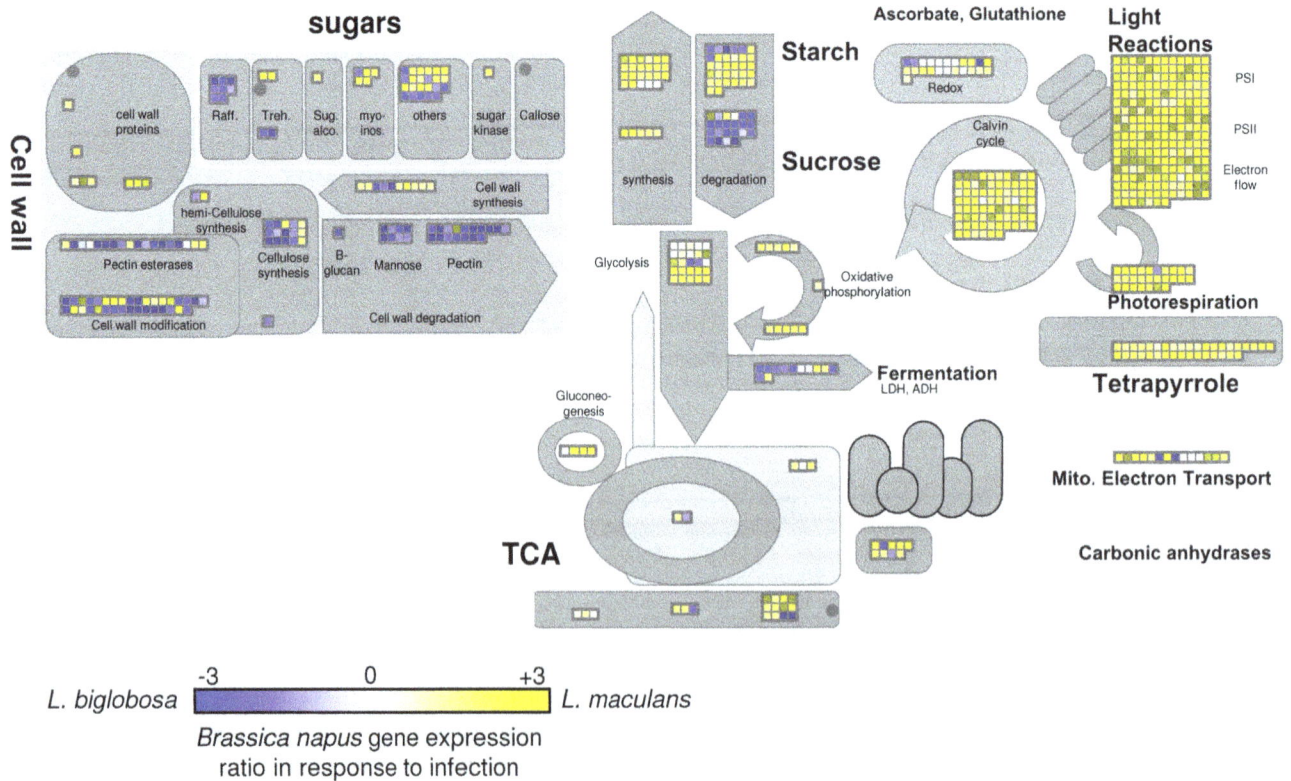

Figure 6. Transcription of *B. napus* genes involved in metabolic processes including photosynthesis seven days after inoculation with *L. maculans* 'brassicae' (Lmb) or *L. biglobosa* 'canadensis' (Lbc). RNA-seq gene expression values for a *B. napus* unigene set [27] were used to calculate a ratio of expression values (log2) for *B. napus* genes after infection by Lmb or Lbc. Ratios were plotted on major metabolic pathways with Mapman software [39]. A yellow square indicates a *B. napus* gene that is expressed more highly during Lmb infection, while a blue square indicates a *B. napus* gene with higher expression during Lbc infection. An expression ratio close to zero is shown with a white square and indicates equivalent expression during infection by either pathogen. Only genes with expression values greater than 10 FPKM were included. Abbreviations: LDH, lactate dehydrogenase; ADH, Alcohol dehydrogenases, TCA, tricarboxylic acid cycle; raff, raffinose; Treh, trehalose; PSI, photosystem one; PSII, photosystem two; ABA, abscisic acid; ET, Ethylene; SA, salicylic acid; JA, jasmonic acid.

Table 9. Top 20 functional categories that are significantly regulated in *B. napus* in response to infection by *L. maculans* 'brassicae' compared to infection by *L. biglobosa* 'canadensis.'

Mapman Function Category	Mapman Function Description	Number of genes	p-value
1	Photosynthesis	313	0.00E+00
1.1	Photosynthesis, light reactions	210	0.00E+00
1.1.1	Photosynthesis, light reactions, photosystem II	105	0.00E+00
26	Miscellaneous	365	4.42E-25
10	Cell wall	116	4.31E-12
1.1.2	Photosynthesis lightreaction photosystem I	44	5.06E-11
1.1.1.2	Photosynthesis lightreaction.photosystem II, PSII polypeptide subunits	63	5.06E-11
1.3	Photosynthesis calvin cycle	76	2.19E-10
1.1.1.1	PS lightreactions photosystem II, LHC-II	42	3.08E-10
1.1.2.2	PS lightreactions photosystem I, PSI polypeptide subunits	39	3.56E-10
20.1	Stress, biotic	123	1.98E-09
10.6	Cell wall, degradation	28	3.88E-09
26.12	Misc peroxidases	26	1.07E-08
26.8	Misc nitrilases, nitrile lyases, berberine bridge enzymes, reticuline oxidases, troponine reductases	42	3.72E-08
26.10	Misc cytochrome P450	46	1.93E-07
26.3	Misc gluco-, galacto- and mannosidases	42	1.35E-06
10.6.3	Cell wall degradation, pectate lyases and polygalacturonases	17	1.74E-05
35	Not assigned	1969	6.69E-05
35.2	Not assigned, unknown	1969	6.69E-05
17	Hormone metabolism	107	8.21E-05
2.2.1	Major CHO metabolism, degradation sucrose	22	6.47E-04
20.1.7.6	Stress biotic, PR-proteins, proteinase inhibitors	12	6.63E-04
1.2	Photosynthesis, photorespiration	27	7.32E-04
17.5	Hormone metabolism.ethylene	30	7.77E-04
34	Transport	208	8.58E-04
33	Development	107	1.62E-03
26.9	Misc glutathione-S-transferases	32	1.92E-03
10.8	Cell wall pectin esterases	18	1.92E-03
17.5.1	Hormone metabolism, ethylene synthesis-degradation	21	1.92E-03
20.1.7.6.1	Stress biotic, PR-proteins, proteinase inhibitors, trypsin inhibitor	11	1.92E-03
1.1.3	Photosynthesis, light reactions, cytochrome b6/f	8	2.25E-03
10.8.1	Cell wall,pectin esterases, PME	13	2.35E-03
31.4	Cell vesicle transport	32	2.41E-03
1.3.6	Photosynthesis, calvin cycle,aldolase	13	2.49E-03
1.1.4	Photosynthesis, lightreaction, ATP synthase	15	2.55E-03
29.5.3	Protein degradation, cysteine protease	18	3.11E-03
29.5.9	Protein degradation, AAA type	15	3.41E-03
2.2.1.3	Major CHO metabolism, degradation, sucrose invertases	13	4.18E-03
13.2.3	Amino acid metabolism, degradation, aspartate family	11	5.62E-03
1.1.5	Photosynthesis, light reaction, other electron carrier (oxidation/reduction)	18	5.77E-03

B. napus genes expressed during infection at 7 dpi were assigned to functional categories and overall pathway regulation was examined. The top 20 categories determined by MapMan analysis are listed. Abbreviations: PS (photosynthesis), LHC (light harvesting complex), CHO (carbohydrate), PR (pathogen response), PME (pectin methyl esterase), ATP (adenosine triphosphate). AAA type (ATPases Associated with diverse cellular Activities). P values indicate the likelihood of the observed pathway regulation being due to chance.

(Lb_j154_P004138) and Lmb (Lema_P082270.1) (Table 2, Table S10, Table S11 in file S2). Its role in disease is unknown. The MEROPS S01A "type" peptidase is bovine chymotrypsin but fungal homologs are usually described as "trypsin-like" because they are similar to both bovine trypsin, and trypsin from *Streptococcus bacteria* [47]. The *in vitro* growth condition resulted in a more distinct expression profile of peptidases, producing low correlation co-efficient values (0.35 and 0.38) between the *in vitro* and 7 dpi *in planta* conditions, for Lbc and Lmb, respectively. Four Lmb peptidases had moderate levels of expression at 7 dpi and were down-regulated during growth *in vitro*, in a similar manner to the trypsin homologs. Two genes, Lema_P058000.1 and Lema_P031600.1, were most similar to C56 cysteine peptidases. The other two genes, Lema_P044030.1 and Lema_P044810.1, were most similar to the S33 serine peptidases, which typically release an N-terminal proline from a peptide substrate. These peptidases may target hydroxyproline-rich glycoproteins such as extensins, which are components of the plant cell wall [48]. Average CAZy class expression for the peptidases showed Lbc expressed its peptidase inhibitors more highly than Lmb did across all treatments (Table S11 in file S2). Glutamyl peptidases were expressed at very low levels in both fungi, and threonine peptidases had the highest average expression of all peptidase classes.

Expression of Brassica genes

The infection stage in a plant-pathogen interaction is often reflected in expression of key genes implicated in host defence signalling pathways. The expression levels for eight defence reporter genes of *B. napus* were determined during invasion by Lmb and Lbc (Figure 5). In general these genes were more highly expressed at 7 dpi during infection by Lbc than by Lmb. During Lbc infection, reporter genes for ethylene signalling (ACS2, CHI, and HEL) were expressed 50- to 138-fold higher than in mock-infected controls at 7 dpi, while Lmb infection resulted in only 2–19 fold induction of these genes. *L. maculans* 'brassicae' triggered a higher expression level of a key salicylic acid reporters, ICS1 and WRKY70, but another salicylic acid reporter gene, PR-1, was not as up regulated compared to Lbc. The salicylic acid reporter gene, PR-1, but not ICS1, were similarly highly induced at 7 dpi in Lbc. The jasmonic acid reporter gene PDF1.2 was induced highly by Lbc infection, but not by Lmb. The upregulation of these genes reflected that jasmonic acid and salicylic acid defence pathways were induced by Lbc, in a pattern consistent with the timing of necrosis.

Sasek et al (2012) showed that interactions of *B. napus* and Lmb, involving the recognition of *AvrLm1* (or the mutant allele, *avrLm1*) and corresponding resistance gene, *Rlm1*, salicylic acid biosynthesis and transcription of SA-associated genes (ICS1, WRKY70 and PR-1) increased as early as 3 dpi. Expression of HEL and CHI, genes involved in ethylene signalling, increased at 7 dpi [49,50]. Although these genes were upregulated during the susceptible response compared to the uninoculated controls, they were much more highly expressed during a resistance response.

The increased level of expression of *B. napus* genes involved in ethylene, jasmonic acid and salicylic acid signalling during Lbc infections prompted further examination of biochemical pathways during infection. A ratio of expression values of *B. napus* genes at 7 days after infection with Lmb compared to Lbc was calculated and analysed by MapMan software to identify pathways co-ordinately responding to early infection of the cotyledon (Figure 6, Table 9). The major difference in host response to infection by the two fungi was in genes with photosynthesis-associated activities. *L. biglobosa* 'canadensis' infection resulted in massive down-regulation of genes involved in photosynthesis (e.g. PSI, PSII), electron transport, photorespiration and chlorophyll (tetrapyrrole) biosynthesis compared to that in Lmb. Similarly, expression of sucrose and starch biosynthesis genes was reduced, possibly as a flow-on from lack of photosynthesis. Levels of sucrose degradation genes were higher during Lbc than during Lmb infection, perhaps due to decreased amounts of photosynthate. Biosynthetic genes for raffinose, a monosaccharide osmoprotectant, were induced by

Lbc, perhaps due to water stress. In contrast, genes involved in starch metabolism (both synthesis and degradation) were transcribed at high levels during Lmb infection at 7 dpi. The cell wall remodelling genes of *B. napus* that modify β-glucans, mannans and pectin were more highly expressed during infection by Lbc than by Lmb [51]. Extensive up regulation of host cell wall remodelling genes occurred at 7 dpi as the necrotic lesion was formed by Lbc, while Lmb infection had much less impact on transcription of these genes. A cohort of genes associated with secondary metabolism in *B. napus* was also plotted using Mapman on the same dataset (Figure S3 in file S1). Isoflavone reductase genes associated with isoflavonoid biosynthesis, were more highly expressed during Lbc infection, while genes associated with carotenoid metabolism (phytoene dehydrogenase, zeta-carotene desaturase, lycopene cyclases and violaxanthin de-epoxidase) were expressed more highly during Lmb infection. These observations are in agreement with the increased necrosis observed during Lbc infection at the early stages of infection. As well as increased expression of secondary metabolism genes during Lbc infection, peroxidase, nitrilase, and cytochrome P450 genes were consistently upregulated by *B. napus* at 7 dpi (Figure S3 in file S1). Twenty four general peroxidases were upregulated suggesting a strong oxidative burst was deployed during infection by Lbc. Forty two cytochrome 450 genes, 15 oxidase genes and 34 nitrilase genes were upregulated by Lbc infection, which also suggest a strong activation of secondary metabolism.

Summary

The hemi-biotroph *L. maculans* 'brassicae' avoids triggering host defence during early infection. This is reflected by the finding that at seven days post inoculation, *L. maculans* 'brassicae' expresses a large number of genes with no known domains, many of them being small secreted proteins. One class of small-secreted protein-encoding genes that is highly expressed at this time is CBM50 (LysM) genes, which suppress chitin-triggered PAMP immunity and evade detection of the fungus by the plant. Also avirulence genes are highly upregulated; at seven days post-inoculation, two avirulence genes are amongst the top 20 most highly upregulated *L. maculans* 'brassicae' genes *in planta*. This pattern is consistent with the relatively asymptomatic growth phase of this fungus at seven days post-inoculation. In contrast, *L. biglobosa* 'canadensis' expresses a high number of cell wall degrading CAZy genes during the first seven days of infection, consistent with extensive necrosis and a high degree of activation of host defence signalling pathways.

Supporting Information

File S1 Supporting figures. Figure S1, Classifications of CBM50 (LysM) domains in *L. maculans* 'brassicae', *L. biglobosa* 'canadensis', *Cladosporium fulvum* and *Zymoseptoria tritici*. LysM domains from Lmb, Lbc and *Z. tritici* (formerly *Mycosphaerella graminicola*) with high sequence similarity to ECP6 of *C. fulvum* were aligned using ClustalW. (A) Domains are numbered by proximity to N-terminus (#1 is closest). Residues are coloured by similarity (black: 100%, dark grey: 80–100%, light grey: 60–80%, white: less than 60%). B) Phylogram based on the amino acid alignment of LysM domains from panel A. Branch numbers show % bootstrap support and scale bar shows amino acid substitutions per site. LysM domains assigned to three Positions A, B and C based on sequence similarity to the *C. fulvum* ECP6 sequence. LysM domain organisation in ECP6-like predicted proteins in Lm, Lb, *Z. tritici* and *C. fulvum* are shown in panel C. Mg LysM genes are from *Z. tritici*. **Figure S2,** Expression of three LysM-

containing genes of *L. maculans* 'brassicae' grown *in vitro* and *in planta*. Quantitative RT-PCR analysis was performed on RNA from Lmb isolate IBCN18 grown in 10% Campbells V8 juice (*in vitro*), and after infection of cotyledons of *B. napus* cv. Westar at 3, 7 and 14 days post inoculation (dpi). Expression levels of Lm2LysM (Lema_P102640.1), Lm4LysM (Lema_P025400.1), Lm5LysM (Lema_P070100.1), were normalised to those of gamma-actin (Lema_P099940.1). Error bars represent one standard error of the mean (n = 2–3 biological replicates). Asterisks indicate values significantly different from *in vitro* levels (p<0.05).
Figure S3, Response of *B. napus* secondary metabolism and large enzyme families to infection by *L. maculans* 'brassicae' or *L. biglobosa* 'canadensis.'RNA-seq gene expression values for a *B. napus* unigene set [27] were used to calculate a ratio of expression values (log2) for *B. napus* genes seven days after infection by Lmb or Lbc. Ratios were plotted on secondary metabolism gene groups using MapMan software on maps for 'Secondary metabolism', and 'Large enzyme families' maps [39]. A yellow square indicates a *B. napus* gene that is expressed more highly during Lmb infection, while a blue square indicates a *B. napus* gene with higher expression during Lbc infection. An expression ratio close to zero is shown with a white square and indicates equivalent expression during infection by either pathogen. Only genes with expression values greater than 10 FPKM were included. MVA is mevalonic acid.

File S2 Supporting tables. Table S1, Total number of RNA-seq reads aligned to reference genomes. **Table S2,** Oligonucleotide Primers. **Table S3,** Top 100 most highly

upregulated *in planta* genes in *L. maculans* 'brassicae' at seven days post-inoculation. **Table S4,** Top 100 most highly upregulated *in planta* genes in *L. biglobosa* 'canadensis' at seven days post-inoculation. **Table S5,** Top 100 most highly upregulated *in planta* genes in *L. maculans* 'brassicae' at 14 days post-inoculation. **Table S6,** Top 100 most highly upregulated *in planta* genes in *L. biglobosa* 'canadensis' at 14 days post-inoculation. **Table S7,** Annotated CAZy domains of *L. maculans* and *L. biglobosa,* and their expression. **Table S8,** Multi-domain CBM18 genes of *L. maculans* 'brassicae' and *L. biglobosa* 'canadensis', and their expression. **Table S9,** The top 10 CAZy families of *L. maculans* 'brassicae' and *L. biglobosa* 'canadensis' based on expression ratio of *in planta* and *in vitro* growth at 7 dpi. **Table S10,** Annotated peptidases of *L. maculans* and *L. biglobosa,* and their expression. **Table S11,** Peptidase expression in *L. maculans* 'brassicae' and *L. biglobosa* 'canadensis'.

Acknowledgments

We thank Sebastian Gornik, the University of Melbourne, for bioinformatics advice.

Author Contributions

Conceived and designed the experiments: RL BH AW. Performed the experiments: RL AW BC JG. Analyzed the data: RL AC JG. Contributed reagents/materials/analysis tools: JG TR. Contributed to the writing of the manuscript: RL BH.

References

1. Oliver RP, Solomon PS (2010) New developments in pathogenicity and virulence of necrotrophs. Curr Opin Plant Biol 13: 415–419.
2. Rouxel T, Grandaubert J, Hane JK, Hoede C, van de Wouw AP, et al. (2011) Effector diversification within compartments of the *Leptosphaeria maculans* genome affected by Repeat-Induced Point mutations. Nat Comm 2: 202.
3. Shoemaker RA, Brun H (2001) The teleomorph of the weakly aggressive segregate of *Leptosphaeria maculans*. Can J Bot 79: 412–419.
4. Voigt K, Cozijnsen AJ, Kroymann J, Pöggeler S, Howlett BJ (2005) Phylogenetic relationships between members of the crucifer pathogenic *Leptosphaeria maculans* species complex as shown by mating type (MAT1-2), actin, and β-tubulin sequences. Mol Phylogenet Evol 37: 541–557.
5. de Gruyter J, Woudenberg JHC, Aveskamp MM, Verkley GJM, Groenewald JZ, et al. (2013) Redisposition of phoma-like anamorphs in Pleosporales. Stud Mycol 75: 1–36.
6. Howlett BJ (2004) Current knowledge of the interaction between *Brassica napus* and *Leptosphaeria maculans*. Can J Plant Pathol 26: 245–252.
7. Van de Wouw A, Thomas V, Cozijnsen A, Marcroft S, Salisbury P, et al. (2008) Identification of *Leptosphaeria biglobosa* 'canadensis' on *Brassica juncea* stubble from northern New South Wales, Australia. Australas Plant D Notes 3: 124–128.
8. Gout L, Fudal I, Kuhn ML, Blaise F, Eckert M, et al. (2006) Lost in the middle of nowhere: the *AvrLm1* avirulence gene of the Dothideomycete *Leptosphaeria maculans*. Mol Microbiol 60: 67–80.
9. Fudal I, Ross S, Gout L, Blaise F, Kuhn ML, et al. (2007) Heterochromatin-like regions as ecological niches for avirulence genes in the *Leptosphaeria maculans* genome: map-based cloning of *AvrLm6*. Mol Plant Microbe Interact 20: 459–470.
10. Balesdent MH, Fudal I, Ollivier B, Bally P, Grandaubert J, et al. (2013) The dispensable chromosome of *Leptosphaeria maculans* shelters an effector gene conferring avirulence towards *Brassica rapa*. New Phytol 198: 887–898.
11. Parlange F, Daverdin G, Fudal I, Kuhn ML, Balesdent MH, et al. (2009) *Leptosphaeria maculans* avirulence gene *AvrLm4-7* confers a dual recognition specificity by the *Rlm4* and *Rlm7* resistance genes of oilseed rape, and circumvents Rlm4-mediated recognition through a single amino acid change. Mol Microbiol 71: 851–863.
12. Van de Wouw AP, Lowe RGT, Elliott CE, Dubois DJ, Howlett BJ (2013) An avirulence gene, *AvrLmJ1*, from the blackleg fungus, *Leptosphaeria maculans*, confers avirulence to *Brassica juncea* cultivars. Mol Plant Pathol doi:10.1111/mpp.12105.
13. Larkan NJ, Lydiate DJ, Parkin IAP, Nelson MN, Epp DJ, et al. (2013) The *Brassica napus* blackleg resistance gene *LepR3* encodes a receptor-like protein

14. Rubartelli A, Lotze MT (2007) Inside, outside, upside down: damage-associated molecular-pattern molecules (DAMPs) and redox. Trends Immunol 28: 429–436.
15. Baxter L, Tripathy S, Ishaque N, Boot N, Cabral A, et al. (2010) Signatures of Adaptation to Obligate Biotrophy in the *Hyaloperonospora arabidopsidis* Genome. Science 330: 1549–1551.
16. Duplessis S, Cuomo CA, Lin Y, Aerts A, Tisserant E, et al. (2011) Obligate biotrophy features unraveled by the genomic analysis of rust fungi. Proc Natl Acad Sci U S A 108: 9166–9171.
17. Spanu PD, Abbott JC, Amselem J, Burgis TA, Soanes DM, et al. (2010) Genome expansion and gene loss in powdery mildew fungi reveal tradeoffs in extreme parasitism. Science 330: 1543–1546.
18. Ohm RA, Feau N, Henrissat B, Schoch CL, Horwitz BA, et al. (2012) Diverse Lifestyles and Strategies of Plant Pathogenesis Encoded in the Genomes of Eighteen *Dothideomycetes* Fungi. Plos Pathog 8: e1003037.
19. Palmer CL, Skinner W (2002) *Mycosphaerella graminicola*: latent infection, crop devastation and genomics. Mol Plant Pathol 3: 63–70.
20. Solomon PS, Lowe RG, Tan KC, Waters OD, Oliver RP (2006) *Stagonospora nodorum*: cause of *stagonospora nodorum* blotch of wheat. Mol Plant Pathol 7: 147–156.
21. Liu Z, Ellwood SR, Oliver RP, Friesen TL (2011) *Pyrenophora teres*: profile of an increasingly damaging barley pathogen. Mol Plant Pathol 12: 1–19.
22. Westermann AJ, Gorski SA, Vogel J (2012) Dual RNA-seq of pathogen and host. Nat Rev Micro 10: 618–630.
23. Kawahara Y, Oono Y, Kanamori H, Matsumoto T, Itoh T, et al. (2012) Simultaneous RNA-seq analysis of a mixed transcriptome of rice and blast fungus interaction. PLoS One 7: e49423.
24. Yazawa T, Kawahigashi H, Matsumoto T, Mizuno H (2013) Simultaneous Transcriptome Analysis of Sorghum and *Bipolaris sorghicola* by Using RNA-seq in Combination with *De Novo* Transcriptome Assembly. PLoS One 8: e62460.
25. Purwantara A, Salisbury PA, Burton WA, Howlett BJ (1998) Reaction of *Brassica juncea* (Indian mustard) lines to Australian isolates of *Leptosphaeria maculans* under glasshouse and field conditions. Eur J Plant Pathol 104: 895–902.
26. Trapnell C, Pachter L, Salzberg SL (2009) TopHat: discovering splice junctions with RNA-Seq. Bioinformatics 25: 1105–1111.
27. Love CG, Graham NS, Ó Lochlainn S, Bowen HC, May ST, et al. (2010) A *Brassica* Exon Array for Whole-Transcript Gene Expression Profiling. PLoS One 5: e12812.

28. Trapnell C, Williams BA, Pertea G, Mortazavi A, Kwan G, et al. (2010) Transcript assembly and quantification by RNA-Seq reveals unannotated transcripts and isoform switching during cell differentiation. Nat Biotech 28: 511–515.

29. Cantarel BL, Coutinho PM, Rancurel C, Bernard T, Lombard V, et al. (2009) The Carbohydrate-Active EnZymes database (CAZy): an expert resource for Glycogenomics. Nucleic Acids Res 37: D233–D238.

30. Yin Y, Mao X, Yang J, Chen X, Mao F, et al. (2012) dbCAN: a web resource for automated carbohydrate-active enzyme annotation. Nucleic Acids Res 40: W445–W451.

31. Petersen TN, Brunak S, von Heijne G, Nielsen H (2011) SignalP 4.0: discriminating signal peptides from transmembrane regions. Nat Meth 8: 785–786.

32. Eddy SR (2011) Accelerated Profile HMM Searches. Plos Comput Biol 7: e1002195.

33. Punta M, Coggill PC, Eberhardt RY, Mistry J, Tate J, et al. (2012) The Pfam protein families database. Nucleic Acids Res 40: D290–D301.

34. Robinson JT, Thorvaldsdottir H, Winckler W, Guttman M, Lander ES, et al. (2011) Integrative genomics viewer. Nat Biotech 29: 24–26.

35. de Jonge R, van Esse HP, Kombrink A, Shinya T, Desaki Y, et al. (2010) Conserved fungal LysM effector Ecp6 prevents chitin-triggered immunity in plants. Science 329: 953–955.

36. Altschul SF, Gish W, Miller W, Myers EW, Lipman DJ (1990) Basic local alignment search tool. J Mol Biol 215: 403–410.

37. Rawlings ND, Barrett AJ, Bateman A (2012) MEROPS: the database of proteolytic enzymes, their substrates and inhibitors. Nucleic Acids Res 40: D343–D350.

38. Epple P, Apel K, Bohlmann H (1997) ESTs reveal a multigene family for plant defensins in Arabidopsis thaliana. Febs Lett 400: 168–172.

39. Thimm O, Bläsing O, Gibon Y, Nagel A, Meyer S, et al. (2004) Mapman: a user-driven tool to display genomics data sets onto diagrams of metabolic pathways and other biological processes. Plant J 37: 914–939.

40. Eckert M, Maguire K, Urban M, Foster S, Fitt B, et al. (2005) Agrobacterium tumefaciens-mediated transformation of Leptosphaeria spp. and Oculimacula spp. with the reef coral gene DsRed and the jellyfish gene gfp. Fems Microbiol Lett 253: 67–74.

41. Zablackis E, Huang J, Muller B, Darvill AG, Albersheim P (1995) Characterization of the cell-wall polysaccharides of Arabidopsis thaliana leaves. Plant Physiol 107: 1129–1138.

42. Ipcho SV, Hane JK, Antoni EA, Ahren D, Henrissat B, et al. (2012) Transcriptome analysis of Stagonospora nodorum: gene models, effectors, metabolism and pantothenate dispensability. Mol Plant Pathol 13: 531–545.

43. Levasseur A, Drula E, Lombard V, Coutinho PM, Henrissat B (2013) Expansion of the enzymatic repertoire of the CAZy database to integrate auxiliary redox enzymes. Biotechnology for biofuels 6: 41.

44. Brunner PC, Torriani SFF, Croll D, Stukenbrock EH, McDonald BA (2013) Coevolution and life cycle specialization of plant cell wall degrading enzymes in a hemibiotrophic pathogen. Mol Biol Evol 30: 1337–1347.

45. Feng J, Zhang H, Strelkov SE, Hwang SF (2014) The LmSNF1 gene is required for pathogenicity in the canola blackleg pathogen Leptosphaeria maculans. PLoS One 9: e92503.

46. O'Connell RJ, Thon MR, Hacquard S, Amyotte SG, Kleemann J, et al. (2012) Lifestyle transitions in plant pathogenic Colletotrichum fungi deciphered by genome and transcriptome analyses. Nat Genet.

47. Rypniewski WR, Hastrup S, Betzel C, Dauter M, Dauter Z, et al. (1993) The sequence and X-ray structure of the trypsin from Fusarium oxysporum. Protein Eng 6: 341–348.

48. Showalter AM, Keppler B, Lichtenberg J, Gu D, Welch LR (2010) A bioinformatics approach to the identification, classification, and analysis of hydroxyproline-rich glycoproteins. Plant Physiol 153: 485–513.

49. Persson M, Staal J, Oide S, Dixelius C (2009) Layers of defense responses to Leptosphaeria maculans below the RLM1- and camalexin-dependent resistances. New Phytol 182: 470–482.

50. Šašek V, Nováková M, Jindřichová B, Bóka K, Valentová O, et al. (2012) Recognition of Avirulence Gene AvrLm1 from Hemibiotrophic Ascomycete Leptosphaeria maculans Triggers Salicylic Acid and Ethylene Signaling in Brassica napus. Mol Plant Microbe Interact 25: 1238–1250.

51. Vogel JP (2002) PMR6, a Pectate Lyase-Like Gene Required for Powdery Mildew Susceptibility in Arabidopsis. Plant Cell 14: 2095–2106.

52. Grandaubert J, Lowe RGT, Soyer JL, Schoch CL, Fudal I, et al. (2014) Transposable Element-assisted evolution and adaptation to host plant within the Leptosphaeria maculans-Leptosphaeria biglobosa species complex of fungal pathogens. Biomed Cent Genomics. In press.

12

Analysis of Papaya Cell Wall-Related Genes during Fruit Ripening Indicates a Central Role of Polygalacturonases during Pulp Softening

João Paulo Fabi[1,3]*, Sabrina Garcia Broetto[1], Sarah Lígia Garcia Leme da Silva[1], Silin Zhong[2], Franco Maria Lajolo[1,3], João Roberto Oliveira do Nascimento[1,3]

1 Department of Food Science and Experimental Nutrition, FCF, University of São Paulo, São Paulo, São Paulo, Brazil, **2** State Key Laboratory of Agrobiotechnology, School of Life Sciences, The Chinese University of Hong Kong, Hong Kong, China, **3** University of São Paulo, – NAPAN – Food and Nutrition Research Center, São Paulo, São Paulo, Brazil

Abstract

Papaya (*Carica papaya* L.) is a climacteric fleshy fruit that undergoes dramatic changes during ripening, most noticeably a severe pulp softening. However, little is known regarding the genetics of the cell wall metabolism in papayas. The present work describes the identification and characterization of genes related to pulp softening. We used gene expression profiling to analyze the correlations and co-expression networks of cell wall-related genes, and the results suggest that papaya pulp softening is accomplished by the interactions of multiple glycoside hydrolases. The polygalacturonase *cpPG1* appeared to play a central role in the network and was further studied. The transient expression of *cpPG1* in papaya results in pulp softening and leaf necrosis in the absence of ethylene action and confirms its role in papaya fruit ripening.

Editor: Sonia Osorio-Algar, University of Malaga-Consejo Superior de Investigaciones Científicas, Spain

Funding: This research was financially supported by grants #2002/12452-9 and #2008/54476-8, scholarships to JPF (#2003/00932-9 and #2007/56515-8), SGB (#2009/53329-4) and SLGLS (#2008/58653-1), São Paulo Research Foundation (FAPESP). The funders had no role in study design, data collection and analysis, decision to publish, or preparation of the manuscript.

Competing Interests: The authors have declared that no competing interests exist.

* Email: jpfabi@usp.br

Introduction

Papaya (*Carica papaya* L.) is a fleshy fruit that undergoes severe pulp softening during ripening. Although softening increases the sensory and nutritional properties of the fruit, it also contributes to post-harvest deterioration and product losses because the fruit becomes susceptible to physical injury and mold growth [1].

The softening of fleshy fruits during ripening is characterized by the degradation of the plant's cell wall, which is mainly formed by polysaccharides organized in the pectic (amorphous) and the hemicellulosic/cellulosic (crystalline) fractions and the loss of the rigidity of cell wall structure is attributed to the solubilization of both fractions [2]. This solubilization is achieved by the interactive action of several enzymes on determined polysaccharides, such as polygalacturonases (PGs), pectinesterases (PMEs) and pectate lyases (PLs) on homogalacturonans, arabinofuranosidases (ARFs) and galactanases (GALs) on heterogalacturonans and xyloglucan endotransglucosylases/hydrolases (XTHs), endoxylanases (EXYs) and cellulases (CELLs) on hemicelluloses [3,4].

In the case of papaya fruit, the investigation of polysaccharide changes during ripening has revealed that the disassembly of the cell wall is predominantly caused by the degradation and solubilization of pectin [5]. In this regard, polygalacturonase (*cpPG1*) [6] and β-galactosidase (*cp_b-GAL*) [7] play a central role in pectin solubilization during papaya ripening. However, other studies have provided evidence that the solubilization of the hemicellulosic fraction by endoxylanases (EXY) is a major contributor to the softening of papaya fruit [8]. This apparent contradiction between the experimental evidence provided by different studies suggests that the disassembly of the cell wall in papaya fruit is a more complex process that involves several enzymes and various cell wall components. In fact, the investigation of several genes or proteins simultaneously may provide better insight on the physiological process, as revealed by data from an XSpecies microarray of papaya ripening that indicated that several cell wall-related genes exhibit similarities to the gene expression profiles observed during the transition of 5- to 11-day-old hypocotyls in *Arabidopsis thaliana* [9,10]. Moreover, evaluating the correlations and co-expression of a set of genes may provide clues to the interactive role of those genes.

Therefore, investigating the network of cell wall-related genes, specifically those encoding depolymerization or degradation enzymes, may further elucidate the mechanism of cell wall disassembly during papaya ripening. Thus, this study describes the qPCR analysis of the mRNA levels of some genes of enzymes that act on homogalacturonans (polygalacturonases, pectinesterases and pectate lyases), heterogalacturonans (arabinofuranosidases and galactanases) and hemicellulose (xyloglucan endotransglucosylases/hydrolases and xylanases), which covers the diversity of polysaccharides from the amorphous and crystalline fractions of cell wall, and the correlations and co-expression network of 25 cell wall-related genes during papaya ripening. In addition, the protein levels and enzymatic activity of endopolygalacturonase and the overexpression of the *cpPG1* gene were measured in agroinfiltrated papaya fruit pulp and leaves.

Results

Cloning cell wall-related genes in papaya fruit

The fruit used for the cloning and expression analysis presented typical softening resulting from the extensive solubilization of cell wall components, as indicated by the climacteric decrease in pulp firmness [1] and cell wall thinning (**Figure 1**).

A BLAST search of the papaya genome allowed the cloning of *cpPG1* (**FJ007644** – Fabi et al., 2009), *cpPG2* (**GQ479794**), *cpPG3* (**GQ479795**), *cpPG4* (**GQ479796**), *cpARF* (**GQ479793**), the previously reported *cp_b-GAL* (**AF064786** – [7]) and *cpPL* (**DQ660903**). As summarized in **Table 1**, the properties of the encoded proteins are similar to those from other plants. The four PGs were differentially expressed and peaked in the climacteric papaya (3DAH) (**Figure 2**). *CpEXY1* was also differentially expressed during ripening, but in a linear increasing manner. Papaya *cp_b-GAL* was also up-regulated, although it was brought to the initial levels after second day after harvesting (2DAH), whereas both *cpPL* and *cpARF* were down-regulated. When the amount of the transcripts of the up-regulated genes was compared, *cpPG1* and *cp_b-GAL* were determined to be the most abundant (**Figure S1 in File S1**), and the amount of *cpPG1* exceeded that of the less abundant *cpPG4* by six orders of magnitude.

Gene expression and correlation analysis

The expression pattern of a collection of 25 cell wall-related genes that were previously described as ripening-related ones [11,12,10] was analyzed for co-expression during ripening. Among the 600 possible correlations, 300 had *p*-values less than 5% and were further analyzed (**Figure 3**), which revealed that PGs, galactanases, PMEs and PLs significantly impact the network assembly (**Figure 4**). The PGs were positively correlated to galactanases, xylanases and, to a lesser degree, expansins and negatively correlated to a second group of genes encompassing PMEs, PLs and ARF. Because the expression of papaya cell wall disassembling genes was similar to *Arabidopsis* from 5 to 11-day-old hypocotyls growth [9,10], a new comparison of papaya data was done with newly *Arabidopsis* data [13]. The comparison (**Table S1 in File S1**) showed changes in the expression of 11 out of 16 genes, which were similar between the ripening fruit and the growing plant. The genes that were most significantly affected in both developmental processes were *cpPG1* and its homologues.

Transient expression of papaya *cpPG1*

We identified four papaya PG genes that have AT-rich intronic regions with two specific patterns, one for *cpPG1* and *cpPG4* and another for *cpPG2* and *cpPG3* (**Figure S2 in File S1**). Notably, the proteins encoded by *cpPG1*/*cpPG4* were separated in a distinct clade from *cpPG2*/*cpPG3* when the PGs from diverse plants were aligned (**Figure S3 in File S1**). The expression of *cpPG1* in prokaryotic cells resulted in a recombinant product with the expected size and identity of the predicted cpPG1 protein, as revealed by MS sequencing (**Figure S4 in File S1**). This protein was used to produce polyclonal antibodies utilized in western blotting (**Figure S5 in File S1**), which showed that the amount of PG protein was correlated to that of the *cpPG1* transcript and its enzymatic activity and with pulp softening (**Figure 5**).

To verify the role played by this PG gene in pulp softening, 1-MCP-treated fruits were transiently transformed to overexpress *cpPG1*. After three days of agroinfiltration, GFP signals were observed in both the control and *cpPG1-GFP*-transformed fruits, which showed transient cpPG1-GFP protein delivered to the extracellular space (**Figure 6A and B**). The transcript analysis of 15 cell wall-related genes showed that *cpPG1* was up-regulated after treatment, and *cpPG2* also appeared to be affected (**Figure 6C**). The absolute quantitation of the transcript revealed a 20% excess of *cpPG1* mRNA (**Figure 6D**), which was correlated with a two-fold increase in PG activity and pulp softening. Papaya leaves that were agroinfiltrated with the same construct to overexpress *cpPG1* in 1-MCP-treated fruit exhibited a 45% increase in *cpPG1* expression, demonstrated increased activity and survived for only two days, whereas the control leaves survived for six days (**Figure 7**).

Discussion

Pulp softening is a fast and marked change that occurs in ripening papayas and is denoted by the severe loss of firmness after just a few days. The thinning of the cell wall, which results from the substantial action of enzymes on the polysaccharide structure, causes a loss in pulp tissue rigidity. Considering its complexity, which is derived from various polysaccharide chains and the arrangement of pectic and hemicellulosic fractions, several enzymes are expected to act as a result of the interactions of co-expressed cell wall-related genes.

According to the evaluation of the gene expression profiles of various cell wall enzymes, the depolymerizing capacity is predominately attributed to the increased expression of PGs, β-gal and endoxylanase. This result suggests that the massive solubilization of pectin may be the predominant event, although a minor contribution of the hemicellulose component should not be disregarded, as indicated by the increase in endoxylanase. Notably, the gene encoding PL, an enzyme responsible for pectin depolymerization through a β-elimination reaction, was down-regulated. In fact, WGCNA analysis revealed both positive and negative correlations among the expression patterns of the genes of pectin-acting enzymes, in contrast from what was observed for strawberries, in which a concomitant action of PL and PG might be essential for pulp softening [14,15]. In addition to PL, the co-expression of PGs was also negatively correlated to that of PMEs. Therefore, the action of papaya PGs does not require the simultaneous removal of pectin methyl groups, which would be consistent with the stable amount of methoxylated pectin during ripening [1,12,10,5].

Another interesting result that emerged from the WGCNA analysis was the co-expression of galactanases, which may act on the heterogalactans of pectins. Alpha-galactosidase genes only marginally contribute to pulp softening [16] and were strongly correlated to PG expression. In contrast, an up-regulated β-galactosidase that is partially responsible for papaya softening [17] demonstrated increased co-expression with genes related to pectin and hemicellulose modifications, such as *GAE6*, *ARF1*, *PL2*, *PL3* and *XTH* [18], [19,20,21,22]. This finding might indicate that *cp_b-GAL* plays a ubiquitous role in papaya plant cell wall mobilization that is not limited to pectin depolymerization during fruit ripening.

The data from the gene expression and co-expression network analysis suggest a cooperative network of cell wall-related genes during papaya pulp softening that is mainly composed of endoPGs, which is consistent with the massive pectin solubilization [5] and the release of uronic acid oligosaccharides [23]. Softening

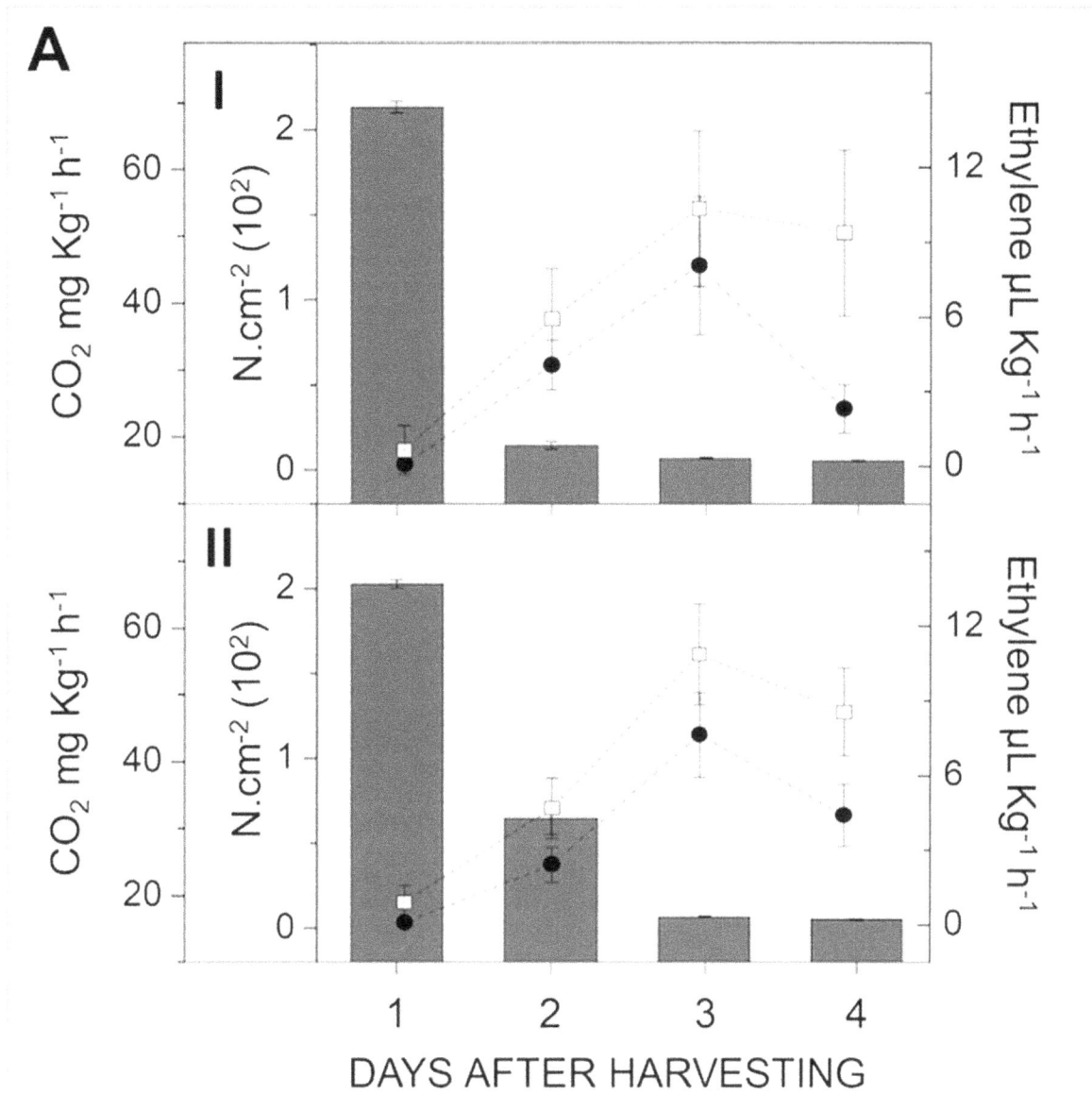

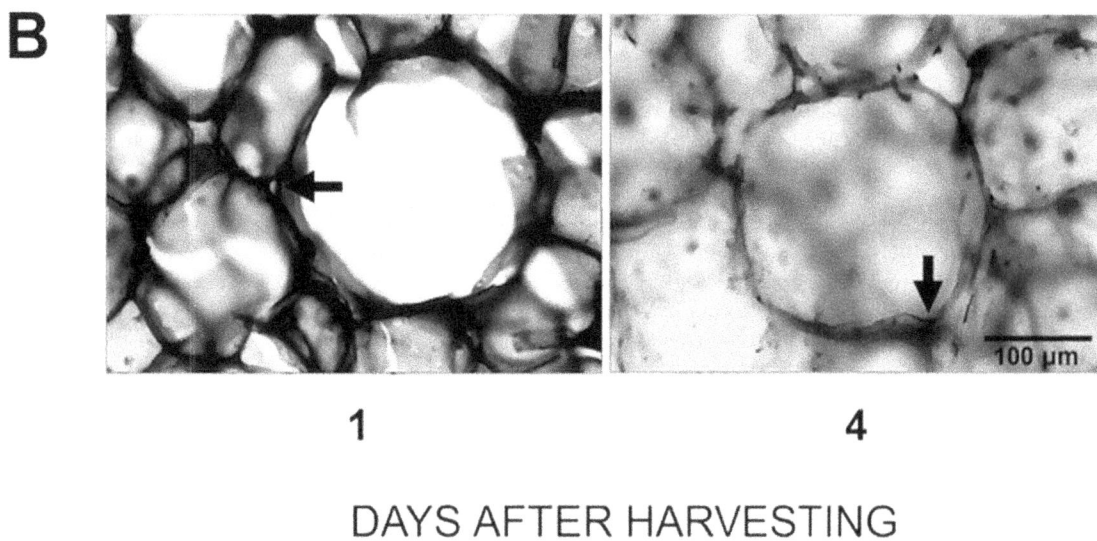

DAYS AFTER HARVESTING

Figure 1. Ripening of papaya fruit. A) The amount of CO_2 produced by respiration (Open squares – mg Kg^{-1} h^{-1}), the production of endogenous ethylene (Black circles - μL Kg^{-1} h^{-1}), and the pulp firmness (Bars – $N.cm^{-2}$) were monitored throughout ripening. Error bars indicate SDs of the mean ($n = 12$) for each sampling (**I** and **II**). **B**) Sections of frozen papaya pulp (12–15 μm) in two different stages (unripe and ripe – 1st and 4th day after harvesting, respectively) were stained with 0.05% of toluidine blue and documented using light microscope. Different colorations indicate cell wall thinning. Arrows indicate the junction zones of cell walls. Bar indicates the scale. The image is a representation of, at least, a triplicate of experiment.

dependence on massive PG action has already been seen in apple fruit, as denoted by the inhibition of softening in transgenic lines with PG suppression [24]. The up-regulation of *cp_EXY1* corroborates previous data reported by Manenoi and Paull [8], and the encoded enzyme accounts for the hydrolysis of β-1,4-D-xylosyl that is bounded to polysaccharides in the cell wall matrix, which are mainly hemicelluloses. However, because cellulase expression decreases during ripening [10], *cp_EXY1* would contribute to the minor disassembly of hemicellulose fractions by the release of heteroglucan chains but not the depolymerization of the cellulose backbone, which is mostly found in a crystalline phase. Therefore, the action of pectinases on less compacted pectin in the amorphous phase would have a more significant effect on cell wall disassembly during ripening.

In addition to a previous observation of the similarities between ripening papayas and *Arabidopsis* hypocotyl gene expression [10], new data from *Arabidopsis* development (young and mature plants – [13] – **Table S1 in File S1**) indicate similar co-expression patterns for several cell wall-related papaya genes. In this regard, the *cpPG1* orthologue from *Arabidopsis* (**AT3G59850**) appears to play a central role in cell wall disassembly because the transgenic expression of this gene and the co-expression of expansin (*EXP2* - AT2G39700 - [25]) markedly changed the pulp firmness and increased the tomato susceptibility to *Botrytis cinerea* [26].

The occurrence of different endoPGs in various fleshy fruit organs has previously been reported [27,28,29] and suggests a degree of redundancy or specialization in pectin depolymerization. In this regard, *cpPG1* appears to play a predominant role in papaya softening as it occurs for apple softening [30], [24]. Thus, the more abundant transcript would account for most of the enzymatic activity observed in the fruit pulp, although it was not so well adjusted to the protein peak abundance at the third day after harvest, probably as a result of the interference of other PG proteins, such as cpPG3 and cpPG4, in the assays. Furthermore, since the performed activity assay measures reducing end formation, we cannot exclude a minor contribution of exopoly-galacturonases. In relation to *cpPG2*, that gene was separated from *cpPG1* in a distinct clade, and the encoded protein had a Lys_{346} catalytic residue instead of a *Glu*, which allowed speculation that this product may act differently on pectin depolymerization (**Table S2 and Figure S6 in File S1**).

The transient transformation of papaya demonstrated the effect of *cpPG1* on pulp softening. Treatment with 1-MCP precluded ethylene synthesis and the triggering of PG genes. Fruit agroinfiltrated with pGBPGWG_cpPG1 transiently expressed cpPG1-GFP proteins that were delivered to the extracellular compartment, and the pulp of the fruit softened compared with those that received the empty vector. Notably, the up-regulation of *cpPG2* may reveal some level of responsiveness or dependence on *cpPG1* and insensitivity to ethylene because the gene was expressed during late ripening in non-transformed fruit (**Figure 2**). At the same time, it is possible that other ethylene-independent cell wall-related genes had contributed to partial softening of the pulp. The results of the transient transformation of papaya leaves with pGBPGWG_cpPG1 were similar to those of tobacco leaves that overexpress polygalacturonases [31], which

cause cell death and leaf necrosis (**Figure 7A**). Moreover, it is likely that the overexpression of the *cpPG1* gene reduced cell wall adhesion causing disruption of leaf organization as was observed in apple tree leaves constitutively expressing a fruit-specific PG [30]. The above-mentioned results demonstrate that the abundantly expressed *cpPG1* gene may exert profound changes in the papaya cell wall structure and play a central role in pectin disassembly during fruit ripening. The softening of fleshy fruit, particularly pectin-rich fruit, may now be studied through a broader approach by considering a key group of cell wall-related genes and their interactions with each other and other cell wall-related genes, as was demonstrated with papayas.

Materials and Methods

Plant material and ripening parameters

Papayas (*Carica papaya* L. cv. Golden) were obtained from a producer in Linhares City, Espírito Santo, Brazil. The fruits were harvested with up to 25% yellow peel (150 days post-anthesis). The fruits were stored in 240 L chambers with controlled temperature and humidity (22°C±0.1°C and 95%, respectively) for seven days. Daily analysis was performed on six fruits (two distinguishable biological replicates from two different seasons). The CO_2, ethylene and pulp firmness were measured according to Fabi et al. [1], and the remaining pulp was N_2-frozen, pooled and stored at −80°C. Sections of frozen tissue (12–15 μm) were cut using a refrigerated microtome, stained with toluidine blue (0.05% in 0.1 M phosphate buffer, pH 6.8) and visualized using a light microscope (Zeiss, Germany).

Bioinformatics and sequence cloning and analysis

The papaya genomic sequences were identified by comparing a partial papaya genome [32] with polygalacturonases from diverse fleshy fruits and *Arabidopsis thaliana* using Blast tools and primers that were designed to flank the coding regions (**Table S3 in File S1**). Papaya pectate lyase (*cpPL1*) and β-galactosidase (*cp_b-GAL*) genes were downloaded from the GenBank database (**DQ660903** and **AF064786**, respectively). For papaya α-L-arabinofuranosi-dase, a pair of degenerated primers was used to identify part of the coding region (**Table S3 in File S1**). The total RNA was extracted according to Fabi et al. [6], and first-strand cDNAs were synthesized from 1 μg of total RNA using an Improm-II Reverse Transcription System kit (Promega, Madison, WI, USA) and oligo-dT primers. The coding regions were PCR-amplified using KOD Hot Start DNA Polymerase (Merck KGaA, Darmstadt, Germany), and sequences with a high GC content (*cpPG2* and *cpPG3*) were PCR-amplified using KAPA HiFi DNA Polymerase (Kappa Biosystems, Boston, MA, USA). The PCR fragments were digested with *Xho*I and ligated on a pET-45b(+) (Merck KGaA, Darmstadt, Germany) plasmid. After cloning and sequencing, the sequences were analyzed with the TargetP 1.1 tool [33] and the SignalP tool [34]. Finally, the DNA was extracted according to Fabi et al. [6], PCR-amplified as described above, cloned and genomically sequenced.

Table 1. Cell wall-related proteins identified in papaya pulp and their probable biological properties.

Gene name	Length[†]	Accession Number	Chromosomes	TAIR	Cleavage Site	Molecular Weight (KDa)[†]	Isoelectric Point[††]	Blast Best Match
cpPG1	397	ACH82233	Un contig 23855 (ABIM01023821)	AT3G59850	Ala^{29}	38.98	9.14	Polygalacturonase [P. communis] CAH18935
cpPG2	494	ACV85695	LG4 contig 16510 (ABIM01016487)	AT3G57510	Ser^{33}	51.03	5.21	Polygalacturonase A [A. chinensis] AAF71160
cpPG3	444	ACV85696	LG4 contig 19176 (ABIM01019150)	AT3G07970	Gly^{31}	45.78	9.22	Polygalacturonase QRT2 [A. thaliana] NP_187454
cpPG4	389	ACV85697	LG1 contig 4651 (ABIM01004644)	AT2G43870	Ala^{31}	38.67	8.54	Polygalacturonase 3 [P. communis] BAF42034
cpARF	710	ACV85694	LG4 contig 9013 (ABIM01009003)	AT3G10740	Ala^{43}	74.16	4.94	α-L-arabinofuranosidase [M. domestica] AAP97437
cp-b-GAL	721	AAC77377	LG7 contig 16621 (ABIM01016598)	AT3G13750	Ala^{21}	78.49	8.26	β-galactosidase 1 [P. persica] ABV32546
cpPL	369	ABG66730	LG6 contig 2998 (ABIM01002995)	AT3G07010	Ser^{19}	40.52	9.14	Pectate lyase [P. persica] BAF43573

[†]Amino acids numbers;
[††]The Molecular Weight and isoelectric points were estimated using the mature proteins.

Quantitative analysis of gene expression by real time-PCR (qPCR)

Gene expression analyses were performed according to [10] and following the 'Minimum Information for Publication of Quantitative Real-Time PCR Experiments – MIQE' [35]. The primer sequences are shown in **Table S4 in File S1**. The gene expression of a previously identified papaya endoxylanase (*cpEXY1* - **AY138968**; [8]) was also analyzed. The actin gene (*ACT*) located on chromosome LG9 contig 1059 (GenBank accession no. ABIM01001059) and the elongation factor 1-alpha gene (*EF1*) located on chromosome LG9 (GenBank accession no. ABIM0101268) were used as internal controls. To evaluate the agroinfiltration experiments using the pGBPGWG vector, a pair of primers enclosing the GFP gene was synthesized (**Table S4 in File S1**). For primer testing and identity confirmation, all of the fragments from the PCR reactions were cloned and sequenced. Real-time PCR was performed using a four-channel Rotor-Gene 3000 multiplexing system (Corbett Research, Sydney, Australia). The non-template controls (NTCs) and melting curve analyses of the amplicons were monitored for all experiments. The threshold cycle (Ct) values (four technical replicates and two biological replicates) were averaged using the Rotor-Gene 3000 software, and the quantification was performed using the relative standard curve method [36]. The results of the standard curves calculations are shown in **Table S5 in File S1**. To assess the gene copy numbers, the corresponding cloned coding regions were used to construct calibration curves (**Table S6 in File S1**). The results are expressed as gene copy numbers/microgram of RNA for each PCR reaction. All of the data were analyzed by one-way ANOVA, and the means were compared using Tukey's test. The statistical analyses were performed using OriginPro version 8 software (OriginLab).

Comparative biology and co-expression patterns

ClustalW phylogenetic trees [37] were calculated from the alignment of the corresponding proteins with proteins from fleshy fruits and *A. thaliana*. To search for co-expression patterns between genes related to papaya cell wall metabolism, the $\log_2$ values of the gene expression data from 25 genes were analyzed using the weighted gene co-expression network analysis (WGCNA) in the R package [38]. In summary, pair-wise gene (Pearson) correlations were calculated using gene expression data from genes identified in this study and three previous studies [10,11,12]. The correlations and associated *p*-values were calculated using the WGCNA R package and the "corAndPvalue" module. The GENE-E software was loaded with values, and corresponding heat map figures were generated (http://www.broadinstitute.org/cancer/software/GENE-E). A weighted adjacency matrix was then constructed by the WGCNA package using default parameters from the "adjacency" module and a soft-threshold parameter equal to 3, which forces the program to emphasize stronger correlations over weaker ones [39]. The obtained co-expression networks were visualized using Cytoscape [40].

Heterologous Protein Expression, Purification and Sequencing

Protein expression was performed by transforming Rosetta 2 (DE3) pLysS (Novagen) *E. coli* with a pET-45b(+) construct containing *cpPG*. Bacteria were grown in LB medium (Cm$^+$/Amp$^+$), and inductions were performed with isopropyl β-D-1-thiogalactopyranoside (IPTG) when the $OD_{600\ nm}$ reached 0.6. A control experiment was performed using a closed pET-45b(+) plasmid. Images of polyacrylamide gels were captured using a

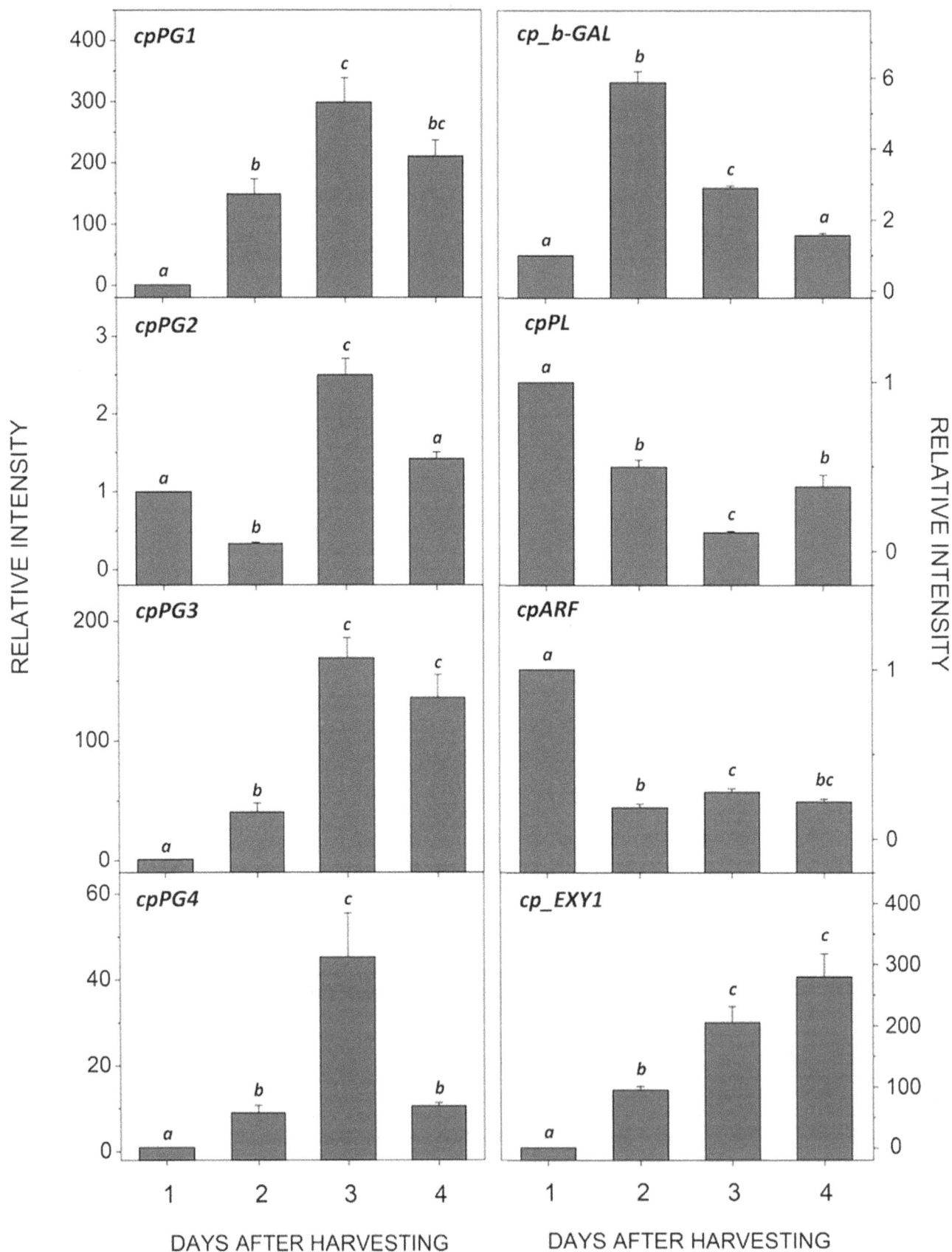

Figure 2. Expression of cell wall-related genes during papaya ripening. Real-time PCR (qPCR) was used to analyze the mRNA levels of various genes during ripening. The column heights indicate the relative mRNA abundance; the expression values for unripened fruit one day after harvest were set to 1. The error bars on each column indicate the SD of four technical replicates from samplings I and II. The different letters represent samples that were significantly different from those collected on other days post-harvest (within the same gene), as determined by one-way ANOVA and Tukey's test ($\alpha < 0.05$, $n = 4$).

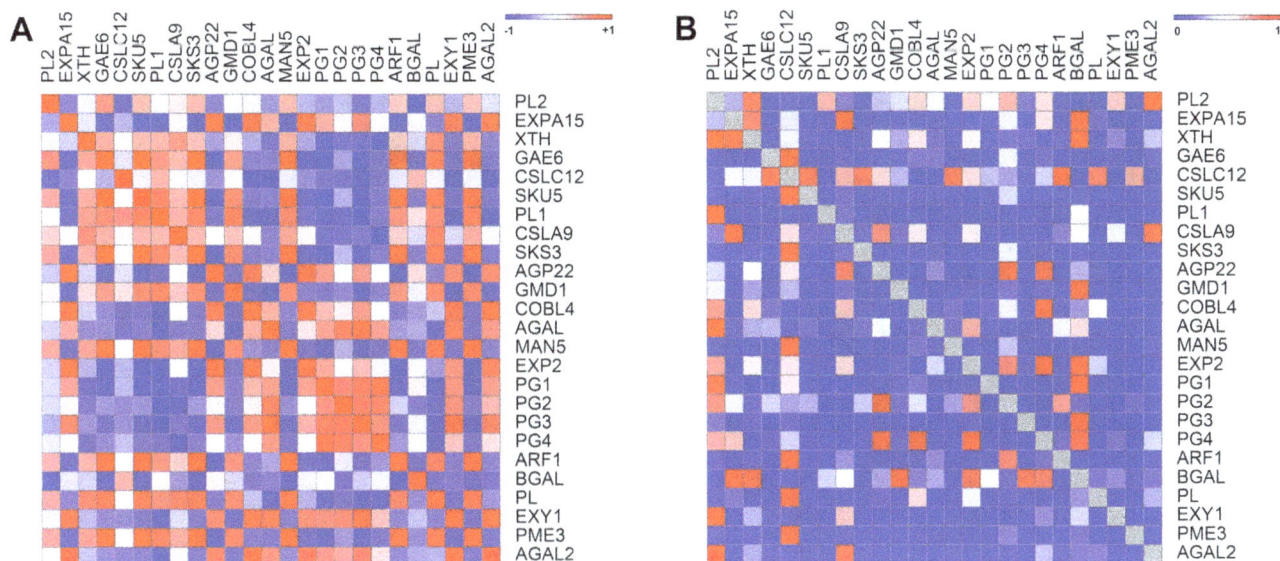

Figure 3. Pearson correlations and associated *p*-values of papaya cell wall-related genes. The R package for weighted correlation network analysis (weighted gene co-expression network analysis - WGCNA) was used to calculate the Pearson correlation (**Figure A**) and the corresponding *p*-values for the 25 papaya cell wall-related genes that were identified by our group (**Figure B**). From the 600 possible gene expression correlations, 300 had *p*-values less than or equal to 0.05 and were further analyzed. **Figure 1 A**: The heat map is described as positive values set to red color and negative values set to blue color. **Figure B**: The heat map is described as values near to one set to red color and values near to zero set to blue color. The twenty-five genes are described in **Table S1 in File S1**.

Versa Doc Gel Imaging System (Bio-Rad, CA, USA). The molecular masses of corresponding bands were estimated by comparing them with standard bands of molecular weight proteins using the Quantity One software (version 4.6.7) (Bio-Rad, CA, USA). The pelleted bacterial suspensions were disrupted using the commercial BugBuster Protein Extraction Reagent (Merck KGaA, Darmstadt, Germany) supplemented with the protease inhibitor cocktail ProteoBlock (Fermentas, Thermo Fisher Scientific Inc., Vilnius, Lithuania). The recombinant proteins were extracted from inclusion bodies using 8 M urea with 24 hours of agitation and purified in a nickel column (Protino NI-TED 1000 - Macherey-Nagel GmbH & Co KG, Düren, Germany) according to the manufacturer's instructions. The samples were analyzed by SDS-PAGE on a 12.5% acrylamide gel [41]. The bands of interest were excised from the gels, and the digestion and protein sequencing were performed according to [42].

Production of Polyclonal Antibodies

After confirming the identities of the recombinant proteins, 1 mg of the purified protein was separated using SDS-PAGE. The band was excised and the rabbits were immunized according to [43].

Western Blotting and Enzymatic assays

The total protein from the pulp of papayas was extracted according to Carpentier et al. [44] by adding ProteoBlock (Fermentas) protease inhibitor. The protein concentrations were determined using a 2-D Quant Kit (GE Healthcare, Piscataway, NJ, USA) according to the manufacturer's instructions. Twenty micrograms of total protein were separated by SDS-PAGE and transferred to nitrocellulose membranes (Hybond-ECL, GE Healthcare, Piscataway, NJ, USA). The membranes were incubated with antiserums and secondary anti-rabbit ECL Plex Goat-the-Rabbit Cy5 antibody (GE Healthcare, Piscataway, NJ, USA) according to the manufacturer's instructions. Monoclonal anti-

actin for plants (Sigma, #A0480, St. Louis, MO, USA - 1:5000 dilution) with a secondary anti-mouse ECL Plex Goat-a-Mouse Cy3 antibody (GE Healthcare, Piscataway, NJ, USA) was used as a control experiment. Images of the membranes were captured with a Versa Doc Gel Imaging System (Bio-Rad) using the corresponding channels for Cy5 and Cy3 detection (multiplexing). Images were taken after 120 seconds of exposure for Cy5 and 30 seconds of exposure for Cy3, and the negative image files were obtained with the Quantity One (Bio-Rad) software. The bands observed were compared with the standard molecular weight ECL Plex Fluorescent Rainbow Marker (GE Healthcare). The proteins for enzymatic assays were extracted as previously described [6]. The polygalacturonase activity was assayed according to Fabi et al. [6].

Transient gene expression of *cpPG1* in papaya fruits and leaves

The *cpPG1* gene was transiently expressed in papaya fruits and leaves by agroinfiltration. The binary vector used in the experiments was pGBPGWG expressing green fluorescent protein (GFP) under the regulation of the 35S promoter (GenBank accession number **AM884372**), and the *cpPG1* ORF was cloned into the vector by recombination, as previously described by Zhong et al. [45]. The agroinfiltration of the papaya fruits was performed according to Spolaori et al. [46]. *Agrobacterium tumefaciens* GV3101 containing a pSoup plasmid (Tet⁺/Gent⁺) was transformed with a binary vector that enclosed *cpPG1* ORF (pGBPGWG_cpPG1 – Cm⁺/Kan⁺) and used in both experiments. Moreover, bacterial suspensions were co-infiltrated (1:1) with *A. tumefaciens* GV2260 that was transformed with the 35S: p19 binary vector (Rif⁺/Kan⁺) to decrease the post-transcriptional gene silencing of exogenous *cpPG1* [47].

One day after harvesting, the papaya fruits were treated with 100 ppb of 1-methylcyclopropene (1-MCP) for 12 hours to reduce *cpPG1* endogenous gene expression, as previously described [1].

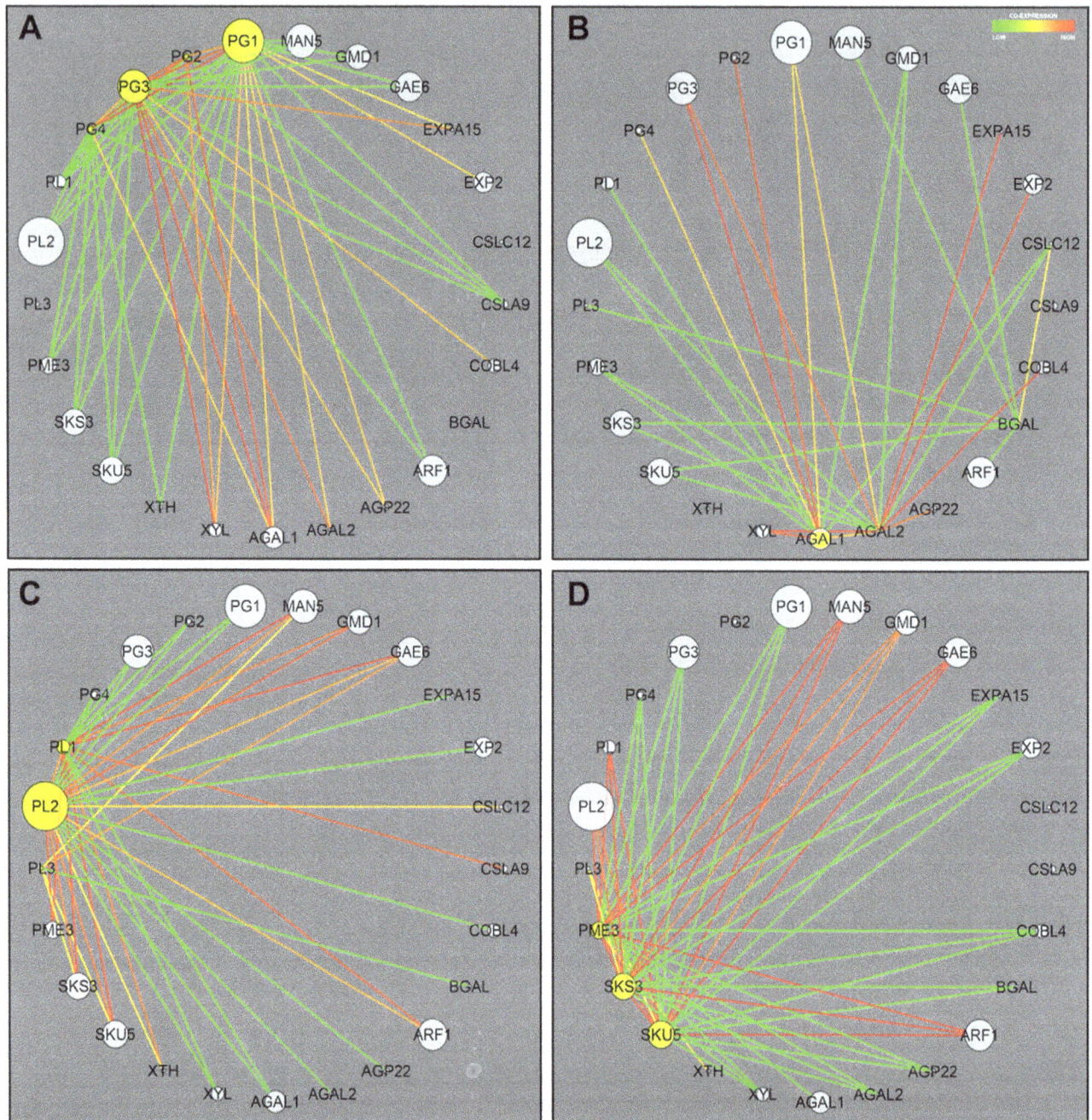

Figure 4. Co-expression network of papaya cell wall-related genes given by the weighted gene co-expression network analysis (WGCNA). The R package of the weighted gene co-expression network analysis (WGCNA) program was used to produce a gene association network for 25 papaya cell wall-related genes. The size of the circle indicates the weight of the gene on the network. In **Figures A, B, C** and **D**, the genes considered for the network assembly are highlighted with yellow backgrounds. **Figure A** describes the network for the papaya *PG* genes. **Figure B** describes the network for the papaya *GAL* genes. **Figure C** describes the network for the papaya *PME* genes. **Figure D** describes the network for the papaya *PL* genes. The color lines are described as positive co-expression values set to red color and negative co-expression values set to green color. The twenty-five genes are described in **Table S1 in File S1**.

After treatment, three fruits $(n = 3)$ were agroinfiltrated using sterile syringes and a short needle (1 mL bacteria suspension per 4 cm^2). The fruits were allowed to ripen at 22°C for a 16-h photoperiod. A previous test using a mixture of *Agrobacterium* and methylene blue was conducted to calculate the extent of agroinfiltration. Three days post-infiltration, the fruits were sliced and photographed using a Versa Doc Gel Imaging System (Bio-

Rad) with parameters for GFP detection. The pulp firmness was measured according to Fabi et al. [1]. The remaining slices were N$_2$-frozen for future gene expression quantification. A control experiment $(n = 3)$ was similarly conducted using the pGBPGWG-closed vector. Sections of pulp were manually cut, stained with FM4-64 vital dye and analyzed using a confocal laser-scanning

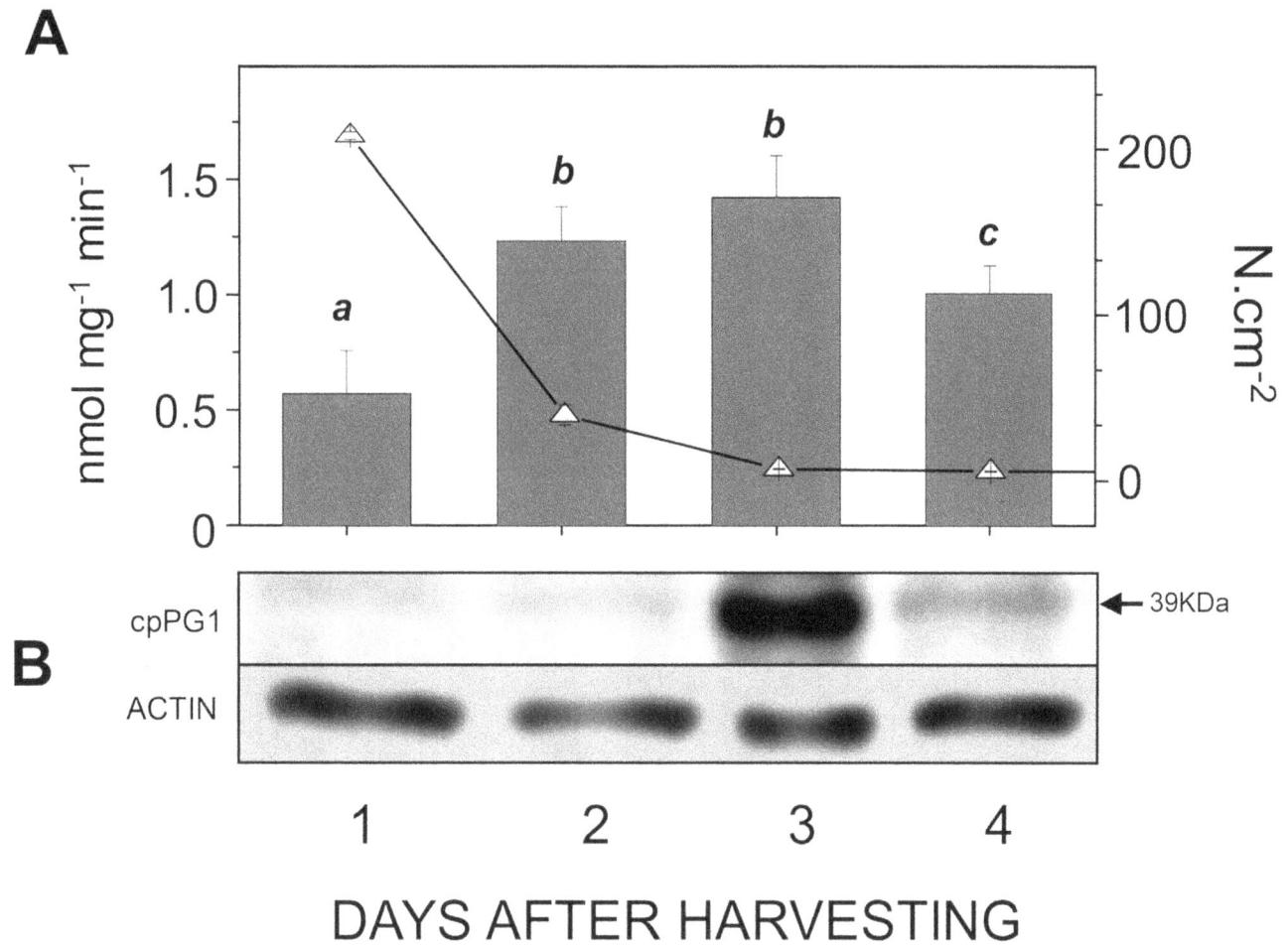

Figure 5. Correlation between pulp softening, enzymatic activity and cpPG1 protein expression during papaya ripening. A) The pulp firmness (open triangles) was monitored during the ripening process using a texturometer (Fabi et al., 2007), and the measurements are given by N.cm^{-2}. PG enzymatic activities was done according to Fabi et al. (2009). The error bars indicate the SDs of the mean. The different letters represent samples that were significantly different from those collected on other days post-harvest, as determined by one-way ANOVA and Tukey's test ($\alpha <$ 0.05, $n = 4$). **B)** Western blotting experiment was done using the monoclonal anti-actin for plants as a control experiment (ACTIN), and the images are representative of at least a triplicate experiment. The arrow indicates the molecular weight of the mature protein cpPG1 with the corresponding molecular weight.

microscope (Zeiss, Jena, Thuringia, Germany, LSM 510). The images were processed using the LSM image browser software.

The agroinfiltration of young papaya leaves (20 days and 50 days after germination) was performed according to Boudart et al. [31]. The abaxial spaces of the leaves ($n = 6$) were agroinfiltrated using sterile syringes (without the needle). The plants were maintained at 22°C for a 16 h photoperiod, and the leaves were collected 2 days post-infiltration (20-day-old leaves). A control experiment (six leaves) was similarly conducted using the pGBPGWG closed vector.

Supporting Information

File S1 These are the legends for Supporting Tables / Figures presented in File S1. Table S1. Cell wall-related genes from ripe papaya and mature *A. thaliana* plant. Table S2. Similarity percentage of amino acid from papaya and other plants PGs. Table S3. Nucleotide sequences used in PCR reactions. Table S4. Nucleotide sequences used in qPCR. Table S5. Calibration curves for relative gene expression. Table S6. Calibration curves for absolute gene expression. Figure S1. Up-

regulation of cell wall-related genes during papaya ripening. Real-time PCR (qPCR) was used to determine the absolute quantitation of the mRNA levels of various genes during papaya ripening. The quantification is represented by the column height. The error bars on each column indicate the SD from four technical replicates from samplings I and II. The different letters represent samples that were significantly different from those collected on other days post-harvest (within the same gene) as determined by one-way ANOVA and Tukey's test ($\alpha < 0.05$, $n = 4$). **Figure B** shows the threshold cycle values (Ct) for the two genes used as internal controls (actin gene – *cpACT* and elongation factor 1-alpha gene - *cp_EF1*). Figure S2. Genomic and mRNA organization of different PGs from papaya fruit. Grey boxes represent coding regions (exons), while black lines represent non-coding regions (introns). White boxes represent the mRNA sequences concatenated from the above compared exons. Figure S3. Unrooted phylogram encompassing PGs from papaya, *Arabidopsis* and several other plant organisms. A phylogenetic tree was calculated using the neighbor-joining method based on the ClustalW alignment of the deduced amino acid sequences. The putative signal peptide from all of the proteins was removed from the

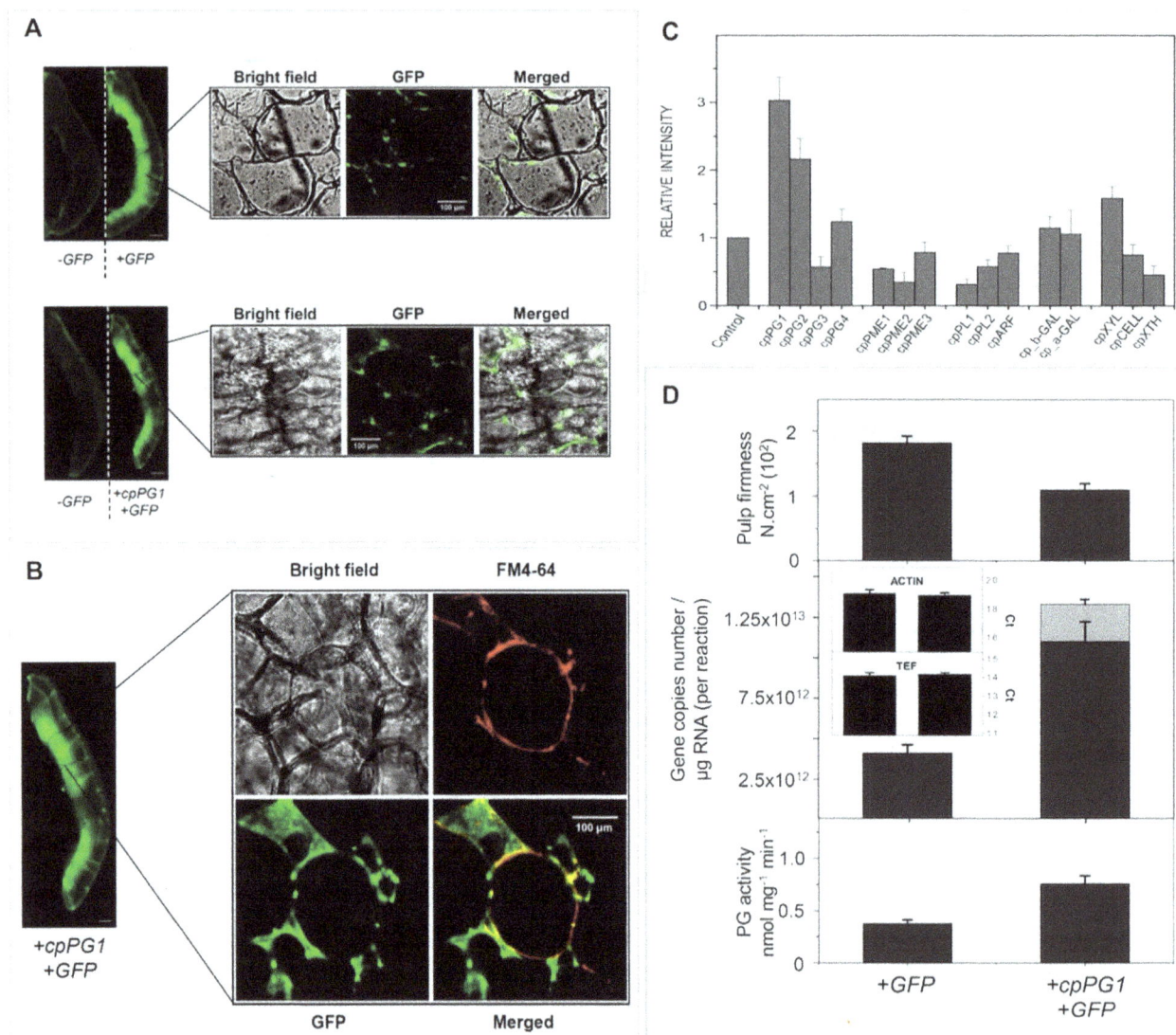

Figure 6. Transient expression of the *cpPG1* gene in the pulp of papaya fruits treated with 1-MCP. Constructs carrying *GFP* or *cpPG1-GFP* genes were agroinfiltrated into papaya fruits treated with 1-MCP, as described in the Materials and Methods. GFP and cpPG1-GFP fusion proteins were transiently expressed and observed with the Versa Doc Gel Imaging System and a confocal laser scanning microscope (**Figures A** and **B**). Contrast with the vital dye FM4-64 shows that the transiently expressed cpPG1-GFP proteins were targeted to the extracellular compartment (**Figure B**). The scale bars in the images of the fruits are 1 cm, and the scale bars in the confocal images indicate the scale. The images are representative of triplicate experiments performed on each agroinfiltrated fruit ($n = 3$). **Figure C** shows the relative mRNA abundance of 15 cell wall-related genes and indicates that *cpPG1* and *cpPG2* may be the main contributors to papaya pulp softening in agroinfiltrated fruits. The expression values for control fruit expressing only GFP (closed vector) were set to 1 for all genes. The absolute quantification of the *cpPG1* mRNA levels in **Figure D** demonstrates that *cpPG1* (gray bar) is transiently expressed. The light gray bar indicates data from transient *cpPG1* expression, the gray bars indicate data from endogenous *cpPG1* expression, and the black bars indicate the threshold cycle values (Ct) for the two genes used as internal controls (*cpACT* and *cp_EF1*).

sequence. The following proteins and their corresponding GenBank IDs were used: *A. thaliana 1, 2 and 3* (**NP_191544**, **NP_191310**, **NP_187454**), *P. persica 1 and 2* (**AAC64184**, **CAA54448**), *P. communis 1 and 2* (**CAH18935**, **BAC22688**), *D. carota* (**BAC87792**), *S. lycopersicum 1 and 2* (**ABW38780**; **AC28947**), *C. melo 1 and 2* (**AAC26510**; **AAC26512**), *V. vinifera 1 and 2* (**XP_002263164**, **XP_002282759**), *B. napus* (**CAA65072**), *G. Max 1 and 2* (**ABC70314**, **AAL30418**), *D. kaki* (**ACJ06506**), *R. communis 1, 2 and 3* (**XP_002529090**, **XP_002529616**, **XP_002517823**), *A. deliciosa* (**P35336**), *O. europaea* (**ACA49228**), *P. trichocarpa 1, 2 and 3* (**XP_002331621**, **XP_002322711**, **XP_002313308**) and *C.*

papaya 1, 2, 3 and 4 (**FJ007644**, **GQ479791**, **GQ479794** e **GQ479795**). The values for the branch lengths are based on the scale bar; there are 0.1 residue substitutions per site. Letters **A** and **B** show the two distinct clades resulting from the phylogenetic analyses. Figure S4. Recombinant proteins sequencing. Figures show the query sequences of cpPG1 (**A**) protein, and the correspondent peptides that were visualized in protein sequencing. Residues highlighted by gray boxes and black letters are from expression vector; residues highlighted by black boxes and white letters are those obtained in protein sequencing. The histidine tags are underlined and asterisks indicate the end of protein. Figure S5. Expression of recombinant proteins from papaya pulp. The figures

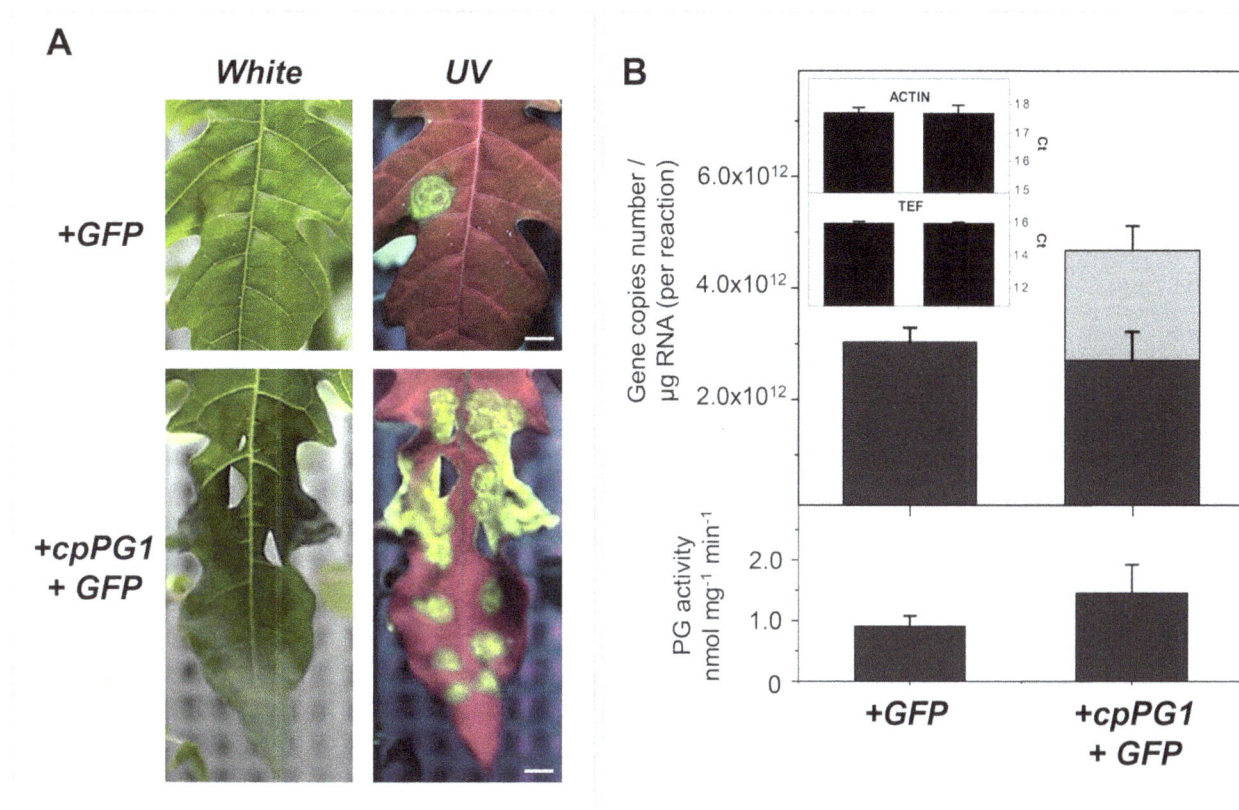

Figure 7. Transient expression of the *cpPG1* gene in papaya leaves. Constructs carrying *GFP* or *cpPG1-GFP* genes were agroinfiltrated into papaya leaves, and the GFP expression in 20-day-old leaves was observed under UV light excitation (**Figure A**). The images are representative of triplicate experiments that were performed on each agroinfiltrated leaf ($n = 3$). The absolute quantification of the *cpPG1* mRNA levels in **Figure B** demonstrate that *cpPG1* (gray bar) is transiently expressed. The light gray bar indicates data from the transient expression of *cpPG1*, the gray bars indicate data from endogenous *cpPG1* expression, and the black bars indicate the threshold cycle values (Ct) for the two genes used as internal controls (*cpACT* and *cp_EF1*). The images are representative of triplicate experiments that were performed on each agroinfiltrated leaf ($n = 3$). The figure bars are scaled to 1 cm.

show representative gels from heterologous expression of *cpPG1* (**A**) gene. **Figure A.** Proteins profiles from bacteria carrying pET45-*cpPG1* before IPTG induction (PG), after 16 hs growth non-induced (PG-NI) and after 16 hs growth IPTG-induced (PG-I); Ni-purified recombinant proteins can be seen after properly elution (E 1 and 2). Proteins markers are highlighted with weight numbers. **Figure B.** The figure displays the hybridization of antiserum for cpPG1 recombinant protein. The arrow indicates the molecular weight for commercially available protein markers. The quantities of proteins are also indicated. Figure S6. Protein alignment of four distinct papaya PGs. The corresponding proteins from *cpPG1* (PG1 - **FJ007644**), *cpPG2* (PG2 - **GQ479791**), *cpPG3* (PG3 - **GQ479794**) and *cpPG4* (PG4 - **GQ479795**) genes were aligned using ClustalW and viewed with BOXSHADE program and then manually edited. Gaps were introduced to optimize alignment. Identic amino acids are highlighted by black boxes and white letters, similar amino acids are highlighted by grey boxes and white letters and different amino

acids are highlighted by white boxes and black letters. The conserved domain of glycosil hydrolases from family 28 (**PS00502**) is indicated by a dotted-line box. The probable catalytic residues are indicated by asterisks, being possible to observe the cpPG2 protein has a Glu_{346} instead of a Lys_{346} as the other three polygalacturonases have.

Acknowledgments

We would like to thank Prof. David Baulcombe, SL and The Gatsby Charitable Foundation for donating the p19 construction.

Author Contributions

Conceived and designed the experiments: JPF JRON. Performed the experiments: JPF SGB SLGLS SZ. Analyzed the data: JPF. Contributed reagents/materials/analysis tools: JPF SZ FML JRON. Contributed to the writing of the manuscript: JPF SZ JRON.

References

1. Fabi JP, Cordenunsi BR, Barreto GPM, Mercadante AZ, Lajolo FM, et al. (2007) Papaya fruit ripening: response to ethylene and 1-methylcyclopropene (1-MCP). J Agric Food Chem 55: 6118–6123.
2. Gapper NE, McQuinn RP, Giovannoni JJ (2013) Molecular and genetic regulation of fruit ripening. Plant Mol Biol 82(6): 575–91.
3. Brummell DA (2006) Cell wall disassembly in ripening fruit. Funct Plant Biol 33: 103–119.
4. Hyodo H, Terao A, Furukawa J, Sakamoto N, Yurimoto H, et al. (2013) Tissue Specific Localization of Pectin-Ca(2+) Cross-Linkages and Pectin Methyl-Esterification during Fruit Ripening in Tomato (*Solanum lycopersicum*). PLoS One 13; 8(11): e78949.

5. Shiga TM, Fabi JP, Nascimento JRO, Petkowicz CLO, Vriesmann LC, et al. (2009) Changes in cell wall composition associated to the softening of ripening papaya: evidence of extensive solubilisation of large molecular mass galactouronides. J Agric Food Chem 57: 7064–7071.

6. Fabi JP, Cordenunsi BR, Seymour GB, Lajolo FM, Nascimento JRO (2009) Molecular cloning and characterization of a ripening-induced polygalacturonase related to papaya fruit softening. Plant Physiol Biochem 47: 1075–1081.

7. Othman R, Choo TS, Ali ZM, Zainal Z, Lazan H (1998) A full-length beta-galactosidase cDNA sequence from ripening papaya. Plant Physiol 118: 1102–1102.

8. Manenoi A, Paull RE (2007) Papaya fruit softening, endoxylanase gene expression, protein and activity. Physiol Plant 131: 470–480.

9. Jamet E, Roujol D, San-Clement H, Irshad M, Soubigou-Taconnat L, et al. (2009) Cell wall biogenesis of *Arabidopsis thaliana* elongating cells: transcriptomics complements proteomics. BMC Gen 10: 505.

10. Fabi JP, Seymour GB, Graham NS, Broadley MR, May ST, et al. (2012) Analysis of Ripening-related Gene Expression in Papaya using an Arabidopsis-based Microarray. BMC Plant Biol 12: 242.

11. Fabi JP, Lajolo FM, Nascimento JRO (2009) Cloning and characterization of transcripts differentially expressed in the pulp of ripening papaya. Sci Hort 121: 159–165.

12. Fabi JP, Mendes LRBC, Lajolo FM, Nascimento JRO (2010) Transcript profiling of papaya fruit reveals differentially expressed genes associated with fruit ripening. Plant Sci 179: 225–233.

13. Chen CC, Fu SF, Lee YI, Lin CY, Lin WC, et al. (2012) Transcriptome analysis of age-related gain of callus-forming capacity in *Arabidopsis* hypocotyls. Plant Cell Physiol 53(8): 1457–69.

14. Santiago-Doménech N, Jiménez-Bermúdez S, Matas AJ, Rose JKC, Muñoz-Blanco J, et al. (2008) Antisense inhibition of a pectate lyase gene supports a role for pectin depolymerization in strawberry fruit softening. J Exp Bot 59: 2769–2779.

15. Quesada MA, Blanco-Portales R, Posé S, García-Gago JA, Jiménez-Bermúdez S, et al. (2009) Antisense down-regulation of the FaPG1 gene reveals an unexpected central role for polygalacturonase in strawberry fruit softening. Plant Physiol 150: 1022–1032.

16. Soh CP, Ali ZM, Lazan H (2006) Characterisation of an alpha-galactosidase with potential relevance to ripening related texture changes. Phytochem 67: 242–254.

17. Lazan H, Ng SY, Goh LY, Ali ZM (2004) Papaya beta-galactosidase/galactanase isoforms in differential cell wall hydrolysis and fruit softening during ripening. Plant Physiol Biochem 42: 847–853.

18. Mandaokar A, Thines B, Shin B, Lange BM, Choi G, et al. (2006) Transcriptional regulators of stamen development in *Arabidopsis* identified by transcriptional profiling. Plant J 46(6): 984–1008.

19. Bassel GW, Fung P, Chow TF, Foong JA, Provart NJ, et al. (2008) Elucidating the germination transcriptional program using small molecules. Plant Physiol 147(1): 143–55.

20. Yadav RK, Girke T, Pasala S, Xie M, Reddy GV (2009) Gene expression map of the *Arabidopsis* shoot apical meristem stem cell niche. Proc Natl Acad Sci U S A 106(12): 4941–4946.

21. Dalchau N, Hubbard KE, Robertson FC, Hotta CT, Briggs HM, et al. (2009) Correct biological timing in *Arabidopsis* requires multiple light-signaling pathways. Proc Natl Acad Sci U S A 107(29): 13171–13176.

22. Ng S, Giraud E, Duncan O, Law SR, Wang Y, et al. (2013) Cyclin-dependent kinase E1 (CDKE1) provides a cellular switch in plants between growth and stress responses. J Biol Chem 288(5): 3449–3459.

23. Sañudo-Barajas JA, Labavitch J, Greve C, Osuna-Enciso T, Muy-Rangel D, et al. (2009) Cell wall disassembly during papaya softening: Role of ethylene in changes in composition, pectin derived oligomers (PDOs) production and wall hydrolases. Postharvest Biol Technol 51: 158–167.

24. Atkinson RG, Sutherland P, Johnston SL, Gunaseelan K, Hallett IC, et al. (2012) Down-regulation of POLYGALACTURONASE1 alters firmness, tensile strength and water loss in apple (*Malus x domestica*) fruit. BMC Plant Biol 12: 129.

25. Zuber H, Davidian JC, Aubert G, Aimé D, Belghazi M, et al. (2010) The seed composition of *Arabidopsis* mutants for the group 3 sulfate transporters indicates a role in sulfate translocation within developing seeds. Plant Physiol 154(2): 913–926.

26. Cantu D, Vicente AR, Greve LC, Dewey FM, Bennett AB, et al. (2008) The intersection between cell wall disassembly, ripening, and fruit susceptibility to *Botrytis cinerea*. Proc Natl Acad Sci U S A 105(3): 859–864.

27. Hiwasa K, Kinugasa Y, Amano S, Hashimoto A, Nakano R, et al. (2003) Ethylene is required for both the initiation and progression of softening in pear (*Pyrus communis* L.) fruit. J Exp Bot 54: 771–779.

28. Morgutti S, Negrini N, Nocito FF, Ghiani A, Bassi D, et al. (2006) Changes in endopolygalacturonase levels and characterization of a putative endo-PG gene during fruit softening in peach genotypes with nonmelting and melting flesh fruit phenotypes. New Phyt 171: 315–328.

29. Sekine D, Munemura I, Gao M, Mitsuhashi W, Toyomasu T, et al. (2006) Cloning of cDNAs encoding cell-wall hydrolases from pear (*Pyrus communis*) fruit and their involvement in fruit softening and development of melting texture. Physiol Plant 126: 163–174.

30. Atkinson RG, Schroder R, Hallett IC, Cohen D, MacRae EA (2002). Overexpression of polygalacturonase in transgenic apple trees leads to a range of novel phenotypes involving changes in cell adhesion. Plant Physiol 129: 122–133.

31. Boudart G, Charpentier M, Lafitte C, Martinez Y, Jauneau A, et al. (2003) Elicitor activity of a fungal endopolygalacturonase in tobacco requires a functional catalytic site and cell wall localization. Plant Physiol 131(1): 93–101.

32. Ming R, Hou S, Feng Y, Yu Q, Dionne-Laporte A, et al. (2008) The draft genome of the transgenic tropical fruit tree papaya (*Carica papaya* Linnaeus) Nature 452: 991–997.

33. Emanuelsson O, Nielsen H, Brunak S, von Heijne G (2000) Predicting subcellular localization of proteins based on their N-terminal amino acid sequence. J Mol Biol 300: 1005–1016.

34. Bendtsen JD, Nielsen H, von Heijne G, Brunak S (2004) Improved prediction of signal peptides: SignalP 3.0. J Mol Bio 340: 783–795.

35. Bustin SA, Benes V, Garson JA, Hellemans J, Huggett J, et al. (2009) The MIQE guidelines: minimum information for publication of quantitative real-time PCR experiments. Clin Chem 55(4): 611–622.

36. Pfaffl MW (2001) A new mathematical model for relative quantification in real-time RT-PCR. Nucleic Acids Res 29 (9): 2002–2007.

37. Thompson JD, Higgis DG, Gibson TJ (1994) CLUSTALW: Improving the sensitivity of progressive multiple sequence alignment through sequence weighting, position specific gap penalties and weight matrix choice. Nucleic Acids Res 22: 4673–4680.

38. Langfelder P, Horvath S (2008) WGCNA: An R package for weighted correlation network analysis. BMC Bio 9: 559.

39. Zhang B, Horvath S (2005) A general framework for weighted gene co-expression network analysis. Stat Appl Genet Mol Biol 4: 17.

40. Shannon P, Markiel A, Ozier O, Baliga NS, Wang JT, et al. (2003) Cytoscape: a software environment for integrated models of biomolecular interaction networks. Genome Res 13(11): 2498–2504.

41. Hames BD, Rickwood D (1990) Gel electrophoresis of proteins – A practical approach. 2nd ed. IRL Press, Oxford University Press. 372p.

42. Nogueira SB, Labate CA, Gozzo FC, Pilau EJ, Lajolo FM, et al. (2012) Proteomic analysis of papaya fruit ripening using 2DE-DIGE. J Proteomics 75(4): 1428–1439.

43. Nascimento JRO, Cordenunsi BR, Lajolo FM (1997) Purification and partial characterization of sucrose-phosphate synthase from pre-climacteric and climacteric bananas (*Musa acuminata*). J Agri Food Chem 45: 1103–1107.

44. Carpentier SC, Witters E, Laukens K, Deckers P, Swennen R, et al. (2005) Preparation of protein extracts from recalcitrant plant tissues: an evaluation of different methods for two-dimensional gel electrophoresis analysis. Proteomics, 5: 2497–2507.

45. Zhong S, Lin Z, Fray RG, Grierson D (2008) Improved plant transformation vectors for fluorescent protein tagging. Transgenic Res 17(5): 985–989.

46. Spolaore S, Trainotti L, Casadoro G (2001) A simple protocol for transient gene expression in ripe fleshy fruit mediated by Agrobacterium. J Exp Bot 52(357): 845–850.

47. Voinnet O, Rivas S, Mestre P, Baulcombe D (2003) An enhanced transient expression system in plants based on suppression of gene silencing by the p19 protein of tomato bushy stunt virus. Plant J 33(5): 949–956.

13

Response of Two Dominant Boreal Freshwater Wetland Plants to Manipulated Warming and Altered Precipitation

Yuanchun Zou[1], Guoping Wang[1], Michael Grace[2], Xiaonan Lou[1], Xiaofei Yu[1], Xianguo Lu[1]*

1 Key Lab of Wetland Ecology and Environment, Northeast Institute of Geography and Agroecology, Chinese Academy of Sciences, Changchun, China, **2** Water Studies Centre and School of Chemistry, Monash University, Clayton, Australia

Abstract

This study characterized the morphological and photosynthetic responses of two wetland plant species when they were subject to 2–6°C fluctuations in growth temperature and ±50% of precipitation, in order to predict the evolution of natural wetlands in Sanjiang Plain of North-eastern China. We investigated the morphological and photosynthetic responses of two dominant and competitive boreal freshwater wetland plants in Northeastern China to manipulation of warming (ambient, +2.0°C, +4.0°C, +6.0°C) and altered precipitation (−50%, ambient, +50%) simultaneously by incubating the plants from seedling to senescence within climate-controlled environmental chambers. Post-harvest, secondary growth of *C. angustifolia* was observed to explore intergenerational effects. The results indicated that *C. angustifolia* demonstrated a greater acclimated capacity than *G. spiculosa* to respond to climate change due to higher resistance to temperature and precipitation manipulations. The accumulated effect on aboveground biomass of post-harvest secondary growth of *C. angustifolia* was significant. These results explain the expansion of *C. angustifolia* during last 40 years and indicate the further expansion in natural boreal wetlands under a warmer and wetter future. Stability of the natural surface water table is critical for the conservation and restoration of *G. spiculosa* populations reacting to encroachment stress from *C. angustifolia* expansion.

Editor: Eric Gordon Lamb, University of Saskatchewan, Canada

Funding: The research was supported by the National Basic Research Program of China (2010CB951301-2), the National Key Technology R & D Program (2012BAC19B05), and the National Natural Science Foundation of China (41271107, 41101195) in the study design, data collection and analysis and preparation of the manuscript.

Competing Interests: The authors have declared that no competing interests exist.

* Email: luxg@iga.ac.cn

Introduction

Wetlands are among the most important terrestrial carbon pools and play an important role in global carbon cycling [1,2]. Boreal wetlands comprise about half of the total global wetland area, store about one third of the world's organic carbon in the form of living or partially decayed vegetation and soil organic carbon and contribute 34% of global atmospheric methane flux [3,4]. Due to their large and dynamic carbon pool, boreal wetlands may not be only significantly impacted by climate change, but may also provide negative (e.g. increasing organic carbon sink) or positive feedback (e.g. increasing greenhouse gas emissions) mechanisms to anthropogenic climate change [5,6] and mitigate climate warming overall.

Boreal wetlands are considered particularly vulnerable to climate change because they depend on specific climatic conditions with low temperature and high water availability [7,8]. To quantify anticipated climate effects on boreal wetland ecosystem processes, field experiments simulating climate change have been undertaken in North America and northern Europe [2,5,9], and these studies suggested that both the performance of a specific plant and the compositions of plant communities will change in

different directions and at different magnitudes as a response to warming and/or changes in the depth of the surface water table that usually connected with temperature and precipitation change. Despite these efforts, species-specific eco-physiological responses of wetland plants to changes in temperature and precipitation, and especially the interactive effects of these factors, remain unclear. Simulations provide a mechanism for exploring the importance of these temperature-precipitation interactions on wetland plants under future climate conditions [10]. Multifactor experimental manipulations of temperature and precipitation in the laboratory can complement field studies by enabling controlled adjustment of covarying climate variables based on the historical data to generate relatively realistic climate change experimental scenarios [11]. The responses then could be scaled up to the level of the ecosystem when the initial structure of the wetland and the environmental driving variables are consequently well coupled.

As typical dominant boreal freshwater wetland plants in Northeast China, Siberia and Far East Russia, *Calamagrostis angustifolia* and *Glyceria spiculosa* both belong to the family Poaceae, and are often distributed as large, dense clumps in palustrine, riverine and lacustrine wetlands [12,13]. Generally, *C. angustifolia* covers saturated microhabitats while *G. spiculosa* covers the flooded microhabitats with greater surface water depth. Both plants play important ecological functions in the wetlands as primary producers for food webs, shelter for some invertebrates and major organic carbon sources for carbon sequestration. In recent years, with climate warming and surface water deficiency in some wetlands of the Sanjiang Plain, the microhabitats once covered by *G. spiculosa* have been almost replaced by *C. angustifolia* due to the drawdown of surface water depth, caused either by climate change or anthropogenic activities such as drainage.

In this paper, we simultaneously considered two major factors (temperature and precipitation) involved in climate change effects on boreal wetland ecosystem functioning. The two plant species, *C. angustifolia* and *G. spiculosa*, were incubated from seedling to senescence in environmental chambers under manipulated warming and altered precipitation conditions. The objectives of this study were: (1) to characterize the morphological and photosynthetic responses of the two plants to 2–6°C warming and ±50% of precipitation changes; and (2) to analyze the resistance and acclimation of the two plants and predict their potential distribution in the future under the projected changed climate conditions in the Sanjiang Plain wetlands of Northeast China.

Materials and Methods

Study site and target species

The study site is located in Xingkai Lake (Khanka) lacustrine wetlands of Sanjiang Plain, the largest expanse of freshwater wetlands in China. Xingkai Lake is the largest lake in Northeast Asia and a transboundary lake shared by China and Russia. According to calculations based on the 54 years of local meteorological records, the annual mean precipitation is 561 mm and the temperature ranges from $-39°C$ in January to 36°C in July with the annual mean of 3.5°C for 1957–2010. According to a linear fitting of local meteorological records in Xingkai Lake lacustrine wetlands (Annual temperature = -52.55 + 0.028*year, $r>0.99$, $p<0.0001$; unpublished), there is a significant warming trend of 0.28°C per decade in Xingkai Lake over the past half century. This is even greater than the average level of 0.22°C per decade for the whole country [14].

The target plant species for the three-month incubation experiments were collected *in situ* from *Calamagrostis angustifolia* (45°20′59″ N, 132°18′52″ E) and *Glyceria spiculosa* (45°21′49″ N, 132°19′26″ E) communities within Xingkai Lake lacustrine wetlands. Representative and homogeneous individuals of *C. angustifolia* and *G. spiculosa* were excavated with intact roots and soils using a planting shovel in May of 2011 and 2012, respectively, when the first new shoots were apparent in springtime. The soil types were silty clay meadow marsh soil and humus marsh soil, respectively. The seedlings and soils with the depth of approximate 20 cm were cut into small blocks with rhizome based, transported to the laboratory in Northeast Institute

of Geography and Agroecology within two days, and transplanted into incubation pots with 10-cm diameter and 15-cm depth with 4–5 seedlings in each pot. To minimize loss of vitality during transplantation, these potted seedlings were first incubated in a greenhouse for one month to germinate new ramets before placement into environmental chambers. The temperature in the greenhouse was set as the average air temperature of June in Xingkai Lake lacustrine wetlands. All the field studies got the permission from administrative authority of Xingkai Lake National Nature Reserve and did not involve any endangered or protected species.

Experimental design

The design between the warming and the precipitation was a two-way randomized block experiment. Four environmental chambers were deployed to control the temperature. Within each chamber, three plastic container (35 cm × 25 cm × 10 cm) were used to control water level fluctuations caused by precipitation manipulation. Each container contained three replicated pots. Before incubation, any pot growing approximate ten vigorous seedlings with similar performance (about 20 cm and 15 cm heights for *C. angustifolia* and *G. spiculosa*, respectively) was selected and randomly separated into 4 groups corresponding to the 4 temperature regimes. The ambient temperature was set as the local monthly average air temperatures in July, August and September, respectively fitted. The ambient temperature was adjusted five times per day based on the local monthly (July, August and September) average air temperatures over the time periods 1:00–4:00, 4:00–10:00, 10:00–16:00, 16:00–19:00 and 19:00–1:00. Table 1 provides the temperature and illuminance values for the plant incubations in the environmental chambers. The illuminance values were based on the field observation. These pots were exchange their positions in the chambers daily to keep the temperature and light intensity similar for a given pot. The warming scenarios were then set according to the ambient temperature regime with +2°C, +4°C and +6°C added for each interval. The lights were plant growth fluorescent lamps that are recognized as suitable for plants' growth [15].

The ambient water depths were set to mimic the natural conditions in Xingkai Lake wetlands, with no net water for *C. angustifolia* and 10 cm of water above the soil surface for *G. spiculosa*. The ambient precipitation regime was set according to the local monthly average precipitation. These seedlings were gently watered over one hour using sprinkler nozzles to simulate natural precipitation every 3 days with one tenth of the amount of monthly precipitation. The altered precipitations were set as -50% and +50% based on the ambient precipitation rates.

After the three month incubation in the environmental chambers and subsequent harvesting of all above-ground biomass, one *C. angustifolia* pot was randomly selected from the 3 replicates of each precipitation treatment container within each environmental chamber. Twelve pots were consequently selected

Table 1. Temperature and illuminance control for plants' incubation in the environmental chambers.

Time period		0:00–7:00	7:00–11:00	11:00–14:00	14:00–16:00	16:00–24:00
Illuminance (lux)		0	10000	20000	15000	5000
Temperature (°C)	July	16	20	28	24	17
	August	17	19	26	23	18
	September	11	14	19	17	12

for a secondary growth test to analyze the potential effect of climate change on the next generation of this plant species. These pots were incubated for one month in the same environmental chamber with the average temperature and precipitation in June of 17.4°C and 74.4 mm, respectively in Xingkai Lake wetlands.

Response measurements

The morphological indicators (height, stem diameter and leaf area) were observed once per month using a meter stick, vernier caliper and portable area meter (LI-3000, Li-Cor Inc., USA) from seedling to senescence, with each indicator measured by 4 replicates for *C. angustifolia* and 3 replicates for *G. spiculosa*. The photosynthetic indicators, including leaf net photosynthetic rate (Pn), leaf stomatal conductance (Gs) and leaf transpiration rate (Ts) were measured by a portable photosynthesis system (LI-6400, Li-Cor, Inc., USA) during the peak growth phase (around August), with each indicator measured by 4 replicates for *C. angustifolia* and 3 replicates for *G. spiculosa*. The chlorophyll content was determined by homogenizing and extracting the leaf tissue in 80% acetone until the green colour turn white completely, then measuring the optical absorbance at 663 and 645 nm for chlorophyll a and b, respectively using a spectrophotometer (UV-2500, Shimadzu, Japan), and calculating concentrations using the specific absorption coefficients [16,17]. The reported leaf chlorophyll content was the sum of chlorophyll a and b, with 3 replicates both for *C. angustifolia* and *G. spiculosa*.

Above-ground biomass was measured after plant senescence by drying to constant weight at 80°C. The total nitrogen content of harvested leaves was determined using a continuous flow analyzer (SAN^{++}, SKALAR, the Netherlands) after digesting 0.3000 g of homogeneous, milled material with H_2SO_4/H_2O_2, with 3 replicates both for *C. angustifolia* and *G. spiculosa*. The below-ground biomass was not measured as it is a mixture of newly developed rhizomes/roots within that growing season and the rhizomes/roots from previous growing seasons. Older material usually accounts for most of the total below-ground biomass [18].

The response measurements for the secondary growth test of *C. angustifolia* were conducted as described above. The measured indicators were height, stem diameter above the soil, leaf area of each leaf on the plant based on measuring a suitable number of leaves, Pn, Gs, Ts and above-ground biomass, with each indicator measured by 3 replicates.

Statistical analyses

Statistical analyses were performed using SPSS Statistics 21.0 (SPSS Inc., USA). The main and the interaction effects of temperature and precipitation on the two plants' morphological indicators were compared by repeated measures analysis of variance (ANOVA). For repeated measures ANOVA, Mauchly's Test of Sphericity was first examined and the Greenhouse-Geisser adjustment was adopted when the null hypothesis was rejected. The main and the interaction effects of temperature and precipitation on the two plants' photosynthetic indicators, leaf chlorophyll contents during the phase of peak growth, the above-ground biomasses and leaf nitrogen contents after senescence were compared by two-way ANOVA. One-way ANOVA was performed to compare the differences of above-ground biomasses of *C. angustifolia* and *G. spiculosa*. Differences in morphology, photosynthesis and above-ground biomass between different secondary grown *C. angustifolia* after the temperature and precipitation manipulation for one month were also compared using one-way ANOVA. The least significant difference (LSD) was used respectively to perform *post hoc* multiple comparisons when

Table 2. Repeated-measures analysis of variance (ANOVA) of morphological indicators for *Calamagrostis angustifolia* and *Glyceria spiculosa*.

Variable		Month		Month × Temperature		Month × Precipitation		Month × Temperature × Precipitation	
		F	p	F	p	F	p	F	p
C. angustifolia	Height	30.864	**<0.001**	0.910	0.482	2.611	0.052	2.025	**0.044**
	Basal stem diameter	0.064	0.860	0.538	0.702	0.343	0.764	0.513	0.835
	Leaf area	151.032	**<0.001**	2.449	**0.047**	0.271	0.857	0.922	0.517
G. spiculosa	Height	111.065	**<0.001**	1.038	0.412	2.203	0.116	2.520	**0.033**
	Basal stem diameter	3.165	0.051	0.412	0.867	1.426	0.240	0.839	0.612
	Leaf area	61.079	**<0.001**	1.739	0.155	3.271	**0.031**	1.363	0.239

Significant effects ($p<0.05$) are shown in bold. When the highest order of interaction effect is significant, the significant main effect is not in bold.

equal variances were assumed; otherwise, Tamhane's T2 method was used.

Results

Morphological indicators

Repeated-measures ANOVA of morphological indicators showed that the interaction effects of temperature and precipitation on the height of *C. angustifolia* and *G. spiculosa* were both significant ($F = 2.025$, $p = 0.044$; $F = 2.520$, $p = 0.033$, respectively). For both species, main effects of temperature or precipitation, and any interaction effects, on basal stem diameter were all insignificant. The main effects of temperature on leaf area of *C. angustifolia* and the effect of precipitation on that of *G. spiculosa* were both significant ($F = 2.449$, $p = 0.047$; $F = 3.271$, $p = 0.031$, respectively) (Table 2).

C. angustifolia's height increased from July to September. The estimated marginal means (as the average of the results for each month) of the height suggested that the greatest heights were observed in the ambient precipitation for each warming manipulation (Fig. 1A). Under ambient precipitation conditions (Fig. 1B), the greatest height occurred at ambient temperature

(except for the early growth of July), while the lowest height was found in the +2°C manipulation (except in September). When the precipitation decreased by 50% from ambient (Fig. 1C), the greatest height occurred in the +4°C manipulation (except for the early growth of July), while the lowest height was in the +6°C manipulation. When the precipitation increased by 50% from ambient, the greatest height occurred in the +4°C manipulation, while the other three treatments produced plants of comparable height (Fig. 1D).

G. spiculosa's height also increased throughout the incubation (Fig. 2B–D). The estimated marginal means of the height suggested that ambient precipitation and 2–4°C of warming could support greater plant height (Fig. 2A). In the ambient precipitation conditions, there were no significant differences in height between the different temperature treatments (Fig. 2B). For both the −50% and +50% precipitation treatments, the greatest height occurred in the +4°C manipulation (Fig. 2C, D).

The leaf area of the two species increased first and then decreased throughout the incubation. For *C. angustifolia*, the leaf area with the +6°C manipulation was smaller than other manipulations in July. There was no significant difference in leaf area between the ambient and the +2°C manipulation in July. In

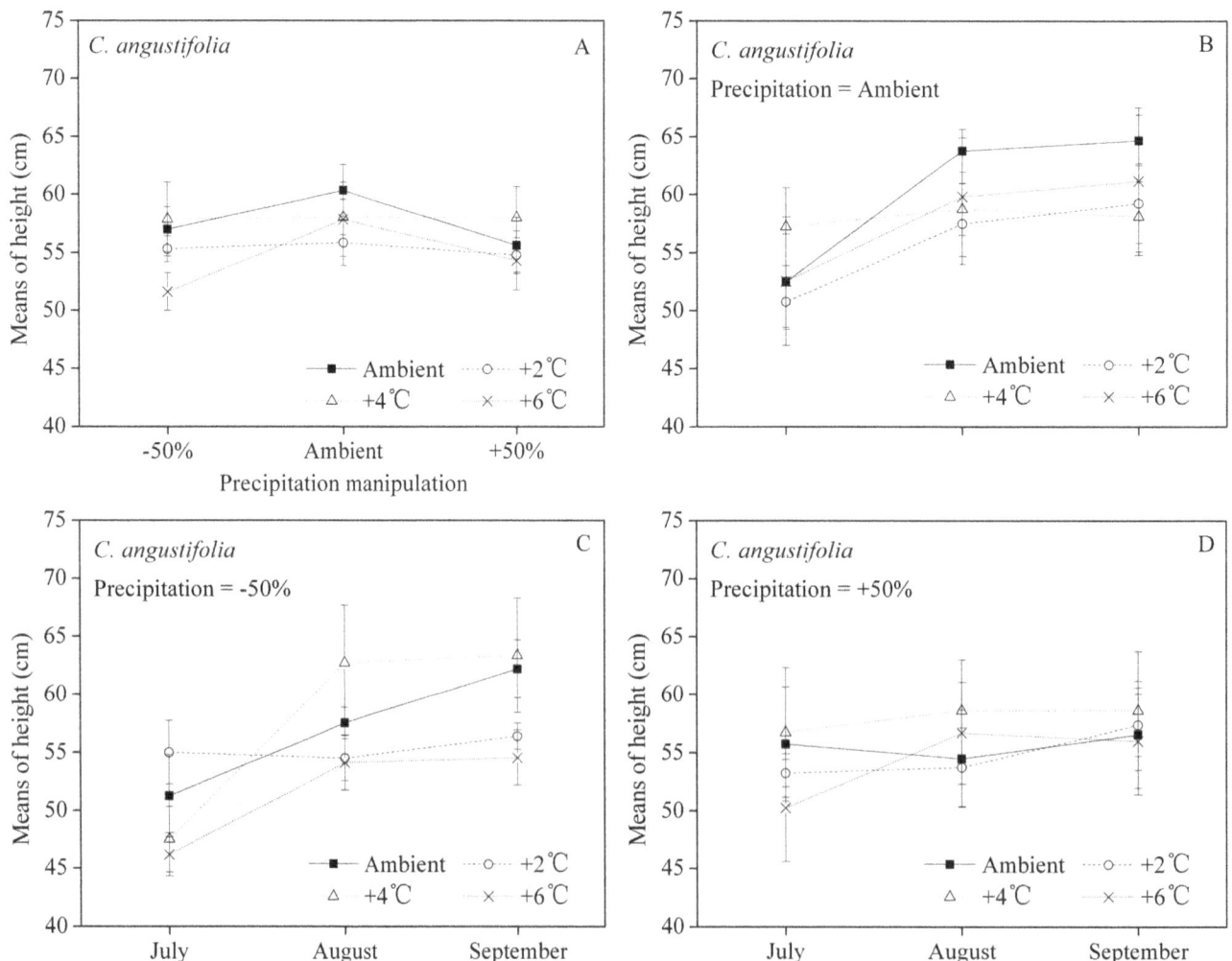

Figure 1. Interaction of temperature and precipitation manipulations on plant height (A) of *Calamagrostis angustifolia***, and the monthly estimated marginal means of height of** *C. angustifolia* **for different temperature treatments when precipitation was manipulated at ambient (B), −50% (C) and +50% (D) levels, respectively.** The error bars represent means ± 1 standard error.

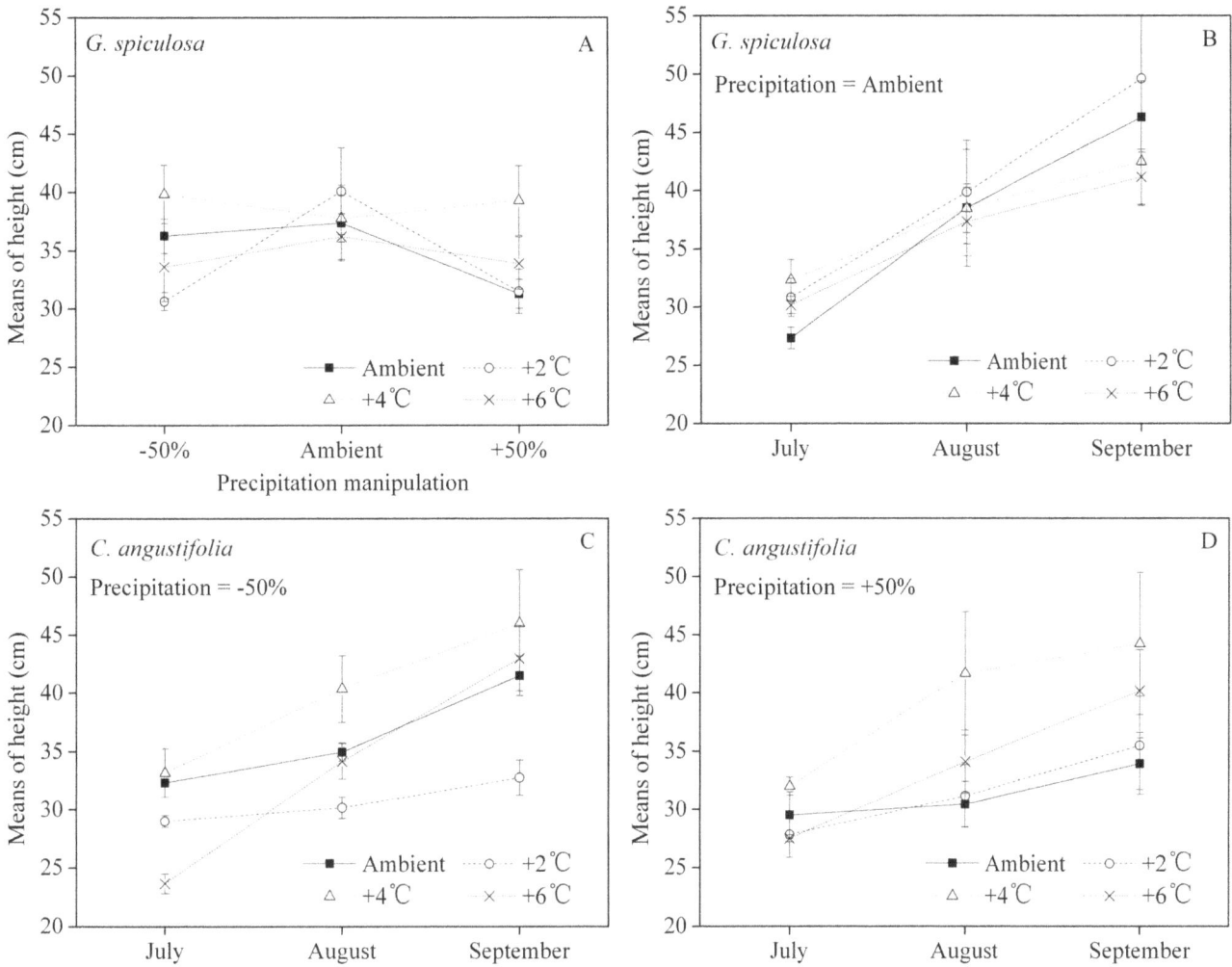

Figure 2. Interaction of temperature and precipitation manipulations on plant height (A) of *Glyceria spiculosa*, and the monthly estimated marginal means of height of *G. spiculosa* for different temperature treatments when the precipitation was manipulated at ambient (B), −50% (C) and +50% (D) levels, respectively. The error bars represent means ± 1 standard error.

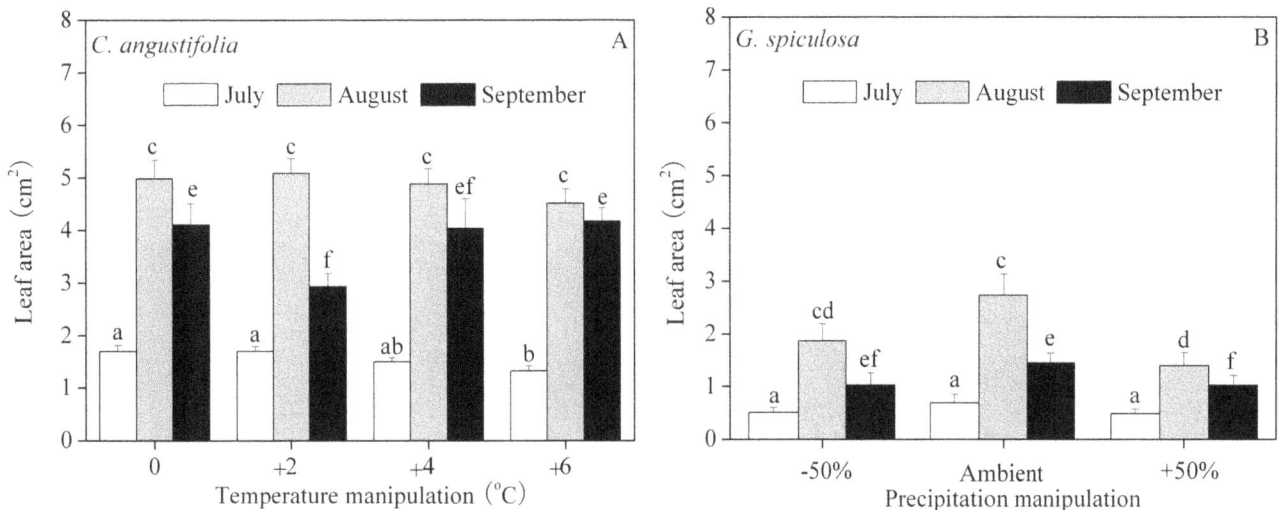

Figure 3. Leaf area of *C. angustifolia* (A) and *G. spiculosa* (B) monthly. For each month, different letters over the bars indicate significant differences ($p < 0.05$) between temperature manipulations for *C. angustifolia* and precipitation manipulations for *G. spiculosa*. The letters 'a' and 'b' are for the July results, 'c' and 'd' for the August results, and 'e' and 'f' for the September results. The error bars represent means ± 1 standard error.

Table 3. Two-way ANOVA of photosynthetic indicators and leaf chlorophyll content for C. angustifolia and G. spiculosa during the phase of peak growth, and above-ground biomass and leaf total nitrogen content after harvest.

Variable		Temperature		Precipitation		Temperature × Precipitation	
		F	p	F	p	F	p
C. angustifolia	Pn	9.878	**<0.001**	7.516	**0.002**	4.974	**0.001**
	Gs	3.691	**0.021**	8.639	**0.001**	6.825	**<0.001**
	Ts	4.410	**0.010**	3.013	0.062	9.208	**<0.001**
	Chlorophyll	0.966	0.419	0.765	0.473	2.545	**0.037**
	Aboveground biomass	4.512	**0.012**	2.349	0.117	0.842	0.550
	Leaf total nitrogen	7.125	**0.001**	3.642	0.052	0.524	0.784
G. spiculosa	Pn	1.663	0.201	11.361	**<0.001**	2.244	0.073
	Gs	17.989	**<0.001**	1.966	0.162	3.400	**0.014**
	Ts	5.091	**0.007**	2.883	0.075	3.170	**0.020**
	Chlorophyll	4.920	**0.008**	7.153	**0.004**	2.322	**0.046**
	Aboveground biomass	4.156	**0.017**	13.637	**<0.001**	1.002	0.447
	Leaf total nitrogen	2.737	0.066	1.134	0.338	2.394	0.059

Significant effects ($p<0.05$) are shown in bold. When the highest order of interaction effect is significant, the significant main effect is not in bold.

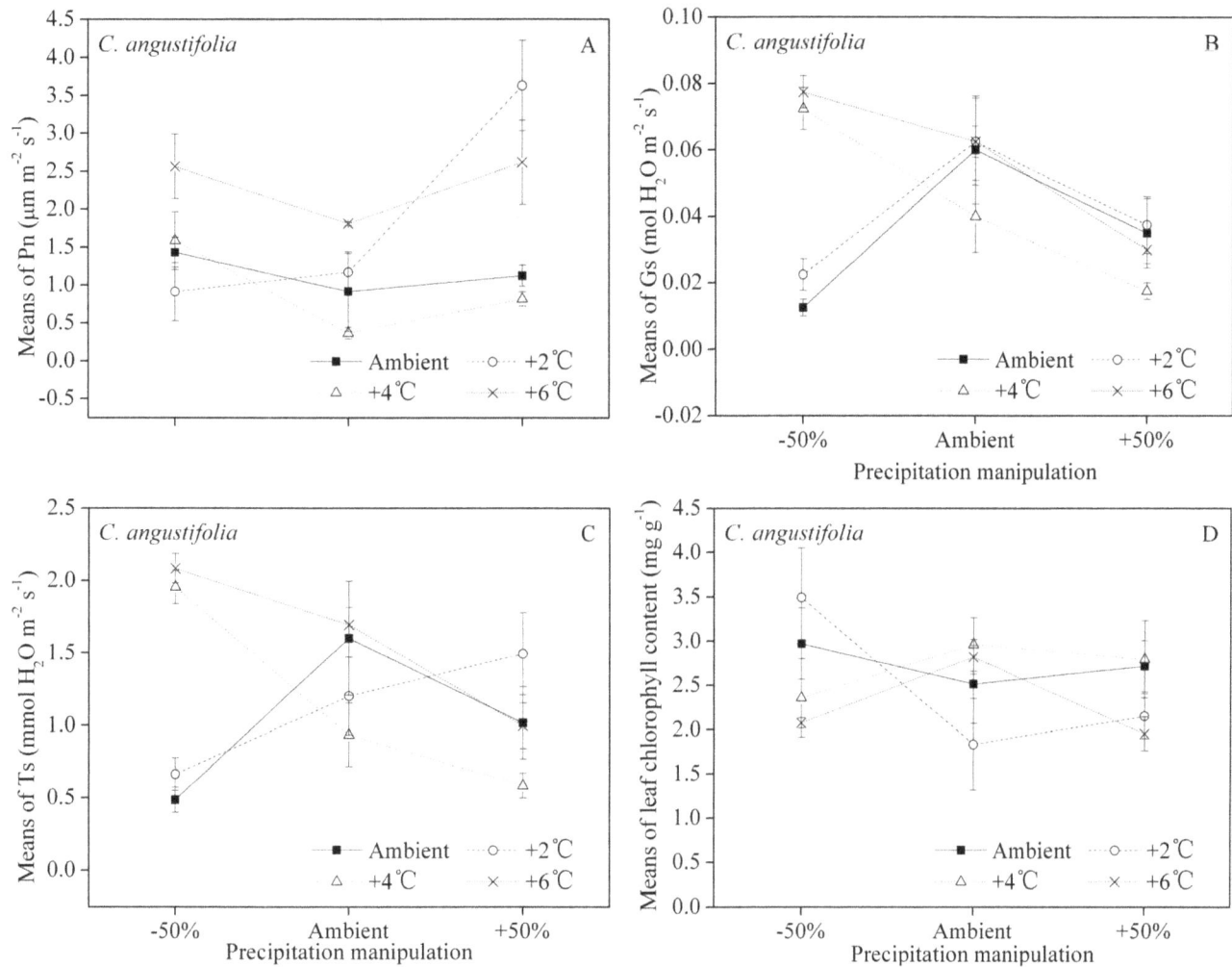

Figure 4. Interactions of temperature and precipitation manipulations on *C. angustifolia*'s leaf net photosynthetic rate (Pn, A), leaf stomatal conductance (Gs, B), leaf transpiration rate (Ts, C) and leaf chlorophyll content (D), respectively. The error bars represent means ± 1 standard error.

September, the leaf area with the +4°C manipulation was not significantly different from either the ambient or the +6°C manipulation. There was no significant difference in leaf area with different temperature manipulations in August (Fig. 3A). For *G. spiculosa*, there was no significant difference between ambient and the −50% precipitation treatments in any of the three months, while the leaf area with the 50% precipitation treatment was smaller than other treatments in August and September (Fig. 3B).

Photosynthetic indicators and leaf chlorophyll content

Two-way ANOVA of photosynthetic indicators and leaf chlorophyll content showed that the interaction effects of temperature and precipitation on these two species were both significant, except for the net photosynthetic rate (Pn) for *G. spiculosa*, which was only significantly affected by precipitation (Table 3).

For *C. angustifolia*, smaller Pn occurred under ambient precipitation except for the +2°C manipulation, and greater Pn was measured in the +6°C manipulation with the exception of the +50% precipitation treatment (Fig. 4A). The leaf stomatal conductance, Gs, decreased with the enhanced precipitation for

the +4°C and +6°C manipulations (Fig. 4B). The greatest leaf transpiration rates, Ts, were found in the +4°C and +6°C manipulations under −50% precipitation (Fig. 4C). Ts rates decreased in both these temperature treatments with increasing precipitation. There was no systematic pattern in Ts at ambient temperature but the +2°C manipulation showed Ts increased with increasing precipitation. There were no consistent trends in leaf chlorophyll content with temperature or precipitation treatments (Fig. 4C) but all concentrations fell within the range 1.83–3.49 mg/g.

For *G. spiculosa*, Pn decreased with precipitation (Fig. 5A). With the exception of the +2°C treatment under the −50% precipitation regimen, the highest Gs results were found under ambient temperature (Fig. 5B). There was no consistent response in Gs to increased precipitation over the temperature treatments. There was no consistent trend in Ts across either temperature or precipitation treatments (Fig. 5C). All temperature treatments showed a decrease in leaf chlorophyll content between −50% and ambient precipitation (Fig. 5D). Increasing precipitation to 50% above ambient resulted in no consistent trend in chlorophyll with increasing temperature regime.

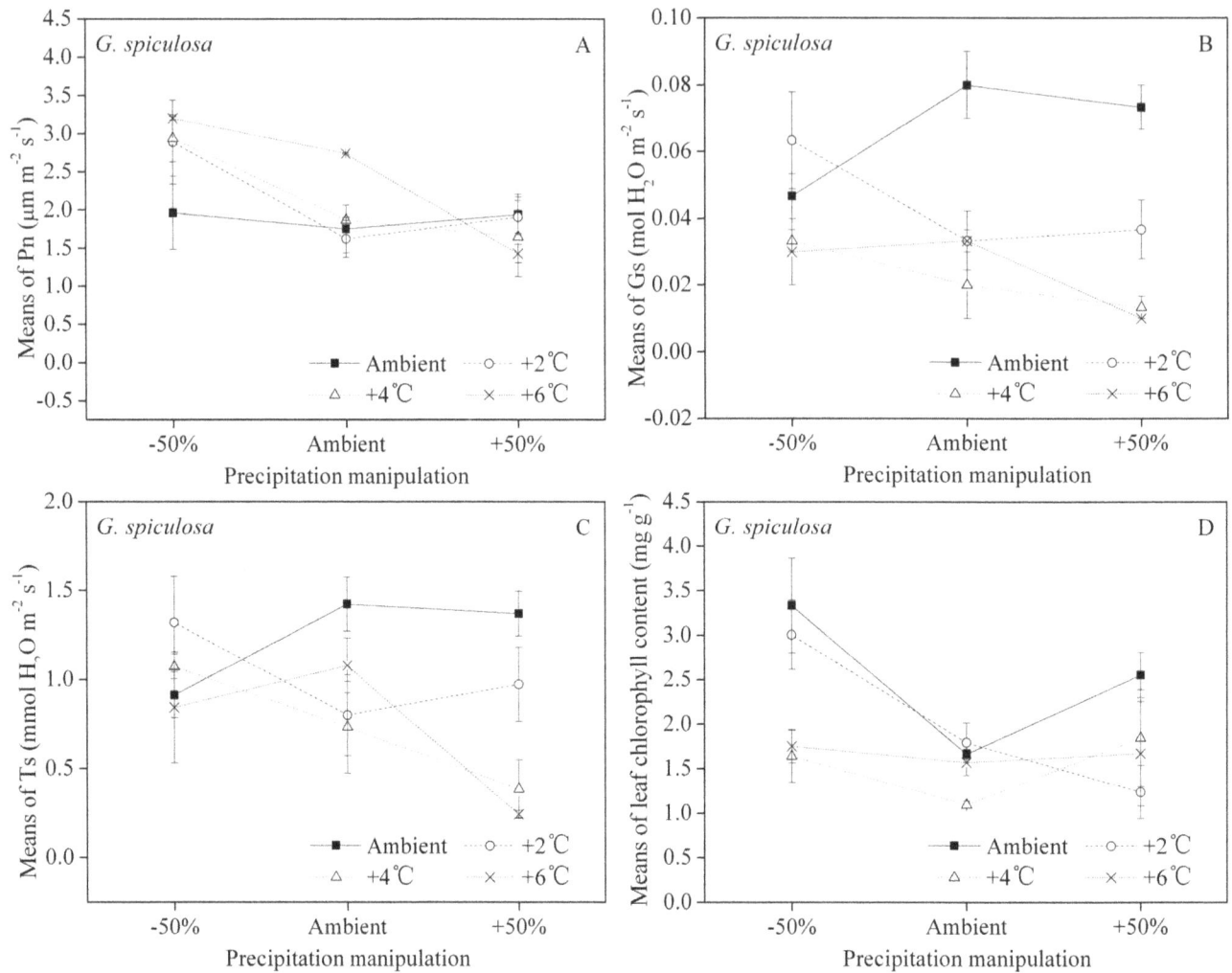

Figure 5. Main effect of precipitation manipulation on *G. spiculosa*'s leaf net photosynthetic rate (Pn, A), interactions of temperature and precipitation manipulations on leaf stomatal conductance (Gs, B), leaf transpiration rate (Ts, C) and leaf chlorophyll content (D), respectively. Different letters shared by the bars indicate significant differences ($p < 0.05$) between precipitation treatments. The error bars represent means $\pm$ 1 standard error.

Above-ground biomass and leaf nitrogen content

Above-ground biomass of *C. angustifolia* was only significantly affected by temperature ($F = 4.512$, $p = 0.012$), while the effect of precipitation and the interaction effect were insignificant (Table 3). Biomass was at a maximum in the +4°C manipulation, increasing by 58% above that at ambient temperature (Fig. 6A). Leaf total nitrogen of *C. angustifolia* was also only significantly affected by temperature ($F = 7.125$, $p = 0.001$; Table 3). All three temperature increments resulted in statistically significant increases above ambient temperature, with a maximum value at +6°C, representing a 39% increase above ambient temperature (Fig. 6B).

Above-ground biomass of *G. spiculosa* was significantly affected both by temperature and precipitation ($F = 4.156$, $p = 0.017$; $F = 13.637$, $p < 0.0001$, respectively), while the interaction effect was insignificant (Table 3). As noted with *C. angustifolia*, biomass was highest in the +4°C manipulation, with an increase of 37% above ambient temperature (Fig. 6C). Both increasing and decreasing precipitation had negative effects on above-ground biomass, with reductions of 29% and 32%, respectively compared to ambient precipitation (Fig. 6D). Any effects of temperature or

precipitation, as well as their interaction effect, on leaf total nitrogen of *G. spiculosa* were insignificant (Table 3).

Secondary growth of *C. angustifolia*

One-way ANOVA results (Table 4) showed that only above-ground biomass displayed a significant difference with the incubation temperature treatment after one month's secondary growth ($F = 4.096$, $p = 0.049$); no other morphological or photosynthetic indicator showed significant differences (Table 4). Above-ground biomass of secondary growth was greatest in the +4°C manipulation, 1.11 times higher than the ambient temperature (Fig. 7).

Discussion

Interaction effects of warming and precipitation alteration

Boreal wetlands are becoming increasingly vulnerable in a warming world [8]. Measuring responses of dominant plant species to warming and precipitation alteration is fundamental to assess and project the vulnerability of boreal wetlands under

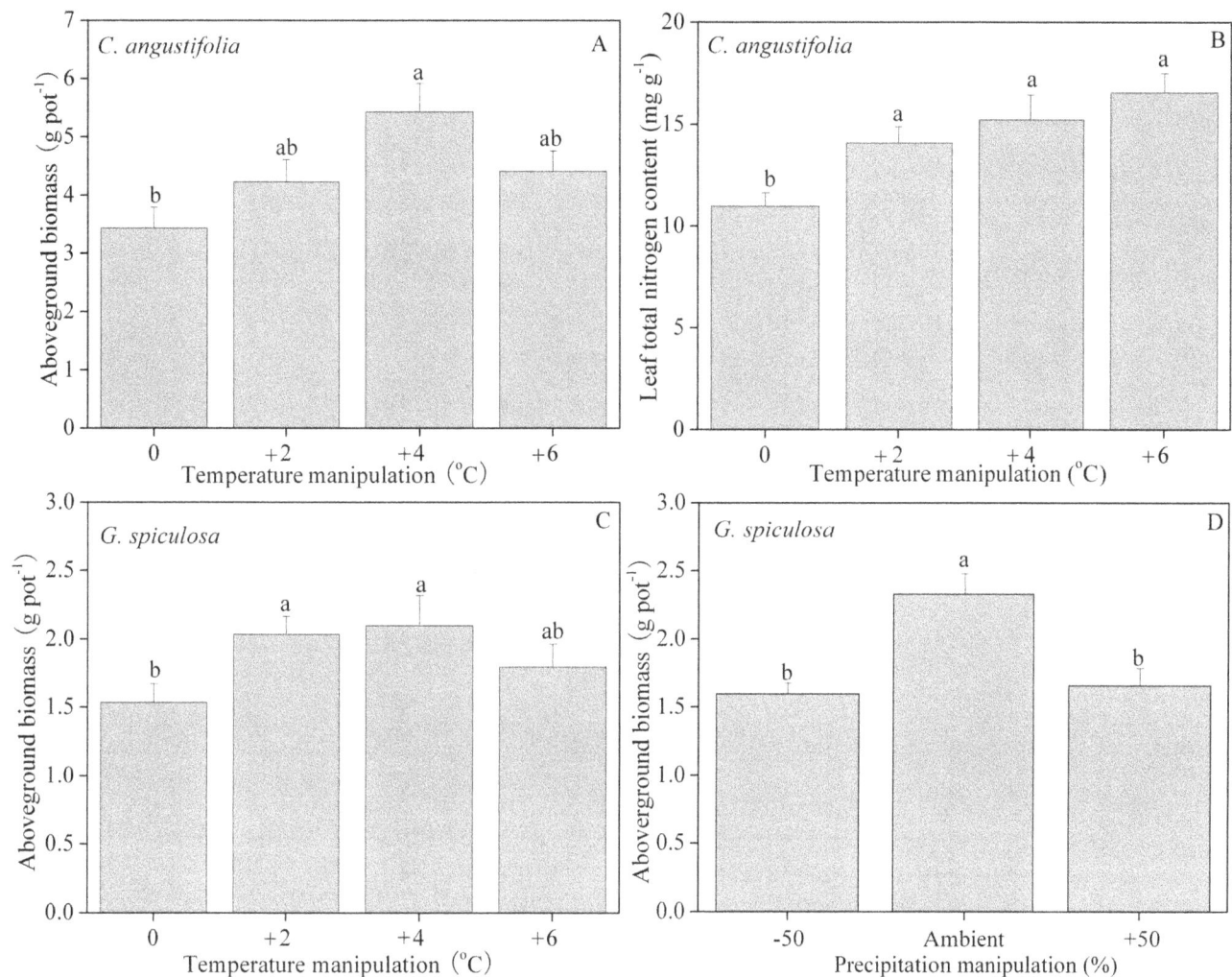

Figure 6. Aboveground biomass (A), leaf total nitrogen (B) of *C. angustifolia*, **and aboveground biomass of** *G. spiculosa* **(C, D) after harvest.** Different letters shared by the bars indicate significant differences ($p<0.05$) between temperature or precipitation manipulations. The error bars represent means $\pm$ 1 standard error.

climate change. The interactions among multiple climate-change factors may cause responses in wetlands that may not be detected when focusing on a single factor, because climate change will include simultaneous changes in temperature and precipitation regimes [19,20]. The above-ground biomass of *C. angustifolia* and *G. spiculosa* initially showed an increasing trend and then decreased with continued increments in mean monthly temperature. The +4°C temperature treatment produced the highest biomass (Fig. 6A, C), which is consistent with the single factor study by Breeuwer et al. [10] and multi-factor meta-analysis results from Lin et al. [21].

The monthly heights in different warming manipulations show intersecting trend lines (Fig. 1A, 2A), indicating that interaction effects of temperature and precipitation should be considered, especially for the +4°C manipulation. In the decreased precipitation treatment (−50%), greater plant heights were at ambient temperature and in the +4°C manipulation for *C. angustifolia* and *G. spiculosa* (Fig. 1C, 2C, respectively). With the exclusion of injury, height is one of the best indicators of an herbaceous plant's growth and competitiveness in light [22,23]. Our results show that any study on species-specific or interspecific plant response to

climate change should also include consideration of the accompanying precipitation regime.

A limited increase of temperature within the optimal range often enhances photosynthesis, depending on the species and its environmental acclimation [24,25]. However, this expected enhancement of Pn can be moderated by concurrent water or drought stress on plants [26]. Our results show that Pn of *C. angustifolia* increased with temperature only in the driest (−50% precipitation) treatment, while ambient precipitation resulted in smaller Pn values in two of the three elevated temperatures (Fig. 4A), indicating that a change in precipitation would counter-intuitively limit the photosynthetic capacity of *C. angustifolia*. The Pn of *G. spiculosa* decreases with increasing precipitation (Fig. 5A), suggesting that soil water availability affects photosynthesis more than temperature variation [27]. Photosynthetic capacity is closely associated with, and approximately proportional to, leaf chlorophyll and nitrogen contents, and all three are favored by increased temperature [28,29,30]. Only the leaf nitrogen content of *C. angustifolia* increased with temperature (Table 3, Fig. 6B), while that of *G. spiculosa* was not significantly affected either by temperature or precipitation (Table 3). Although the leaf chlorophyll contents of two species were altered both by

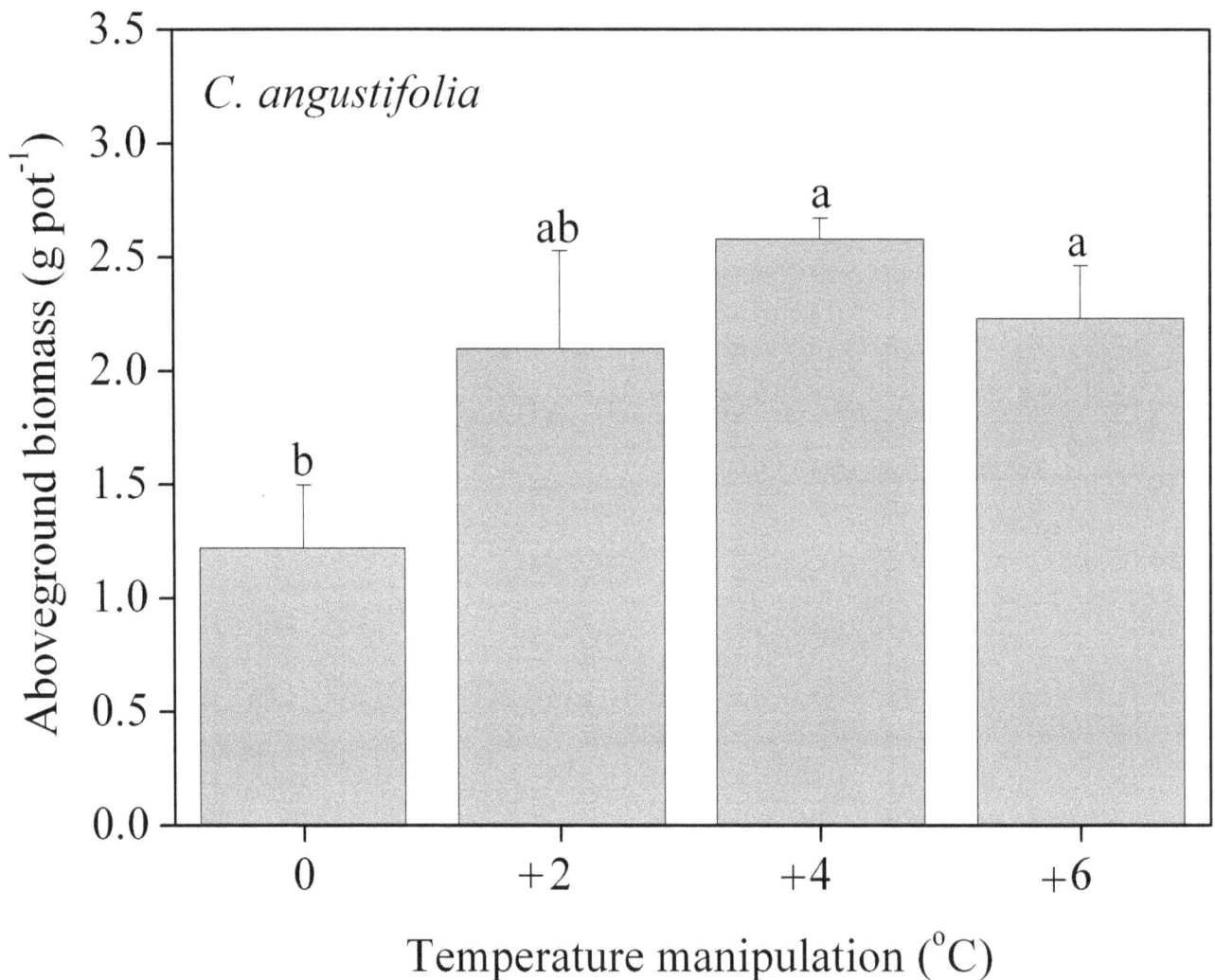

Figure 7. Aboveground biomass of *C. angustifolia* after secondary growth for one month. Different letters shared by the bars indicate significant differences ($p < 0.05$) between the temperature manipulations. The error bars represent means $\pm$ 1 standard error.

temperature and precipitation (Table 3), the temperature responses under different precipitation treatments differed between species (4D, 5D).

These differential, and even opposite, responses of two species to our manipulated temperature and precipitation treatments are consistent with the widely supported observations and modeling that show species respond individualistically to changes in environmental conditions [5].

Species-specific responses to temperature and precipitation change

The effects of climate change on boreal wetland ecosystems ranges from regional, ecosystem, vegetation type, functional type to species [21,31,32,33]. Our results suggest that within the same functional plant type, species-specific response was a significant determinant in height (Table 3, Fig. 1, 2), leaf area (Fig. 3), photosynthetic indicators, leaf chlorophyll contents (Table 3, Fig. 4, 5), and above-ground biomass (Table 3, Fig. 6).

The most marked species-specific difference was leaf area and above-ground biomass (Table 2, 3). Leaf area is only affected by temperature for *C. angustifolia* (Fig. 3A) and precipitation for *G.*

spiculosa (Fig. 3B). Above-ground biomass of *C. angustifolia* is only affected by temperature (Fig. 6A), while both temperature and precipitation influence *G. spiculosa* (Fig. 6C, D). Given the direct relationship with net primary productivity, above-ground biomass is considered as one of the plant characteristics responding strongly to climate change [34]. The $+2°C$ manipulation increased above-ground biomass 23% and 31% compared with ambient conditions for *C. angustifolia* (Fig. 6A) and *G. spiculosa* (Fig. 6C), respectively. According to the linear fitting of local meteorological records in Xingkai Lake lacustrine wetlands, a $2°C$ warming is likely to take place by the 2080s. Consequently, this species-specific difference in predicted biomass would affect carbon and nutrient budgets in boreal wetland ecosystems responding to future temperature change [31].

Previous species-specific comparisons under simulated climate change are mostly based on only one growth season, and the intergenerational response is not readily apparent [35,36]. The significant difference in above-ground biomass of secondary growth (Fig. 7) with the different temperature treatments indicates that effects of climate change would continue to the next generation, because *C. angustifolia* mainly reproduces clonally by rhizomes [12]. These rhizomes would store a large amount of

Table 4. One-way ANOVA of aboveground biomass, morphological and photosynthetic indicators of *C. angustifolia* after secondary growing for one month.

ANOVA	Height	Basal stem diameter	Leaf area	Pn	Gs	Ts	Aboveground biomass
F	0.281	0.874	1.849	2.963	2.650	2.802	4.096
p	0.838	0.494	0.217	0.097	0.120	0.108	**0.049**

Significant effects ($p < 0.05$) are shown in bold.

biomass below-ground, translocated from above-ground when the shoot is harvested. Greater biomass results in better performance during secondary growth [37]. Considering the biomass accumulation of biomass year by year, *C. angustifolia* is more acclimated to temperature and precipitation change. This acclimated capacity supports the continued presence of *C. angustifolia* under climate change and perhaps even expansion, because a competitor is not as evolutionarily successful in adapting to change and hence will decline leaving an ecological niche to be filled [31].

Inter-specific interaction to temperature and precipitation change

Interspecies interactions play a key role in regulating the distribution and composition of plant communities. The impacts of inter-species interactions can be altered by external drivers such as climatic conditions or anthropogenic activities [38]. For well conserved boreal wetlands, climate change may affect wetland plant communities by directly limiting or fostering the performance of particular species or by altering species competition and abundance [10,39].

As the two dominant species in Sanjiang Plain wetlands of Northeast China, *C. angustifolia* competes with *G. spiculosa* both in mesic and hydric habitats. Based on sequentially documented surveys of plant communities in a mesic habitat from 172 sampling plots from 1973 to 2003, the "importance values", as the average of relative density, frequency and coverage which gives an overall estimate of the influence of importance of a plant species in the community, of *C. angustifolia* decreased from 0.55 in the 1970s to 0.50 in the 2000s and that of *G. spiculosa* from 0.30 in the 1970s to 0.13 in the 2000s, accompanied by an increase in mesophytes. Although importance values for both hygrophytes declined, the former decreased by 9%, while the latter decreased 57%. Another sequentially documented *Carex lasiocarpa* community in a hydric habitat showed that the importance value of *G. spiculosa* decreased from 0.19 in the 1970s to 0.13 in the 2000s, while that of *C. angustifolia* increased from 0.021 to 0.023 [40]. Consequently, *C. angustifolia* is more competitive than *G. spiculosa* and *G. spiculosa* is more environmentally sensitive than *C. angustifolia*, which might be partly attributed to the greater acclimated capacity of *C. angustifolia* responding to precipitation and/or temperature change (excluding anthropogenic activities). Firstly, the greatest above-ground biomass of *C. angustifolia* presents in the +4°C treatment (Fig. 6A), while that of *G. spiculosa* is in the +2°C and +4°C manipulations (Fig. 6C), suggesting that *C. angustifolia* might be more acclimated to greater warming, even occasional extreme high temperatures. Secondly, there is no significant difference in the precipitation manipulations for *C. angustifolia*'s leaf area (Table 3) or above-ground biomass (Table 3), while either -50% or +50% of precipitation significantly decreased the leaf area (Fig. 3B) in August and September and above-ground biomass (Fig. 6D) of *G. spiculosa*.

Considering the warming trend of 2.5–5.4°C in the Sanjiang Plain within this century [41], it is predicted that the distribution of *C. angustifolia* will expand while that of *G. spiculosa* will shrink. However, the composition dynamics of these communities will be determined both by the intrinsic biological characteristics of each species and the external environment including climate change [5,42]. The species-specific responses described in this research coupled with inter-species interactions demonstrate the complexity of multi-factor effects. In addition to climate, other factors, e.g. soil nutrient concentrations, will modify or attenuate temperature and precipitation effects [35]. Consequently, minimization of other external (anthropogenic) factors is required in order to assess climate change impacts and this can best be achieved using well-conserved boreal wetlands.

Implications for boreal wetland conservation and restoration

Boreal wetland ecosystems, especially those that are well conserved, have an innate resistance and acclimated self-regulation potential, allowing them to respond to climate change and variation in habitat conditions [7,8]. However, long-term stress caused by high temperature, extreme drought or flooding, or the degradation and eventual loss of dominant species or functional groups will alter fundamental ecosystem properties and processes, because changes in the abundance, production, or distribution of species are often sufficient to alter both structure and function of the ecosystem [43]. Therefore, investigating the thresholds of irreversible climate change effects is both urgent and critical to the conservation and restoration of boreal wetland ecosystems.

Stability and acclimation can by delineated by ecological thresholds. Within these thresholds, the species interactions of freshwater wetland plants tend to shift from competitive to facilitative with increased stress [44]; however, when the environment changes beyond these thresholds, perhaps through a combination of habitat alteration by climate change and direct damage by humans, boreal wetlands may shift to other ecosystems such as grassland or shrubland [45]. This may result in loss of the organic carbon sequestered in these wetlands for hundreds or thousands of years, ultimately reducing the net cooling effect of a carbon sink on global warming [7,45]. To avoid this consequence, conservation policies and restoration practices will need to be revised in the face of potential thresholds for irreversible change to key wetland plant species under climate change. For example, hydrologically damaged wetlands will experience greater vulnerability to climate change effects compared to wetlands with an intact hydrological regime [8]. According to our results (Fig. 6D), the stability of the natural surface water table is critical for the conservation and restoration of the *G. spiculosa* population, which indicates special actions may be required at both the management and policy level if this species is to be safeguarded in the future.

Acknowledgments

The authors express gratitude to the reviewers and editors for their critical comments on an earlier version of the manuscript. We gratefully acknowledge Jiawei Guo and Haiyang Zhao for seedling collection, and Yuxia Zhang and Bokun Lou for sample analyses.

Author Contributions

Conceived and designed the experiments: YZ GW. Performed the experiments: X. Lou. Analyzed the data: XY. Contributed reagents/materials/analysis tools: X. Lu. Contributed to the writing of the manuscript: YZ MG.

References

1. Gorham E (1991) Northern peatlands, role in the carbon cycle and probable responses to climatic warming. Ecol Appl 1:182–195.
2. Aerts R, Callaghan TV, Dorrepaal E, van Logtestijn RSP, Cornelissen JHC (2012) Seasonal climate manipulations have only minor effects on litter decomposition rates and N dynamics but strong effects on litter P dynamics of sub-arctic bog species. Oecologia 170:809–819.
3. Bartlett KB, Harriss RC (1993) Review and assessment of methane emissions from wetlands. Chemosphere 26:261–320.
4. Bridgham SD, Pastor J, Updegraff K, Malterer TJ, Johnson K, et al. (1999) Ecosystem control over temperature and energy flux in northern peatlands. Ecol Appl 9:1345–1358.
5. Weltzin JF, Bridgham SD, Pastor J, Chen J, Harth C (2003) Potential effects of warming and drying on peatland plant community composition. Global Change Biol 9:141–151.
6. Flanagan LB, Syed KH (2011) Stimulation of both photosynthesis and respiration in response to warmer and drier conditions in a boreal peatland ecosystem. Global Change Biol 17:2271–2287.
7. Dise N (2009) Peatland response to global change. Science 326:810–811.
8. Essl F, Dullinger S, Moser D, Rabitsch W, Kleinbauer I (2012) Vulnerability of mires under climate change, implications for nature conservation and climate change adaptation. Biodivers Conserv 21:655–669.
9. Charles H, Dukes JS (2009) Effects of warming and altered precipitation on plant and nutrient dynamics of a New England salt marsh. Ecol Appl 19:1758–1773.
10. Breeuwer A, Heijmans MMPD, Robroek BJM, Berendse F (2008) The effect of temperature on growth and competition between *Sphagnum* species. Oecologia 156:155–167.
11. Thompson RM, Beardall J, Beringer J, Grace M, Sardina P (2013) Means and extremes, building variability into community-level climate change experiments. Ecol Lett 16:799–806.
12. Editorial Committee of Flora Republicae Popularis Sinicae (1987) Flora Republicae Popularis Sinicae 9(3). Beijing, China: Science Press. 221–223 p.
13. Editorial Committee of Flora Republicae Popularis Sinicae (2002) Flora Republicae Popularis Sinicae 9(2). Beijing, China: Science Press. 327–328 p.
14. Ren G, Guo J, Xu M, Chu Z, Zhang L, et al. (2005) Climate changes of China's mainland over the past half century. Acta Meteorol Sin 63:942–956.
15. Guo R, Shi L, Ding X, Hu Y, Tian S, et al. (2010) Effects of saline and alkaline stress on germination, seedling growth, and ion balance in wheat. Agron J 102:1252–1260.
16. Arnon DI (1949) Copper enzymes in isolated chloroplasts Polyphenoxidase in *Beta vulgaris*. Plant Physiol 24:1–15.
17. Porra RJ (2002) The chequered history of the development and use of simultaneous equations for the accurate determination of chlorophylls a and b. Photosynth Res 73:149–156.
18. Zou YC, Jiang M, Yu XF, Lu XG, David JL, et al. (2011) Distribution and biological cycle of iron in freshwater wetlands of Sanjiang Plain, Northeast China. Geoderma 164:238–248.
19. Fenner N, Freeman C, Lock MA, Harmens H, Reynolds B, et al. (2007) Interactions between elevated CO_2 and warming could amplify DOC exports from peatland catchments. Environ Sci Technol 41:3146–3152.
20. Kardol P, Cregger MA, Campany CE, Classen AT (2010) Soil ecosystem functioning under climate change, plant species and community effects. Ecology 91:767–781.
21. Lin DL, Xia JY, Wan SQ (2010) Climate warming and biomass accumulation of terrestrial plants, a meta-analysis. New Phytol 188:187–198.
22. Wilson SD, Tilman D (1993) Plant competition and resource availability in response to disturbance and fertilization. Ecology 74:599–611.
23. Gough L, Gross KL, Cleland EE, Clark CM, Collins SL, et al. (2012) Incorporating clonal growth form clarify the role of plant height in response to nitrogen addition. Oecologia 169:1053–1062.
24. Berry J, Bjorkman O (1980) Photosynthetic response and adaptation to temperature in higher plants. Annu Rev Plant Physiol 31:491–543.
25. Rustad LE, Campbell JL, Marion GM, Norby RJ, Mitchell MJ, et al. (2001) A meta-analysis of the response of soil respiration, net nitrogen mineralization, and aboveground plant growth to experimental ecosystem warming. Oecologia 126:543–562.
26. De Boeck HJ, Lemmens CMHM, Zavalloni C, Gielen B, Malchair S, et al. (2008) Biomass production in experimental grasslands of different species richness during three years of climate warming. Biogeosciences 5:585–594.
27. Ge Z, Zhou X, Kellomäki S, Zhang C, Peltola H, et al (2012) Acclimation of photosynthesis in a boreal grass (*Phalaris arundinacea* L) under different temperature, CO_2, and soil water regimes. Photosynthetica 50:141–151.
28. Farquhar GD, Sharkey TD (1982) Stomatal conductance and photosynthesis. Annu Rev Plant Physiol 33:317–345.
29. Evans JR (1989) Photosynthesis and nitrogen relationship in leaves of C_3 plants. Oecologia 78:9–19.
30. Mae T (1997) Physiological nitrogen efficiency in rice, nitrogen utilization, photosynthesis, and yield potential. Plant Soil 196:201–210.
31. Weltzin JF, Pastor J, Harth C, Bridgham SD, Updegraff K, et al. (2000) Response of bog and fen plant communities to warming and water-table manipulations. Ecology 81:3464–3478.
32. Erwin KL (2009) Wetlands and global climate change, the role of wetland restoration in a changing world. Wetl Ecol Manage 17:71–84.
33. Liancourt P, Spence LA, Song D, Lkhagva A, Nharkhuu A, et al. (2013) Plant response to climate change varies with topography, interactions with neighbors, and ecotype. Ecology 94:444–453.
34. Van Minnen JG, Onigkeit J, Alcamo J (2002) Critical climate change as an approach to assess climate change impacts in Europe, development and application. Environ Sci Policy 5:335–347.

35. Luo WB, Xie YH (2009) Growth and morphological responses to water level and nutrient supply in three emergent macrophyte species. Hydrobiologia 624:151–160.

36. Gedan KB, Bertness MD (2010) How will warming affect the salt marsh foundation species *Spartina patens* and its ecological role? Oecologia 164:479–487.

37. Qiu ZC, Wang M, Lai WL, He FH, Chen ZH (2011) Plant growth and nutrient removal in constructed monoculture and mixed wetlands related to stubble attributes. Hydrobiologia 661:251–260.

38. Brooker RW (2006) Plant-plant interactions and environmental change *New Phytol*, 171, 271–284.

39. Pastor J, Peckham B, Bridgham S, Weltzin J, Chen J (2002) Plant community dynamics, nutrient cycling and alternative stable equilibria in peatlands. Am Nat 160:553–568.

40. Lou YJ (2008) Species biodiversity spatial patterns of typical wetland plants and their changes within 30 years in the Sanjiang Plain. Graduate University of Chinese Academy of Sciences, Beijing, China.

41. Yin XM (2013) Simulation Study of Climate Change impacts on Wetlands Productivity in Sanjiang Plain. Graduate University of Chinese Academy of Sciences, Beijing, China.

42. Hoeppner SS, Dukes JS (2012) Interactive responses of old-field plant growth and composition to warming and precipitation. Global Change Biol 18:1754–1768.

43. Berry PM, Dawson TP, Harrison PA, Pearson RG (2002) Modelling potential impacts of climate change on the bioclimatic envelope of species in Britain and Ireland. Global Ecol Biogeogr 11:453–462.

44. He Q, Bertness MD, Altieri AH (2013) Global shifts towards positive species interactions with increasing environmental stress. Ecol Lett 16:695–706.

45. Hobbs RJ, Higgs E, Harris JA (2009) Novel ecosystems, implications for conservation and restoration. Trends Ecol Evol 24:599–605.

14

Alfalfa Cellulose Synthase Gene Expression under Abiotic Stress: A Hitchhiker's Guide to RT-qPCR Normalization

Gea Guerriero*, Sylvain Legay, Jean-Francois Hausman

Department Environment and Agro-biotechnologies (EVA), Centre de Recherche Public, Gabriel Lippmann, Belvaux, Luxembourg

Abstract

Abiotic stress represents a serious threat affecting both plant fitness and productivity. One of the promptest responses that plants trigger following abiotic stress is the differential expression of key genes, which enable to face the adverse conditions. It is accepted and shown that the cell wall senses and broadcasts the stress signal to the interior of the cell, by triggering a cascade of reactions leading to resistance. Therefore the study of wall-related genes is particularly relevant to understand the metabolic remodeling triggered by plants in response to exogenous stresses. Despite the agricultural and economical relevance of alfalfa (*Medicago sativa* L.), no study, to our knowledge, has addressed specifically the wall-related gene expression changes in response to exogenous stresses in this important crop, by monitoring the dynamics of wall biosynthetic gene expression. We here identify and analyze the expression profiles of nine cellulose synthases, together with other wall-related genes, in stems of alfalfa plants subjected to different abiotic stresses (cold, heat, salt stress) at various time points (e.g. 0, 24, 72 and 96 h). We identify 2 main responses for specific groups of genes, i.e. a salt/heat-induced and a cold/heat-repressed group of genes. Prior to this analysis we identified appropriate reference genes for expression analyses in alfalfa, by evaluating the stability of 10 candidates across different tissues (namely leaves, stems, roots), under the different abiotic stresses and time points chosen. The results obtained confirm an active role played by the cell wall in response to exogenous stimuli and constitute a step forward in delineating the complex pathways regulating the response of plants to abiotic stresses.

Editor: Olga A. Zabotina, Iowa State University, United States of America

Funding: Financial support was obtained through the Fonds National de la Recherche Luxembourg (FNR) Project CANCAN C13/SR/5774202 and through internal funding sources. The funders had no role in study design, data collection and analysis, decision to publish, or preparation of the manuscript.

Competing Interests: The authors have declared that no competing interests exist.

* Email: guerrier@lippmann.lu

Introduction

The study of biological phenomena requires several sensitive analytical techniques, which can convey detailed information at different depths of organismal complexity, namely tissular, metabolic, genomic. One such type of information is represented by gene expression changes, which provide clues about transcripts dynamics, e.g. in response to exogenous stimuli.

Currently one of the most reliable and reproducible methods to perform differential gene expression profiling is quantitative reverse transcription PCR (hereafter referred to as RT-qPCR), a method which is robust enough to quantify challenging targets, as microRNAs (miRNAs) e.g. [1]. However, accurate gene expression analyses rely on several critical aspects and experimental steps (namely RNA purity and integrity, genomic DNA contamination, reverse transcription) and, in the case of relative quantification, on the identification of suitable reference genes for data normalization [2–3]. Those are genes whose expression is stable and not subject to fluctuations across the different conditions tested. This feature is particularly critical, as the choice of inappropriate reference genes can significantly bias the results obtained and therefore lead to misinterpretations of biological events.

The use of RT-qPCR is particularly suitable to study the response of a set of genes in plants after the application of specific stresses e.g. [4]: being sessile organisms, plants are not capable of escaping from adverse environmental conditions and are therefore characterized by a very responsive transcriptional regulation, which results in phenotypic plasticity [5–7]. Abiotic stresses constitute serious threats for plants, as they can affect not only their development, growth, reproduction and productivity, but can be so detrimental to cause their death. Exogenous stresses unleash a cascade of reactions, which lead to plant response and resistance, usually by means of wall fortification.

Many studies in the literature have provided a comprehensive view of gene expression changes in different plant species in response to abiotic stresses and identified a list of suitable reference genes for data normalization e.g. [8–13]. These studies have also shown how the expression of reference genes can vary in different plant species and conditions and how important it is to validate their stability in the specific experimental set-ups used.

Despite the agricultural and economical importance of the legume crop *Medicago sativa* L. (a.k.a alfalfa, or lucerne), no study has so far tested suitable reference genes for expression analysis using RT-qPCR in this plant. Suitable reference genes have been identified in *Medicago truncatula* [14] and potential reference genes in alfalfa have been proposed by Yang et al. [15], however their suitability for RT-qPCR studies has, to our knowledge, never been validated so far.

Alfalfa is an experimentally valuable model: it is not only suitable for the study of symbiotic interactions e.g. [16], but has also been proposed as an excellent model system to study dicot cell wall development [17]. Its stem shows indeed 2 clearly-defined regions characterized by active elongation and lignification/thickening, which provide "snapshots" of the cell wall maturation process. Although the genome of alfalfa has not yet been sequenced, several studies have shown the suitability of using the genome of the closely related barrel medic (*M. truncatula*) [18] for molecular analyses. These studies have delivered valuable information concerning the regulation of wall polysaccharide biosynthesis in cultivars with contrasting cell wall composition [19].

The aim of the present study is to provide a time-course analysis of cell wall-related gene expression in response to different abiotic stresses in alfalfa stems. In particular we analyzed nine cellulose synthases (hereafter named *MsCesAs*), identified on the basis of the sequence homology with the orthologs from *M. truncatula* (*MtCesAs*), together with other genes linked to wall biogenesis (namely sucrose synthase, *SuSy*; phenylalanine ammonia lyase, *PAL*; cinnamyl alcohol dehydrogenase, *CAD*; cellulose synthase-like gene, *CslD4*). To perform a reliable gene expression analysis, accurate data normalization is mandatory, which prompted us to identify the most suitable reference genes for expression analysis. We chose as candidate reference genes a set of known and widely used genes which have been tested on *M. truncatula* [14], together with candidates proposed by Yang et al. [15] and Huis et al. [20].

We decided to extend our survey not only to stems, but also to other tissues (namely roots and leaves), in order to provide a list of genes to be used in tissue- and/or growth condition-specific studies in alfalfa and which can be tested in other legume crops too. Ten candidate reference genes were chosen and their reliability for RT-qPCR studies tested at different time-points in different tissues of *M. sativa* plants exposed to abiotic stresses. To further validate their suitability, we studied the expression of a stress-associated kinase (*SK1*) in the different tissues and growth conditions.

Despite the unanimously recognized role of plant walls as cellular structures sensing and responding to stimuli [21–24] and despite the economic significance of alfalfa, to our knowledge no study is yet available on the expression analysis of key wall biosynthetic genes (as *CesAs*) in response to different abiotic stresses in *M. sativa*. We here provide such a study and identify the main trends characterizing the response to abiotic stresses in alfalfa stems.

Materials and Methods

Plant growth and abiotic stress treatments

Medicago sativa L. seeds, variety Giulia (Italy), were inoculated with a peat-based inoculant (HiStick, Becker Underwood) according to the manufacturer's instructions. Five seeds were sown per pot in 1 L containers filled with soil (50% topsoil, 25% potting soil, 25% sand). After 4 weeks of cultivation under controlled greenhouse conditions (photoperiod of 13 h light/11 h darkness, minimum temperature of 20°C, maximum 27°C), plants were moved to incubators (programmed to provide exactly the same light/dark cycles) for moderate cold and heat stress treatments, while those subjected to moderate salt stress were left in the greenhouse. For the cold stress condition, plants were grown at a constant temperature of 5°C; for the heat stress condition, they were grown at 28°C/32°C (night/day); for the salt stress treatment, plants were supplemented with 100 mM NaCl. A total of 3 biological replicates (each consisting of a pool of 15 plants) were used per treatment. For each time point studied (0- 24- 72-

96 h) a control group was always kept (i.e. plants grown without any treatment for 24- 72- 96 h), for appropriate comparisons.

RNA extraction and cDNA synthesis

Sampled tissues (roots, leaves and whole stems) were ground to a fine powder in liquid nitrogen, using a mortar and a pestle. One hundred mg of finely ground sample were weighed on a balance and total RNA was extracted using the RNeasy Plant Mini Kit with the on-column DNase I treatment (Qiagen). The integrity of the extracted RNA was checked with an Agilent Bioanalyzer (all the RINs were >8) and the purity/concentration measured using a NanoDrop ND-1000 spectrophotometer (A260/280 and A260/230 ratios between 1.9 and 2.2). Subsequently, 1 μg of extracted RNA was retro-transcribed using the Superscript II cDNA Synthesis kit (Invitrogen), according to the manufacturer's instructions.

Identification of CesA genes from alfalfa

To identify and amplify putative *CesA* genes from *M. sativa*, initial data mining was performed on *M. truncatula* [25], a closely related species for which the genome is available. A total of nine putative *M. truncatula* cellulose synthase proteins (hereafter indicated MtCESAs) were identified. BLASTp searches were performed against non-redundant protein databases of *Arabidopsis thaliana* and *Populus trichocarpa* from the National Centre for Biotechnology [26] to check the percentage of identity of the identified sequences. To amplify the orthologous genes from alfalfa, primers were designed on the identified *MtCesAs* genes (listed in Table S1). Three full-length *CesA* genes from alfalfa were identified and designated *MsCesA3*, *MsCesA4* and *MsCesA7-A* [GenBank: KJ398155, KJ398156, KJ398157; Fig. S1], on the basis of their phylogenetic kinship, while partial sequences were obtained for the other *MsCesAs* (Figs. S1 and S2). The phylogenetic tree was built by aligning the amino acid regions of CESAs from *M. truncatula*, *M. sativa*, *P. trichocarpa* and *A. thaliana* encompassing the U1–U4 regions, the QXXRW motif and the HVR2 region, which allows class discrimination [27], using MUSCLE [28]. Phylogeny was analyzed using PhyML [29]. The maximum-likelihood phylogenetic tree was rendered using TreeDyn [30]. Microarray data for *M. truncatula CesAs* were retrieved at [31] and electronic fluorescent pictographic (eFP) representations at [32].

Quantitative real-time PCR and statistical analysis

For quantitative real-time PCR analysis, 10 ng cDNA were used as template. The cDNA was amplified using the MESA GREEN qPCR MasterMix Plus.

Low ROX (Eurogentec) on a ViiA 7 Real-Time PCR System (Applied Biosystems) in a final volume of 25 μl.

The reactions were performed in technical triplicates and repeated on the above-described 3 biological independent replicates. The PCR conditions consisted of an initial denaturation at 95° for 10 min, followed by 45 cycles of denaturation at 95° for 15 sec, annealing/extension at 60° for 60 sec.

A dissociation kinetics analysis was performed at the end of the experiment to check the specificity of the annealing.

Ten candidate reference genes were analyzed, namely actin, tubulin, ubiquitin-conjugating protein 13 (UBC13), cyclophilin (cyclo), elongation initiation factor 4A (eif4A), elongation initiation factor 5A (eif5A), translation initiation factor IIA (TFIIA), glyceraldehyde-3P dehydrogenase (GAPDH), actin-depolymerizing protein (ADF1), poly(A) binding protein 4 (PAB4). Their stability was evaluated using NormFinder [33] and geNorm[PLUS] [2], two of the most commonly used software, which rank candidate reference genes on the basis of their stability. The software geNorm[PLUS] performs a pairwise comparison and computes the M-value, i.e. the variation of a gene compared to all the remaining candidates, while NormFinder computes first the intra-group and subsequently the inter-group expression variability of a candidate reference gene [33–34]. NormFinder calculates both a single best gene (best gene) and an optimal gene pair (best pair); the best pair might display compensating expression in the different experimental groups. The candidate reference genes primers for actin, tubulin, GAPDH were designed using the sequences from *M. truncatula* [GenBank: XM_003621971, XM_003603622, XM_003595990]. The primers for the other reference genes were designed using the sequences of the candidate housekeeping genes reported by Yang et al. [15], which show an average RPKM-normalized value higher than 10 and the lowest coefficient of variation identified with RNAseq. The list of primers used to perform RT-qPCR analyses is shown in Table S2. The RT-qPCR primers for *CAD* and *CslD4* (Table S2) were designed on the sequences from *M. truncatula* genes (probesets Mtr.8985.1.S1_at and Mtr.45005.1.S1_at, respectively) [19], while those for *SuSy* and *PAL* (Table S2) were designed on the reported sequences from alfalfa (probeset Msa.2902.1.S1_at) [19] and [GenBank: CAA41169]. Primers were designed using Primer3Plus [35] and analyzed with OligoAnalyzer 3.1 [36]. The primers size was 20 bp, the amplicon sizes were between 70–150 bp (Table S2), the %GC was between 40–60% and Tm 60°C. The primers used were not intron spanning. Primer efficiencies were tested and are reported in Table S2. All the amplicons were verified by sequencing on an Applied Biosystems 3500 Genetic Analyser using the BigDye Terminator v3.1 Cycle Sequencing and the BigDye XTerminator Purification kits, according to the manufacturer's instructions.

The results relative to the expression of the target genes were analyzed using the software qBase[PLUS] version 2.5 (Biogazelle, [37]) and normalized taking into account the most stable reference genes (as indicated in the text). The expression levels of the genes detected in the different tissues and conditions analyzed are here expressed as "Normalized relative expression". A one-way ANOVA (with Tukey's HSD post-hoc test) was performed on the log_2 transformed calibrated normalized relative quantities (CNRQs), using IBM SPSS Statistics (version 19), after having checked the normal distribution of the data with a Kolmogorov-Smirnov test.

Hierarchical clustering was generated with Cluster 3.0 [38] and visualized with Java TreeView [39], available at [40].

Results

Stability of putative reference genes in different tissues of *M. sativa* subjected to abiotic stresses

Analyses with geNorm[PLUS] were performed to rank the expression stability of the 10 candidate reference genes in the tissues and conditions analyzed (Fig. 1). According to geNorm[PLUS], TFIIA ranks among the most stable genes in roots, leaves and stems, but interestingly this gene is not among the most stably expressed when all the tissues are grouped together (Fig. 1). However NormFinder ranks TFIIA among the 4 most stable genes when all the tissues are taken into account (Table 1). The gene eIF4A is very stable in leaves and stems, while it is among the least stable in roots (Fig. 1). The most stably expressed genes when all the tissues are grouped together are ADF1 and PAB4 (Fig. 1). Actin is among the least stable genes in all the conditions tested (Fig. 1). These results show that care should be taken when choosing candidate reference genes for expression

analysis in different plant tissues, as stable genes in a specific tissue might not be suitable for normalization of expression data in another one. In the literature, several studies have shown the importance of determining the stability of the reference genes in the different plant tissues, in order to use the most reliable ones in the condition examined e.g. [11,20]. The stability data obtained with geNorm[PLUS] have been compared to the rankings generated via the other widely used software, i.e. NormFinder. Ranking lists were generated for each single tissue under all the conditions tested, for all the tissues together under each single condition (which can be of particular interest when a specific abiotic stress is studied), as well as for all the tissues and conditions together (Table 1). From the rankings it is possible to confirm the expression stability of TFIIA in the single tissues under the different stress treatments; in particular the pairs ADF1/TFIIA and eIF4a/TFIIA are confirmed as the most suitable genes for normalization in roots and stems respectively (Table 1). PAB4 ranks always among the 5 most stable genes when all the tissues are analyzed together in each of the conditions tested, while the high stability of eIF5A and PAB4 is confirmed when all the tissues and conditions are taken into consideration (Table 1). Both geNorm[PLUS] and NormFinder show how unsuitable tubulin and actin are for data normalization in our experimental system: this is quite important, as these genes, although suitable for normalization in some instances [41–42], might not be ideal in others [8,20,43]. The best gene pairs identified by NormFinder are different in some tissues from those identified by geNorm[PLUS], a finding which has already been reported in other studies e.g. [20] and might be due to the different ranking methodology used by the two softwares: normalization in the leaves requires TFIIA and eIF4A according to geNorm[PLUS], while NormFinder suggests ADF1 and PAB4, which in the geNorm[PLUS] ranking are the 4th and 7th least stable gene, respectively (Fig. 1). Similarly, if all the tissues and stresses are considered, the best gene pair is GAPDH/PAB4 according to NormFinder, however geNorm[PLUS] ranks GAPDH as the most unstable gene in this configuration (Fig. 1).

Nevertheless, taking into account the rankings of NormFinder and geNorm[PLUS], it emerges that for expression studies in alfalfa tissues (and possibly in other legume crops) actin and tubulin are not ideal, whereas a suitable panel of reference genes should include eIF4A, PAB4, ADF1 and TFIIA, as they rank among the most stable genes according to the two softwares. This result can be of particular interest when studying gene expression in different plant tissues subjected to a specific treatment: if a tissue-maximization strategy is selected in the experimental design, it is helpful to know *a priori* which panel of candidates to include for stability test.

Optimal number of reference genes for normalization in *M. sativa* tissues using geNorm[PLUS] ranking

In order to calculate the appropriate number of reference genes for data normalization in alfalfa, we used geNorm[PLUS] to compute the pairwise variation (Vn/Vn+1) between two consecutive normalization factors (NFn and NFn+1). The analysis shows that for accurate normalization in roots, stems and leaves, 2 reference genes are required: the addition of a third gene is indeed not necessary, as the V value relative to 2 reference genes is already below the cut-off threshold of 0.15 (Fig. 2). However, if all the tissues are grouped together, the number of genes required for accurate data normalization increases to 3, since the V value relative to 2 genes is above the cut-off threshold (0.159) (Fig. 2).

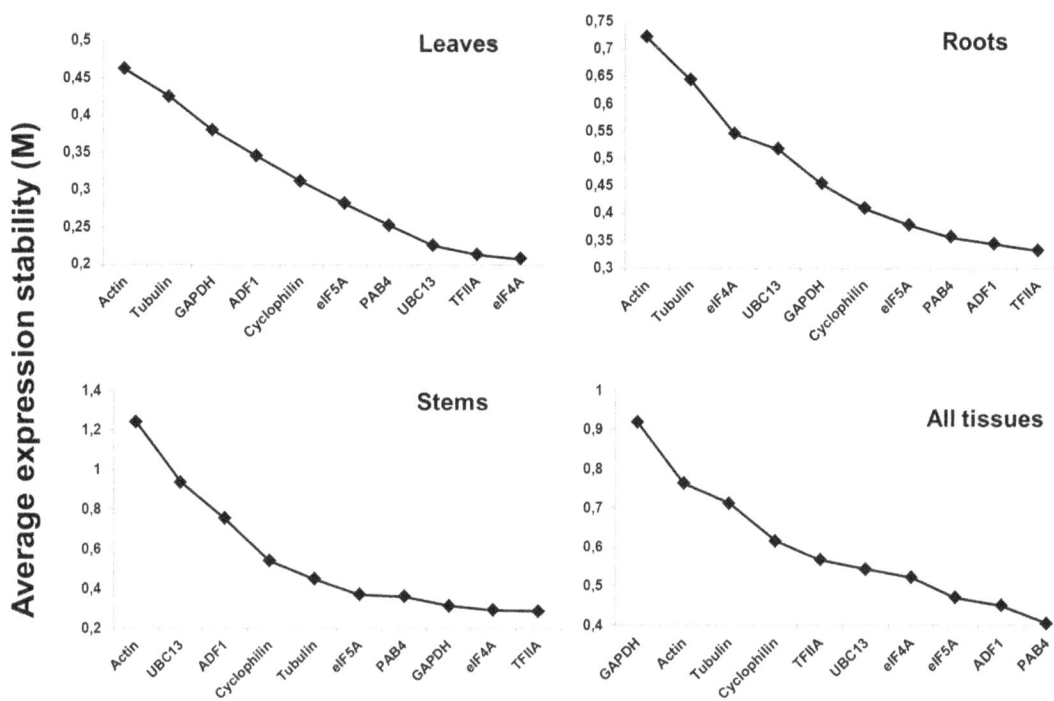

Figure 1. Candidate reference genes in alfalfa. Ranking of ten candidate reference genes in different tissues of *M. sativa* according to the parameter M calculated by geNorm[PLUS]. Increasing stability of the candidate genes is determined by a decrease in the M value.

Validation of the selected reference genes in different tissues

The validity of the candidate reference genes identified via the geNorm[PLUS] and NormFinder analyses was tested in the different tissues and conditions by studying the expression profiles of a stress-associated kinase orthologous to *MtSK1* [GenBank: XP_003592980] [44]. This gene is a member of the SnRK group of plant kinases and was shown to be induced upon wounding in cultured tissues [44].

Since SnRKs are involved in stress response in plants e.g. [45], we decided to use this gene both to validate the identified reference genes in the different conditions and to study its expression profile in response to different abiotic stresses in alfalfa tissues. It was assumed that the experimental treatment would not alter the expression of the reference genes, but would instead affect the expression of the stress-associated kinase. The data were analyzed with qBASE[PLUS] and normalized using ADF1/.

TFIIA and eIF4A/TFIIA for the roots and stems respectively, since these candidates were selected by both geNorm[PLUS] and NormFinder, then a comparison of normalization strategies was performed for the leaves (Figs. 3 and 4), since the two softwares chose different candidates (namely ADF1/PAB4 by NormFinder and TFIIA/eIF4A by geNorm[PLUS]; Fig. 1 and Table 1). As can be seen in Fig. 3, the stresses which triggered the most significant changes were cold and heat: in all the tissues examined, a significant decrease in expression could indeed be observed during cold stress treatment, while heat stress induced expression, where the highest increase was present in roots. Salt stress, on the other hand, did not appreciably change the expression of the stress-associated kinase, apart from a mild increase at 24 and 72h in the stems (Fig. 3). This result was unexpected, as it was previously shown that the expression of the ortholog from *M. truncatula* increased in the leaves after salt stress treatment [44], however it should be noted that the analysis was here performed on another

species and that fluctuations in expression were observed in control condition over the different time-points (Fig. 3). These fluctuations contribute to make the expression changes not significant.

In order to compare the normalization strategies using the gene pairs recommended by NormFinder and geNorm[PLUS], we chose to perform a test on the leaves, since for the roots and the stems the two softwares agreed on the best gene pairs (Fig. 1 and Table 1). As can be seen in Fig. 4, the expression trend in response to the different stresses did not change: different normalized relative expression values could be observed for a same time point between the NormFinder and geNorm[PLUS] normalization (Fig. 4). In particular, higher error bars could be observed at some time points (e.g. 24 h heat, 72 h heat, 96 h heat) for the expression values obtained with NormFinder-based normalization (Fig. 4): this is most likely a reflection of the intrinsic computing differences of the two algorithms. However the Student's t-test did not show statistically significant differences between the magnitude changes calculated by NormFinder and geNorm[PLUS] (not shown).

Identification and phylogenetic analysis of CesAs from *M. truncatula* and *M. sativa*

In silico analysis of *M. truncatula* genome led to the identification of 9 putative *CesA* genes (Table 2). On the basis of the amino acid sequence identity with the orthologs from *A. thaliana* and poplar, a nomenclature is here proposed (Table 2) which follows the one recently proposed for *Populus* [46]. *M. truncatula* CESAs are between 981 and 1098 amino acids long and show from 6 to 8 transmembrane domains (TMDs; Table S3) according to the parameters of TMHMM [47]. The CESAs showing 6 TMDs actually display the occurrence of 2 additional potential TMDs, which however do not reach the critical threshold of the software (not shown). Therefore, since the CESAs so far described typically show the occurrence of 8 TMDs, the

Table 1. Ranking of candidate reference genes according to NormFinder.

	Leaves (all treatments)		Roots (all treatments)		Stems (all treatments)		Control (all tissues)		Cold (all tissues)		Heat (all tissues)		Salt (all tissues)		All tissues and treatments	
	Gene	Stability	Gene	Stability	Gene	Stability	Gene	Stability	Gene	Stability	Gene	Stability	Gene	Stability	Gene	Stability
	Act	0.150	Act	0.251	Act	0.306	Act	0.117	Tub	0.148	Tub	0.280	ADF1	0.154	Tub	0.168
	Tub	0.126	Tub	0.211	ADF1	0.153	TFIIA	0.109	eIF5A	0.125	Act	0.202	Act	0.105	Act	0.133
	eIF5A	0.103	Cyclo	0.139	Tub	0.143	Tub	0.108	Act	0.106	eIF5A	0.118	Cyclo	0.080	UBC13	0.084
	GAPDH	0.091	UBC13	0.138	Cyclo	0.128	eIF5A	0.104	TFIIA	0.093	Cyclo	0.116	Tub	0.068	eIF4A	0.076
	UBC13	0.087	eIF5A	0.110	UBC13	0.127	GAPDH	0.059	GAPDH	0.081	UBC13	0.112	eIF5A	0.073	Cyclo	0.076
	Cyclo	0.086	PAB4	0.102	eIF5A	0.102	Cyclo	0.053	ADF1	0.064	GAPDH	0.109	TFIIA	0.050	ADF1	0.074
	ADF1	0.072	eIF4A	0.094	PAB4	0.081	eIF4A	0.053	eIF4A	0.054	PAB4	0.085	GAPDH	0.049	TFIIA	0.067
	eIF4A	0.056	GAPDH	0.089	eIF4A	0.067	UBC13	0.051	UBC13	0.051	eIF4A	0.080	PAB4	0.046	eIF5A	0.058
	TFIIA	0.051	TFIIA	0.071	GAPDH	0.062	ADF1	0.040	PAB4	0.047	ADF1	0.074	UBC13	0.041	GAPDH	0.055
Best gene	PAB4	0.043	ADF1	0.036	TFIIA	0.046	PAB4	0.033	Cyclo	0.041	TFIIA	0.040	eIF4A	0.029	PAB4	0.050
Best pair	ADF1/PAB4	0.033	ADF1/TFIIA	0.040	TFIIA/eIF4A	0.040	Cyclo/eIF4A	0.019	Cyclo/PAB4	0.033	Cyclo/GAPDH	0.042	GAPDH/UBC13	0.020	GAPDH/PAB4	0.010

The best gene and the best combination of genes are shown. The analysis has been carried out to find the most stable reference genes in the different tissues under all the treatments tested, in all the tissues under different treatments and when all the tissues and treatments studied are grouped together. Abbreviations here used: Act (actin), Tub (tubulin), GAPDH, Cyclo (cyclophilin).

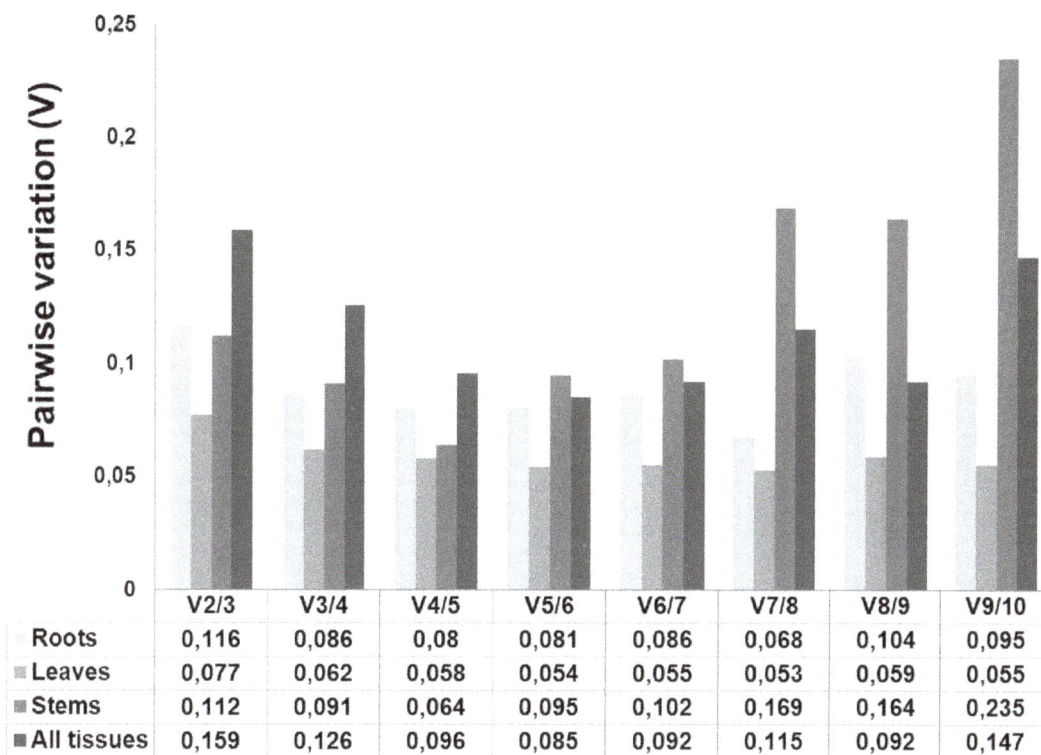

	V2/3	V3/4	V4/5	V5/6	V6/7	V7/8	V8/9	V9/10
Roots	0,116	0,086	0,08	0,081	0,086	0,068	0,104	0,095
Leaves	0,077	0,062	0,058	0,054	0,055	0,053	0,059	0,055
Stems	0,112	0,091	0,064	0,095	0,102	0,169	0,164	0,235
All tissues	0,159	0,126	0,096	0,085	0,092	0,115	0,092	0,147

Figure 2. Determination of the appropriate number of reference genes for data normalization in *M. sativa* tissues under abiotic stress conditions, as computed by geNorm^PLUS. The pairwise variation (Vn/Vn+1) was calculated between the normalization factors NFn and NFn+1. The recommended cut-off threshold of 0.15 was kept in the present study.

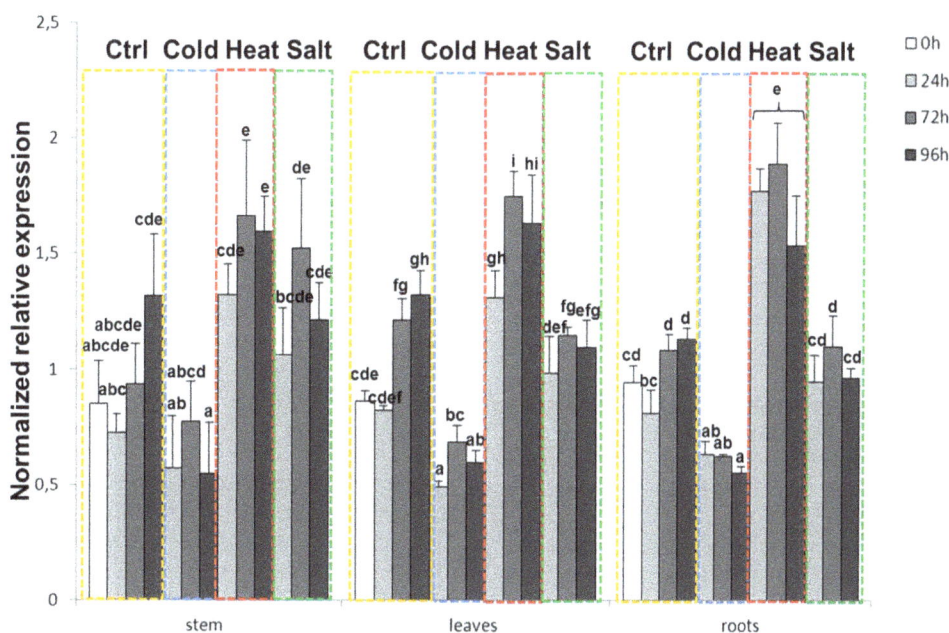

Figure 3. *MsSK1* expression in alfalfa tissues under abiotic stresses. Expression profiles of *MsSK1* in the different tissues under abiotic stresses (yellow dotted frame is control; blue dotted frame is cold stress; red dotted frame is heat stress; green dotted frame is salt stress). The Y-axis indicates NRE (Normalized Relative Expression of *MtSK1*). Data were normalized using ADF1/TFIIA and eIF4A/TFIIA for the roots and stems respectively and TFIIA/eIF4A for the leaves. Means sharing a letter are not significantly different at α = 0.05.

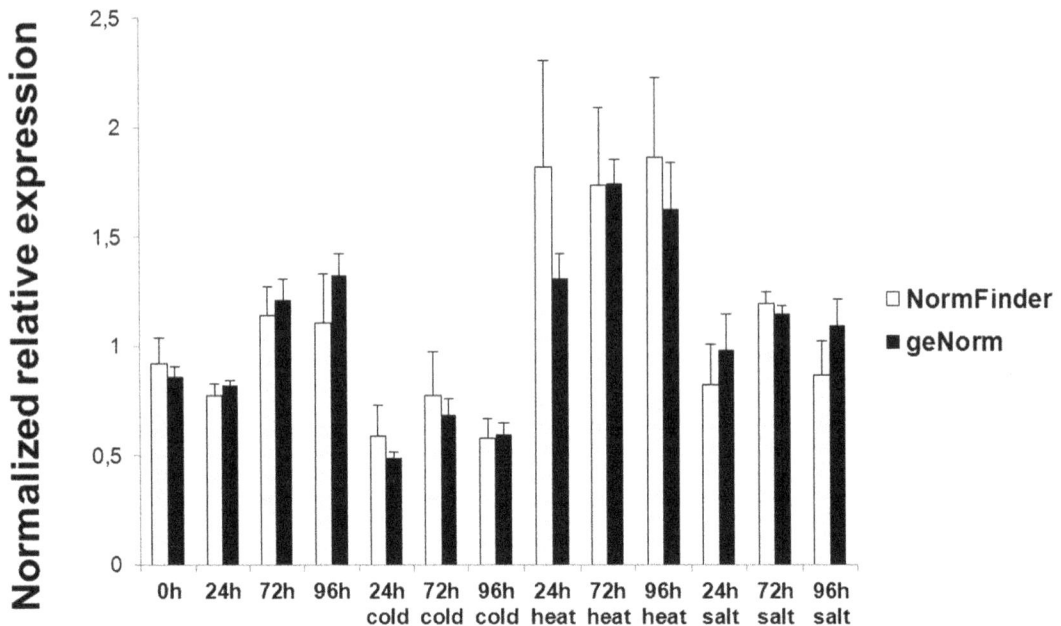

Figure 4. Comparison of NormFinder and geNorm^PLUS normalization methods. Comparison of *MsSK1* expression profiles in leaves when normalization is performed using ADF1/PAB4 (according to NormFinder), or TFIIA/eIF4A (according to geNorm^PLUS).

alfalfa proteins might as well all share the same feature. All the proteins show the occurrence of the signature motif typical of processive glycosyltransferases from family 2 (GT2s), i.e. D, D, DxD, QxxRW; MtCESA6-F, however, shows amino acid substitutions in the conserved motif (Fig. S3). The genes also have the zinc-finger domain (CxxC)4 (Fig. S3). Other genes with amino acid substitutions in the processive GT2s motif have been classified as CESAs (i.e. in *Cicer aretinum* and *Phaseolus vulgaris*) [GenBank: XP_004499618.1, ESW20735.1] (Fig. S3), moreover phylogenetic and blast analyses both classify MtCESA6-F as a putative CESA and assign it to the primary CESAs clade (Fig. 5). Therefore this gene was assigned to the *CesA6* branch and retained for expression analysis.

Phylogenetic analysis showed the occurrence of 6 CESA clades with proteins involved in primary and secondary cell wall biosynthesis (Fig. 5). MtCESA1, MtCESA3, MtCESA6-B, MtCESA6-C, MtCESA6-F belong to the primary cell wall clade, while MtCESA4, MtCESA7-A, MtCESA7-B and MtCESA8 belong to the secondary cell wall clade (Fig. 5). Although MtCESA7-B and MtCESA6-F show the lowest % identity (Table 2), the phylogenetic tree classifies them as representatives of the CESA6 and CESA7 clade respectively (Fig. 5). The branches relative to these genes correspond to higher evolutionary distance (Fig. 5), a finding, which might indicate different roles with respect to their paralogs. Nevertheless the branch support values for the CESA6-E/F and CESA7-A/B clades are high (98 and 100%, respectively; Fig. 5).

The phylogenetic analysis shows, as expected, that orthologous genes from different species are more related than homologs from the same species [48]. Some of the identified CESAs are represented by different genes in *M. truncatula*. Three orthologs of AtCESA6 are present: these genes might display specific roles in primary cell wall biosynthesis, but it is possible that they participate in secondary cell wall biosynthesis too, since poplar CESA6-E and CESA6-F were shown to be part of one of the two types of complexes found in differentiating xylem [49]. The

CESA6 members group together with the *A. thaliana* CESA2, CESA5, CESA9: this finding reflects their possible interchangeability in the primary CESA complex [50–51]. In the secondary cell wall clade, the occurrence of 2 AtCESA7 orthologs is observed. This is especially interesting if one considers that the presence of 2 *CesA7* and *CesA8* genes is a reported feature for woody angiosperms such as poplar, where the biosynthesis of wood represents an important process. Further functional characterizations are necessary to unveil the role of the 2 CESA7 in *M. truncatula*. However, in the light of the specialization and promiscuity that the different CESAs display, e.g. mucilage or seed coat biosynthesis [52–54], involvement in both primary and secondary cell wall biosynthesis or formation of mixed complexes [51,55–56], it is plausible to hypothesize that these 2 proteins co-participate in the assembly of secondary wall complexes and/or possess specific functions in cell wall biosynthesis. The tissue-specific expression of *M. truncatula CesAs* obtained from publicly available microarray data [31] confirmed the annotation of the genes into the primary and secondary clades: as can be seen from Fig. 6, the primary *CesAs* show a homogeneous expression in roots, leaves and stems, while the secondary display a higher expression in the stems. Notably the primary *CesAs MtCesA3* and *MtCesA1* show a high level of expression in the stems (Fig. 6 and Fig. S4), a finding which suggests a role for these genes in alfalfa stem cell wall biosynthesis.

Three full-length *CesA* sequences from *M. sativa* have been here obtained [GenBank: KJ398155, KJ398156, KJ398157; Fig. S1]; the phylogenetic analysis classifies them as MsCESA3, MsCESA4 and MsCESA7-A (Fig. 5). Partial sequences have been obtained for the other *CesAs* of *M. sativa* (Figs. S1 and S2).

Cell wall-related genes from *M. sativa* show two main trends in response to abiotic stresses

Variations in the expression pattern in response to abiotic stresses can be observed among the different cell wall-related genes. From the Heat Map visualization, it is possible to discern

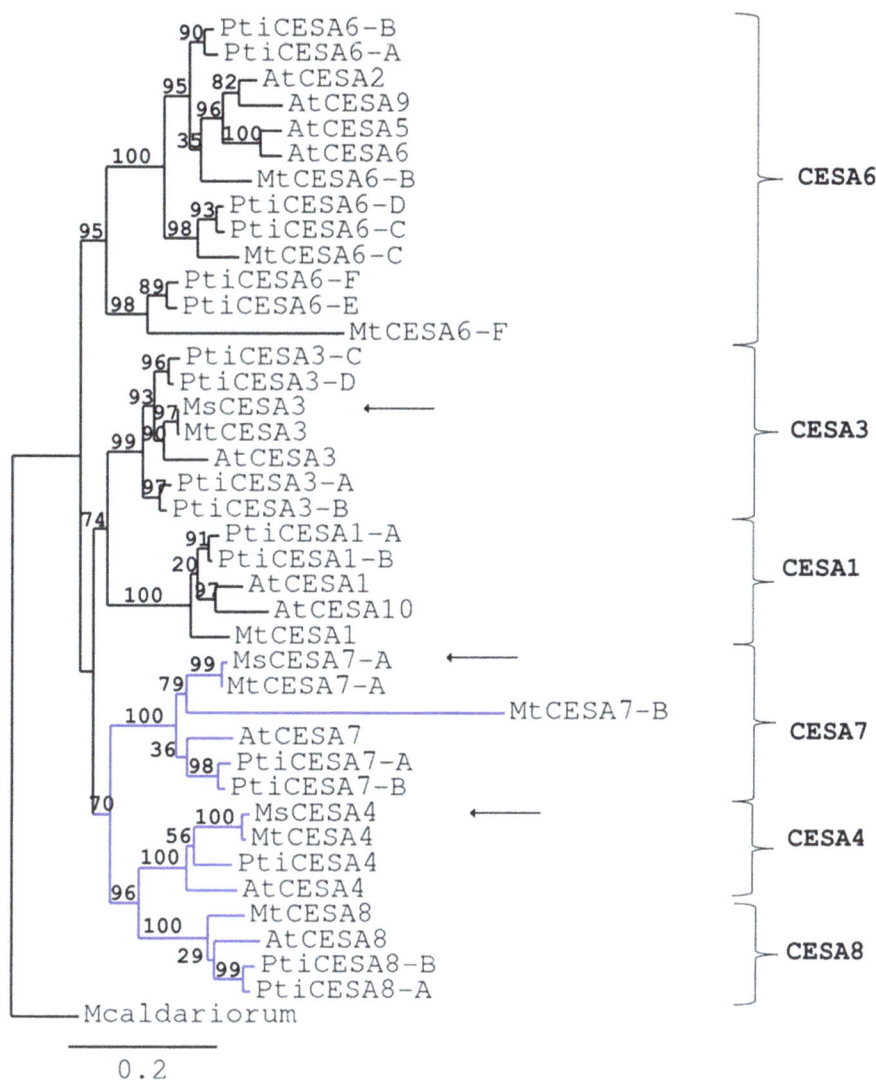

Figure 5. Phylogenetic relationships of CESAs from *M. truncatula, M. sativa, P. trichocarpa* and *A. thaliana* by maximum-likelihood analysis. Bootstrap = 100. Numbers indicate percentage of branch support values. The *scale bar indicates* an evolutionary distance of 0.2 amino acid substitutions per positions. *Mesotaenium caldariorum* CESA1 [GenBank: AAM83096] was used as outgroup to root the tree. The branch of secondary CESAs is indicated in blue. Arrows point to the three full-length CESAs identified in *M. sativa.*

two main groups: a heat/salt-induced and a cold/heat-repressed group of cell wall genes (Fig. 7). Salt/heat-induced genes are represented by the primary *CesAs MsCesA1, MsCesA3, MsCesA6-B* (with a Pearson correlation coefficient of 0.883) and to this group *CAD* belongs too (although with a lower correlation coefficient of 0.690). *CslD4* and *PAL* are also assigned to this group, although they cluster in a different branch, as their trend is less sharp than the one observed for primary *CesAs* (Fig. 7). The cold/heat-repressed group is represented by the secondary *CesAs*, together with *SuSy* (correlation coefficient of 0.91 for the cluster *SuSy, MsCesA4* and *MsCesA7-A*, and of 0.93 for *MsCesA4* and *MsCesA7-B*). The hierarchical clustering assigns to this group also *MsCesA6-C* and *MsCesA6-F* (Fig. 7). The statistical analyses carried out on the RT-qPCR data (Fig. S5) reveal that the changes in expression for *MsCesA1* and *MsCesA6-F* are statistically not significant; however their expression patterns can be interpreted as an overall trend which enables their classification in the heat/salt-induced and cold/heat-repressed group, respectively (Fig. 7).

A more detailed analysis of *MsCesAs* expression profiles shows mild but significant change for *MsCesA3*, with respect to the control, in response to salt stress after 96 h (Fig. S5; Table S4). *MsCesA6-B* displays an increase in expression at late stages of heat and salt application, which reaches a maximum after 96 h of treatment (Fig. S5; Table S4). *MsCesA6-C* shows a noteworthy decrease after 24 and 72 h of cold stress treatment (Fig. S5; Table S4).

MsCesA4, MsCesA7-A and *MsCesA7-B* show a trend towards decrease in expression already after 24 h of heat stress, while *CesA8* responds later, 72 h after the application of the stress (Fig. S5; Table S5). The correlation analysis of the wall-related genes performed with qBase^PLUS revealed a strong correlation between *MsCesA4* and *MsCesA7-A* in all the conditions tested (Fig. S6). This is not surprising, since these two genes belong both to the secondary CESAs clade, they are necessary for secondary cell wall biosynthesis together with CESA8 [57] and have been shown to interact in *Arabidopsis* [56].

Table 2. Proposed nomenclature for the *CesA* genes from *M. truncatula* based on amino acid identities with the orthologous proteins from *A. thaliana* and *P. trichocarpa*.

Populus/Medicago %identity	P. trichocarpa	M. truncatula	A. thaliana	Arabidopsis/Medicago %identity
87	PtiCesA1-A estExt_fgenesh4_pm.C_LG_XVIII0125	MtCesA1 Medtr3g107520	AtCesA1/RSW1 AT4G32410	84
87	PtiCesA1-B fgenesh4_pg.C_LG_VI001789	MtCesA1 Medtr3g107520	AtCesA1/RSW1 AT4G32410	84
83	PtiCesA3-A eugene3.00060479	MtCesA3 Medtr3g030040	AtCesA3/CEV1 AT5G05170	85
84	PtiCesA3-B eugene3.00160483	MtCesA3 Medtr3g030040	AtCesA3/CEV1 AT5G05170	85
88	PtiCesA3-C estExt_fgenesh4_pg.C_LG_IX0979	MtCesA3 Medtr3g030040	AtCesA3/CEV1 AT5G05170	85
88	PtiCesA3-D estExt_Genewise1_v1.C_LG_I1792	MtCesA3 Medtr3g030040	AtCesA3/CEV1 AT5G05170	85
84	PtiCesA4 eugene3.00002636	MtCesA4 Medtr2g035780	AtCesA4/IRX5 AT5G44030	77
87	PtiCesA6-B estExt_fgenesh4_pg.C_LG_VII0650	MtCesA6-B Medtr8g092590	AtCesA6/IXR2/PRC1 AT5G64740	82
81	PtiCesA6-C estExt_fgenesh4_pg.C_LG_V1107	MtCesA6-C Medtr1g098550	AtCesA6/IXR2/PRC1 AT5G64740	74
68	PtiCesA6-F fgenesh4_pg.C_scaffold_133000012	MtCesA6-F Medtr3g007770	AtCesA6/IXR2/PRC1 AT5G64740	62
87	PtiCesA7-A estExt_Genewise1_v1.C_LG_VI2188	MtCesA7-A Medtr4g130510	AtCesA7/IRX3 AT5G17420	85
63	PtiCesA7-B gw1.XVIII.3152.1	MtCesA7-B Medtr8g063270	AtCesA7/IRX3 AT5G17420	64
80	PtiCesA8-A gw1.XI.3218.1	MtCesA8 Medtr8g086600	AtCesA8/IRX1 AT4G18780	76
77	PtiCesA8-B eugene3.00040363	MtCesA8 Medtr8g086600	AtCesA8/IRX1 AT4G18780	76

Loci are as reported in the Phytozome web portal [25].

Discussion

The use of RT-qPCR for gene expression studies is a tool of unanimously recognized value, even in the current scientific era marked by the next generation sequencing revolution. Its utility is indeed unquestionable and necessary for validation of results massively produced via high-throughput methods.

For relative gene expression studies using RT-qPCR, the selection of suitable reference genes is a factor of paramount importance. Several studies in the literature have already undertaken the analysis of a set of candidate reference genes for normalization strategies in different plant species and conditions. Lists of stable genes are already available for relative RT-qPCR studies in plant tissues; however it is important to check their suitability in the experimental set-up adopted.

We have here identified and validated the use of reference genes for expression studies in alfalfa plants under different abiotic stresses. Two well-known and widely-used softwares, geNorm^{PLUS} [2] and NormFinder [33], have been chosen to rank the stability of the selected genes and we show that for some tissues, the best gene pairs identified differed between the 2 methods (Fig. 1 and Table 1). However, we were able to identify and propose a set of reference genes, ranked among the most stable by both softwares, namely eIF4A, PAB4, ADF1 and TFIIA. These genes can

therefore be included in a panel of candidates to be tested for RT-qPCR studies in alfalfa and, potentially, in other leguminous plants.

For the validation phase, we have used as a model gene a plant kinase, *SK1*, known for its susceptibility to stresses [44] and we show that the response pattern is similar in the different tissues, where cold and heat stress cause the most pronounced responses, namely reduction and increase of expression, respectively (Fig. 3).

We have subsequently extended our RT-qPCR study to cell wall biosynthetic genes in stems, since our efforts are currently devoted towards understanding the regulation of cell wall biosynthesis dynamics in stems of alfalfa plants. In particular we here show the expression of nine putative *CesAs*, belonging to both primary and secondary wall clades (Fig. 5), together with other wall-related genes (Fig. 7). Although several reports in the literature have shown a link between cell wall biosynthesis/ modification and abiotic stresses [58–61], a detailed investigation of cell wall gene expression changes in response to different abiotic stresses is lacking.

The main finding of our investigation is the elucidation of the wall-related gene dynamism in alfalfa plants subjected to abiotic stresses. The hierarchical clustering analysis identified two main trends in response to abiotic stresses: a salt/heat-induced and a cold/heat-repressed group of genes. Interestingly, a gene known to

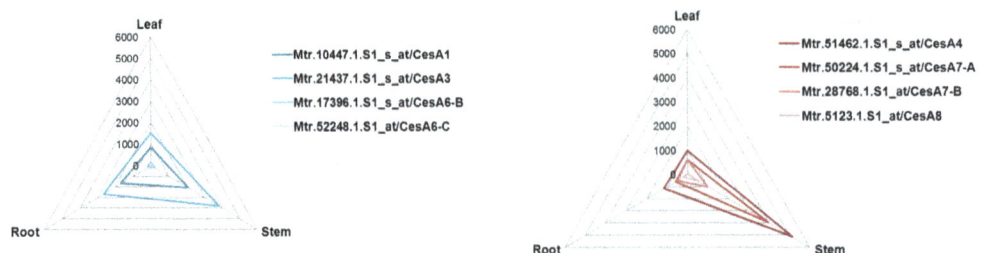

Figure 6. Radar plots of *M. truncatula* *CesAs* obtained plotting the microarray data retrieved at [31].

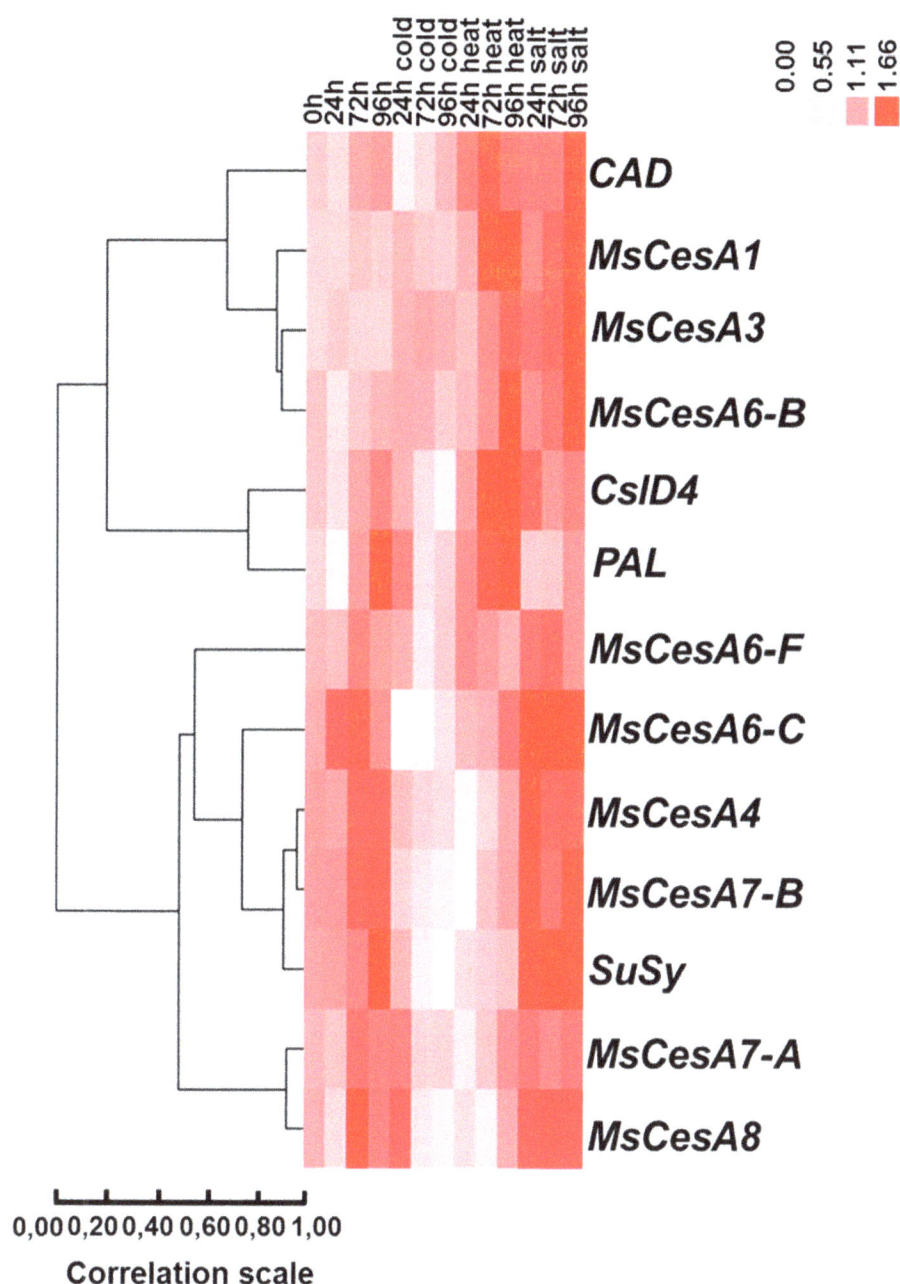

Figure 7. Heat Map representation of the data in Fig. S5 showing the hierarchical clustering of cell wall-related genes in response to the abiotic stresses at the different time points in alfalfa stems. The data collected refer to 3 independent biological replicates, each consisting of a pool of 15 plants. For each stress treatment a control group was always kept for the 24 h-48 h-96 h time points, for appropriate comparisons. The group clustering was generated with Cluster 3.0 [38] and visualized with Java TreeView [39], as described in Material and Methods.

be involved in lignin biosynthesis, *CAD*, grouped together with the primary *CesA*s *MsCesA1*, *MsCesA3* and *MsCesA6-B* (Fig. 7): this indicates that these genes, although not strictly related, show a common response mechanism to abiotic stress. In this respect it should be noted that induction of a peroxidase, triggering in its turn an increase in lignin and suberin deposition, has been reported in tomato plants exposed to salt stress [62] and that tomato plants under salt stress show an increased number of lignified cells [63]. In addition to this, a link between miRNAs, abiotic stresses and lignification has been unveiled in *A. thaliana*, as miR397b, a miRNA targeting a laccase (and consequently

affecting lignification), was shown to be up-regulated in response to salt stress [64–65]. *PAL* and *CslD4* also clustered with the *CAD*-primary *CesA*s group, although with a lower correlation: both genes display a heat and salt-stress responsive trend at later stages of treatment (Fig. 7; Fig. S5; Table S6). Cellulose synthase-like genes belong to the CESA superfamily and several members involved in wall glycan biosynthesis have been identified [66]. Many members of the *Csl* group of genes have not yet been functionally characterized, however representatives of the *CslD* clade are required for tip-growing cells [67] and *CslD1* and *CslD4* have been shown to affect cellulose biosynthesis in pollen tubes

[68]. Moreover, another member of the *CslD* clade, *CslD5*, was shown to be required for osmotic stress tolerance in *A. thaliana* [60]. The results shown by the hierarchical clustering (Fig. 7) suggest that *PAL* and *CslD4* might be involved in cell wall remodeling in response to heat stress in alfalfa stems in a pathway likely involving increased lignin biosynthesis and cellulose deposition to strengthen the wall under the adverse condition. Heat stress triggers substantial modifications in plants: changes in ultrastructural anatomy and cell wall polysaccharide composition have been observed in coffee leaves subjected to heat stress, with an increase in monolignol content [69].

The susceptibility of primary *CesAs* to exogenous stresses is a known feature: the *A. thaliana cev1/CesA3* mutant shows constitutive expression of stress responsive genes, together with an increased resistance to fungal attack [70].

The second group of genes identified by the hierarchical clustering is represented by the secondary *CesAs* together with *SuSy*, *MsCesA6-C* and *MsCesA6-F* (Fig. 5). *MsCesA6-C* and *MsCesA6-F* belong to the primary *CesAs* and it is interesting that these genes cluster with secondary *CesAs*. This might indicate that, as already discussed for *CAD*, a similar response mechanism exists between these genes and the secondary *CesAs*, or it can indirectly show that they are more functionally related to secondary *CesAs*. This needs verification, however the presence of multiple *CesA6* genes in alfalfa might indicate overlapping and/or distinct roles in cell wall biosynthesis.

The second group of genes identified by the hierarchical clustering shows down-regulation in response to cold and heat stress. Heat stress is known to cause a decreased expression of *SuSy* in pollen grains (accompanied by a decrease in the expression of other wall-related genes and vacuolar invertases; [71]) and in chickpea leaves [72]. The RT-qPCR analysis performed on alfalfa stems suggests that the decrease observed in secondary *CesAs* expression upon cold and heat stress treatment might be related to an impaired fueling of UDP-glucose by *SuSy*, which, despite not strictly required for cellulose synthesis [73–74], might contribute to increase the rate of synthesis, by concentrating the substrate [74].

Conclusions

The present work constitutes a useful guide for the identification of appropriate reference genes in expression studies on alfalfa, which can be extended to other legume crops for analysis. Through analyses using NormFinder and geNorm[PLUS], we have identified a set of suitable candidates, which can be included in a panel of reference genes to be tested for differential expression analysis.

The results concerning *CesAs* and a few other wall-related genes confirm an active role played by the cell wall in response to exogenous stimuli and constitute a step forward in delineating the complex pathways fine-tuning the response of plants to abiotic stresses.

Supporting Information

Figure S1 Nucleotide sequences of *MsCesAs*. Sequence details of the *CesAs* identified in alfalfa.

Figure S2 Alignment of alfalfa partial *CesA* sequences with *M. truncatula CesAs*. Alignment of the *CesAs* from alfalfa with the respective orthologs from *M. truncatula*.

Figure S3 Sequence details of *MtCESA6-F*. Alignment of *MtCESA6-F* with *CesAs* from *C. aretinum* [GenBank: XP_004499618.1] and *P. vulgaris* [GenBank: ESW20735.1] showing the amino acid substitutions in the processive GT2s motif (bold and underlined). The zinc-finger domain (CxxC)4 is highlighted in yellow.

Figure S4 Electronic Fluorescence Pictographic (eFP) representations of *M. truncatula CesA1*, *CesA3*, *CesA6-B*, *CesA6-C*, *CesA4*, *CesA7-A*, *CesA7-B*, *CesA8*.

Figure S5 Gene expression profiles of cell wall-related genes in stems of alfalfa plants subjected to abiotic stress. Data were normalized using eif4A/TFIIA. Means sharing a letter are not significantly different at $\alpha = 0.05$. NRE indicates Normalized Relative Expression.

Figure S6 *MsCesA7-A* and *MsCesA4* relationship. Correlation between *MsCesA7-A* and *MsCesA4* in stems under abiotic stress conditions. Pearson (log) r = 0.962; Spearman (log) r = 0.961.

Table S1 List of primers used to amplify *MsCesAs*. Name of the primers, with the respective sequences, used to amplify the *CesAs* from *M. sativa*.

Table S2 List of primers used for the RT-qPCR study. Name of the primers used for the RT-qPCR study, with the respective sequences. Details concerning the amplicons details (length, Tm), PCR efficiencies and regression coefficients are included.

Table S3 CESAs from *M. truncatula*. Details concerning number of predicted transmembrane helices (TMHs, according to [47]) and the length of the putative CESAs from *M. truncatula*.

Table S4 Normalized Relative Expression for primary *CesAs*. Normalized Relative Expression values ± standard deviation and significance (Sig.) for the primary *CesAs*. Data were normalized using eif4A/TFIIA.

Table S5 Normalized Relative Expression for secondary *CesAs*. Normalized Relative Expression values ± standard deviation and significance (Sig.) for the secondary *CesAs*. Data were normalized using eif4A/TFIIA.

Table S6 Normalized Relative Expression for for *CAD*, *CslD4*, *PAL* and *SuSy*. Normalized Relative Expression values ± standard deviation and significance (Sig.) for *CAD*, *CslD4*, *PAL* and *SuSy*. Data were normalized using eif4A/TFIIA.

Acknowledgments

The authors wish to thank Dr. Lucien Hoffmann for critical reading of the manuscript. Laurent Solinhac is acknowledged for technical assistance.

Author Contributions

Conceived and designed the experiments: GG SL JFH. Performed the experiments: GG SL JFH. Analyzed the data: GG SL JFH. Contributed reagents/materials/analysis tools: GG SL JFH. Contributed to the writing of the manuscript: GG SL JFH.

References

1. Schmittgen TD, Lee EJ, Jiang J, Sarkar A, Yang L, et al. (2008) Real-time PCR quantification of precursor and mature microRNA. Methods 44: 31–38.
2. Vandesompele J, De Preter K, Pattyn F, Poppe B, Van Roy N, et al. (2002) Accurate normalization of real-time quantitative RT-PCR data by geometric averaging of multiple internal control genes. Genome Biol 3: RESEARCH0034.
3. Bustin SA, Benes V, Garson JA, Hellemans J, Huggett J, et al. (2009) The MIQE guidelines: minimum information for publication of quantitative real-time PCR experiments. Clin Chem 55: 611–622.
4. Prasch CM, Sonnewald U (2013) Simultaneous application of heat, drought, and virus to *Arabidopsis* plants reveals significant shifts in signaling networks. Plant Physiol 162: 1849–1866.
5. Sultan SE (2000) Phenotypic plasticity for plant development, function and life history. Trends Plant Sci 5: 537–542.
6. Sultan SE (2003) Phenotypic plasticity in plants: a case study in ecological development. Evol Dev 5: 25–33.
7. Puijalon S, Bornette G (2006) Phenotypic plasticity and mechanical stress: biomass partitioning and clonal growth of an aquatic plant species. Am J Bot, 93: 1090–1099.
8. Nicot N, Hausman JF, Hoffmann L, Evers D (2005) Housekeeping gene selection for real-time RT-PCR normalization in potato during biotic and abiotic stress. J Exp Bot 56: 2907–2914.
9. Wan H, Zhao Z, Qian C, Sui Y, Malik AA, et al. (2010) Selection of appropriate reference genes for gene expression studies by quantitative real-time polymerase chain reaction in cucumber. Anal Biochem 399: 257–261.
10. Gu C, Chen S, Liu Z, Shan H, Luo H, et al. (2011) Reference gene selection for quantitative real-time PCR in *Chrysanthemum* subjected to biotic and abiotic stress. Mol Biotechnol 49: 192–197.
11. Le DT, Aldrich DL, Valliyodan B, Watanabe Y, Ha CV, et al. (2012) Evaluation of candidate reference genes for normalization of quantitative RT-PCR in soybean tissues under various abiotic stress conditions. PLoS One 7: e46487.
12. Selim M, Legay S, Berkelmann-Löhnertz B, Langen G, Kogel KH, et al. (2012) Identification of suitable reference genes for real-time RT-PCR normalization in the grapevine-downy mildew pathosystem. Plant Cell Rep 31: 205–216.
13. Zhu J, Zhang L, Li W, Han S, Yang W, et al. (2013) Reference gene selection for quantitative real-time PCR normalization in *Caragana intermedia* under different abiotic stress conditions. PLoS One 8: e53196.
14. Kakar K, Wandrey M, Czechowski T, Gaertner T, Scheible WR, et al. (2008) A community resource for high-throughput quantitative RT-PCR analysis of transcription factor gene expression in *Medicago truncatula*. Plant Methods 4: 18.
15. Yang SS, Tu ZJ, Cheung F, Xu WW, Lamb JF, et al. (2011) Using RNA-Seq for gene identification, polymorphism detection and transcript profiling in two alfalfa genotypes with divergent cell wall composition in stems. BMC Genomics 12: 199.
16. Pini F, Frascella A, Santopolo L, Bazzicalupo M, Biondi EG, et al. (2012) Exploring the plant-associated bacterial communities in *Medicago sativa* L. BMC Microbiol 12: 78.
17. Tesfaye M, Yang SS, Lamb JFS, Jung H-JG, Samac DA, et al. (2009) *Medicago truncatula* as a model for dicot cell wall development. Bioenergy Res 2: 59–76.
18. Young ND, Debellé F, Oldroyd GE, Geurts R, Cannon SB, et al. (2011) The *Medicago* genome provides insight into the evolution of rhizobial symbioses. Nature 480: 520–524.
19. Yang SS, Xu WW, Tesfaye M, Lamb JF, Jung HJ, et al. (2010) Transcript profiling of two alfalfa genotypes with contrasting cell wall composition in stems using a cross-species platform: optimizing analysis by masking biased probes. BMC Genomics 11: 323.
20. Huis R, Hawkins S, Neutelings G (2010) Selection of reference genes for quantitative gene expression normalization in flax (*Linum usitatissimum* L.). BMC Plant Biol 10: 71.
21. Humphrey TV, Bonetta DT, Goring DR (2007) Sentinels at the wall: cell wall receptors and sensors. New Phytol 176: 7–21.
22. Seifert GJ, Blaukopf C (2010) Irritable walls: the plant extracellular matrix and signaling. Plant Physiol 153: 467–478.
23. Hamann T, Denness L (2011) Cell wall integrity maintenance in plants: lessons to be learned from yeast? Plant Signal Behav 6: 1706–1709.
24. Hamann T (2012) Plant cell wall integrity maintenance as an essential component of biotic stress response mechanisms. Front Plant Sci 3: 77.
25. Phytozome website. Available: www.phytozome.com. Accessed 2014 July 10.
26. National Center for Biotechnology Information (NCBI) website. Available: http://www.ncbi.nlm.nih.gov/. Accessed 2014 July 10.
27. Carroll A, Specht CD (2011) Understanding plant cellulose synthases through a comprehensive investigation of the cellulose synthase family sequences. Front Plant Sci 2: 5.
28. Edgar RC (2004) MUSCLE: multiple sequence alignment with high accuracy and high throughput. Nucleic Acids Res 32: 1792–1797.
29. Guindon S, Gascuel O (2003) A simple, fast, and accurate algorithm to estimate large phylogenies by maximum likelihood. Syst Biol 52: 696–704.
30. Chevenet F, Brun C, Bañuls AL, Jacq B, Christen R (2006) TreeDyn: towards dynamic graphics and annotations for analyses of trees. BMC Bioinformatics 7: 439.

31. *Medicago truncatula* Gene Expression Atlas website. Available: http://mtgea.noble.org/v3/. Accessed 2014 July 10.
32. *Medicago* electronic fluorescent pictographic (eFP) representations website. Available: http://bar.utoronto.ca/efpmedicago/cgi-bin/efpWeb.cgi. Accessed 2014 July 10.
33. Andersen CL, Jensen JL, Ørntoft TF (2004) Normalization of real-time quantitative reverse transcription-PCR data: a model-based variance estimation approach to identify genes suited for normalization, applied to bladder and colon cancer data sets. Cancer Res 64: 5245–5250.
34. Sørby LA, Andersen SN, Bukholm IR, Jacobsen MB (2010) Evaluation of suitable reference genes for normalization of real-time reverse transcription PCR analysis in colon cancer. J Exp Clin Cancer Res 29: 144.
35. Untergasser A, Nijveen H, Rao X, Bisseling T, Geurts R, et al. (2007) Primer3Plus, an enhanced web interface to Primer3. Nucleic Acids Res 35: W71–W74.
36. OligoAnalyzer 3.1 website. Available: http://eu.idtdna.com/analyzer/Applications/OligoAnalyzer/. Accessed 2014 July 10.
37. Hellemans J, Mortier G, De Paepe A, Speleman F, Vandesompele J (2007) qBase relative quantification framework and software for management and automated analysis of real-time quantitative PCR data. Genome Biol 8: R19.
38. Eisen MB, Spellman PT, Brown PO, Botstein D (1998) Cluster analysis and display of genome-wide expression patterns. Proc Natl Acad Sci U S A 95: 14863–14868.
39. Saldanha AJ (2004) Java Treeview-extensible visualization of microarray data. Bioinformatics 20: 3246–3248.
40. Java TreeView website. Available: http://jtreeview.sourceforge.net/. Accessed 2014 July 10.
41. Jian B, Liu B, Bi YR, Hou WS, Wu CX, et al. (2008) Validation of internal control for gene expression study in soybean by quantitative real-time PCR. BMC Mol Biol, 9: 59.
42. González-Verdejo CI, Die JV, Nadal S, Jiménez-Marín A, Moreno MT, et al. (2008) Selection of housekeeping genes for normalization by real-time RT-PCR: analysis of Or-MYB1 gene expression in *Orobanche ramosa* development. Anal Biochem 379: 176–181.
43. Guénin S, Mauriat M, Pelloux J, Van Wuytswinkel O, Bellini C, et al. (2009) Normalization of qRT-PCR data: the necessity of adopting a systematic, experimental conditions-specific, validation of references. J Exp Bot 60: 487–493.
44. Nolan KE, Saeed NA, Rose RJ (2006) The stress kinase gene MtSK1 in *Medicago truncatula* with particular reference to somatic embryogenesis. Plant Cell Rep 25: 711–722.
45. Coello P, Hey SJ, Halford NG (2011) The sucrose non-fermenting-1-related (SnRK) family of protein kinases: potential for manipulation to improve stress tolerance and increase yield. J Exp Bot 62: 883–893.
46. Kumar M, Thammannagowda S, Bulone V, Chiang V, Han KH, et al. (2009) An update on the nomenclature for the cellulose synthase genes in *Populus*. Trends Plant Sci 14: 248–254.
47. TMHMM website. Available: http://www.cbs.dtu.dk/services/TMHMM/. Accessed 2014 July 10.
48. Handakumbura PP, Matos DA, Osmont KS, Harrington MJ, Heo K, et al. (2013) Perturbation of *Brachypodium distachyon* CELLULOSE SYNTHASE A4 or 7 results in abnormal cell walls. BMC Plant Biol 13: 131.
49. Song D, Shen J, Li L (2010) Characterization of cellulose synthase complexes in *Populus* xylem differentiation. New Phytol 187: 777–790.
50. Persson S, Paredez A, Carroll A, Palsdottir H, Doblin M, et al. (2007) Genetic evidence for three unique components in primary cell-wall cellulose synthase complexes in *Arabidopsis*. Proc Natl Acad Sci U S A 104: 15566–15571.
51. Betancur L, Singh B, Rapp RA, Wendel JF, Marks MD, et al. (2010) Phylogenetically distinct cellulose synthase genes support secondary wall thickening in *Arabidopsis* shoot trichomes and cotton fiber. J Integr Plant Biol 52: 205–220.
52. Stork J, Harris D, Griffiths J, Williams B, Beisson F, et al. (2010) CELLULOSE SYNTHASE9 serves a nonredundant role in secondary cell wall synthesis in *Arabidopsis* epidermal testa cells. Plant Physiol 153: 580–589.
53. Sullivan S, Ralet MC, Berger A, Diatloff E, Bischoff V, et al. (2011) CESA5 is required for the synthesis of cellulose with a role in structuring the adherent mucilage of *Arabidopsis* seeds. Plant Physiol 156: 1725–1739.
54. Mendu V, Griffiths JS, Persson S, Stork J, Downie AB, et al. (2011) Subfunctionalization of cellulose synthases in seed coat epidermal cells mediates secondary radial wall synthesis and mucilage attachment. Plant Physiol 157: 441–453.
55. Carroll A, Mansoori N, Li S, Lei L, Vernhettes S, et al. (2012) Complexes with mixed primary and secondary cellulose synthases are functional in *Arabidopsis* plants. Plant Physiol 160: 726–737.
56. Li S, Lei L, Gu Y (2013) Functional analysis of complexes with mixed primary and secondary cellulose synthases. Plant Signal Behav 8: e23179.
57. Taylor NG, Howells RM, Huttly AK, Vickers K, Turner SR (2003) Interactions among three distinct CesA proteins essential for cellulose synthesis. Proc Natl Acad Sci U S A 100: 1450–1455.
58. Pilling E, Höfte H (2003) Feedback from the wall. Curr Opin Plant Biol 6: 611–616.

59. Moura JC, Bonine CA, de Oliveira FVJ, Dornelas MC, Mazzafera P (2010) Abiotic and biotic stresses and changes in the lignin content and composition in plants. J Integr Plant Biol 52: 360–376.

60. Zhu J, Lee BH, Dellinger M, Cui X, Zhang C, et al. (2010) A cellulose synthase-like protein is required for osmotic stress tolerance in *Arabidopsis*. Plant J 63: 128–140.

61. Wolf S, Mravec J, Greiner S, Mouille G, Höfte H (2012) Plant cell wall homeostasis is mediated by brassinosteroid feedback signaling. Curr Biol 22: 1732–1737.

62. Quiroga M, Guerrero C, Botella MA, Barceló A, Amaya I, et al. (2000) A tomato peroxidase involved in the synthesis of lignin and suberin. Plant Physiol 122: 1119–1127.

63. Sánchez-AguayoI, Rodríguez-Galán JM, García R, Torreblanca J, Pardo JM (2004) Salt stress enhances xylem development and expression of S-adenosyl-L-methionine synthase in lignifying tissues of tomato plants. Planta 220: 278–285.

64. Sunkar R, Zhu JK (2004) Novel and stress-regulated microRNAs and other small RNAs from *Arabidopsis*. Plant Cell 16: 2001–2019.

65. Martin RC, Liu PP, Goloviznina NA, Nonogaki H (2010) microRNA, seeds, and Darwin?: diverse function of miRNA in seed biology and plant responses to stress. J Exp Bot 61: 2229–2234.

66. Yin Y, Huang J, Xu Y (2009) The cellulose synthase superfamily in fully sequenced plants and algae. BMC Plant Biol 9: 99.

67. Bernal AJ, Yoo CM, Mutwil M, Jensen JK, Hou G, et al. (2008) Functional analysis of the cellulose synthase-like genes CSLD1, CSLD2, and CSLD4 in tip-growing *Arabidopsis* cells. Plant Physiol 148: 1238–1253.

68. Wang W, Wang L, Chen C, Xiong G, Tan XY, et al. (2011) *Arabidopsis* CSLD1 and CSLD4 are required for cellulose deposition and normal growth of pollen tubes. J Exp Bot 62: 5161–5177.

69. Lima RB, dos Santos TB, Vieira LG, Ferrarese Mde L, Ferrarese-Filho O, et al. (2013) Heat stress causes alterations in the cell-wall polymers and anatomy of coffee leaves (*Coffea arabica* L.). Carbohydr Polym 93: 135–143.

70. Ellis C, Karafyllidis I, Wasternack C, Turner JG (2002) The *Arabidopsis* mutant cev1 links cell wall signaling to jasmonate and ethylene responses. Plant Cell 14: 1557–1566.

71. Bita CE, Gerats T (2013) Plant tolerance to high temperature in a changing environment: scientific fundamentals and production of heat stress-tolerant crops. Front Plant Sci 4: 273.

72. Kaushal N, Awasthi R, Gupta K, Gaur P, Siddique KHM, et al. (2013) Heat-stress-induced reproductive failures in chickpea (*Cicer arietinum*) are associated with impaired sucrose metabolism in leaves and anthers. Funct Plant Biol 40: 1334–1349.

73. Barratt DHP, Derbyshire P, Findlay K, Pike M, Wellner N, et al. (2009) Normal growth of *Arabidopsis* requires cytosolic invertase but not sucrose synthase. Proc Natl Acad Sci U S A 106: 13124–13129.

74. Carpita NC (2011) Update on mechanisms of plant cell wall biosynthesis: how plants make cellulose and other (1->4)-β-D-glycans. Plant Physiol 155: 171–184.

15

On the Use of Leaf Spectral Indices to Assess Water Status and Photosynthetic Limitations in *Olea europaea* L. during Water-Stress and Recovery

Pengsen Sun[1], Said Wahbi[2], Tsonko Tsonev[3], Matthew Haworth[4], Shirong Liu[1], Mauro Centritto[4]*

1 Institute of Forest Ecology, Environment and Protection, Chinese Academy of Forestry, Beijing, P. R. China, 2 Laboratoire de Biotechnologie et Physiologie Végétale, Faculté des Sciences Semlalia, Université Cadi Ayyad, Marrakech, Morocco, 3 Institute of Plant Physiology and Genetics, Bulgarian Academy of Sciences, Sofia, Bulgaria, 4 Trees and Timber Institute, National Research Council, Sesto Fiorentino, Florence, Italy

Abstract

Diffusional limitations to photosynthesis, relative water content (RWC), pigment concentrations and their association with reflectance indices were studied in olive (*Olea europaea*) saplings subjected to water-stress and re-watering. RWC decreased sharply as drought progressed. Following rewatering, RWC gradually increased to pre-stress values. Photosynthesis (A), stomatal conductance (g_s), mesophyll conductance (g_m), total conductance (g_t), photochemical reflectance index (PRI), water index (WI) and relative depth index (RDI) closely followed RWC. In contrast, carotenoid concentration, the carotenoid to chlorophyll ratio, water content reflectance index (WCRI) and structural independent pigment index (SIPI) showed an opposite trend to that of RWC. Photosynthesis scaled linearly with leaf conductance to CO_2; however, A measured under non-photorespiratory conditions ($A_{1\%O2}$) was approximately two times greater than A measured at 21% [O_2], indicating that photorespiration likely increased in response to drought. $A_{1\%O2}$ also significantly correlated with leaf conductance parameters. These relationships were apparent in saturation type curves, indicating that under non-photorespiratory conditions, CO_2 conductance was not the major limitations to A. PRI was significant correlated with RWC. PRI was also very sensitive to pigment concentrations and photosynthesis, and significantly tracked all CO_2 conductance parameters. WI, RDI and WCRI were all significantly correlated with RWC, and most notably to leaf transpiration. Overall, PRI correlated more closely with carotenoid concentration than SIPI; whereas WI tracked leaf transpiration more effectively than RDI and WCRI. This study clearly demonstrates that PRI and WI can be used for the fast detection of physiological traits of olive trees subjected to water-stress.

Editor: Pilar Hernandez, Institute for Sustainable Agriculture (IAS-CSIC), Spain

Funding: Financial support was provided by the National Natural Science Foundation of China (Grant No. 31290223), the Key Laboratory of Forest Ecology and Environment of State Forestry Administration of China, and the Ministero dell'Istruzione dell'Università della Ricerca of Italy: PRIN 2010-2011 PRO-ROOT and Progetto Premiale 2012 Aqua. The funders had no role in study design, data collection and analysis, decision to publish, or preparation of the manuscript.

Competing Interests: The authors have declared that no competing interests exist.

* Email: mauro.centritto@cnr.it

Introduction

The effective utilisation of renewable water resources in agriculture is a significant challenge in many areas of the globe. To mitigate the effects of increasing chronic water shortages, there is a need to develop and expand irrigation management practices. However, considering the unprecedented pressure on water resources for agriculture caused by rapidly growing water demand for urban and industrial uses, the expansion of irrigation may only become possible through the development of precision water saving irrigation techniques [1]. These are based on the real time detection of crop physiological status, and require the development of advanced, non-invasive phenotyping methods to monitor water relations and photosynthetic status in plants experiencing water-stress [2]. To this end, remotely sensed vegetation indices are increasingly being used as reliable cost-effective plant-based indicators to assess physiological traits associated with plant water status [3]. Furthermore, the real-time detection of plant physiological changes in regions subjected to drought is important for precision crop management and the estimation of terrestrial productivity. Therefore, improved knowledge of the relationship between leaf spectral and physiological responses under variable water conditions is of crucial importance.

Water deficit constrains all physiological processes involved in plant growth and development. These changes are part of a cascade of responses to drought that affect primary processes including tissue water relations and gas exchange mechanisms. It is well known that one of the earliest responses to water deficit is diminished stomatal conductance (g_s), because stomata act as control valves in the exchange of water vapor between leaf and the atmosphere to match soil water uptake rate with transpiration rate (E) to maintain plant water balance [4]. However, as g_s declines, CO_2 diffusion into the leaves also decreases, leading to reduced substomatal CO_2 concentration (C_i) [5]. This diffusional limitation to photosynthesis (A) in the gas phase is frequently associated with a coordinated change limiting CO_2 transport through the mesophyll (i.e., mesophyll conductance, g_m) [6,7]. As a consequence, during the early stages of drought stress, photosynthetic limitations have been shown to be predominantly caused by decreased total diffusive conductance (g_t, which is related to g_s and

g_m) leading to low chloroplastic CO_2 concentration (C_c) [2,7]. Alteration of photosynthetic metabolism generally becomes more prominent as water-stress progresses [5]. Nevertheless, there is still a degree of controversy on the relative importance of g_s, g_m or metabolic impairments in the limitation of A under drought (Lawlor and Cornic 2002) and during recovery upon re-watering [5].

Alterations in leaf water status, photosynthetic pigment concentrations and photosynthetic activity in turn lead to changes in spectral reflectance properties [3]. Reflectance in specific wavelength bands in the visible and near-infrared (NIR) region have potential applications in the estimation of plant water status [3]. Many spectral reflectance indices have been proposed in the direct monitoring of plant water-status; expressed as relative water content (RWC), leaf water potential and photosynthetic status. Among the water spectral indices, the NIR-based water index (WI) [3] is increasingly employed in monitoring water status; as recently demonstrated by Gutierrez et al., [8] to assess water relations in contrasting wheat genotypes, and by both Serrano et al., [9] and Marino et al., [10] to estimate vine water status at leaf and canopy levels. Furthermore, Sun et al., [11] developed a new index, the water content reflectance index (WCRI), which was inversely related to RWC. However, this relationship was only clear at low RWC values, suggesting that this water content-based spectral index may not be effective in detecting moderate stress.

The photochemical reflectance index (PRI) was originally developed to estimate rapid changes in the epoxidation state of xanthophyll pigments [12], a major component of non-photochemical quenching. However, PRI has been increasingly used to detect photosynthetic activity at different scales from plant biochemistry to the canopy level (see the review by Garbulsky et al., [13]. Dobrowsky et al., [14] found that heat and water-stress induced changes in steady-state chlorophyll fluorescence that were tracked by PRI; suggesting that PRI was a more effective real time indicator of photosynthetic function than indices based upon leaf water content and pigment concentration. Consistent with these findings, Ripullone et al., [15] observed strong relationships between photosynthetic activity and PRI in individual forest tree species subjected to water-stress. Similarly, significant linear relationships between PRI and A, and between PRI and g_s, were also demonstrated in *Solanum lycopersicum* [16] and *Quercus ilex* [17] subjected to water-stress, and at the tree canopy-level in irrigated and rainfed adult olive trees grown under field condition [10]. The PRI is not sensitive to sudden drought, but did track A, g_s and leaf water content during slow developing drought stress in *Olea europaea* [11,18]. However, to the best of our knowledge there is only one study that investigates the possibility of using PRI to track photosynthesis diffusional limitations (g_s and g_m) during the early stages of drought stress and during recovery upon re-watering [17].

In this study, diffusional limitations to photosynthesis, leaf water status, pigment concentrations and spectral reflectance indices were evaluated in olive (*Olea europaea* L.) plants during water-stress and recovery. The objective was to understand the links between simultaneous variations in photosynthetic diffusional limitations and PRI alongside leaf water relation parameters and WI by examining time course changes in response to water availability. Additionally, the constancy of other water content and pigment related spectral indices were also evaluated in response to water-stress and recovery.

Materials and Methods

Growing conditions and drought treatment

Three-year-old plants of olive (*Olea europaea* L., cv Leccino) were grown in 15 dm^3 pots filled with commercial soil in the National Research Council greenhouse in Montelibretti, Rome, Italy. Temperature within the greenhouse was maintained at 25–27°C, photosynthetically active radiation followed the natural light regime and relative humidity ranged between 60–70%. Large volume 15 dm^3 pots relative to the *Olea europaea* plants were chosen to avoid any potential effects of root restriction generating root to shoot signals that may interfere with the response of the plants to drought [19,20]. Furthermore, the g_s and PRI values of well-watered and drought treated plants were indistinguishable in the early stages of drought treatment and the values recorded in the drought treatment plants subsequently returned to pre-stress levels following re-watering in the recovery period, indicating that potential root restriction did not influence the effect of drought on simultaneous measurement of reflectance and gas exchange under conditions of water-deficit. The plants were regularly watered to pot water capacity and fertilized with Hoagland solution once a week to supply mineral nutrients at free access rates [21,22] for the first two months after the onset of the growing season. On the afternoon preceding the initiation of the experiment (Day 0), plants were fully irrigated and allowed to drain the excess water over-night. Then half of the plants were water-stressed by withholding water until the seedlings showed symptoms of severe water-stress, while the remaining half of the seedlings continued to be well-watered to pot capacity. On the afternoon of day 18 water-stressed plants were irrigated with 500 cm^3 of water as supplementary irrigation. On the evening of the Day 23, the water-stressed seedlings were then re-watered daily to pot capacity over a 7-day recovery period. Drought stress treatment was performed during May - June. The plants were fully randomized and five replicate plants per treatment were analyzed for all measurements.

Gas exchange and fluorescence measurements

Simultaneous measurement of gas exchange and fluorescence were conducted on five leaves per plant (these values were then averaged to produce a mean value for the individual plant, and then the mean of five replicate plants taken for a given treatment value) using a LI-6400-40 portable infrared gas-analyser (Li-Cor, Lincoln, NE, USA). All gas exchange measurements were made between 11.00 and 15.00 h, with the leaf chamber set to a saturating photosynthetic photon flux density (PPFD) (1300 µmol $m^{-2} s^{-1}$), relative humidity of the air ranging between 45–55% and at a leaf temperature of 25°C. To reduce diffusion leaks through the chamber gasket [23], a supplementary external chamber gasket composed of the same polymer foam was added to create an interspace between the two gaskets (i.e. a double-gasket design with a 5 mm space separating the internal and external gaskets). Then the CO_2 and H_2O gradients between the in-chamber air and pre-chamber air were minimized by feeding the IRGA exhaust air into the interspace between the chamber and the pre-chamber gaskets [24]. Instantaneous measurement of steady-state photosynthesis (A), transpiration (E), stomatal conductance (g_s), and $\Delta F/F_m'$ (i.e. the quantum yield of PSII in the light) were made on five plants per treatment after removal of stomatal limitation of A by lowering the external atmospheric [CO_2] (C_a) to 50 ppm as described by Centritto et al., [6]. Measurements of dark respiration (R_d) were performed at ambient [CO_2] concentration in the dark on the same leaves by switching off the light in the leaf cuvette; when CO_2 release from the leaf had become stable for approximately five to ten minutes this was recorded and considered to represent R_d. Mesophyll conductance to CO_2 diffusion (g_m), the inverse of the total resistance encountered by CO_2 across the leaf mesophyll, was calculated using the variable J method. One percent oxygen for the measurement of A under non-photorespiratory conditions was produced using a Brooks Instruments mass flow controllers (Brooks Instruments, Hatfield, PA, USA) and a gas mixing system connected to cylinders of compressed pure nitrogen and oxygen gas, this was then connected to the gas input line of the LiCor Li6400. Then g_m in olive leaves was calculated following Aganchich et al., [30], whereas total diffusion conductance (g_t) was calculated as: $g_t = g_s g_m/(g_s+g_m)$.

Determination of relative water content and pigment concentrations

Leaf samples for relative water content (RWC) and pigment concentrations were taken immediately after the gas exchange and spectral measurements. Fully-developed leaves were detached and weighed to determine leaf fresh mass (F_M). After measuring leaf area, using a leaf area meter (LI3100, LI-COR Inc., Lincoln, NE, USA), leaves were then covered with a plastic bag and allowed to rehydrate with the cut-end under water in a dark cold room at $5°C$ for 18 h. Following rehydration, each leaf was weighed to determine saturated mass (S_M), and then oven-dried at $80°C$ for 48 hours to determine dry mass (D_M). RWC was finally calculated as follows: $RWC = (F_M - D_M)/(S_M - D_M)$.

The carotenoid and total chlorophyll (*Chl*, a and b) concentration was measured in intact leaf tissues by immersion in N,N-dimethylformamide (DMF). Leaf discs with a total area of 1.5 cm^2 were placed into glass vials containing 2 cm^3 DMF and immediately placed in darkness at $4°C$ in an orbital shaker set to 100 rpm for 4 h. The absorbance (A_B) of the solution was then read on a spectrophotometer (Perkin Elmer, Norwalk, CT) at 663.8, 646.8, and 480 nm, using DMF as a blank. The pigment concentrations were calculated according to the following equations [25]:

$$Total\ chlorophyll\ (Chl_a + Chl_b)$$
$$= (12A_{B663.8} - 3.11A_{B646.8}) + (20.78A_{B646.8} - 4.88A_{B663.8})$$

$$Carotenoid = (1000A_{B480} - 1.12Chl_a - 34.7Chl_b)/245.$$

Spectral measurements and indices

Spectral measurements were carried out in the laboratory, by a portable spectroradiometer (Field spec FR 350–2500 nm, ASD Inc., Boulder, CO, USA). As the reflectance properties of leaves alter with age [26], all of the leaves analysed were of approximately the same age, less than one year-old from the uppermost part of the canopy. The instrument measures spectral reflectance between 350 and 2500 nm to a resolution of 3 nm. The instrument automatically calculates the reflectance value as the ratio between the incident radiation reflected from the surface target and the incident radiation reflected by a reference spectral panel. The reference spectral panel material can be regarded as a Lambertian reflector. The Field spec FR may be operated with different lenses that control the field of view, in this study an $8°$ lens was used to restrict the field of view of the instrument. Reflectance spectra (R) were collected from a distance of 5.0 cm from the sample, which was fixed on a dark platform. To reduce

instrument noise each spectral signature was calculated from the average of 100 scans. A further check of the stability of the reflected signal was performed by taking the white reflectance at the beginning and end of each sample. Reflectance spectra were preprocessed using the View Spec Pro (version 5.6, ASD, Inc.) software. A summary of the spectral measurements used within the study are presented in Appendix S1.

Statistics

Data were tested using a simple factorial ANOVA (two-way maximum interactions) and Tukey *post-hoc* test. The statistical analysis was performed by Sigma Plot 11.0 (Systat software, Inc., San Jose, CA, USA). Linear and non-linear curve-fitting (Sigma Plot 11.0), which minimizes the difference between observed and predicted values, were used for regression analyses.

Ethics Statement

No permits were required for the experimental analysis of *Olea europaea* plants that were bought commercially and complied with all relevant regulations and did not involve an endangered or protected species.

Results

Figure 1 shows the time-course of RWC, PRI, A, g_s, g_m and g_t during the study period of soil drying and subsequent recovery. As expected, relative water content (Fig. 1a) was strongly affected by the water-regime $(P<0.001)$, decreasing first slowly (Days 1–8) and then sharply (Days 10–16) as drought progressed. RWC was significantly increased $(P<0.05)$ by supplementary irrigation (Day 20) and then following re-watering to pot water capacity (Days 23–30). The whole experimental process resulted in a "W-shaped" time course in RWC. In fact, upon relief of the water-stress, RWC increased very rapidly to pre-stress levels. The response of PRI to the drought and recovery cycle (Fig. 1b) closely followed the pattern of RWC.

Photosynthesis (Fig. 1c), g_s, g_m, and g_t (Fig. 1d) were also strongly affected $(P<0.001)$ by water availability. Water-stressed saplings exhibited a significant decrease in A even during the early phase of the drying cycle (Fig. 1c). When water-stress became very severe (Day 16), A measured in ambient air was almost completely inhibited and only slightly, but not significantly, increased following the supplementary irrigation (Day 20). At low $[O_2]$, A measured under non-photorespiratory conditions $(A_{1\%O2})$, was increased by ~75% in well-watered plants at the beginning and at the end of the experiment. In comparison to control values of $A_{1\%O2}$, significant decreases in $A_{1\%O2}$ were only observed in severely stressed plants. It is interesting to note that the ratio between $A_{1\%O2}$ and A increased during water-stress, reaching a value of about two on Day 20 following the supplementary irrigation, and a similar value one day after re-watering on Day 24. Mild water-stress also induced significant decreases in g_s (Fig. 1d); in a similar pattern to A, g_s approached zero when soil water decreased to a level where there was no longer water available within the pot to sustain transpiration, and then did not significantly respond to supplementary irrigation. Upon relief of water-stress, g_s and A increased rapidly to pre-stress levels; but their recovery was less rapid than that of RWC. Interestingly, seven days after re-watering, A and g_s significantly exceeded pre-stress values by about 25% and 31%, respectively. Mesophyll conductance to CO_2 diffusion and, in turn, g_t mirrored the trend in g_s during the drought and recovery cycle (Fig 1d).

Stomatal and mesophyll conductance values were not only similar in magnitude but also linearly correlated $(P<0.001)$

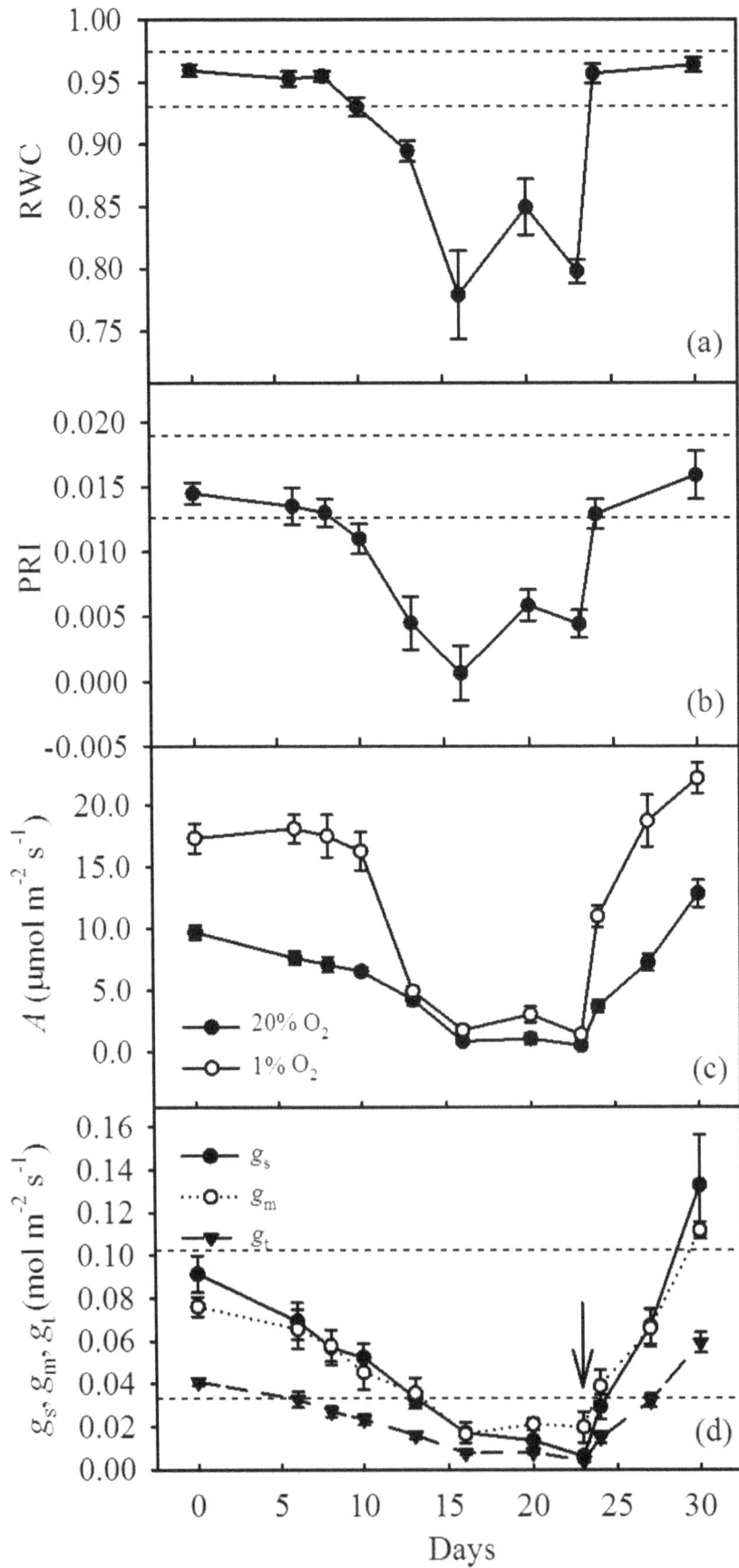

Figure 1. Time courses of leaf (a) relative water content (RWC), dashed horizontal lines indicate the range of values (mean ± one standard deviation) recorded in the well-watered plants over the duration of the experiment; (b) photochemical reflectance index (PRI) (dashed horizontal lines as in 1a); (c) photosynthesis (A) (measured in both ambient air and in air with 1% [O_2] ($A1\%O2$), and; (d) stomatal conductance (g_s), mesophyll conductance (g_m) and total conductance (g_t) of olive saplings grown during and after the drought cycle (days 1–23), dashed horizontal lines indicate the range of g_s values recorded in the well-watered plants. ↓ = end of the drying cycle. Data points are means of five plants (five leaves per plant) ±1 SEM.

(Fig. 2). There was a significant linear correlation between A and g_s ($r^2 = 0.966$, $P<0.001$), when measurements conducted on control and water-stressed leaves were pooled (Fig. 3a). Similar linear relationships were also found between A and g_m ($r^2 = 0.959$, $P<0.001$) (Fig. 3b) and A and g_t ($r^2 = 0.969$, $P<0.001$) (Fig. 3c). Furthermore, $A_{1\%O2}$ scaled significantly with leaf conductance parameters (Fig. 3d,e,f). In contrast to A, the responses of $A_{1\%O2}$ to g_s ($r^2 = 0.945$, $P<0.001$), g_m ($r^2 = 0.965$, $P<0.001$), and g_t ($r^2 = 0.955$, $P<0.001$) took the form of a saturating curve.

After pooling together of all measurements during the drought and recovery cycle, there were significant exponential relationships between the gas-exchange parameters A (Fig. 4a), g_s, g_m and g_t (Fig. 4b), when plotted against RWC. The pre-stress and recovery (i.e., high RWC) gas-exchange parameters showed higher variation than those recorded during water-stress (Fig.4a,b). Stomatal conductance increased more rapidly than g_m and g_t during the late water-stress relief period (i.e., at leaf RWC ranging from 0.9 to 0.96) (Fig. 1d, Fig. 4b). The relationship between PRI and RWC was also non-linear ($r^2 = 0.935$, $P<0.001$), with RWC varying from 0.77 to 0.96 during the study period (Fig. 4c). The PRI showed less variability than the gas exchange parameters as RWC recovered to pre-stress levels. The PRI was also significantly correlated with to A (Fig. 4d), g_m (Fig. 4e) and g_t (Fig. 4e).

The water content-related reflectance indices, RDI (Fig. 5a) and WI (Fig 5c), showed a similar time-course response to that of RWC (Fig. 1a); they decreased significantly as water-stress progressed and significantly increased following the relief of water-stress. In contrast, WCRI (Fig. 5b) increased significantly as water-stress became severe, and recovered quickly and completely upon re-watering to pot water capacity. The supplementary irrigation (Day 20) altered RDI significantly ($P<0.05$), but had a less apparent effect on both WI and WCRI. Overall, RDI (Fig. 5d), WCRI (Fig. 5e) and WI (Fig. 5f) showed significant linear relationships with RWC. However, WCRI differed from the other two indices by being inversely related to RWC. The RDI and WCRI were found to exhibit comparatively stronger correlations with RWC than WI.

The RDI showed significant non-linear relationships with g_s (Fig. 6a) and E (Fig. 6b); whereas, WI scaled linearly with g_s (Fig. 6c) and E (Fig. 6d). In contrast, there was an inverse curvilinear relationship between WCRI and both g_s (Fig. 6e) and E (Fig. 6f). It is noteworthy that RDI and WCRI were highly sensitive to both g_s and E under severe stress conditions ($g_s < 0.06$ mol m^{-2} s^{-1}), whereas this sensitivity decreased sharply in mild-stress or optimal conditions ($g_s > 0.06$ mol m^{-2} s^{-1}).

Carotenoid concentration (Fig. 7a) was significantly affected ($P<0.05$) by water-stress. Carotenoid concentration increased as water-stress progressed; resulting in a 34% increase when water-stress was most severe (Days 16 and 23), and then decreased significantly following both the supplementary irrigation (Day 20) and re-watering to pot water capacity (Day 24). Chlorophyll concentration was inversely associated to that of carotenoids (data not shown), and this led to a progressive decrease in the carotenoid to chlorophyll ratio as water-stress progressed, and to a correspondingly rapid increase in response to the supplementary irrigation and relief of water-stress (Fig. 7b). The time-course of structural independent pigment index (SIPI, i.e. a measure of carotenoid to chlorophyll a ratio) (Fig. 7c) mirrored that of the overall carotenoid to chlorophyll ratio. Finally, significant linear relationships to carotenoid concentration were found between both PRI (Fig. 8a) and SIPI (Fig. 8b).

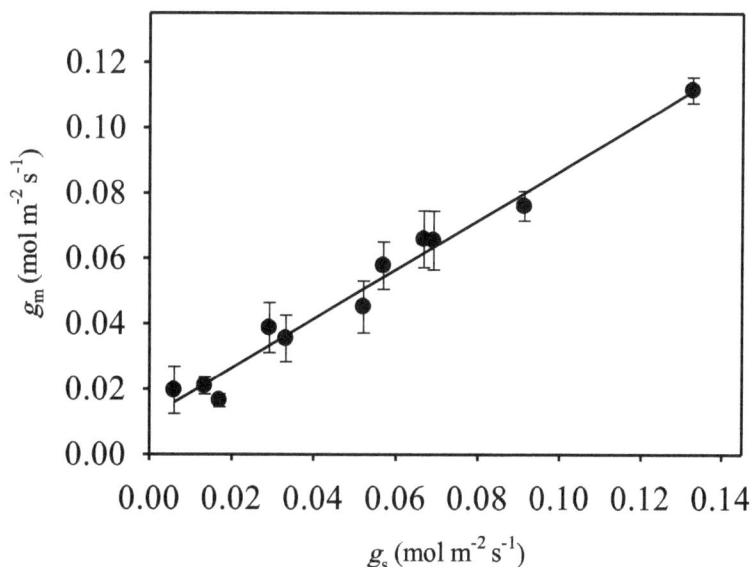

Figure 2. Linear relationships between mesophyll conductance (g_m) and stomatal conductance (g_s) ($r^2 = 0.978$, $P<0.001$) in olive saplings grown during and after the drought cycle.

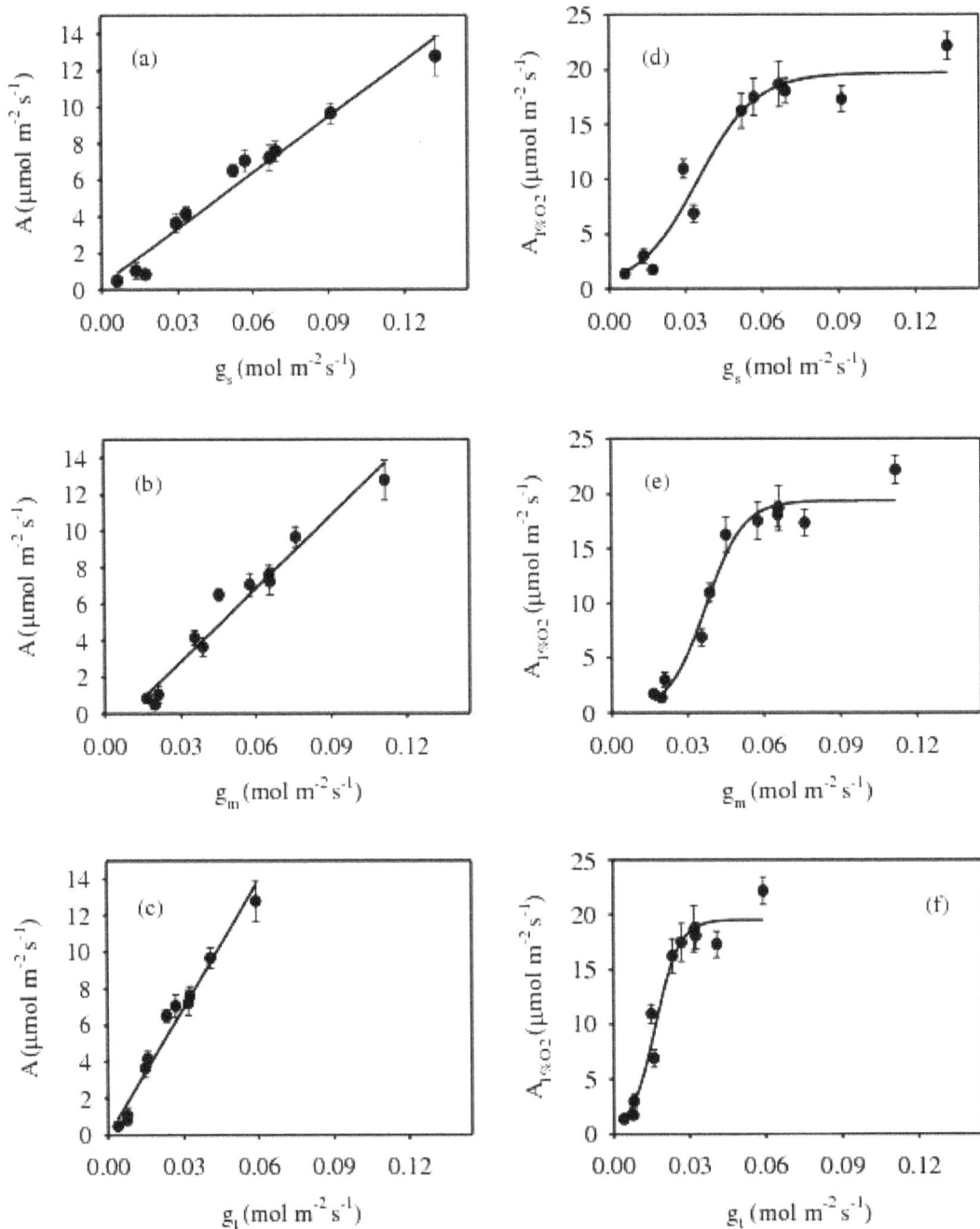

Figure 3. Relationships of photosynthesis (A), measured in ambient air (a, b, c) and in air with 1% [O₂] (A1%O2) (d, e, f), and (a, d) stomatal conductance (g_s), (b, e) mesophyll conductance (g_m), and (c, f) total conductance (g_t) in olive saplings grown during and after the drought cycle.

Discussion

Water shortage is a crucial problem limiting plant productivity across large areas of the globe. Changes in biogeophysical cycles (that is increased evaporative demand and water holding capacity of the air) associated with global warming are expected to further worsen the water crisis in arid and semi-arid zones, and to also increasingly affect temperate regions [27]. To mitigate the effects of drought on plant growth it essential to improve our knowledge of plant functional responses to soil drying. In this study, we assessed a cascade of changes in leaf water relations, pigment

concentrations, photosynthetic CO₂ diffusional limitations in olive saplings subjected to water-stress and recovery by using spectral reflectance indices to monitor the water-status and photosynthetic function of olive trees.

The time-course in leaf water status, as expressed by RWC (Fig. 1a) and PRI (Fig. 1b), in response to water availability was mirrored by that of the photosynthetic parameters. Furthermore, as observed in other studies [2], the decline of both A (Fig. 4a) and leaf conductance parameters (Fig. 4b) was significantly correlated to RWC. We observed a gradual decrease of A (Fig. 1c) in water-stressed saplings during the early period of the drought treatment

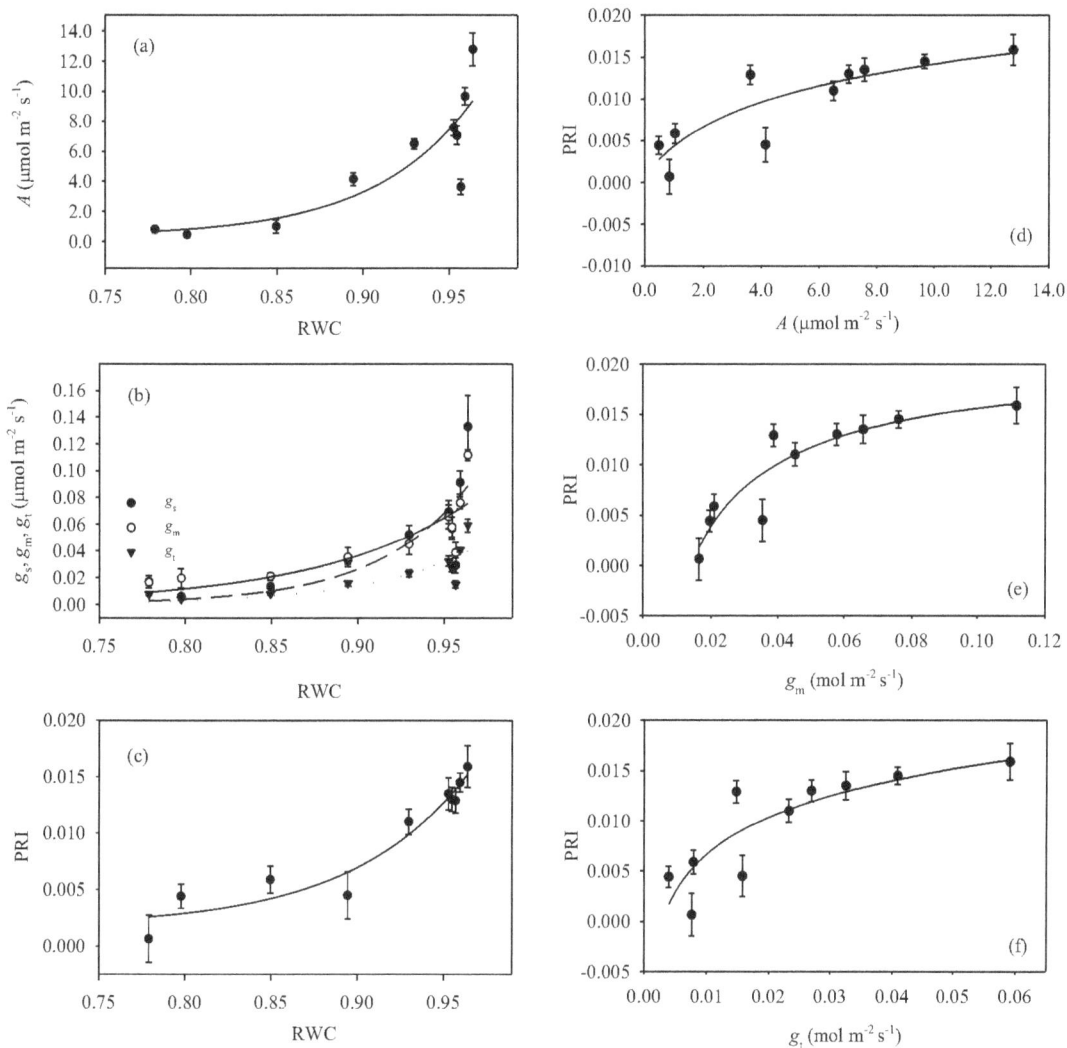

Figure 4. Relationships between relative water content (RWC) and (a) photosynthesis (A) ($r^2 = 0.742$, $P = 0.008$), (b) stomatal conductance (g_s, ——) ($r^2 = 0.667$, $P = 0.005$), mesophyll conductance (g_m, ----) ($r^2 = 0.748$, $P = 0.003$) and total conductance (g_t, ·····) ($r^2 = 0.653$, $P < 0.02$), and (c) photochemical reflectance index (PRI) ($r^2 = 0.935$, $P < 0.001$), and between PRI and (d) A ($r^2 = 0.762$, $P = 0.001$), (e) g_m ($r^2 = 0.844$, $P < 0.001$), and (f) g_t ($r^2 = 0.725$, $P = 0.002$) in olive saplings grown during and after the drought cycle.

(Day 10). The reduction in conductance values of CO_2 diffusion (g_s, g_m and g_t) (Fig. 1d) followed similar patterns, suggesting that during the early phase of the drought stress A was limited by reduced C_c [5,28,29]. There are numerous studies suggesting that under water-stress, A is mostly limited by reduced C_i due to stomatal closure [5]. However, information about the contribution of g_m during water-stress is still lacking [28,29]; although new evidence indicates that g_m is linearly correlated to g_s (Fig. 2) [2,6], and an important component in the determination of water use efficiency [7]. When water-stress reached severe levels (Day 16), A was almost fully inhibited, and upon relief of the water-stress A gradually recovered to the pre-stress level in parallel with g_s, g_m and g_t. The A and g_s values observed in this study, in both well-watered and stressed conditions, are consistent with the findings of previous work performed on young pot-grown olive plants [11,30] and mature olive trees grown under field conditions [10,31,32,33]. Moriana et al., [31] and Marino et al., [10] observed that some degree of metabolic limitation to A occurred in olive trees exposed to severe summer drought. However, the rapid increase in A and its complete restoration to pre-stress levels after one week of stress-

relief, in addition to the linear relationships between A and leaf conductance parameters (Fig. 3a,b,c), may indicate that A was likely mainly inhibited by impaired CO_2 uptake [34].

Photosynthesis measured under non-photorespiratory conditions ($A_{1\%O2}$) changed in parallel with A (Fig. 1c). Assuming that the difference between $A_{1\%O2}$ and A is an approximate measure for photorespiration, it is evident that this parameter was stimulated dramatically at the onset of water stress. This is may have been caused by reductions in C_c as a result of decreases in g_t (Fig. 1d). Reduced C_c may then lead to increased oxygenation of ribulose-1,5-bisphosphate (RuBP) by Rubisco, thus increasing photorespiration, not only relative to photosynthesis, but also in absolute terms [35,36]. For example, photorespiration in four grape varieties increased under moderate water-stress and helped maintain a relatively high photochemical efficiency of PSII and rapid recovery of A after re-watering [37]. However, as photorespiration depends directly on RuBP recycling in the Benson-Calvin cycle, which in turns depends on C_c, severe drought may also result in reduced rates of photorespiration [35]. This may explain the progressive reduction of $A_{1\%O2}$ under

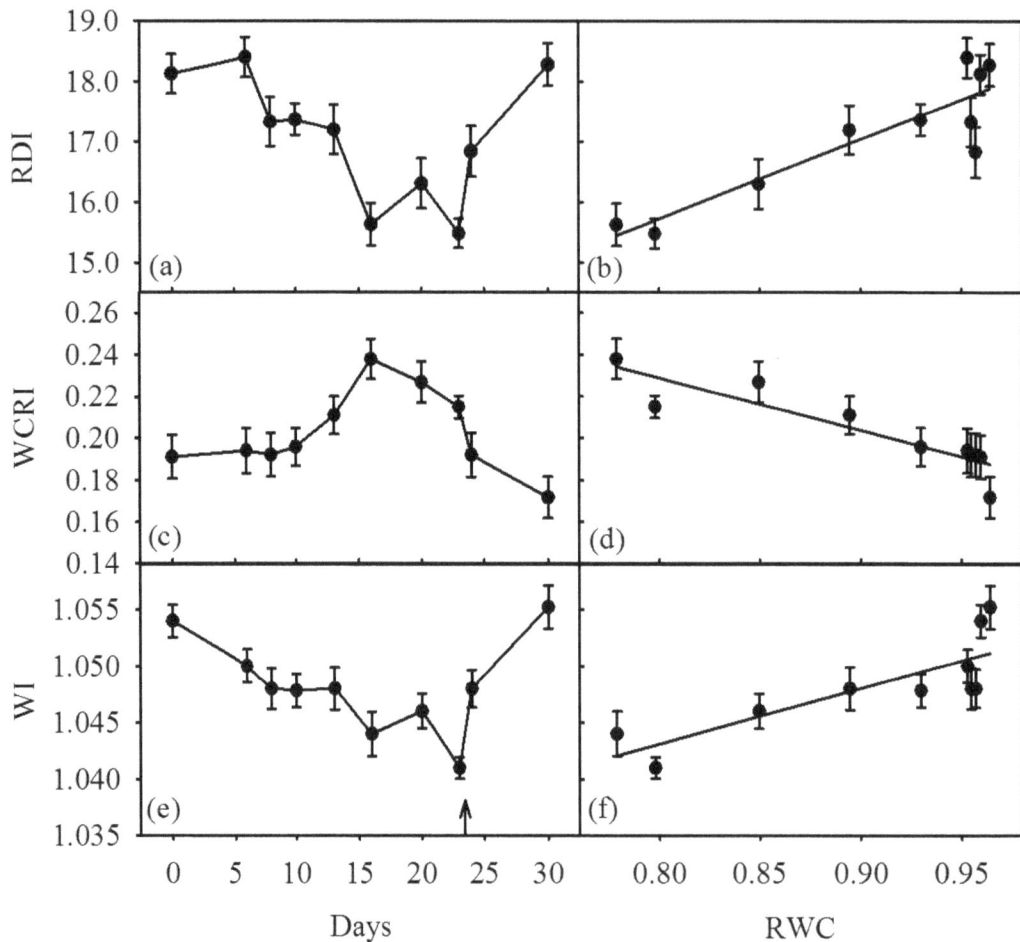

Figure 5. Time courses of (a) relative depth index (RDI), (b) water correlated reflectance index (WCRI) and (c) water index (WI), and relationships between relative water content (RWC) and (d) RDI (r² = 0.803, P<0.001), (e) WCRI (r² = 0.814, P<0.001) and (f) WI (r² = 0.675, P= 0.004) in olive saplings grown during and after the drought cycle (days 1–23). ↑ = end of the drying cycle. Data points are means of five plants (five leaves per plant) per measurement ±1 SEM.

conditions of severe water-stress. Conversely, the increase in $A_{1\%O2}$ observed after the supplementary irrigation and relief of water-stress (Days 20 and 24, respectively), when $A_{1\%O2}$ was approximately two-fold higher than A, may likely be attributed to the corresponding increase in g_m and C_c. Furthermore, the saturating responses of $A_{1\%O2}$ to g_m (Fig. 3d), g_s (Fig. 3e), and g_t (Fig. 3f) imply that under non-photorespiratory conditions, conductance of CO_2 is not the major limitation to A. Thus, because broad-leaved sclerophyll plants, such as olive, possess inherently low leaf conductance, they may have a competitive advantage over mesophyllous vegetation in the future as atmospheric [CO_2] rises [38].

As CO_2 transport from the ambient air to the active site of Rubisco decreases in response to progressive water-stress, A and the quantum yield of photosystem II (PSII) become progressively inhibited. In turn, the proportion of the PPFD energy absorbed by the photosynthetic antenna and dissipated as heat increases, while that used for photochemistry declines. This photo-protective reaction is associated with the de-epoxidation state of xanthophylls [39]. PRI is based on a normalized difference of the 531 and 570 nm bands where xanthophyll pigments absorption occurs and, thus can be used as a robust proxy of radiation use efficiency [12] and photosynthetic activity [13]. The PRI, was dramatically

affected by water availability (Fig. 1b), and exhibited close correlation to RWC (Fig. 4c) and A (Fig. 4d). The strong relationship between PRI and RWC (P<0.001) observed in this study is in keeping with previous studies showing that PRI is a good indicator of leaf water status [15]. Furthermore, PRI detected at tree canopy level, has been recently found to be correlated to A in field-grown mature olive plants under rainfed and well-watered conditions [10]. Significant correlations between PRI and A have also been found in *Arbutus unedo* [15], *Ceratonia siliqua* [40], *Solanum lycopersicum* [16] and *Quercus ilex* [15,17] subjected to water-stress in different environmental conditions. Photosynthesis is mainly limited by diffusive limitations to CO_2 uptake [7,28,29], unless water-stress becomes particularly severe to the point where there is virtually no water available to support transpiration [4,34]; consequently, the direct detection of diffusional limitations by spectral indices may be of great importance in the detection of plant function under stress conditions. In this study, significant relationships between g_t (Fig. 4f) and g_m (Fig. 4e) were established. Similar results were also observed in a study of *Q. ilex* [17] response to water-stress and recovery. These relationships, which were similar to that observed between A and PRI (Fig. 4d), can be viewed as supportive indications of the suitability of PRI in tracking photosynthetic activity.

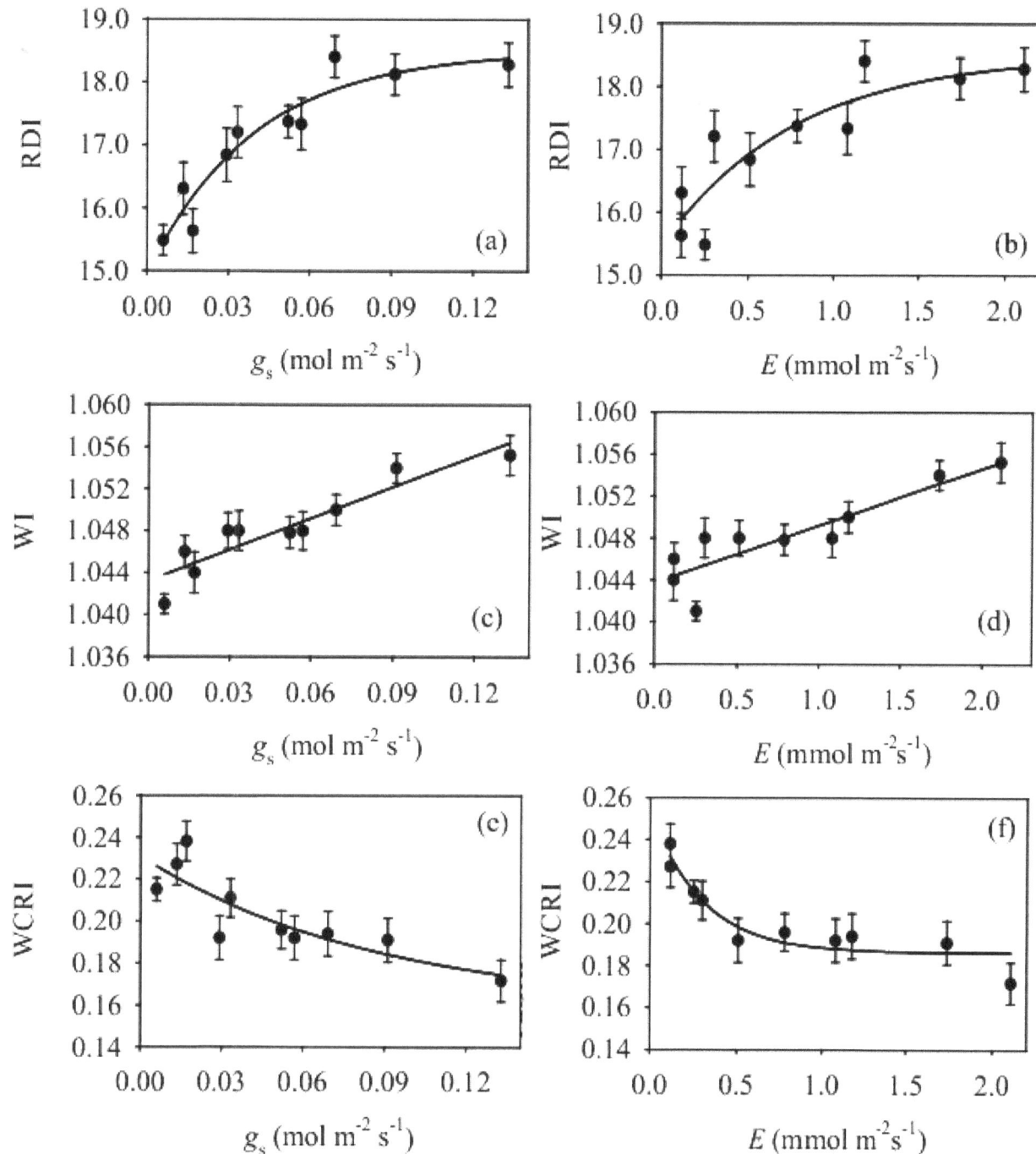

Figure 6. Relationships between relative depth index (RDI) and (a) leaf stomatal conductance (g_s) ($r^2 = 0.908$, $P<0.001$) and (b) leaf transpiration (E) ($r^2 = 0.796$, $P<0.01$), between WI and (c) g_s ($r^2 = 0.863$, $P<0.001$) and (d) E ($r^2 = 0.801$, $P<0.001$), and between WCRI and (e) g_s ($r^2 = 0.717$, $P<0.012$) and (f) E ($r^2 = 0.902$, $P<0.002$) in olive saplings grown during and after the drought cycle (days 1–23).

Variation in PRI can be affected by short-term changes in xanthophyll epoxidation state and over longer time-scales by shifts in the size of carotenoid and chlorophyll pools [41,42,43]. Similarly, time-course variations in SIPI, a pigment-based reflectance index [3], in response to water status (Fig. 7c) may reflect corresponding changes in the ratio of carotenoids to chlorophyll. In this study, carotenoid concentration (Fig. 7a) and the carotenoid to chlorophyll ratio (Fig. 7b) were responsive to water availability, as they significantly increased as the water-stress

progressed and then decreased following supplementary irrigation and re-watering to pot water capacity. This may indicate the ability of olive trees to up- and down-regulate photo-protective thermal dissipation mechanisms by xanthophyll cycle carotenoids in response to water status [39]. The PRI (Fig. 8a) and SIPI (Fig. 8b) both inversely reflected changes in carotenoid concentrations. Interestingly, SIPI was less effective than PRI in gauging carotenoid concentration. Unlike PRI, which is estimated in the green-yellow region of the radiation spectrum (i.e., within the

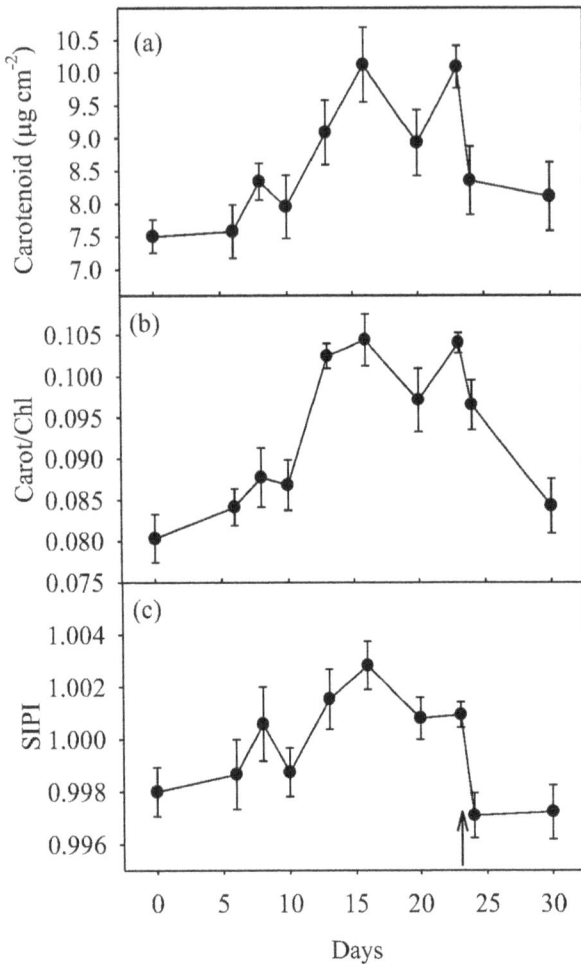

Figure 7. Time courses of (a) leaf carotenoid concentration (μg cm^{-2}), (b) carotenoid to chlorophyll ratio (Carot/Chl), and (c) structural independent pigment index (SIPI) in olive saplings grown during and after the drought cycle (days 1–23). ↑ = end of the drying cycle. Data points are means of five plants (five leaves per plant) per measurement ±1 SEM.

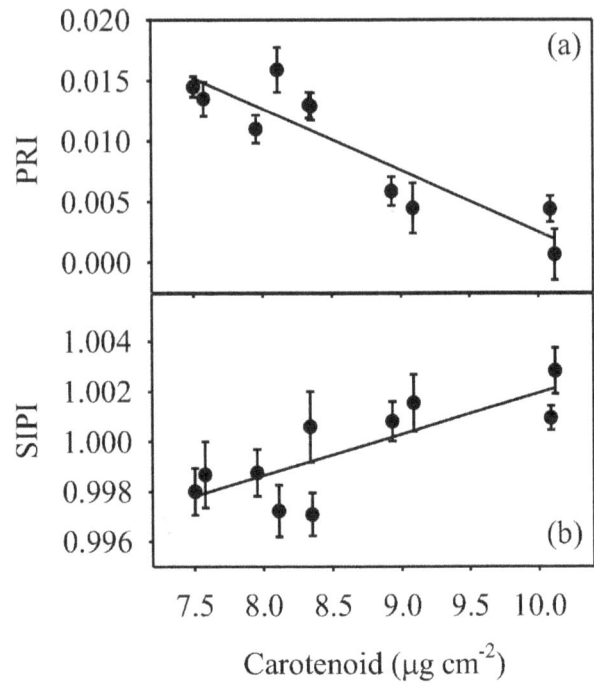

Figure 8. Linear relationships between carotenoid concentration (μg cm^{-2}) and (a) photochemical reflectance index (PRI) ($r^2 = 0.813$, $P < 0.001$) and (b) structural independent pigment index (SIPI) ($r^2 = 0.622$, $P < 0.007$) in olive saplings grown during and after the drought cycle.

carotenoid absorption spectra), SIPI is estimated using blue and red bands that are sensitive to both carotenoid and chlorophyll pigment changes. Therefore, the combined variations in both carotenoid and chlorophyll pools may have contributed to the higher variability of SIPI in comparison to that of PRI. However, taken together, these results reinforce the growing body of literature showing that PRI [10,13,41], and to a lesser extent SIPI, [10,44] are correlated to photosynthetic parameters.

The three water-related reflectance indices, RDI, WI and WCRI, were significantly affected during the water-stress cycle and re-watering. The RDI (Fig. 5a) and WI (Fig. 5c) mirrored the time course of RWC (Fig.1a); whereas WCRI (Fig. 5b) increased as water-stress progressed and decreased upon rewatering. These differential responses resulted in significant direct positive relationships between RWC and both RDI (Fig. 5d) and WI (Fig. 5f), and a significant negative relationship between RWC and WCRI (Fig. 5e). As all these indices are formulated by several water absorption wavelengths, these linear relationships indicate the consistent sensitivity of RDI, WI and WCRI to mild or severe drought conditions. Furthermore, these three spectral indices were also significantly correlated to g_s and E (Fig. 6), due to the close

dependence of stomata behavior and transpiration upon leaf water status. The RDI (Fig. 6a,b) and WCRI (Fig. 6e,f) had curvilinear relationships with g_s and E, and were consequently more sensitive as E and g_s decreased, and less sensitive under mild-stress and optimal conditions. Conversely, WI was linearly related to both g_s (Fig. 6c) and E (Fig. 6d), and as a consequence was more sensitive to these two physiological parameters than RDI and WCRI. This differential sensitivity may result from the different wavelengths used in calculating RDI, WI and WCRI, as middle infrared wavelengths are used to estimate RDI and WCRI while near-infrared wavelengths are used to calculated WI. It is noteworthy that Marino et al., [10] demonstrated in a recent study of mature olive trees that WI, estimated at tree canopy level, was the most accurate predictive index of plant water status and whole-plant transpiration. Similarly, Serrano et al., [9] found that WI was a good indicator of vineyard water status. We therefore suggest that WI is the most reliable tool in predicting parameters of plant water-status.

Conclusion

Drought is a complex syndrome affecting several leaf biophysical and biochemical properties that subsequently influence leaf reflectance spectra. In the present study, we examined the reliability of water absorption and pigment based spectral indices to assess changes in leaf water status, pigment concentration and photosynthetic traits during water-stress and recovery. In general, PRI was very sensitive to leaf water status, pigments concentrations and A. It is noteworthy that PRI also significantly tracked leaf CO_2 conductance. These relationships may reflect the tight control that diffusional limitations exert on A under water-stress and then following relief from water-stress. The WI was the most reliable index in the prediction of plant water-status. Finally, A was likely mainly inhibited by leaf CO_2 conductance parameters during water-stress; whereas under non-photorespiratory conditions, CO_2 conductance was not the major limitation to photosynthesis. In consideration of a predicted future elevated atmospheric $[CO_2]$, it is possible that sclerophyll plants will likely out-perform mesophyllous vegetation.

Acknowledgments

The comments of two anonymous reviewers significantly improved this manuscript. The authors are grateful to the Scientific Cooperation Agreements between CNR and the Chinese Academy of Forestry; CNR and the Centre National pour la Recherche Scientifique et Technique of Morocco.

Author Contributions

Conceived and designed the experiments: PS SW TT SL MC. Performed the experiments: PS SW TT. Analyzed the data: MC MH. Contributed reagents/materials/analysis tools: MC TT. Wrote the paper: PS MC MH.

References

1. Fereres E, Orgaz F, Gonzalez-Dugo V (2011) Reflections on food security under water scarcity. Journal of Experimental Botany 62: 4079–4086
2. Centritto M, Lauteri M, Monteverdi MC, Serraj R (2009) Leaf gas exchange, carbon isotope discrimination, and grain yield in contrasting rice genotypes subjected to water deficits during the reproductive stage. Journal of Experimental Botany 60: 2325–2339
3. Peñuelas J, Filella L (1998) Visible and near-infrared reflectance techniques for diagnosing plant physiological status. Trends in Plant Science 3, 151–156
4. Centritto M, Brilli F, Fodale R, Loreto F (2011a) Different sensitivity of isoprene emission, respiration, and photosynthesis to high growth temperature coupled with drought stress in black poplar (Populus nigra). Tree Physiology 31: 275–286
5. Lawlor DW, Cornic G (2002) Photosynthetic carbon assimilation and associated metabolism in relation to water deficits in higher plants. Plant, Cell and Environment 25: 275–294
6. Centritto M, Loreto F, Chartzoulakis K (2003) The use of low $[CO_2]$ to estimate diffusional and non-diffusional limitations of photosynthetic capacity of salt-stressed olive saplings. Plant Cell and Environment 26: 585–594
7. Flexas J, Niinemets U, Gallé A, Barbour MM, Centritto M, et al. (2013) Diffusional conductances to CO_2 as a target for increasing photosynthesis and photosynthetic water-use efficiency. Photosynthesis Research, 117: 45–59
8. Gutierrez M, Reynolds MP, Klatt AR (2010) Association of water spectral indices with plant and soil water relations in contrasting wheat genotypes. Journal of Experimental Botany 61: 3291–3303
9. Serrano L, González-Flor C, Gorchs G (2010) Assessing vineyard water status using the reflectance based water index. Agriculture, Ecosystems and Environment 139: 490–499
10. Marino G, Pallozzi E, Cocozza C, Tognetti R, Giovannelli A, et al. (2014) Assessing gas exchange, sap flow and water relations using tree canopy spectral reflectance indices in irrigated and rainfed Olea europaea L. Environmental and Experimental Botany 99: 43–52
11. Sun P, Grignetti A, Liu S, Casacchia R, Salvatori R, et al. (2008) Associated changes in physiological parameters and spectral reflectance indices in olive (Olea europaea L.) leaves in response to different levels of water stress. International Journal of Remote Sensing 29: 1725–1743
12. Gamon JA, Peñuelas J, Field CB (1992) A narrow-waveband spectral index that tracks diurnal changes in photosynthetic efficiency. Remote Sensing of Environment 41: 35–44
13. Garbulsky MF, Peñuelas J, Gamon JA, Inoue Y, Filella I (2011) The photochemical reflectance index (PRI) and the remote sensing of leaf, canopy and ecosystem radiation use efficiencies: a review and meta-analysis. Remote Sensing of Environment 115: 281–297
14. Dobrowski SZ, Pushnik JC, Zarco-Tejada PJ, Ustin SL (2005) Simple reflectance indices track heat and water stress-induced changes in steady-state chlorophyll fluorescence at the canopy scale. Remote Sensing of Environment 97: 403–414
15. Ripullone F, Rivelli AR, Baraldi R, Guarini R, Guerrieri R, et al. (2011) Effectiveness of the photochemical reflectance index to track photosynthetic activity over a range of forest tree species and plant water statuses. Functional Plant Biology 38: 177–186
16. Sarlikioti V, Driever SM, Marcelis LFM (2010) Photochemical reflectance index as a mean of monitoring early water stress. Annals of Applied Biology 157: 81–89
17. Tsonev T, Wahbi S, Sun P, Sorrentino G, Centritto M (2014) Gas exchange, water relations and their relationships with photochemical reflectance index in Quercus ilex plants during water stress and recovery. International Journal of Agriculture and Biology 16: 335–341
18. Suárez L, Zarco-Tejada PJ, Sepulcre-Cantó G, Pérez-Priego O, Miller JR, et al. (2008) Assessing canopy PRI for water stress detection with diurnal airborne imagery. Remote Sensing of Environment 112: 560–575
19. Arp WJ (1991) Effects of source-sink relations on photosynthetic acclimation to elevated CO_2. Plant, Cell & Environment 14: 869–875
20. Ismail A, Hall A, Bray E (1994) Drought and pot size effects on transpiration efficiency and carbon isotope discrimination of cowpea accessions and hybrids. Functional Plant Biology 21: 23–35.
21. Magnani F, Centritto M, Grace J (1996) Measurement of apoplasmic and cell-to-cell components of root hydraulic conductance by a pressure-clamp technique. Planta 199: 296–306
22. Centritto M, Lee HSJ, Jarvis PG (1999) Long-term effects of elevated carbon dioxide concentration and provenance on four clones of Sitka spruce (Picea sitchensis) I. Plant growth, allocation and ontogeny. Tree Physiology 19: 799–806
23. Flexas J, Díaz-Espejo A, Berry J, Cifre J, Galmés J, et al. (2007) Analysis of leakage in IRGA's leaf chambers of open gas exchange systems: quantification and its effects in photosynthesis parameterization. Journal of Experimental Botany 58: 1533–1543
24. Rodeghiero M, Niinemets U, Cescatti A (2007) Major diffusion leaks of clamp-on leaf cuvettes still unaccounted: how erroneous are the estimates of Farquhar, et al. model parameters? Plant, Cell and Environment 30: 1006–1022
25. Wellburn AR (1994) The spectral determination of Chlorophylls a and b, as well as total carotenoids, using various solvents with spectrophotometers of different resolution. Journal of Plant Physiology 144: 307–313
26. Peñuelas J, Filella I, Gamon JA (1995) Assessment of photosynthetic radiation-use efficiency with spectral reflectance. New Phytologist 131: 291–296
27. Dai A (2010) Drought under global warming: a review. WIREs Climate Change 2: 45–65
28. Flexas J, Ribas-Carbó M, Diaz-Espejo A, Galmés J, Medrano H (2008) Mesophyll conductance to CO_2: current knowledge and future prospects. Plant, Cell and Environment 31: 602–621
29. Loreto F, Centritto M (2008) Leaf carbon assimilation in a water-limited world. Plant Biosystems 142: 154–161
30. Aganchich B, Wahbi S, Loreto F, Centritto M (2009) Partial root zone drying: regulation of photosynthetic limitations and antioxidant enzymatic activities in young olive (Olea europaea) saplings. Tree Physiology 29: 685–696
31. Moriana A, Villalobos FJ, Fereres E (2002) Stomatal and photosynthetic responses of olive (Olea europaea L.) leaves to water deficits. Plant, Cell and Environment 25: 395–405
32. Centritto M, Wahbi S, Serraj R, Chaves MM (2005) Effects of partial rootzone drying (PRD) on adult olive tree (Olea europaea) in field conditions under arid climate. II. Photosynthetic responses. Agriculture, Ecosystems and Environment 106: 303–311
33. Diaz-Espejo A, Nicolas E, Fernández JE (2007) Seasonal evolution of diffusional limitations and photosynthetic capacity in olive under drought. Plant, Cell and Environment 30: 922–933
34. Brilli F, Tsonev T, Mahmood T, Velikova V, Loreto F, et al. (2013) Ultradian variation of isoprene emission, photosynthesis, mesophyll conductance and optimum temperature sensitivity for isoprene emission in water-stressed Eucalyptus citriodora saplings. Journal of Experimental Botany 64: 519–528
35. Wingler A, Quick WP, Bungard RA, Bailey KJ, Lea PJ, et al. (1999) The role of photorespiration during drought stress: an analysis utilizing barley mutants with reduced activities of photorespiratory enzymes. Plant, Cell and Environment 22: 361–373
36. Sage RF (2013) Photorespiratory compensation: a driver for biological diversity. Plant Biology 15: 624–638
37. Guan XQ, Zhao SJ, Li DQ, Shu HR (2004) Photoprotective function of photorespiration in several grapevine cultivars under drought stress. Photosynthetica 42: 31–36
38. Centritto M, Tognetti R, Leitgeb E, Střelcová K, Cohen S (2011). Above ground processes - Anticipating climate change influences. In: Bredemeier M, Cohen S, Godbold DL, Lode E, Pichler V, Schleppi P, eds. Forest Management and the Water Cycle: An Ecosystem-Based Approach. Ecological Studies 212, Springer, pp 31–64
39. Demmig-Adams B, Adams WWIII (2000) Harvesting sunlight safely. Nature 403: 371–374

40. Osório J, Osório ML, Romano A (2012) Reflectance indices as nondestructive indicators of the physiological status of *Ceratonia siliqua* seedlings under varying moisture and temperature regimes. Functional Plant Biology, 39: 588–597

41. Sims DA, Gamon JA (2002) Relationships between leaf pigment content and spectral reflectance across a wide range of species, leaf structures and developmental stages. Remote Sensing of Environment 81: 337–354

42. Guo J, Trotter CM (2004) Estimating photosynthetic light-use efficiency using the photochemical reflectance index: variations among species. Functional Plant Biology 31: 255–265

43. Filella I, Porcar-Castell A, Munne-Bosch S, Back J, Garbulsky MF, et al. (2009) PRI assessment of long-term changes in carotenoids/chlorophyllratio and short-term changes in de-epoxidation state of the xanthophyllcycle. International Journal of Remote Sensing 30: 4443–4455

44. Ollinger SV (2011) Sources of variability in canopy reflectance and the convergent properties of plants. New Phytologist 189: 375–394.

16

Identification of Candidate Genes Associated with Leaf Senescence in Cultivated Sunflower (*Helianthus annuus L.*)

Sebastian Moschen[1,2], Sofia Bengoa Luoni[3], Norma B. Paniego[1,2], H. Esteban Hopp[1,4], Guillermo A. A. Dosio[2,5], Paula Fernandez[1,2,3¶], Ruth A. Heinz[1,2,4*¶]

1 Instituto de Biotecnología, Centro de Investigaciones en Ciencias Agronómicas y Veterinarias, Instituto Nacional de Tecnología Agropecuaria, Hurlingham, Buenos Aires, Argentina, 2 Consejo Nacional de Investigaciones Científicas y Técnicas, Ciudad Autónoma de Buenos Aires, Argentina, 3 Escuela de Ciencia y Tecnología, Universidad Nacional de San Martín, San Martín, Buenos Aires, Argentina, 4 Facultad de Ciencias Exactas y Naturales, Universidad de Buenos Aires, Ciudad Autónoma de Buenos Aires, Argentina, 5 Laboratorio de Fisiología Vegetal, Unidad Integrada Universidad Nacional de Mar del Plata, Estación Experimental Agropecuaria INTA Balcarce, Balcarce, Buenos Aires, Argentina

Abstract

Cultivated sunflower (*Helianthus annuus L.*), an important source of edible vegetable oil, shows rapid onset of senescence, which limits production by reducing photosynthetic capacity under specific growing conditions. Carbon for grain filling depends strongly on light interception by green leaf area, which diminishes during grain filling due to leaf senescence. Transcription factors (TFs) regulate the progression of leaf senescence in plants and have been well explored in model systems, but information for many agronomic crops remains limited. Here, we characterize the expression profiles of a set of putative senescence associated genes (SAGs) identified by a candidate gene approach and sunflower microarray expression studies. We examined a time course of sunflower leaves undergoing natural senescence and used quantitative PCR (qPCR) to measure the expression of 11 candidate genes representing the NAC, WRKY, MYB and NF-Y TF families. In addition, we measured physiological parameters such as chlorophyll, total soluble sugars and nitrogen content. The expression of *Ha-NAC01*, *Ha-NAC03*, *Ha-NAC04*, *Ha-NAC05* and *Ha-MYB01* TFs increased before the remobilization rate increased and therefore, before the appearance of the first physiological symptoms of senescence, whereas *Ha-NAC02* expression decreased. In addition, we also examined the trifurcate feed-forward pathway (involving *ORE1*, *miR164*, and *ETHYLENE INSENSITIVE 2*) previously reported for Arabidopsis. We measured transcription of *Ha-NAC01* (the sunflower homolog of *ORE1*) and *Ha-EIN2*, along with the levels of *miR164*, in two leaves from different stem positions, and identified differences in transcription between basal and upper leaves. Interestingly, *Ha-NAC01* and *Ha-EIN2* transcription profiles showed an earlier up-regulation in upper leaves of plants close to maturity, compared with basal leaves of plants at pre-anthesis stages. These results suggest that the *H. annuus* TFs characterized in this work could play important roles as potential triggers of leaf senescence and thus can be considered putative candidate genes for senescence in sunflower.

Editor: Cynthia Gibas, University of North Carolina at Charlotte, United States of America

Funding: This research was supported by ANPCyT/FONCYT, Préstamo BID PICT 15-32905 and PAE 37100-PICT 019, INTA-PE AEBIO 241001 and 245001, INTA-PE AEBIO 245711, INTA-AEBIO 243532, INTA PN CER 1336 and UNMdP, AGR212, AGR260. The funders had no role in study design, data collection and analysis, decision to publish, or preparation of the manuscript.

Competing Interests: The authors have declared that no competing interests exist.

* Email: rheinz@cnia.inta.gov.ar

¶ These authors contributed equally to this work.

¶ These authors are joint last authors on this work.

Introduction

As the last stage of leaf development, the genetically determined and highly ordered process of senescence involves characteristic changes in gene expression that result in decreased photosynthetic activity, active degradation of cellular structures, nutrient recycling, lipid peroxidation and, ultimately, cell death [1,2]. Multiple variables control the complex mechanism of senescence; these genetic and environmental variables have a strong effect on crop yield [3]. Annual plants, such as grain and oil crops, undergo visible senescence towards the end of the reproductive stage, accompanied by nutrient remobilization from leaves to developing seeds [4]. In monocarpic species, the development of the reproductive structure controls leaf senescence [5]. Prematurely induced senescence, caused by biotic or abiotic stress, can reduce crop yield. Thus, leaf senescence has an economic impact, affecting the gap between potential and real yields.

Sunflower (*Helianthus annuus L.*) is the third most important source of edible vegetable oil worldwide, and the second in Argentina. It also provides an important source of biodiesel [6] (Sunflower Statistics NSA 2007–2009, USA) [7]. Recent work has produced some genomic information for this crop [8], but the complete genome sequence remains unavailable. However, functional genomics tools for cultivated sunflower have been developed, including transcriptional and metabolic profiling tools [9–21].

In sunflower, adverse environmental conditions and foliar diseases abruptly trigger senescence [22], resulting in seriously limited production. In different crops, including sunflower, a delay in leaf senescence has an important impact on yield, by maintaining photosynthetic leaf area during the reproductive stage [23–26]. The maintenance of functional leaves for longer periods could increase the intercepted radiation and thus favour seed weight and oil content during the grain filling period [27,28].

Sunflower is an annual, monocarpic species in which the reproductive organs exert a strong control on leaf senescence and nutrient remobilization, and final grain mass is affected by the source:sink ratio [29]. However, the age of a leaf, and its position on the stem affect the triggering of senescence and the rate of the remobilization of nutrients [30]. Given the high extinction coefficient of light in the canopy for sunflower, shading of lower leaves affects not only the amount of incident radiation, but also the quality of the light, thereby causing senescence [31].

Two main stages can be distinguished in the senescence process: an initial stage involving up- or down-regulation of expression of a set of genes involved in nutrient export and degradation of cell structures, and a second stage involving progression of senescence, which proceeds at different rates under different conditions. The first stage of senescence can be assessed by measuring gene expression and the second stage can be assessed by measuring physiological traits. For example, chlorophyll content, the most commonly measured parameter, is directly associated with turnover of CO_2-fixing Rubisco [30,32]. However, other physiological parameters, such as the drop in soluble sugars, can also be measured to evaluate the progression of senescence [33,34]. Nitrogen content represents an important leaf senescence-associated variable, given its central role in key cellular processes such as photosynthesis. Previous studies examined several physiological parameters related to leaf senescence in sunflower [9,35–37], but the relationship of these parameters to gene expression, particularly the induction of transcription factors at an early stage of leaf development, has not been reported yet.

NAC transcription factors related to senescence have been identified in model (Arabidopsis thaliana, Medicago sativa, Orzya sativa L., Brassina napus) and non-model species (Chrysanthemum lavandulifolium, Gossipium hirsutum L., Malus domestica, Solanum tuberosum L., Triticum aestivum) and described as relevant players in the regulation of leaf senescence, particularly related to programmed cell death [38–56]. In Arabidopsis thaliana, the expression of a NAC gene family member, ORE1, occurs under the control of the ethylene signaling pathway gene EIN2 (ETHYLENE INSENSITIVE2) [45]. EIN2 encodes a central protein of ethylene signaling that is located in the endoplasmic reticulum membrane [57,58] and is down-regulated by miR164. In young leaves, which have high levels of miR164, ORE1 expression remains low, but during the leaf aging process, as miR164 expression gradually decreases, ORE1 expression increases [38]. Despite evidence of ORE1 function in senescence, our knowledge about the possible targets of this transcription factor remains limited. Recent work identified the preferred binding sequence of ORE1, which directly activates BIFUNC-TIONAL NUCLEASE1 (BFN1) [43].

Previous work on leaf senescence in sunflower identified Hahb-4, a transcription factor related to the ethylene signaling pathway [59]. Ethylene positively regulates Hahb-4 during normal leaf senescence; once induced, Hahb-4 negatively regulates the biosynthesis of ethylene and the expression of genes related to this signaling pathway.

In this work, we evaluated the transcription profiles of NAC, AP2/EREBP, WRKY, MYB and NF-Y family TF as potential candidate genes, as well as miR164, in sunflower leaves at different developmental stages during natural senescence. Concomitantly, we also measured physiological parameters such as chlorophyll, total soluble sugars and nitrogen content, to assess the triggering time of the different functional variables along the onset and evolution of senescence in sunflower, a relevant oil crop with limited available genomic information.

Materials and Methods

Plant material and experimental conditions

A field experiment was carried out at the INTA Balcarce Experimental Station (37°45′ S, 58°18′ W) during the 2010/11 growing season. Sunflower hybrid VDH 487 (Advanta Seeds) was sown at a 7.2 plants/m^2. Emergence occurred 10 days later. Diseases, weeds and insects were adequately controlled. Soil fertility assured maximum yields under non-limiting water conditions. Soil volumetric humidity was measured periodically using the time domain reflectometry technique (Trase System, Model 6050X1, Soil moisture Equipment Corp., Santa Barbara, CA, USA). Soil water was maintained by irrigation above 50% of volumetric humidity in the first 0.60 m of soil during the entire growing season.

Time was expressed on a thermal time basis by daily integration of air temperature with a threshold temperature of 6°C and with plant emergence as thermal time origin [60].

The experiment was conducted as a randomized complete block design with three replicates (plant–plots). Each biological replicate consisted of three randomly selected plants from each plot. Both molecular and physiological measurements were performed in tissue obtained from leaves 10 and 20. Both leaves were sampled after they reached about 15 cm width and with an accumulation of degree days from emergence (°Cd) between 432 to 861 and 670 to 1180, for leaves 10 and 20 respectively.

Chlorophyll measurements

Chlorophyll content of the sampled leaves was measured by chemical extraction with N, N dimethylformamide [61]. Two 0.5 cm-diameter discs were taken from the base of each sampled leaf. A total of 6 disks for each biological replicate (3 plants sampled in each plot) were dried with tissue paper and incubated in vials containing 6 ml of N, N dimethylformamide overnight at room temperature in darkness. Absorbance of each sample was measured using a spectrophotometer (Spectronic 20, Bausch & Lomb, Bausch & Lomb Place, Rochester, New York, USA). Chlorophyll content was calculated as:

Chlorophyll $(mg.l^{-1}) = 17.9 * abs\ (647) + 8.08 * abs\ (664)$
Chlorophyll $(mg.cm^{-2}) = Cl\ (mg.l^{-1})/1.1775\ cm^2$

Where: $abs\ (647)$ = absorbance at 647 nm, corresponding to the maximum absorption peak of chlorophyll B; $abs\ (664)$ = absorbance at 664 nm, corresponding to the maximum absorption peak of chlorophyll A and the values 17.9 and 8.08 are the extinction coefficients of chlorophylls A and B, respectively [61]

Total soluble carbohydrates (TSC)

Three 1.2 cm-diameter discs were taken from the base of each leaf sampled. A total of 9 disks for each biological replicate (3 plants sampled in each plot) were weighed and dried at 60°C for 48 hours. Five ml of distilled water was added to a test tube containing 50 mg of dry tissue. Samples were incubated in a water bath at 100°C for 10 min and centrifuged at 2500 rpm for 5 min; this step was repeated three times. A 100 µl volume of the

Table 1. Primer sequences for qPCR analysis.

Sunflower TF	Primer sequence Left	Primer sequence Right
Ha-NAC01	AAGAAGTACCCGACCGGATT	TCACCCAATTCGTCTTTCC
Ha-NAC02	GACTTCCGTTGATCCGACAT	AATGGGTCGGTTGTGCTTAG
Ha-NAC03	ATTTCTGGCGGTCAAACAAC	CCGGTCTTGTATCTCGGGTA
Ha-NAC04	TGGTGTGAAGAAGGCACTTG	TTCGACACAAAACCCAATCA
Ha-NAC05	ATGTTTGGCGAGAAGGAATG	TCCTTTCGGAGCTTTACCAA
Ha-WRKY01	CATAATGCCCCATTCAATCC	CATTTTGCTTGGTTGGAGGT
Ha-MYB01	TACTTGCCCGAAAATCCAGT	ATTGCTCGTCCAATCAATCC
Ha-MYB02	GAAGTAATGCCCCTGGATCA	TTCTTGCCATTGAACTGCTG
Ha-RAV01	CCTCATCGCGATACAAAGGT	GCCTTTGCAGCTTCATCTTC
Ha-NF-YB3	ACGTCAGGGTGGTGAAAAAG	ATAACCCGAACCGATTTTCC
Ha-EIN2	AGCTGGCGTTCTCTAAACCA	TGCAATCTCCACATCCTTCA
Ha-CAB2	TTGATCCATGCACAAAGCAT	AGCTACCACCGGGGTAAAGT
Ha-EF-1α	ACCAAATCAATGAGCCCAAG	GAGACTCGTGGTGCATCTCA

supernatant was placed in a test tube and distilled water was added to a volume of 1 ml. TSC were quantified by a colorimetric method using a phenol and sulfuric acid reaction [62]. One volume of phenol 5% and five volumes of sulfuric acid were added to the solution and incubated for 30 min at 25°C. After a color reaction with a mixture of phenol / sulfuric acid, optical density was measured in a spectrophotometer at 490 nm (Spectronic 20, Bausch & Lomb, Rochester, New York, USA). Total soluble carbohydrates were deduced from a standard curve constructed with glucose standard solutions $(0–15 \ \mu g.\mu l^{-1})$.

Total Nitrogen (%)

The percentage of total nitrogen was measured according to the Dumas method from 60 mg of dry tissue [63]. Briefly, the Dumas method consists of a dry combustion at 950°C, using oxygen as the combustion accelerator. Combustion products (H_2O, NO, N_2) are filtered and dried. NO is reduced to N_2 by copper, and is swept by helium gas to a thermal conductivity cell where the concentration is measured (TrunSpec CN, Leco, Michigan, USA).

RNA isolation and quality controls

Total RNA isolation was performed on healthy leaf samples starting from 430°Cd after emergence in order to assure RNA integrity. Samples were immediately frozen in liquid nitrogen and saved at −80°C until processing. High quality total RNA was isolated from 100 mg of frozen tissue using TRIzol, following the manufacturers instructions (Invitrogen, Argentina). Genomic DNA was eliminated after treatment with DNase I for 20 min at room temperature using DNase I (Invitrogen, Argentina).

RNA concentration was measured using a Nanodrop ND-1000 spectrophotometer (NanoDrop Technologies, Wilmington, Delaware USA). Purity and integrity of total RNA was determined by 260/280 nm ratio and the integrity was checked by electrophoresis in 1.5% agarose gel.

Selection of Transcription Factors

Eleven transcription factors were selected by a literature search for putative orthologs of candidate genes associated with leaf senescence that have been reported in model species, as well as TFs identified based on differential expression levels, from a customized 4×44 K microarray analysis conducted on the Agilent

platform [15], in which leaf samples at different development stages and growing conditions were assessed (Table S1 and S2) (unpublished data). Statistical analysis was performed using in house routines to fit, gene by gene, a linear mixed-effects model. The Sunflower Custom Oligo Microarray includes 4 arrays per chip; therefore, the chip effect (incomplete block) was included as a random effect. The set of routines mentioned above were based on the *lme* function of the *nlme* library of R [64] implemented in InfoStat statistical software [65]. Differential gene expression analysis was carried out using the limma package. Gene set analysis was carried out according to the Gene Ontology terms using FatiScan [66] integrated in the Babelomics suite [67].

Primer design for Reference Genes and SAGs

Different reference genes were assessed in a previous gene expression study in sunflower leaf senescence [68]. In this study, Elongation Factor 1-α (Ha-EF-1α) was selected as a reference gene, as it showed stable expression in the different samples for both leaves.

Sunflower Unigene Resource (SUR v1.0) database available at http://atgc-sur.inta.gob.ar// was used to search candidate TFs [15].

Specific primer pairs for qPCR were designed based on selected sequences using Primer3 software [69] with the default parameters (Table 1).

Quantitative RT-PCR analysis

For each sample, 500 ng DNase treated RNA was reverse-transcribed using Superscript III first strand synthesis system (Invitrogen, USA) and random hexamer primers according to the manufacturer's instructions. qPCR was carried out in a 25-μl reaction mix containing 200 nM of each primer, 1 μl of cDNA sample and FastStart Universal SYBR Green Master (Roche Applied Science). Negative controls (no RT added and non-template control) were incorporated in the assays. qPCRs were performed using a 96-well plate thermocycler (ABI Prism 7000 Sequence Detection System and software, PE Applied Biosystems, USA). The thermal profile was set to 95°C for 10 min, and 40 cycles of 95°C for 15 s, and hybridization temperature for 1 min. Amplicon specificity was verified by melting curve analysis (60 to 95°C) after 40 PCR cycles. The qPCR assay was carried out using three biological replicates for each condition, and two technical replicates from two independent cDNA synthesis reactions.

A qPCR assay was performed to quantify the expression of candidate genes in leaves 10 and 20 (numbered from the bottom to the top of the plant) at different sampling times during plant development. Expression of these genes was estimated in relation to Elongation Factor -1α (*EF-1α*) previously selected as a reference gene [68]. Amplification efficiencies and Ct values were determined for each gene and each tested condition, with the slope of a linear regression model using the LinRegPCR [70]. These profiles were estimated in relation to first sampling (S-1) and reference genes using fgStatistic software (Figure 1A and B) [71], based on previously published algorithms [72].

miRNA Northern blot

High quality total RNA was isolated from leaves by using TRIzol, following the manufacturers instructions (Invitrogen, Argentina), repeating the chloroform extraction two times. Fifteen micrograms of RNA was resolved in 17% polyacrylamide gels containing 7 M urea. After electrophoresis, RNA was blotted to GeneScreen Plus membrane (PerkinElmer Life Science, Waltham, MA). Sunflower microRNA164 probe (5′TTCATGTGCC-CTGCTTCTCCA3′) [73] was end-labelled using γATP and T4 Polynucleotide Kinase. The labelled probe was purified using the QIAquick Nucleotide Removal kit (QIAGEN, Argentina). The eluted radiolabeled oligonucleotide was incubated with the membrane in 5× SSC, 7% SDS, 20 nM Na_2HPO_4 (pH7.2), 2×

Denhardt's solution and 1 mg of sheared salmon sperm DNA at 50°C overnight.

The membrane was washed two times with washing solution containing 3× SSC, 5% SDS, 25 nM NaH_2PO_4 (pH7.5), 10× Denhardt's solution for 15 minutes and exposed overnight. The intensity of each band was quantified using a Typhoon Trio (Amersham Biosciences, Piscataway, NJ). Radioactivity intensity of each band was normalized based on ethidium bromide rRNA labelled loaded in each well.

Results

Measurement of Physiological Parameters

We assessed the progress of senescence by measuring changes in chlorophyll content, nitrogen and soluble carbohydrates. These three physiological variables showed similar profiles in leaves 10 and 20, numbered from the bottom of the plant (Figure 2A, B and C).

The maximum chlorophyll content in leaf 10 reached 0.03 mg.cm^{-2} at more than 200°Cd before flowering time. From 700°Cd after emergence, chlorophyll began to decline until it became zero, at close to 900°Cd (Figure 2A). In leaf 20, the maximum chlorophyll content was slightly higher than in leaf 10 (close to 0.035 mg.cm^{-2}), and the maximum was reached 900°Cd after emergence, about 100°Cd after flowering, when leaf 10 had already senesced. In leaf 20, chlorophyll began to decline at 1,100°Cd, and leaf 20 senesced 400°Cd later (Figure 2A). Leaf 20

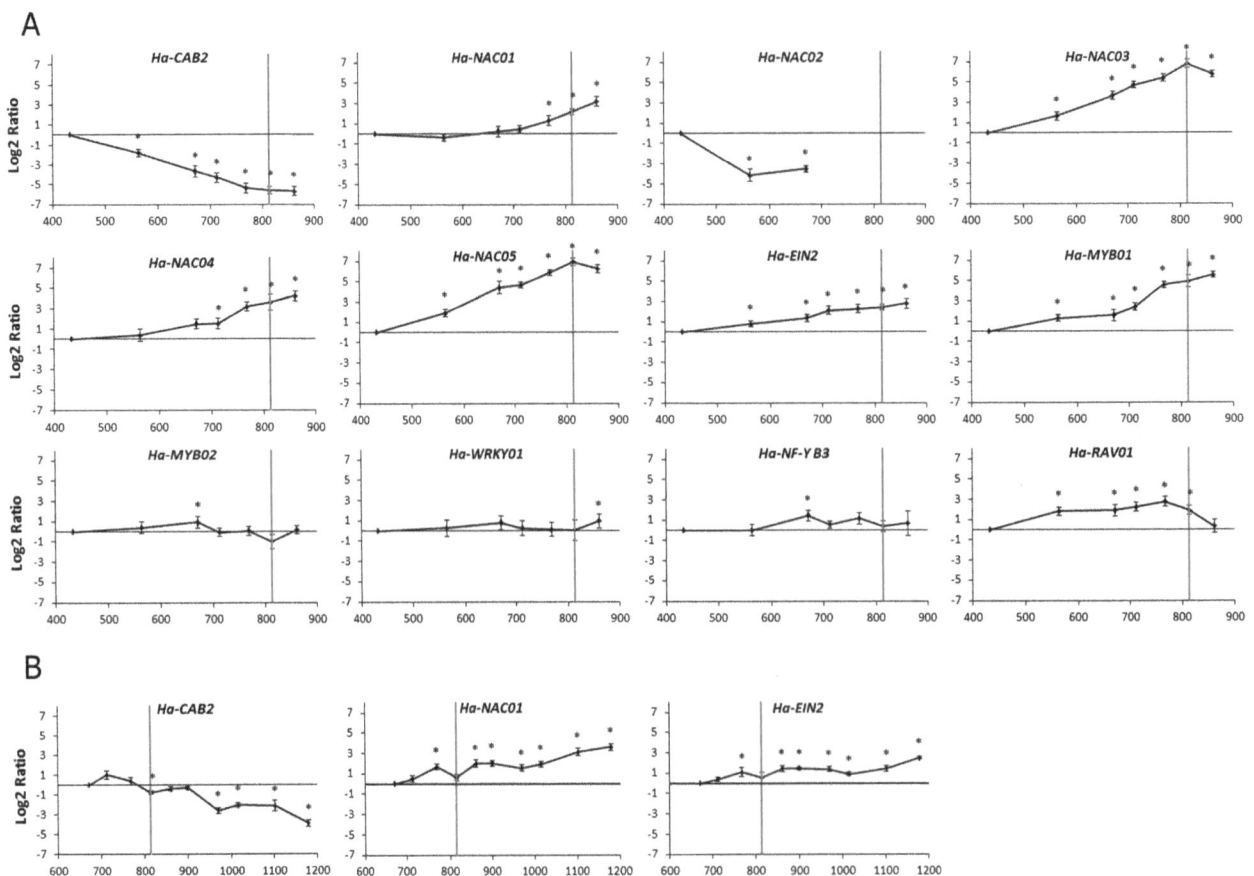

Figure 1. qPCR expression data. (A) Leaf 10. (B) Leaf 20. Relative transcript level for each candidate gene at different sampled points determined by qPCR. The x –axis showed the thermal time after emergence (°Cd). Asterisks indicate significant difference between each sampling in relation to first sampling point (S-1) and reference genes (p-value<0.05). The red line indicates anthesis time. Error bars correspond to standard errors.

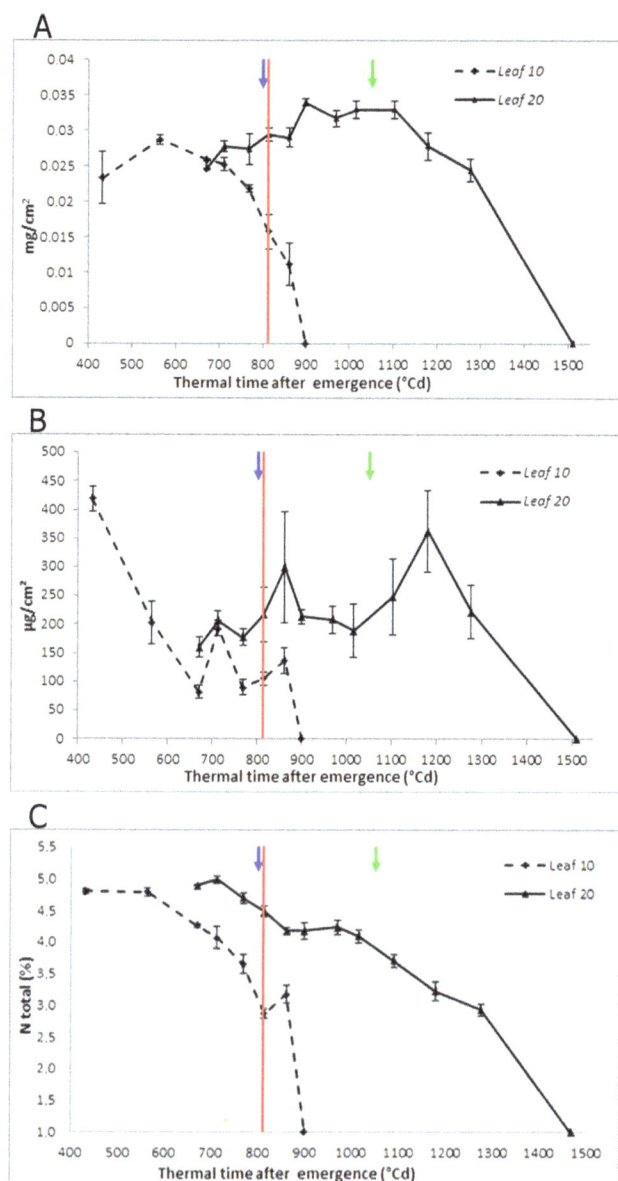

Figure 2. Physiological measurements of the progression of senescence. (A) Chlorophyll content in mg.cm^{-2}. (B) Total soluble carbohydrates in μg/cm^2 and (C) Total nitrogen percentage. The red line indicates anthesis time. Blue and green arrows show the date when the *Ha-NAC01* expression ratio reaches a value of 4 (Log2 ratio = 2) relative to first sampling point in leaf 10 and 20, respectively, determined by qPCR (*Ha-EF1α* as RG). Error bars correspond to standard errors.

had a longer lifespan; despite having been initiated at the apex 115°Cd later (data not shown), its active lifespan was lengthened by 600°Cd compared to leaf 10.

After a high initial value above 400 μg.cm^{-2}, total soluble carbohydrates in leaf 10 decreased to between 100 and 150 μg.cm^{-2}, from 600 to 850°Cd after emergence, just after flowering. Then, they dropped abruptly until leaf death at 900°Cd (Figure 2B). In leaf 20, carbohydrates remained relatively stable between 200 and 250 μg.cm^{-2} from 650 to 1,000°Cd after emergence. Nearly 1,100°Cd after emergence, soluble carbohydrates increased up to 350 μg.cm^{-2}, and thereafter started to drop

at 400°Cd after flowering, reaching zero at senescence, around 1,500°Cd (Figure 2B).

The maximum N (%) content was close to 5% in both leaves (Figure 2C). In leaf 10, it started to decline at 600°Cd and in leaf 20 at 700°Cd after emergence. In both leaves the content of N started to decline before flowering; at flowering, the N content in leaf 10 had already diminished to 50% of its maximum content, and the N content in leaf 20 had decreased to 80% of maximum. Furthermore, the decline was markedly slower in leaf 20 than in leaf 10 (Figure 2C).

Measurement of Transcription Factor Gene Expression

We used BLASTX [74] to identify putative TFs highly similar to Arabidopsis SAGs. Searches of the *Sunflower Unigene Resource (SUR v1.0)* [15] identified 42 genes with a significant score. Out of these, we selected 11 TFs by a candidate gene strategy based on literature searches in model species and on gene expression levels measured in a sunflower microarray analysis (unpublished data). The selected genes included NAC, WRKY, MYB, RAV, NF-Y, and AP2 TFs (Table S1 and S2).

NAC TF transcript levels increased significantly during leaf development (Figure 3A). All of the NAC TFs tested showed an up-regulation at early stages of leaf development, except *Ha-NAC02*, which showed down-regulated expression at an early stage. *Ha-NAC01* and *Ha-NAC04* transcript levels gradually increased from emergence, with high transcript levels in later developmental phases, close to anthesis. Moreover, *Ha-NAC03* and *Ha-NAC05* transcript levels increased at an early stage of leaf development, with continued increases in expression until anthesis, when they reached their highest levels. Transcript levels of *Ha-EIN2* gradually increased at an early stage and then showed uniform expression through leaf development (Figure 3B). In contrast, *Ha-MYB01* transcript levels gradually increased in early stages, then strongly increased from 700°Cd until the end of leaf development. *Ha-RAV01* transcript levels also gradually increased up to 750°Cd and then declined until the last stage of development. Transcript levels of TFs *Ha-MYB02*, *Ha-NF-YB3* and *Ha-WRKY01* did not change during leaf development (Figure 3B).

Arabidopsis *ORE1* and *EIN2* have been reported as triggers of senescence [38,39]. Therefore, we measured the expression of *Ha-NAC01*, the putative sunflower ortholog of Arabidopsis *ORE1*, in both leaves and found that it progressively increased from emergence to approximately 700°Cd for leaf 10 and to 1,000°Cd for leaf 20, with a strong increase in the expression in the last phases of development in both leaves (Figure 4). *Ha-EIN2*, which shows high sequence similarity to Arabidopsis *EIN2*, showed different expression profiles in the two leaves. In leaf 10, *Ha-EIN2* transcript levels increased strongly in early developmental stages, even earlier than *Ha-NAC01* and prior to anthesis. By contrast, in leaf 20, *Ha-EIN2* transcript levels showed a mild and constant increase, with a significant increase towards the last stages of leaf development (Figure 5A and C).

To assess the abundance of *miR164*, previously reported as a negative regulator of *ORE1* in Arabidopsis [38], we measured its abundance by Northern blot. The abundance of *miR164* changed inversely with the expression of *Ha-NAC01* and *Ha-EIN2* in leaf 10 and 20 (Figure 5B and D). The young leaves had low levels of *Ha-NAC01* transcript levels and high levels of *miR164*; as leaf development progressed, the amount of *miR164* gradually decreased and *Ha-NAC01* expression increased.

Finally, we also measured the transcript levels of *Ha-CAB2* (chlorophyll a/b-binding protein) [75] in the same samples, to follow the progress of senescence at the molecular level (Figure 6).

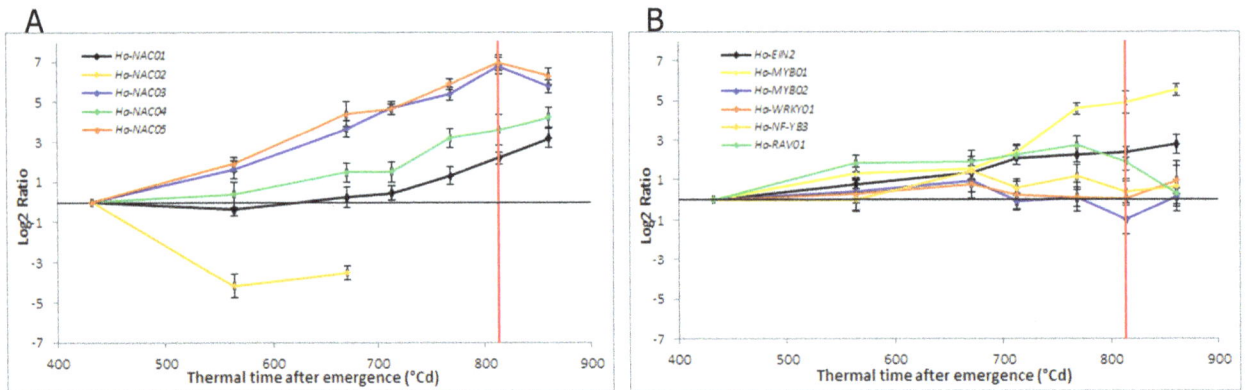

Figure 3. Expression profiles of transcription factors. (A) NAC TFs relative transcript level at each sampled point in relation to the level at the first sampling point determined by qPCR in leaf 10 (*Ha-EF1α* as RG). (B) AP2/EREBP, MYB, WRKY, NF-Y and RAV TFs relative transcript level at each sampled point in relation to the level at the first sampling point determined by qPCR in leaf 10 (*Ha-EF1α* as RG). The red line indicates anthesis time. Error bars correspond to standard errors.

Ha-CAB2 expression decreased during leaf development, with a steeper decrease in leaf 10 than in leaf 20. Hence, these results were consistent with those observed for physiological parameters, as expected (Figure 2).

Discussion

The complex and highly coordinated mechanism of plant senescence may have substantial effects on agricultural production. In sunflower, as in many other crops, leaf senescence triggers abruptly, coinciding with adverse environmental conditions and foliar diseases, and limiting production. The onset and progression of senescence involves global changes in gene expression, regulated by internal and external factors. Multiple pathways respond to various stimuli and interconnect, leading to a complex regulatory network for senescence [76].

In this work, we quantified the transcription profiles of eleven candidate genes previously reported as potential regulators of senescence in the model plant *A. thaliana*. We evaluated these transcripts in association with physiological parameters, quantifying transcript levels over time in senescing sunflower leaves, to monitor the natural progression of senescence in this crop.

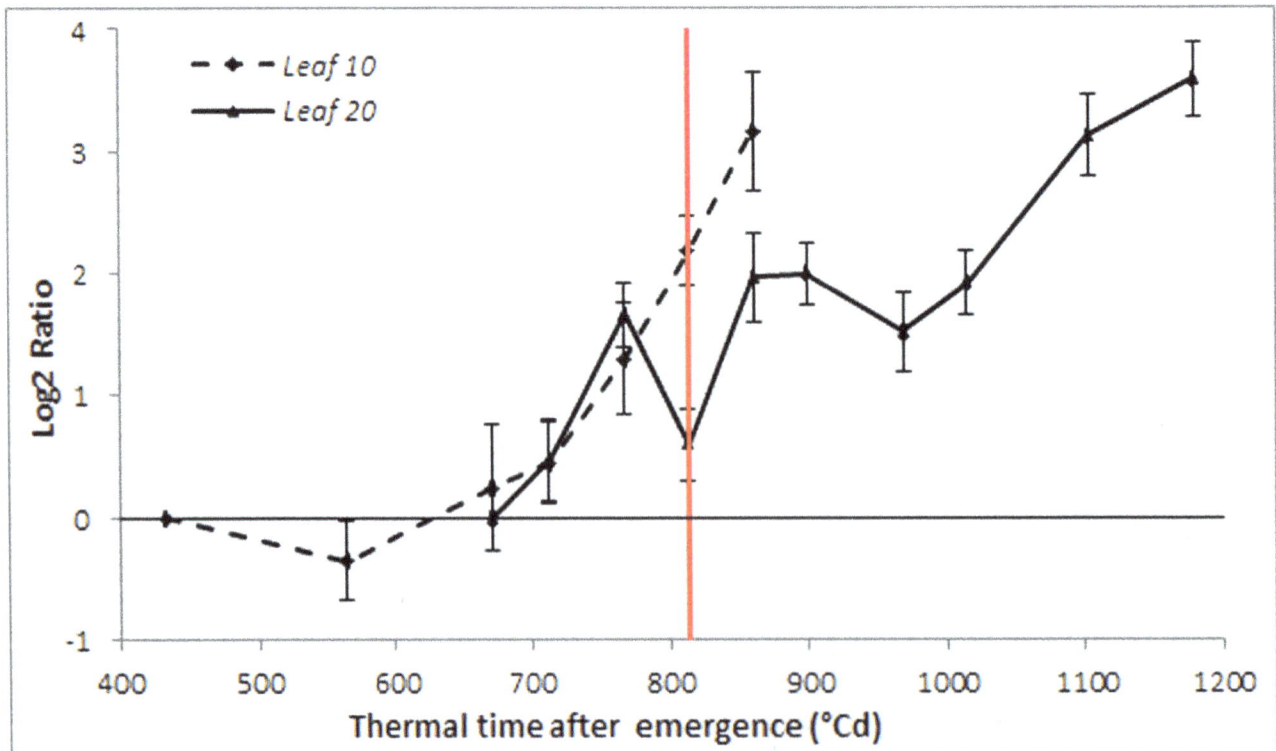

Figure 4. *Ha-NAC01* expression profile. Relative *Ha-NAC01* transcript level at each sampled point in relation to the level at the first sampling point determined by qPCR in leaf 10 and 20 (*Ha-EF1α* as RG). The red line indicates anthesis time. Error bars correspond to standard errors.

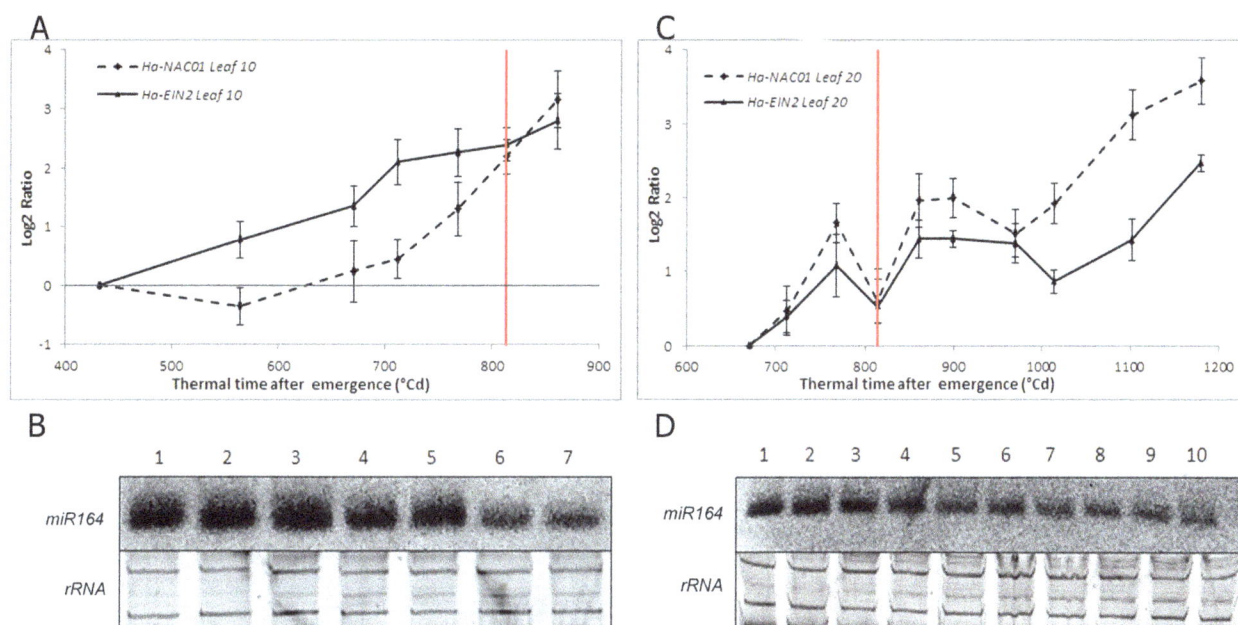

Figure 5. *Ha-EIN2, Ha-NAC01* and *miR164* expression profiles. (A) and (C) Transcript levels of *Ha-EIN2* and *Ha-NAC01* in relation to the first sampling point, using *Ha-EF1α* as RG in leaf 10 and 20 respectively. The red line indicates anthesis time. (B) and (D) Northern blot analysis of *miR164* levels in leaf 10 and 20, RNA from leaves 10 and 20 with an accumulation of degree days (°Cd) between 432 to 861 (B: lanes 1–7) and 670 to 1180 (D:1–10), respectively, was blotted to nylon membrane and hybridized to radioactively labelled sunflower microRNA164 probe. rRNA bands stained with ethidium bromide are shown as loading control. Error bars correspond to standard errors.

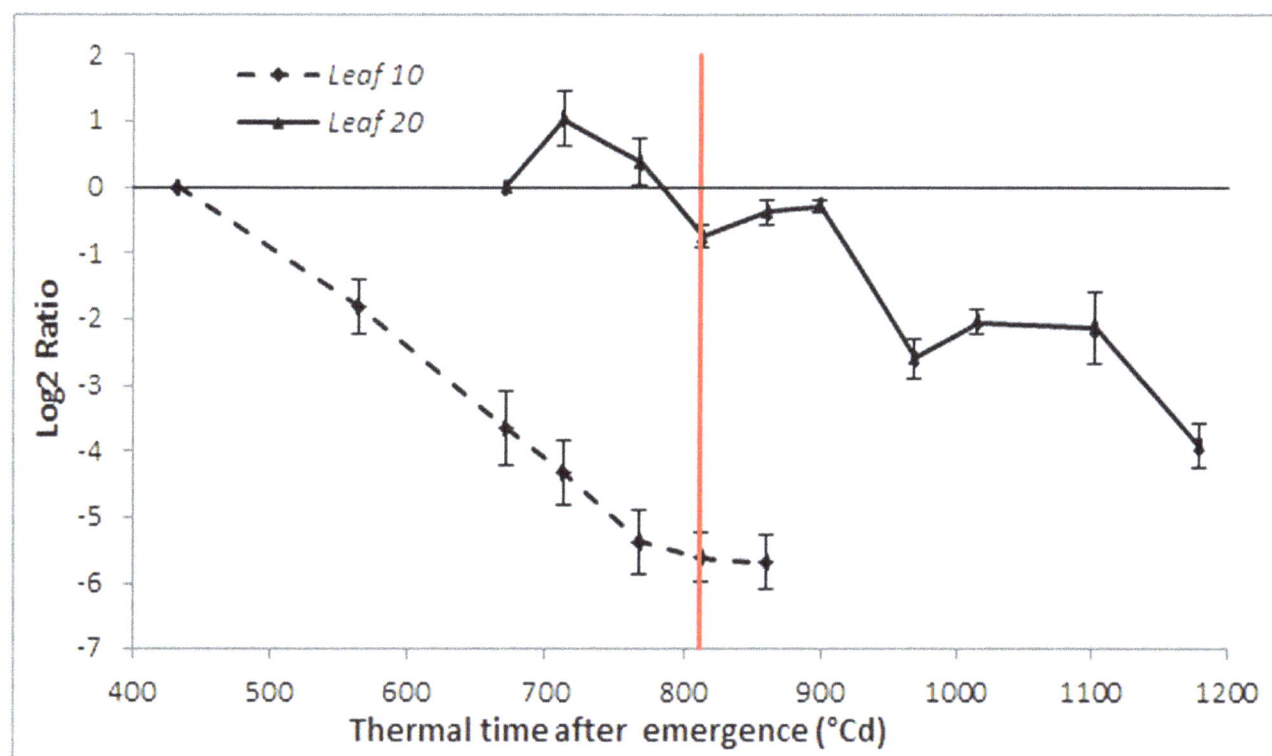

Figure 6. *Ha-CAB2* expression profile. Relative values of *Ha-CAB2* transcript levels referred to first sampling point, determined by qPCR (leaf 10 and 20) (*Ha-EF1α* as RG). The red line indicates anthesis time. Error bars correspond to standard errors.

We found that *Ha-NAC02*, *Ha-NAC03* and *Ha-NAC05*, which are highly similar to Arabidopsis *ANAC072*, *ANAC055* and *ANAC019*, respectively, showed contrasting expression profiles. In Arabidopsis, *ANAC072*, *ANAC055* and *ANAC019* belong to the same clade of NAC genes and have overlapping expression patterns [77]. In sunflower, the expression of *Ha-NAC03* and *Ha-NAC05* rapidly increased toward anthesis, but *Ha-NAC02* showed an opposite expression pattern. These results suggest that different mechanisms might regulate *Ha-NAC02*, compared to *Ha-NAC03* and *Ha-NAC05*. In addition, expression analysis of the *anac019* and *anac055* mutants during senescence also indicated involvement of different signalling pathways for these genes [77].

We also found that *Ha-NAC04*, which is highly similar to Arabidopsis *ANAC047*, was up-regulated during leaf development in sunflower. In Arabidopsis, *ANAC047* was up-regulated during leaf senescence and down-regulated in mutants with defective jasmonic acid, salicylic acid or ethylene pathways, indicating a putative role for this protein during leaf senescence related to hormone signalling [78].

The Arabidopsis TF *MYB62* functions in the response to phosphate starvation and affects gibberellic acid (GA) biosynthesis. Overexpression of *MYB62* results in a typical GA-deficient phenotype, with reduced apical dominance, delayed flowering and late senescence, suggesting that MYB62 acts as a transcriptional repressor of GA biosynthetic genes [79]. In sunflower, we detected up-regulation of *Ha-MYB01* (highly similar to Arabidopsis *MYB62*) starting close to anthesis, at 700°Cd after emergence and increasing thereafter, in a period likely related to nutrient starvation, concomitant with the critical period of grain filling.

In contrast to *Ha-MYB01*, we found that *Ha-RAV01* showed a similar expression pattern to Arabidopsis *RAV1*, which increased in expression at an early stage, before the appearance of senescence symptoms and started to decrease towards the last senescence stages. Woo et al. [75] showed that RAV1 plays a regulatory role during the initiation of leaf senescence and suggested that it might control senescence by the transcriptional activation and/or repression of genes involved in the execution of leaf senescence. *RAV1* and *RAV2* were also induced by several external factors, such as pathogen attack, low temperature, drought, salt stress, darkness, and wounding [80–82].

In Arabidopsis, the *ORE1* transcription factor can induce leaf senescence [39], suggesting that this gene functions as a regulator of senescence. In addition, it was postulated that the micro-RNA *miR164* suppresses *ORE1* transcript levels and both elements are regulated in a loop that also involves *EIN2*, where EIN2 promotes the expression of *ORE1* and inhibits *miR164*. In this work, we evaluated the trifurcate pathway *Ha-NAC01*, *HaEIN2* and *miR164* [38] in two different leaves, 10 and 20 to identify differences in the expression profile associated with nutrient remobilization, during different developmental stages. We found similar expression profiles to those observed in Arabidopsis, indicating potencial conservation of this signaling pathway.

During pre-anthesis senescence, in leaf 10, the *Ha-EIN2* transcript accumulated earlier than the *Ha-NAC01* transcript, showing a significant increase in expression at the last stages, just before anthesis. At this time, the first leaf senescence symptoms appeared in leaf 10, coinciding with an increase in nutrient remobilization rate, indicating that this leaf is an important source of nutrients for the flower and younger leaves (Figure 5A). Chlorophyll and nitrogen contents in leaf 10, and (with some differences) total soluble carbohydrates, showed a rapid decrease, which was inversely proportional to *Ha-EIN2* and *Ha-NAC01*

transcript accumulation. Meanwhile, *miR164* showed high levels in the first samplings, until 700–800°Cd after emergence, and thereafter *miR164* levels decreased towards the last stages, when *Ha-NAC01* expression increased (Figure 5A and B). *miR164* and *Ha-EIN2* showed opposite expression patterns, indicating a potential role of *Ha-EIN2* in the negative regulation of *miR164*. These results are consistent with those observed in Arabidopsis, where *miR164* negatively regulates *ORE1* and *EIN2* is up-regulated earlier in the pathway, activating *ORE1* and inhibiting *miR164* [38]. In Arabidopsis, gene expression and miRNA levels were assessed in the third and fourth foliar leaves before flowering time. In the present study, leaf 10 represents an "old leaf" from a "young sunflower plant". Hence, this can be considered as a comparable physiological stage to those reported in Arabidopsis, showing similar transcriptional patterns for *Ha-NAC01*, *Ha-EIN2* and *miR164*.

Hahb4, a sunflower HD-Zip transcription factor previously reported as related to leaf senescence and in response to biotic and abiotic stress, is also under the control of ethylene signaling pathway [59,83,84]. These authors found that *Hahb-4* transcript levels were elevated in mature/senescent leaves, being its expression induced by ethylene. In addition, transient transformation of sunflower leaves demonstrated the action of Hahb-4 in the regulation of ethylene-related genes [59]. This information concomitant with the results derived from the present work, focused in the study of members of other TF families, highlight the relevance of the ethylene signaling pathways in the initiation and/or regulation of leaf senescence process in sunflower.

Opposite to leaf 10, a mid-upper leaf such as leaf 20 maintains the photosynthetic leaf area, representing an important source of photoassimilates during grain filling. When we measured post-anthesis senescence, we observed a delay in physiological symptoms for this recently expanded leaf, in comparison to leaf 10, regardless of the time of initiation on the apex (data not shown) (Figure 2). In contrast to our observations in leaf 10, *Ha-EIN2* expression increased in leaf 20, then remained stable throughout later stages with a mild increase toward the end of development. *Ha-NAC01* transcript levels in leaf 20 also increased prior to anthesis, similar to leaf 10, but then remained stable until the last stages, when its transcript level increased, before the detection of the first senescence symptoms (Figure 5C). In this leaf, chlorophyll and sugar contents remained stable during the leaf lifespan. Once *Ha-NAC01* reached high expression levels (1100-1200°Cd after emergence), these physiological parameters decreased (Figure 2A and B). Nitrogen contents showed a slow and constant decrease along leaf developmental phases (Figure 2C).

The *miR164* accumulation profile was similar to that observed in leaf 10, with a decrease of transcript levels in the last sampling when *Ha-NAC01* was up-regulated, indicating a potential role in the regulation of *Ha-NAC01* (Figure 5C and D). *miR164* levels were also opposite to *Ha-EIN2* expression, indicating a putative association between them.

These results indicate that the accumulation of Ha-EIN2 could induce expression of *Ha-NAC01* and inhibit *miR164*. In leaf 10, an earlier up-regulation of *Ha-EIN2* could lead to rapid accumulation of Ha-EIN2 protein, but in leaf 20, *Ha-EIN2* transcript levels were low during the leaf's lifespan. However, *Ha-EIN2* transcripts reached high levels later in plant development. The expression of *Ha-EIN2* could be regulated by different internal factors (such as age) and external factors (such as the incidence of radiation and red/far-red ratio), indicating a complex regulatory network for senescence.

During pre-anthesis senescence, when *Ha-NAC01* expression shows a 4-fold increase compared to the first sampling (Log2 ratio = 2) at 800°Cd after emergence, chlorophyll and nitrogen contents had decreased to 61% and 65% of the initial values, respectively. By contrast, during post-anthesis senescence, *Ha-NAC01* expression reaches a 4-fold increase at 1050°C after emergence. At this point, chlorophyll and nitrogen content showed a slight decrease, to 96% and 79%, respectively, of the initial values (Figure 2). Differences in the senescence profiles of leaf 10 and 20 in sunflower might be due to the short lifespan of leaf 10 and to light quality, accelerating senescence. Thus, *Ha-NAC01* transcript accumulation evolves slower in this leaf than in leaf 20 in relation to physiological indicators. Interestingly, in leaf 20, the *Ha-NAC01* expression profile was detected as an early senescence candidate gene.

Based on our results, leaf senescence in sunflower has two markedly different molecular profiles according to the plant developmental stage: leaf senescence during the vegetative stage, prior to anthesis and leaf senescence during the reproductive period, after anthesis. As an example of pre-anthesis senescence, leaf 10 is a developed leaf in a young plant, and the onset of senescence in leaf 10 takes place near anthesis. In this stage, nutrient remobilization from senescent organs supplies developing organs like the flower and young leaves. Different levels and quality of photosynthetically active radiation (PAR) received by the leaves can affect leaf lifespan during both pre- and post-anthesis phases [31]. Shading of the upper leaves, which decreases the incidence of PAR and the red/far-red ratio, accelerates senescence.

As an example of post-anthesis senescence, leaf 20 is a young leaf on a mature plant, and its senescence is delayed until the last plant developmental stages. In this case, the remobilized nutrients are mainly delivered to the grains. The greater incidence and quality of PAR received in this mid-upper leaf leads to a high photosynthetic activity and, consequently, to a delay of senescence, as detected by physiological measurements (Figure 2A, B and C).

In conclusion, this work identified and characterized, for the first time in cultivated sunflower, senescence-associated candidate genes previously reported in model species. Most of the putative NAC TFs transcripts evaluated, as well as *Ha-MyB1*, *Ha-RAV01* and *Ha-EIN2*, exhibited an early up-regulation, with increased transcript levels detected before the first senescence symptoms appeared. Interestingly, *Ha-NAC02* showed an opposite profile, with a sharp decrease in transcript levels at a very early stage of plant development. In addition, *Ha-NAC01* expression showed an inverse relationship with *miR164*, showing high expression levels before the maximum rate of decreases of chlorophyll, carbohydrate, and nitrogen content, mainly in leaf 20, suggesting that these TFs may play a role as potential triggers of senescence in

sunflower plants. However, this study also showed differences of sunflower candidate genes with the Arabidopsis model, since the plant developmental stage strongly affects SAG transcriptional profiles in sunflower. In particular, *Ha-NAC01* showed an earlier up-regulation in leaves of plants close to maturity compared to younger plants.

These results allowed us to detect early-induced candidate genes. However, further functional validation will help us to clarify their role in sunflower senescence. On-going integrated analyses using the sunflower oligonucleotide expression 4×44 K Agilent chip [15] as well as functional studies involving the isolation of full-length sequences of the candidate gene, their overexpression and/or silencing in transgenic plants, both model species and sunflower, will produce new evidence to support our present results. This work opens new avenues for further studies leading to decipher metabolic pathways involved in the triggering or progression of senescence, as the same leaves were measured over the course of the full crop season to obtain physiological and molecular candidates under field conditions. To date, only a few transcriptomic studies have been published describing assays under field conditions [85,86].

Taking these results together, our work provides a new and integrated contribution to the current knowledge of senescence in sunflower, allowing the identification of early candidate genes associated with senescence opening new avenues to explore senescence in this strategic oil crop.

Supporting Information

Table S1 BlastX results and Transcription Factors expression levels derived from a customized 4×44 K microarray analysis conducted on Agilent platform [15]. Three different developmental stages and growing conditions were assessed (T1 = 630°Cd, T2 = 860°Cd and T3 = 970°Cd after emergence).

Table S2 BlastX results and Transcription Factor domain detected using Conserved Domain Database (CDD) (http://www.ncbi.nlm.nih.gov/cdd).

Author Contributions

Conceived and designed the experiments: SM NBP GAAD PF RAH. Performed the experiments: SM SBL. Analyzed the data: SM SBL. Contributed reagents/materials/analysis tools: NBP HEH. Wrote the paper: SM NBP GAAD PF RAH.

References

1. Srivalli B, Khanna-Chopra R (2004) The developing reproductive 'sink' induces oxidative stress to mediate nitrogen mobilization during monocarpic senescence in wheat. Biochem Biophys Res Commun 325: 198–202.

2. Agüera E, Cabello P, de la Haba P (2010) Induction of leaf senescence by low nitrogen nutrition in sunflower (Helianthus annuus) plants. Physiol Plant 138: 256–267.

3. Noodén LD, Guiamet JJ, Isaac J (1997) Senescence mechanisms. Physiologia plantarum 101: 746–753.

4. Buchanan-Wollaston V, Earl S, Harrison E, Mathas E, Navabpour S, et al. (2003) The molecular analysis of leaf senescence -a genomic approach. Plant Biotechnol J 1: 3–22.

5. Nooden LD (1988) Whole plant senescence. In: AC L, editor. Senescence and aging in plants: Academic Press, San Diego. pp. 392–439.

6. Fernandez P, Di Rienzo JA, Fernandez L, Hopp H, Paniego N, et al. (2008) Transcriptomic identification of candidate genes involved in sunflower responses to chilling and salt stresses based on cDNA microarray analysis. BMC Plant Biology 8.

7. Schuster I, Cruz C (2004) Estatística genômica aplicada a populações derivadas de cruzamentos controlados. 568.

8. Kane NC, Gill N, King MJ, Bowers JE, Berges H, et al. (2011) Progress towards a reference genome for sunflower. Botany 89: 429–437.

9. Agüera E, Cabello P, de la Mata L, Molina E, de la Haba P (2012) Metabolic Regulation of Leaf Senescence in Sunflower (Helianthus annuus L.) Plants. Senescence: InTech Open Access Publisher.

10. Bachlava E, Taylor Ca, Tang S, Bowers JE, Mandel JR, et al. (2012) SNP discovery and development of a high-density genotyping array for sunflower. PLoS ONE 7: e29814.

11. Cabello P, Agüera E, Haba PDl (2006) Metabolic changes during natural ageing in sunflower (Helianthus annuus) leaves: expression and activity of glutamine synthetase isoforms are regulated differently during senescence. Physiol Plant 128: 175–185.

12. Fernandez P, Paniego N, Lew S, Hopp HE, Heinz RA (2003) Differential representation of sunflower ESTs in enriched organ-specific cDNA libraries in a small scale sequencing project. BMC Genomics 4: 40.

13. Fernandez P, Di Rienzo J, Fernandez L, Hopp HE, Paniego N, et al. (2008) Transcriptomic identification of candidate genes involved in sunflower responses to chilling and salt stresses based on cDNA microarray analysis. BMC Plant Biology 8: 11.

14. Fernandez P, Moschen S, Paniego N, Heinz RA (2012) Functional approaches to study leaf senescence in sunflower. Senescence: InTech Open Access Publisher.

15. Fernandez P, Soria M, Blesa D, DiRienzo J, Moschen S, et al. (2012) Development, characterization and experimental validation of a cultivated sunflower (Helianthus annuus L.) gene expression oligonucleotide microarray. PLoS ONE 7.

16. Lai Z, Kane N, Kozik A, Hodgins K, Dlugosch K, et al. (2012) Genomics of Composi tae weeds: EST libraries, microarrays, and e vidence of introgression. Am J Bot 99: 209–218.

17. Paniego N, Heinz R, Fernandez P, Talia P, Nishinakamasu V, et al. (2007) Sunflower. In: Kole C, editor. Genome Mapping and Molecular Breeding in Plants. Berlin Heidelberg: Springer-Verlag. pp. 153–177.

18. Peluffo L, Lia V, Troglia C, Maringolo C, Norma P, et al. (2010) Metabolic profiles of sunflower genotypes with contrasting response to Sclerotinia sclerotiorum infection. Phytochemistry 71: 70–80.

19. Moschen S, Radonic LM, Ehrenbolger GF, Fernández P, Lía V, et al. (2014) Functional Genomics and Transgenesis Applied to Sunflower Breeding. In: Arribas JI, editor. Sunflowers: Growth and Development, Environmental Influences and Pests/Diseases: Nova Science Publishers. pp. 131–164.

20. Bazin J, Langlade N, Vincourt P, Arribat S, Balzergue S, et al. (2011) Targeted mRNA oxidation regulates sunflower seed dormancy alleviation during dry after-ripening. Plant Cell 23: 2196–2208.

21. Chan RL (2009) The use of sunflower transcription factors as biotechnological tools to improve yield and stress tolerance in crops. FYTON 78: 5–10.

22. Dosio GAA, Quiroz FJ (2010) Enfermedades foliares en girasol y su relación con la formación del rendimiento y el contenido de aceite. In: Miralles D.J., Aguirrezábal L.A.N., Otegui M.E., Kruk B.C., Izquierdo N.G., editors. Avances en ecofisiología de cultivos de granos. Buenos Aires, Argentina.: Editorial Facultad de Agronomía UBA. pp. 237–254.

23. Sadras VO, Echarte L, Andrade FH (2000) Profiles of Leaf Senescence During Reproductive Growth of Sunflower and Maize. Annals of Botany 85: 187–195.

24. Sadras VO, Quiroz F, Echarte L, Escande A, Pereyra VR (2000) Effect of Verticillium dahliae on Photosynthesis, Leaf Expansion and Senescence of Field-grown Sunflower. Annals of Botany 86: 1007–1015.

25. De la Vega A, Cantore MA, Sposaro NN, Trapani N, Lopez Pereira M, et al. (2011) Canopy stay green and yield in non stressed sunflower. Fields Crop Researchs 121: 175–185.

26. Kusaba M, Tanaka A, Tanaka R (2013) Stay-green plants: what do they tell us about the molecular mechanism of leaf senescence. Photosynth Res.

27. Dosio GAA, Aguirrezábal LAN, Andrade FH, Pereyra VR (2000) Solar radiation intercepted during seed filling and oil production in two sunflower hybrids. Crop Science 40: 1637–1644.

28. Aguirrezábal LAN, Lavaud Y, Dosio GAA, Izquierdo NG, Andrade FH, et al. (2003) Weight per seed and oil concentration in a sunflower hybrid are accounted for by intercepted solar radiation during a definite period of seed filling. Crop Science 43: 152–161.

29. Lopez Pereira M, Trapania N, Sadras VO (2000) Genetic improvement of sunflower in Argentina between 1930 and 1995. Part III. Dry matter partitioning and grain composition. Field Crop Res 67: 215–221.

30. Thomas H (2013) Senescence, ageing and death of the whole plant. New Phytol 197: 696–711.

31. Rousseaux MC, Hall AJ, Sanchez RA (1996) Far-red enrichment and photosynthetically active radiation level influence leaf senescence in field-grown sunflower. Physiol Plant 96: 217–224.

32. Breeze E, Harrison E, McHattie S, Hughes L, Hickman R, et al. (2011) High-Resolution Temporal Profiling of Transcripts during Arabidopsis Leaf Senescence Reveals a Distinct Chronology of Processes and Regulation. The Plant Cell Online 23: 873–894.

33. Wingler A, Masclaux-Daubresse C, Fischer AM (2009) Sugars, senescence, and ageing in plants and heterotrophic organisms. J Exp Bot 60: 1063–1066.

34. van Doorn WG (2008) Is the onset of senescence in leaf cells of intact plants due to low or high sugar levels? J Exp Bot 59: 1963–1972.

35. de la Mata L, Cabello P, de la Haba P, Agüera E (2013) Study of senescence process in primary leaves of sunflower (Helianthus annuus L.) plants under two different lights intensities. Photosynthetica 51: 85–94.

36. Hewezi T, Leger M, Gentzbittel L (2008) A comprehensive analysis of the combined effects of high light and high temperature stresses on gene expression in sunflower. Ann Bot 102: 127–140.

37. Sadras V, Echarte L, Andrade F (2000) Profiles of leaf senescence during reproductive growth of sunflower and maize. Annals of Botany 85: 187–195.

38. Kim JH, Woo HR, Kim J, Lim PO, Lee IC, et al. (2009) Trifurcate feed-forward regulation of age-dependent cell death involving miR164 in Arabidopsis. Science 323: 1053–1057.

39. Balazadeh S, Siddiqui H, Allu AD, Matallana-Ramirez LP, Caldana C, et al. (2010) A gene regulatory network controlled by the NAC transcription factor ANAC092/AtNAC2/ORE1 during salt-promoted senescence. Plant J 62: 250–264.

40. Hu R, Qi G, Kong Y, Kong D, Gao Q, et al. (2010) Comprehensive analysis of NAC domain transcription factor gene family in Populus trichocarpa. BMC Plant Biol 10: 145.

41. Nuruzzaman M, Manimekalai R, Sharoni AM, Satoh K, Kondoh H, et al. (2010) Genome-wide analysis of NAC transcription factor family in rice. Gene 465: 30–44.

42. Balazadeh S, Kwasniewski M, Caldana C, Mehrnia M, Zanor MI, et al. (2011) ORS1, an HO-responsive NAC transcription factor, controls senescence in Arabidopsis thaliana. Mol Plant 4: 346–360.

43. Matallana-Ramirez LP, Rauf M, Farage-Barhom S, Dortay H, Xue GP, et al. (2013) NAC Transcription Factor ORE1 and Senescence-Induced BIFUNCTIONAL NUCLEASE1 (BFN1) Constitute a Regulatory Cascade in Arabidopsis. Mol Plant.

44. Guo Y, Gan S (2006) AtNAP, a NAC family transcription factor, has an important role in leaf senescence. The Plant Journal 46: 601–612.

45. He XJ, Mu RL, Cao WH, Zhang ZG, Zhang JS, et al. (2005) AtNAC2, a transcription factor downstream of ethylene and auxin signaling pathways, is involved in salt stress response and lateral root development. Plant J 44: 903–916.

46. Jensen MK, Kjaersgaard T, Nielsen MM, Galberg P, Petersen K, et al. (2010) The Arabidopsis thaliana NAC transcription factor family: structure-function relationships and determinants of ANAC019 stress signalling. Biochem J 426: 183–196.

47. Huang H, Wang Y, Wang S, Wu X, Yang K, et al. (2012) Transcriptome-wide survey and expression analysis of stress-responsive NAC genes in Chrysanthemum lavandulifolium. Plant Sci 193–194: 18–27.

48. Wu A, Allu AD, Garapati P, Siddiqui H, Dortay H, et al. (2012) JUNGBRUNNEN1, a reactive oxygen species-responsive NAC transcription factor, regulates longevity in Arabidopsis. Plant Cell 24: 482–506.

49. de Zelicourt A, Diet A, Marion J, Laffont C, Ariel F, et al. (2012) Dual involvement of a Medicago truncatula NAC transcription factor in root abiotic stress response and symbiotic nodule senescence. Plant J 70: 220–230.

50. Shah ST, Pang C, Fan S, Song M, Arain S, et al. (2013) Isolation and expression profiling of GhNAC transcription factor genes in cotton (Gossypium hirsutum L.) during leaf senescence and in response to stresses. Gene 531: 220–234.

51. Su H, Zhang S, Yuan X, Chen C, Wang XF, et al. (2013) Genome-wide analysis and identification of stress-responsive genes of the NAM-ATAF1,2-CUC2 transcription factor family in apple. Plant Physiol Biochem 71: 11–21.

52. Singh AK, Sharma V, Pal AK, Acharya V, Ahuja PS (2013) Genome-wide organization and expression profiling of the NAC transcription factor family in potato (Solanum tuberosum L.). DNA Res 20: 403–423.

53. Wang YX (2013) Characterization of a novel Medicago sativa NAC transcription factor gene involved in response to drought stress. Mol Biol Rep 40: 6451–6458.

54. Zhou Y, Huang W, Liu L, Chen T, Zhou F, et al. (2013) Identification and functional characterization of a rice NAC gene involved in the regulation of leaf senescence. BMC Plant Biol 13: 132.

55. Ying L, Chen H, Cai W (2014) BnNAC485 is involved in abiotic stress responses and flowering time in Brassica napus. Plant Physiol Biochem 79C: 77–87.

56. Mao X, Chen S, Li A, Zhai C, Jing R (2014) Novel NAC transcription factor TaNAC67 confers enhanced multi-abiotic stress tolerances in Arabidopsis. PLoS ONE 9: e84359.

57. Alonso JM, Hirayama T, Roman G, Nourizadeh S, Ecker JR (1999) EIN2, a bifunctional transducer of ethylene and stress responses in Arabidopsis. Sciencie 284: 2148–2152.

58. Bisson MM, Bleckmann A, Allekotte S, Groth G (2009) EIN2, the central regulator of ethylene signalling, is localized at the ER membrane where it interacts with the ethylene receptor ETR1. Biochem J 424: 1–6.

59. Manavella PA, Arce AL, Dezar CA, Bitton F, Renou JP, et al. (2006) Cross-talk between ethylene and drought signalling pathways is mediated by the sunflower Hahb-4 transcription factor. Plant J 48: 125–137.

60. Kiniry JR, Blanchet R, Williams JR, Texier V, Jones K, et al. (1992) Sunflower simulation using the EPIC and ALMANAC models. Field Crops Research 30: 403–423.

61. Inskeep WP, Bloom PR (1985) Extinction coefficients of chlorophyll a and B in n,n-dimethylformamide and 80% acetone. Plant Physiol 77: 483–485.

62. Dubois M, Gilles KA, Hamilton JK, Rebus PA, Smith F (1956) Colorimetric method for the determination of sugars and related substances. Analytical Chemistry 28: 350–356.

63. Dumas A (1826) Annales de chimie. 33,342.

64. Pinheiro J, Bates D, DebRoy S, Sarkar D, Team tRDC (2012) nlme: Linear and Nonlinear Mixed Effects Models. R package. 3.1–105 ed.

65. Di Rienzo JA, Casanoves F, Balzarini MG, Gonzalez L, Tablada M, et al. InfoStat. In: Grupo InfoStat F, Universidad Nacional de Córdoba, Argentina. Available: http://www.infostat.com.ar, editor. 2013 ed.

66. Al-Shahrour F, Arbiza L, Dopazo H, Huerta-Cepas J, Minguez P, et al. (2007) From genes to functional classes in the study of biological systems. BMC Bioinformatics 8: 114.

67. Al-Shahrour F, Minguez P, Vaquerizas JM, Conde L, Dopazo J (2005) BABELOMICS: a suite of web tools for functional annotation and analysis of groups of genes in high-throughput experiments. Nucleic Acids Res 33: W460–464.

68. Fernandez P, Di Rienzo JA, Moschen S, Dosio GA, Aguirrezabal LA, et al. (2011) Comparison of predictive methods and biological validation for qPCR reference genes in sunflower leaf senescence transcript analysis. Plant Cell Rep 30: 63–74.

69. Rozen S, Skaletsky H (2000) Primer3 on the WWW for general users and for biologist programmers. Methods Mol Biol 132: 365–386.

70. Ruijter JM, Ramakers C, Hoogaars WM, Karlen Y, Bakker O, et al. (2009) Amplification efficiency: linking baseline and bias in the analysis of quantitative PCR data. Nucleic Acids Res 37: e45.

71. fgStatistics (2009) In: Di Rienzo JA, editor. pp. Facultad de Ciencias Agropecuarias, Universidad Nacional de Córdoba, Argentina. https://sites.google.com/site/fgstatistics/.

72. Pfaffl MW (2001) A new mathematical model for relative quantification in real-time RT-PCR. Nucleic Acid Research 29: e45 2001.

73. Barozai MY, Baloch IA, Din M (2012) Identification of MicroRNAs and their targets in Helianthus. Mol Biol Rep 39: 2523–2532.

74. Altschul S, Gish W, Miller W, Myers E, Lipman D (1990) Basic local alignment search tool. Journal of Molecular Biology 215: 403–410.

75. Woo HR, Kim JH, Kim J, Lee U, Song IJ, et al. (2010) The RAV1 transcription factor positively regulates leaf senescence in Arabidopsis. J Exp Bot 61: 3947–3957.

76. He Y, Tang W, Swain JD, Green AL, Jack TP, et al. (2001) Networking senescence-regulating pathways by using Arabidopsis enhancer trap lines. Plant Physiol 126: 707–716.

77. Hickman R, Hill C, Penfold CA, Breeze E, Bowden L, et al. (2013) A local regulatory network around three NAC transcription factors in stress responses and senescence in Arabidopsis leaves. Plant J 75: 26–39.

78. Buchanan-Wollaston V, Page T, Harrison E, Breeze E, Lim PO, et al. (2005) Comparative transcriptome analysis reveals significant differences in gene expression and signalling pathways between developmental and dark/starvation-induced senescence in Arabidopsis. Plant J 42: 567–585.

79. Devaiah BN, Madhuvanthi R, Karthikeyan AS, Raghothama KG (2009) Phosphate starvation responses and gibberellic acid biosynthesis are regulated by the MYB62 transcription factor in Arabidopsis. Mol Plant 2: 43–58.

80. Fowler S, Thomashow MF (2002) Arabidopsis transcriptome profiling indicates that multiple regulatory pathways are activated during cold acclimation in addition to the CBF cold response pathway. Plant Cell 14: 1675–1690.

81. Lee D, Polisensky DH, Braam J (2005) Genome-wide identification of touch- and darkness-regulated Arabidopsis genes: a focus on calmodulin-like and XTH genes. New Phytol 165: 429–444.

82. Sohn KH, Lee SC, Jung HW, Hong JK, Hwang BK (2006) Expression and functional roles of the pepper pathogen-induced transcription factor RAV1 in bacterial disease resistance, and drought and salt stress tolerance. Plant Mol Biol 61: 897–915.

83. Manavella PA, Dezar CA, Bonaventure G, Baldwin IT, Chan RL (2008) HAHB4, a sunflower HD-Zip protein, integrates signals from the jasmonic acid and ethylene pathways during wounding and biotic stress responses. The Plant Journal 56: 376–388.

84. Manavella PA, Dezar CA, Ariel FD, Drincovich MF, Chan RL (2008) The sunflower HD-Zip transcription factor HAHB4 is up-regulated in darkness, reducing the transcription of photosynthesis-related genes. Journal of Experimental Botany 59: 3143–3155.

85. Schafleitner R, Gutierrez Rosales RO, Gaudin A, Alvarado Aliaga CA, Martinez GN, et al. (2007) Capturing candidate drought tolerance traits in two native Andean potato clones by transcription profiling of field grown plants under water stress. Plant Physiol Biochem 45: 673–690.

86. Sato Y, Antonio B, Namiki N, Motoyama R, Sugimoto K, et al. (2011) Field transcriptome revealed critical developmental and physiological transitions involved in the expression of growth potential in japonica rice. BMC Plant Biol 11: 10.

PERMISSIONS

LIST OF CONTRIBUTORS

Jesse M. Wilson, Rodney Severson and J. Michael Beman
Life and Environmental Sciences, Environmental Systems, and Sierra Nevada Research Institute, University of California Merced, Merced, California, United States of America

Youxiong Que, Yachun Su, Jinlong Guo, Qibin Wu and Liping Xu
Key Laboratory of Sugarcane Biology and Genetic Breeding, Ministry of Agriculture, Fujian Agriculture and Forestry University, Fuzhou, Fujian, China

Jing Zhang and Qifa Zhou
College of Life Sciences, Zhejiang University, Hangzhou, China

Wenjiang Huang
Key Laboratory of Digital Earth Science, Institute of Remote Sensing and Digital Earth, Chinese Academy of Sciences, Beijing, China

Qing Liu
Key Laboratory of Plant Resources Conservation and Sustainable Utilization, South China Botanical Garden, Chinese Academy of Sciences, Guangzhou, China

Huan Liu
Key Laboratory of Plant Resources Conservation and Sustainable Utilization, South China Botanical Garden, Chinese Academy of Sciences, Guangzhou, China University of Chinese Academy of Sciences, Beijing, China

Jun Wen and Paul M. Peterson
Department of Botany, National Museum of Natural History, Smithsonian Institution, Washington, D.C., United States of America

Abdellatif Bahaji, Edurne Baroja-Fernández, Ángela María Sánchez-López, Francisco JoséMuñoz, Jun Li, Goizeder Almagro, Javier Pozueta-Romero and Manuel Montero
Instituto de Agrobiotecnología, Universidad Pública de
Navarra/Consejo Superior de Investigaciones Científicas / Gobierno de Navarra, Mutiloabeti, Nafarroa, Spain

Pablo Pujol and Regina Galarza
Servicio de Apoyo a la Investigación, Universidad Pública de Navarra, Campus de Arrosadia, Iruña, Nafarroa, Spain

Kentaro Kaneko, Kazusato Oikawa, Kaede Wada and Toshiaki Mitsui
Department of Applied Biological Chemistry, Niigata University, Niigata, Japan

Till Meineke, Chithra Manisseri and Christian A. Voigt
Phytopathology & Biochemistry, Biocenter Klein Flottbek, University of Hamburg, Hamburg, Germany

Ni Huang, Li Wang, Pengyu Hao and Zheng Niu
The State Key Laboratory of Remote Sensing Science, Institute of Remote Sensing and Digital Earth, Chinese Academy of Sciences, Beijing, China

Yiqiang Guo
Land Consolidation and Rehabilitation Center, Ministry of Land and Resources, Beijing, China

Vanessa González-Ortiz, Luis G. Egea, Rocio Jiménez-Ramos, Francisco Moreno-Marín, JoséL. Pérez-Lloréns and Fernando G. Brun
Department of Biology, Faculty of Marine and Environmental Sciences of University of Cadiz, Puerto Real, Cadiz, Spain,

Tjeed J. Bouma
Department of Spatial Ecology, Netherlands Institute for Sea Research, Yerseke, The Netherlands

David Pinzon-Latorre and Michael K. Deyholos
Department of Biological Sciences, University of Alberta, Edmonton, Alberta, Canada

Emelie Lindquist, Mohamed Alezzawi and Henrik Aronsson
Department of Biological and Environmental Sciences, University of Gothenburg, Gothenburg, Sweden

Rohan G. T. Lowe, Bethany L. Clark, Angela P. Van de
Wouw and Barbara J. Howlett
School of Botany, The University of Melbourne, Parkville, Victoria, Australia

Andrew Cassin
ARC Centre of Excellence in Plant Cell Walls, School of Botany, The University of Melbourne, Parkville, Victoria, Australia

Jonathan Grandaubert and Thierry Rouxel
INRA-Bioger, UR1290, Thiverval-Grignon, France

João Paulo Fabi, Franco Maria Lajolo and João Roberto Oliveira do Nascimento
Department of Food Science and Experimental Nutrition, FCF, University of São Paulo, São Paulo, São Paulo, Brazil
University of São Paulo, – NAPAN – Food and Nutrition Research Center, São Paulo, São Paulo, Brazil

Sabrina Garcia Broetto and Sarah Lígia Garcia Leme da Silva
Department of Food Science and Experimental Nutrition, FCF, University of São Paulo, São Paulo, São Paulo, Brazil

Silin Zhong
State Key Laboratory of Agrobiotechnology, School of Life Sciences, The Chinese University of Hong Kong, Hong Kong, China

Yuanchun Zou, Guoping Wang, Xiaonan Lou, Xiaofei Yu and Xianguo Lu
Key Lab of Wetland Ecology and Environment, Northeast Institute of Geography and Agroecology, Chinese Academy of Sciences, Changchun, China

Michael Grace
Water StudiesCentre and School of Chemistry, Monash University, Clayton, Australia

Gea Guerriero, Sylvain Legay and Jean-Francois Hausman
Department Environment and Agro-biotechnologies (EVA), Centre de Recherche Public, Gabriel Lippmann, Belvaux, Luxembourg

Index

www.ingramcontent.com/pod-product-compliance
Lightning Source LLC
Chambersburg PA
CBHW080252230326
41458CB00097B/4304